Lecture Notes in Computer Science   11085

Commenced Publication in 1973
Founding and Former Series Editors:
Gerhard Goos, Juris Hartmanis, and Jan van Leeuwen

More information about this series at http://www.springer.com/series/7407

Zvi Lotker · Boaz Patt-Shamir (Eds.)

Structural Information and Communication Complexity

25th International Colloquium, SIROCCO 2018
Ma'ale HaHamisha, Israel, June 18–21, 2018
Revised Selected Papers

 Springer

Editors
Zvi Lotker
Ben-Gurion University of the Negev
Beer-Sheva, Israel

Boaz Patt-Shamir
Tel Aviv University
Tel Aviv, Israel

ISSN 0302-9743   ISSN 1611-3349   (electronic)
Lecture Notes in Computer Science
ISBN 978-3-030-01324-0   ISBN 978-3-030-01325-7   (eBook)
https://doi.org/10.1007/978-3-030-01325-7

Library of Congress Control Number: 2018955971

LNCS Sublibrary: SL1 – Theoretical Computer Science and General Issues

This Springer imprint is published by the registered company Springer Nature Switzerland AG
The registered company address is: Gewerbestrasse 11, 6330 Cham, Switzerland

Preface

This volume contains the papers presented at the 25th International Colloquium on Structural Information and Communication Complexity (SIROCCO 2018). This year was particularly special for SIROCCO, as it was the half-jubilee of SIROCCO. The conference and celebration were held during June 18–21, in Ma'ale HaHamisha, Israel.

This year we received 46 submissions in response to the call for papers. Each submission was reviewed by at least three reviewers; we had a total of 21 Program Committee members and 49 external reviewers. The Program Committee decided to accept 23 papers for regular presentations, and eight papers for brief announcements. All these papers are included in this volume.

In addition the conference program included five additional talks: four keynote talks, and one talk by the winner of the SIROCCO Prize for Innovation in Distributed Computing. The invited speakers were Kurt Melhorn, David Peleg, Claire Mathieu, and Seth Pettie. Additionally, there was a talk by Zvi Lotker, the recipient of the 2018 SIROCCO Prize for Innovation in Distributed Computing. Papers representing these talks are also included in this volume.

The Program Committee selected the following paper as the winner of the SIROCCO 2018 Best Student Paper Award: "Mixed Fault Tolerance in Server Assignment: Combining Reinforcement and Backup," by Tal Navon and David Peleg. Selected papers will also appear in a special issue of the *Theoretical Computer Science* journal devoted to SIROCCO 2018.

We would like to thank all of the authors for their high-quality submissions and all of the speakers for their excellent talks. We are grateful to the Program Committee and all external reviewers for their efforts in putting together a great conference program, to the Steering Committee, chaired by Andrzej Pelc, for their help and support, and to everyone who was involved in the local organization for making it possible to have SIROCCO 2018 in lovely Israel. Thanks also to Michael Borokhovich for his work as the proceedings chair.

August 2018

Zvi Lotker
Boaz Patt-Shamir

Organization

Conference Committee

General Chair

Boaz Patt-Shamir   Tel Aviv University, Israel

Web

Chen Avin   Ben Gurion University, Israel

Proceedings

Michael Borokhovich

Jubilee Celebration

Pierre Fraigniaud   CNRS, France

Publicity

Moti Medina   Ben Gurion University, Israel

Steering Committee

Shantanu Das	Aix-Marseille University, France
Andrzej Pelc (Chair)	UQO
Nicola Santoro	Carleton University, Canada
Christian Scheideler	University of Paderborn, Germany
Sébastien Tixeuil	University of Paris 6, France
Jukka Suomela	Aalto University, Finland

Program Committee

Ittai Abraham	Hebrew University, Israel
Amotz Bar-Noy	CUNY, USA
Jurek Czyzowicz	University of Québec Outaouais, Canada
Taisuke Izumi	Nagoya IT, Japan
Paola Flocchini	University of Ottawa, Canada
Luisa Gargano	University of Salerno, Italy
Leszek Gąsieniec	University Liverpool, UK
Danny Hendler	Ben-Gurion University, Israel
Qiang-Sheng Hua	HUST
Adrian Kosowski	Inria, France
Zvi Lotker (Program Co-chair)	Ben-Gurion University, Israel

Euripides Markou	University of Thessaly, Greece
Gopal Pandurangan	University of Houston, USA
Merav Parter	Weizmann Institute of Science, Israel
Boaz Patt-Shamir (Program Co-chair)	Tel Aviv University, Israel
Sriram Pemmaraju	University of Iowa, USA
Harald Räcke	TU Munich, Germany
Sergio Rajsbaum	UNAM
Ivan Rappaport	University of Chile
Adi Rosén	CNRS and University of Paris Diderot, France
Ladislav Stacho	Simon Fraser University, Canada
Lewis Tseng	Boston College, USA

Additional Reviewers

James Aspnes
Przemysław Uznański
Yossi Azar
Lelia Blin
Lucas Boczkowski
Benedikt Bollig
Costas Busch
Soumyottam Chatterjee
Huda Chuangpishit
Gennaro Cordasco
Shantanu Das
Giuseppe Di Luna
Gabriele Di Stefano
Nguyen Dinh Pham
Michal Dory
Arun Ganesh
Vijay Garg
Konstantinos Georgiou
Sukumar Ghosh
Robert Gmyr
Emmanuel Godard
Maurice Herlihy
Riko Jacob
Eleni Kanellou
Ryan Killick

David Kirkpatrick
Christian Konrad
Kishori Konwar
Rastislav Kralovic
Arnaud Labourel
Nikos Leonardos
Reut Levi
Pedro Montealegre
William Moses
Yoram Moses
Alfredo Navarra
Shreyas Pai
Paolo Penna
Joseph Peters
Franck Petit
Adele Rescigno
Talal Riaz
Eric Ruppert
Dimitris Sakavalas
Stefan Schmid
Paulo Sérgio Almeida
Gokarna Sharma
Giovanni Viglietta
Mengchuan Zou

Sponsoring Institutions

We gratefully acknowledge the generous support that was provided to the SIROCCO 2018 Conference by the Israel Science Foundation, the Israel Ministry of Science Technology and Space, and Springer.

Invited Talks (Abstracts)

The Distributed Lovász Local Lemma Problem

Seth Pettie

University of Michigan
seth@pettie.net

Abstract. The Lovász Local Lemma (LLL) is a well known tool to prove the *existence* of a combinatorial object, by showing that a randomly chosen object satisfies some property with positive (but small) probability. The LLL has been applied in numerous areas, e.g., to compute graph colorings, packet-routing schedules, and satisfying assignments to CNF-SAT formulae. *Algorithmic* versions of the LLL can compute such objects efficiently, in polynomial time.

In this talk I will define the *Distributed LLL* problem and survey its role in algorithm design and complexity theory in the LOCAL model. Among the take-away messages from this talk are the following:

- The LLL is instrumental for designing fast algorithms for edge-coloring, defective coloring, frugal coloring, and other problems.
- There is an exponential gap between randomized and deterministic complexity in the LOCAL model, and the Distributed LLL is the foremost problem realizing this gap.
- The randomized Distributed LLL is *complete* for sublogarithmic randomized time. In particular, any sublogarithmic time algorithm for a locally checkable labeling problem can be automatically sped up to match the time of the Distributed LLL.
- The *deterministic* complexity of the Distributed LLL is inextricably linked to computing network decompositions deterministically. On the one hand, network decompositions are the basis of the fastest Distributed LLL algorithms. Conversely, a deterministic polylog(n) LLL algorithm implies a deterministic (polylog(n), polylog(n))-network decomposition algorithm. (The Distributed LLL is PSLOCAL-hard.)

Keywords: LOCAL model · Probabilistic method · Graph coloring Lovász local lemma

Supported by NSF grants CCF-1514383 and CCF-1637546.

On Fair Division for Indivisible Goods

Kurt Mehlhorn

Max Planck Institute for Informatics, Saarland Informatics Campus (SIC),
66123 Saarbrücken, Germany

We consider the task of dividing indivisible goods among a set of n agents in a fair manner. More precisely, we consider the following scenario. We have m distinct goods. Goods are available in several copies or items; there are k_j items of good j. The agents have decreasing utilities for the different items of a good, i.e., for all i and j

$$u_{i,j,1} \geq u_{i,j,2} \geq \ldots \geq u_{i,j,k_j}.$$

An allocation assigns the items to the agents. For an allocation x, x_i denotes the multi-set of items assigned to agent i, and $m(j, x_i)$ denotes the multiplicity of j in x_i. The total utility of bundle x_i under valuation u_i is given by

$$u_i(x_i) := \sum_j \sum_{1 \leq \ell \leq m(j,x_i)} u_{i,j,\ell}.$$

Each agent has a utility cap c_i. The utility of bundle x_i for agent i is defined as

$$\bar{u}_i(x_i) = \min(c_i, u_i(x_i)).$$

Our notion of fairness is *Nash social welfare* (NSW) [Nas50], i.e., the goal is to maximize the geometric mean

$$\text{NSW}(x) = \left(\prod_{1 \leq i \leq n} \bar{u}_i(x_i) \right)^{1/n}$$

of the capped utilities. All utilities and caps are assumed to be integers.

The problem has a long history. For divisible goods, maximizing Nash Social Welfare (NSW) for any set of valuation functions can be expressed via an Eisenberg-Gale program [EG59]. For *additive valuations* ($c_i = \infty$ for each agent i and $k_j = 1$ for each good j) this program is equivalent to a Fisher market with identical budgets and maximizing NSW is achieved via the well-known fairness notion of competitive equilibrium with equal incomes (CEEI) [Mou03].

For indivisible goods, the problem is NP-complete [NNRR14] and APX-hard [Lee17]. Several constant-factor approximation algorithms are known for the case of additive valuations. They use different approaches.

The first one was pioneered by Cole and Gkatzelis [CG15] and uses spending-restricted Fisher markets. Each agent comes with one unit of money to the market. Spending is restricted in the sense that no seller wants to earn more than one unit of money. If the price p of a good is higher than one in equilibrium, only a fraction $1/p$ of the good is sold. Cole and Gkatzelis showed how to compute a spending restricted equilibrium in polynomial time and how to round its allocation to an integral allocation with good NSW. In the original paper they obtained an approximation ratio of $2e^{1/e} \approx 2.889$. Subsequent work [CDG+17] improved the ratio to 2.

The second approach is via stable polynomials. Anari et al. [AGSS17] obtained an approximation factor of e.

The third approach is via integral allocations that are Pareto-optimal and envy-free up to one good introduced by Barman et al. [BMV17]. Let x_i be the set of goods that are allocated to agent i. An allocation is envy-free up to one good if for any two agents i and k, there is a good j such that $u_i(x_k - j) \leq u_i(x_i)$, i.e., after removal of one good from k's bundle its value for i is no larger than the value of i's bundle for i. Caragiannis et al. [CKM+16] have shown that an allocation maximizing NSW is Pareto-optimal and envy-free up to one good. Barman et al. [BMV17] studied allocations that are Pareto-optimal and almost envy-free up to one good (ε-EF1), i.e., $u_i(x_k - g) \leq (1 + \varepsilon)u_i(x_i)$, where ε is an approximation parameter. They showed that a Pareto-optimal and ε-EF1 allocation approximates NSW up to a factor $e^{1/e} + \varepsilon \approx 1.445 + \varepsilon$. They also showed how to compute such an allocation in polynomial time.

There are also constant-factor approximation algorithms beyond additive utilities.

Garg et al. [GHM18] studied budget-additive utilities ($k_j = 1$ for all goods j and arbitrary c_i). They showed how to generalize the Fisher market approach and obtained an $2e^{1/2e} \approx 2.404$-approximation.

Anari et al. [AMGV18] investigated multi-item concave utilities ($c_i = \infty$ for all i and k_j arbitrary). They generalized the Fisher market and the stable polynomial approach and obtained approximation factors of 2 and e^2, respectively.

In [CCG+18] is shown that the envy-free allocation approach can handle both generalizations combined and yields an approximation ratio of $e^{1/e} + \varepsilon \approx 1.445 + \varepsilon$. The approach via envy-freeness does not only yield better approximation ratios, it is also easier to state and to analyse.

References

[AGSS17]   Anari, N., Gharan, S.O., Saberi, A., Singh, M.: Nash social welfare, matrix permanent, and stable polynomials. In: ITCS, pp. 36:1–36:12 (2017)

[AMGV18]   Anari, N., Mai, T., Gharan, S.O., Vazirani, V.V.: Nash social welfare for indivisible items under separable, piecewise-linear concave utilities. In: SODA, pp. 2274–2290 (2018)

[BMV17]   Barman, S., Murthy, S.K.K., Vaish, R.: Finding fair and efficient allocations. CoRR, abs/1707.04731 (2017). To appear in EC 2018

[CCG+18]   Cheung, Y.K., Chaudhuri, B., Garg, J., Garg, N., Hoefer, M., Mehlhorn, K.: On Fair Division of Indivisible Items. CoRR, abs/1805.06232 (2018)

[CDG+17]   Cole, R., Devanur, N.R., Gkatzelis, V., Jain, K., Mai, T., Vazirani, V.V., Yazdanbod, S.: Convex program duality, fisher markets, and Nash social welfare. In: EC, pp. 459–460 (2017)

[CG15]   Cole, R., Gkatzelis, V.: Approximating the Nash social welfare with indivisible items. In: STOC, pp. 371–380 (2015)

[CKM+16]   Caragiannis, I., Kurokawa, D., Moulin, H., Procaccia, A.D., Shah, N., Wang, J.: The unreasonable fairness of maximum Nash welfare. In: EC, pp. 305–322 (2016)

[EG59]   Eisenberg, E., Gale, D.: Consensus of subjective probabilities: the pari-mutuel method. Ann. Math. Statist. **30**, 165–168 (1959)

[GHM18]   Garg, J., Hoefer, M., Mehlhorn, K.: Approximating the nash social welfare with budget-additive valuations. In: SODA 2018, pp. 2326–2340 (2018)

[Lee17]   Lee, E.: APX-hardness of maximizing Nash social welfare with indivisible items. Inf. Process. Lett. **122**, 17–20 (2017)

[Mou03]   Moulin, H.: Fair Division and Collective Welfare. MIT Press (2003)

[Nas50]   Nash, J.: The bargaining problem. Econometrica **18**, 155–162 (1950)

[NNRR14]   Nguyen, N.-T., Nguyen, T.T., Roos, M., Rothe, J.: Computational complexity and approximability of social welfare optimization in multiagent resource allocation. Autonom. Agents Multi-Agent Syst. **28**(2), 256–289 (2014)

College Admissions in Practice

Claire Mathieu

CNRS, Paris, France

Abstract. The Gale-Shapley algorithm is the standard method in practice for stable marriage in large matching markets, but must be adapted to the constraints of each situation. We study the design of college admissions in a setting with the following features and constraints:

- Lack of trust in the platform: students worry that the rankings of students by schools will factor in their own ranking of schools
- Simplicity: the general public must be able to understand the method
- Transparency: the final result must not be given as a black box but come with an "explanation" that helps rebuild trust
- Quotas: schools have a legal obligation to respect certain quotas of student types. The types and quotas vary from school to school
- Housing: schools provide need-based housing to some of their students. Some students can only afford to attend if housing is provided. The offers must thus take into account both the students' academic ranking and their ranking according to need.

I will present some preliminary work to address such issues, with an application to the French higher education admissions problem.

This is ongoing joint work with Hugo Gimbert.

Taking Turing to the Theater
(Abstract of Award Lecture)

Zvi Lotker

Ben Gurion University in Israel
zvilo@bgu.ac.il

Abstract. Computer science has grown out of the seed of imitation. From von Neumann's machine to the famous Turing test, which sparked the field of AI, algorithms have always tried to imitate humans and nature. Examples of such "imitation algorithms" are simulated annealing which imitates thermodynamics, genetic algorithms which imitate biology, or deep learning which imitates human learning.

In this talk, I describe an algorithm which imitates human psychology. Specifically, I discuss M algorithms, which serve as a simple example of psychology-based imitation algorithms. The M algorithm is one of the simplest natural language processing (NLP) algorithms.

Respecting the long tradition of imitation algorithms, the M algorithm is simple yet powerful. Like other imitation algorithms, the M algorithm is able to efficiently solve difficult problems. The M algorithm pinpoints critical events in films, theater productions, and other scripts, revealing the rhythm of the texts.

At first glance, when trying to design an algorithm which pinpoints critical events of a text, it seems necessary for the algorithm to understand the complete text. Additionally, it would be expected that all layers of the narrative, background information, etc., would also be necessary. In short, it would be expected that the algorithm would imitate the human process of comprehending a text.

Surprisingly, the M algorithm utilizes the structure of the complete text itself without understanding even a *single* word, sentence, or character in order to discover critical events. The content of the narrative is not necessary for the algorithm to work. Other than an awareness of the illusion of time, borrowed from psychology, the M algorithm circumvents the human process of reading.

In the link below, we can see the computerized summary of several movies and relevant data. The M algorithm extracted the critical points on all those movies. As you can see these synopsis provides an "executive" summary of the movies. https://zvilotker.myportfolio.com/psychological-alg.

This talk is based on my upcoming book (in process).

Contents

Invited Talks and Brief Announcments

Realizability of Graph Specifications: Characterizations and Algorithms

Amotz Bar-Noy[1], Keerti Choudhary[2], David Peleg[2], and Dror Rawitz[3(✉)]

[1] City University of New York (CUNY), New York, USA
amotz@sci.brooklyn.cuny.edu
[2] Weizmann Institute of Science, Rehovot, Israel
{keerti.choudhary,david.peleg}@weizmann.ac.il
[3] Bar Ilan University, Ramat-Gan, Israel
dror.rawitz@biu.ac.il

Abstract. The study of graphs and networks often involves studying various parameters of the graph vertices, capturing different aspects of the graph structure, such as the vertex degrees or the distances between the vertices. Given an n-vertex graph G and a parameter of interest f, one may associate with G a vector $\mathcal{F}(G) = \langle f_1, \ldots, f_n \rangle$ giving the value of f for each vertex. This vector can be thought of as the f-profile of the graph. This paper concerns the dual problem, where given an n-entry f-specification vector $F = \langle f_1, \ldots, f_n \rangle$, we need to decide whether it is possible to find a graph G realizing this specification, namely, whose f-profile $\mathcal{F}(G)$ conforms to F. The paper introduces the notion of graph realiziations and illustrates a number of example problems related to finding graph realiziations for given specifications.

1 Introduction

A common theme in the theory of graphs and networks involves extracting and studying a variety of graph parameters that are useful for understanding the graph properties. Over the years, numerous types of graph parameters and measures became the object of attention of graph theorists and network researchers. As a colloquial running example let us pick *vertex degrees*. Given an n-vertex graph G, we denote its *degree sequence* by $\mathcal{DEG}(G) = \langle d_1, \ldots, d_n \rangle$, where d_i denotes the degree of vertex i. It is easy to extract the degree sequence from a given graph, and one may use this information in many different ways, depending on the desired application.

An interesting branch of research, on which we focus here, concerns the *dual* problem where, rather than being given the *graph*, we are given a *sequence* of integers $D = \langle d_1, \ldots, d_n \rangle$. Thinking of this sequence as a *specification* for a desired graph, it is natural to ask whether it is possible to find a graph *realizing*

A. Bar-Noy—Research was sponsored by the Army Research Laboratory and was accomplished under Cooperative Agreement Number W911NF-09-2-0053 (the ARL Network Science CTA).

© Springer Nature Switzerland AG 2018
Z. Lotker and B. Patt-Shamir (Eds.): SIROCCO 2018, LNCS 11085, pp. 3–13, 2018.
https://doi.org/10.1007/978-3-030-01325-7_1

this specification, namely, whose degree sequence conforms to D. Formally, given D, we would like to decide whether there exists a graph G such that $\mathcal{DEG}(G) = D$. Such a sequence (for which there exists a realization) is sometimes called a *graphic* sequence. Note that this problem encapsulates (at least) two separate questions. The first concerns the principal existence of a realizing graph, namely it seeks a characterization (or, a necessary and sufficient condition) for a sequence to be graphic. The second question concerns the practical aspect of the problem, namely, the existence of an effective (and hopefully efficient) algorithm for finding such a realizing graph, if exists. Indeed, both problems were studied in the past. Erdös and Gallai gave a necessary and sufficient condition (which also implies an $O(n)$ decision algorithm) for a sequence to be graphic [8]. However, it is unclear how to efficiently construct a graph that has a given graphic sequence using their method. Havel and Hakimi (independently) gave another algorithm for graphic sequences [10,11], which also implies an efficient $O(m)$ method for constructing a realizing graph for a given graphic sequence, where m is the number of edges in the graph. Their work was later extended in various ways, cf. [19].

In fact, a number of related questions present themselves as well, including the following: (a) Given a degree sequence, find all the (non-isomorphic) graphs that realize it. (b) Given a degree sequence, count all its (non-isomorphic) realizing graphs. (c) Given a degree sequence, sample a random realization as uniformly as possible. (d) Determine the conditions under which a given degree sequence defines a unique realizing graph. (This may be referred to as the *graph reconstruction* problem.) These realization and reconstruction questions are well-studied in the literature, cf. [5,8,10–12,15–18], and have found several interesting applications, most notably in the study of social networks, cf. [4,6,13]. Sampling questions were studied extensively as well, for instance for regular graphs (cf. [20]). In particular, they are used as a component in algorithms for sampling the universe of all graphs with the same degree sequence and estimating its size.

The current paper is motivated by the key observation, made already in [1], that similar questions may be asked for many *other* types of graph specifications or profiles, based on a variety of other graph parameters and measures, and catering to a host of significant applications. For example, for each vertex i, let m_i denote the maximum vertex degree in i's neighborhood. Then $\mathcal{MAX}(G) = \langle m_1, \ldots, m_n \rangle$ is the maximum neighborhood degree profile of G. The same realizability questions asked above for degree sequences can be asked for the maximum neighborhood degree profile as well.

This observation paves the road to a rich field of investigation, which to the best of our knowledge has so far been relatively little explored. The only examples that we are aware of for a study of a realization problem (other than for degree sequences) are the results of [1] on the neighborhood list problem and the closely related *shotgun assembly* problem, where the characteristic f_i associated with the vertex i is the full description of its neighborhood upto radius r, $\Gamma_r(i)$. This problem and some related variants were studied in [14]. Our main aim in the current paper is to look at a number of illustrative example profiles and discuss some of the issues that arise, focusing more on questions than on answers, in the hope of promoting this interesting research direction.

2 Specifications and Realizations

2.1 Basic Notions

Assume that each vertex i $(i = 1, \ldots, n)$ in a graph G is associated with a characteristic f_i. We call the vector $\mathcal{F}(G) = \langle f_1, \ldots, f_n \rangle$ the f-profile of G. The profile can be composed of Boolean variables, integers, real numbers, or even pairs or vectors of numbers. In the degree sequence case, $f_i = d_i$, the degree of vertex i.

We consider situations where we are given an f-specification vector $F = \langle f_1, \ldots, f_n \rangle$ of the right form. Note that this vector might not correspond to (i.e., be the f-profile of) any n-vertex graph. We say that a length n f-specification vector F is a *realizable f-profile* if there exists an n-vertex graph G whose f-profile satisfies $\mathcal{F}(G) = F$.

Hereafter, we consider the following problems:

1. **Profile realizability:** Find a necessary and sufficient condition for an f-specification vector F to be a realizable f-profile.
2. **Profile realization:** For a given realizable f-specification F, construct a realizing graph G (namely, one whose f-profile is $\mathcal{F}(G) = F$).
3. **Approximate realization:** For a given non-realizable f-specification F, construct a graph G realizing the largest possible fraction of F (namely, one whose f-profile $\mathcal{F}(G)$ matches F for as many vertices as possible).

Let us return to the $\mathcal{MAX}$ profile example discussed earlier.

Example 1. **Max neighbor degrees:** Consider the $\mathcal{MAX}$-profile that contains, for every vertex i, the value $m_i = \max_{j \in \Gamma(i)} d_j$, where $\Gamma(i)$ denotes the set of neighbors of i. (This could be the "closed" or "open" neighborhood, i.e., including or excluding i itself; for concreteness let us consider closed neighborhoods.) For the realizability problem, it is convenient to represent the input m-specification vector M alternatively in compressed form, as $\tilde{M} = \langle n_1, n_2, \ldots, n_k \rangle$, where n_i's are non-negative integers with $\sum_{i=1}^{k} n_i = n$; here the specification requires that G contains exactly n_i vertices whose maximum neighbouring degree is i. We may also assume that n_k is non-zero.

The realizability of a $\mathcal{MAX}$-specification was studied in [2], where it was shown that a necessary and sufficient condition for the compressed vector $\tilde{M} = \langle n_1, n_2, \ldots, n_k \rangle$ to be a realizable $\mathcal{MAX}$-profile is that $n_k \geq 1 + k$ and n_1 is even.

2.2   Boolean Profiles

When the characteristic f_i is Boolean, it can be thought of as a vertex *property* P_f, such that

$$f_i = \begin{cases} 1, \text{ vertex } i \text{ has property } P_f, \\ 0, \text{ otherwise.} \end{cases}$$

In many cases, the profile $\mathcal{F}(G)$ can be represented more compactly by a pair of numbers $\langle n, \ell \rangle$, representing the fact that the graph consists of n vertices and ℓ of them satisfy the property P_f.

Example 2. **Degree threshold:** The *degree threshold* k profile $\mathcal{DT}^k$ is defined as follows. For $0 \le k \le n - 1$, let

$$dt_i^k = \begin{cases} 1, d_i \ge k, \\ 0, \text{ otherwise.} \end{cases}$$

If $k = 0$, then the only realizable $\mathcal{DT}^0$-specification is $\langle n, n \rangle$. If $k \ge 1$, then every $\mathcal{DT}^k$-specification $\langle n, \ell \rangle$ where $0 \le \ell \le n$ and $\ell \ne k$ is realizable. To see this, consider two cases. If $\ell \ge k + 1$, then the realizing graph G is a split graph composed of a clique K_ℓ of ℓ vertices and an independent set I of $n - \ell$ isolated vertices, where there are no edges between K_ℓ and I. Otherwise ($\ell \le k - 1$), the graph G is a split graph consisting of a clique K_ℓ of ℓ vertices and an independent set I of $n - \ell$ vertices, where all edges in $K_\ell \times I$ are contained in the edge set. In this case the degree of each of the ℓ vertices in K_ℓ is $n - 1 \ge k$, and the degree of each vertex in I is $\ell < k$.

In the remaining case where $\ell = k$, different situations arise. For example, $\ell = k = 1$ yields an unrealizable pair, and so does $\ell = k = n - 1$. For $2 \le \ell = k \le n - 2$, a possible realizing graph consists of two vertices u and v and a clique K_ℓ of ℓ vertices such that every vertex of K_ℓ is connected to exactly one of u or v, while the degree of u and v is at least 1. The remaining vertices (if exist) remain isolated. Examples of the constructions are given in Fig. 1.

2.3   Notions of Vertex Happiness

In certain contexts in social networks, research has focused on comparisons between peers. For example, people often compare their number of friends with the number of friends of their friends. Consequently, various notions by which a vertex may compare itself with its neighbors were considered in the literature. We may say that a vertex i is *happy* if its degree satisfies a certain condition compared to its neighbors. Here are two possible definitions of happiness.

Example 3. **Low relative loneliness:** For any vertex i, the relative loneliness of i is defined as the ratio $rl_i = \frac{\text{avg}_{j \in \Gamma(i)} d_j}{d_i}$, where avg denotes the average. The relative loneliness profile $\mathcal{RL}$ can be used to define a Boolean profile of

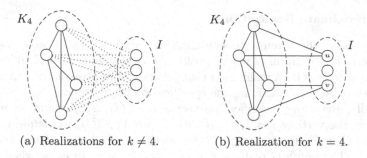

(a) Realizations for $k \neq 4$.   (b) Realization for $k = 4$.

Fig. 1. $\mathcal{DT}^k$ realizations of $\langle 7, 4 \rangle$ for $k < 4$, $k > 4$, and $k = 4$. In all cases it is a split graph which consists of K_4 and an independent set I with three vertices. In (a), the dotted edges are used only when $k > 4$.

happiness by considering a vertex to be happy if its relative loneliness ratio is small, say, $rl_i < 1$. Denote the resulting happiness profile by $\mathcal{H}^{RL}$, where for every i, $h_i^{RL} = 1$ if and only if $rl_i < 1$, or,

$$h_i^{RL} = \begin{cases} 1, & d_i > \text{avg}_{j \in \Gamma(i)} d_j, \\ 0, & \text{otherwise}. \end{cases}$$

Example 4. **Not lowest:** According to this definition of happiness, the vertex i is happy if d_i is greater than the degree of *some* neighbor of i, i.e.,

$$h_i^{NL} = \begin{cases} 1, & d_i > d_j \text{ for some neighbor} j \text{ of } i, \\ 0, & \text{otherwise}. \end{cases}$$

The resulting (compressed) profile $\mathcal{H}^{NL} = \langle n, \ell \rangle$ implies that exactly ℓ of the n vertices have a neighbor of a lower degree. It is clear that for $n = 2$, the only realizable $\mathcal{H}^{NL}$-specification is $\langle 2, 0 \rangle$. So consider $n \geq 3$. For $\ell = 0$, the specification $\langle n, 0 \rangle$ is trivially realizable by a complete graph. It is also easy to verify that the "all-happy" $\mathcal{H}^{NL}$-specification where $\ell = n$ is unrealizable, since the lowest degree vertex in the graph is inevitably unhappy. For $1 \leq \ell \leq n - 2$, one can realize the $\mathcal{H}^{NL}$-specification $\langle n, \ell \rangle$ by a split graph construction similar to that presented earlier for the degree threshold profile. The only remaining case is $\ell = n - 1$. One can verify that the $\mathcal{H}^{NL}$-specification $\langle n, n - 1 \rangle$ is realizable for every $n \geq 7$, and direct case analysis reveals that it is unrealizable for $n = 3, 4, 5, 6$.

For the happiness profile based on relative loneliness, $\mathcal{H}^{RL}$, the well-known *friendship paradox* states that in most graphs, and in particular in most social networks, the relative loneliness ratio of most vertices is greater than 1 (these are the "sad" vertices, whose friends have more friends on average) [7,9]. Nevertheless, note that there are graphs for which most of the vertices are happy. In particular, the graph $K_n - \{1, 2\}$, i.e., the complete graph minus one edge, has $n - 2$ happy vertices, and only two slightly sad vertices.

2.4   Approximate Realizations

In certain cases, the given f-specification vector F is unrealizable, i.e., it is impossible to find a graph G whose profile $\mathcal{F}(G)$ coincides with F. It may still be of interest to look for a realization that (exactly or approximately) maximizes the number of vertices satisfying the specification requirements.

Formally, given an f-specification vector $F = \langle f_1, \ldots, f_n \rangle$ and an n-vertex graph G on the vertices $i = 1, \ldots, n$ with $\mathcal{F}$-profile $\mathcal{F}(G)$, we define the *compatibility* of G to F, denoted $comp(G, F)$, as the number of vertices i such that $\mathcal{F}(G)_i = f_i$. The graph G realizes F if $comp(G, F) = n$. In cases when finding a realizing graph is hard or impossible, a more modest goal may be to find a graph G with as large compatibility $comp(G, F)$ as possible.

For example, consider the profile of happiness based on the "not lowest" property h^{NL} defined above, namely, having a neighbor of lower degree. As mentioned earlier, it is impossible to realize the "all-happy" profile when $\ell = n$. Suppose our goal is to find an optimal realization, namely, one maximizing the number of happy vertices (under this definition). Then we can show that there are graphs with $n - 1$ happy vertices.

3   Three Additional Examples

3.1   The Clique Profile

Let G be a simple (with no parallel edges or self loops) undirected connected graph over the vertex set $V = \{1, 2, \ldots, n\}$. The *clique profile* of G, denoted by $\mathcal{CLIQUE}(G) = \langle k_1, k_2, \ldots, k_n \rangle$ is defined by setting k_i to be the size of the largest clique that includes vertex i, for $1 \leq i \leq n$. Without loss of generality assume that $k_1 \geq k_2 \geq \cdots \geq k_n$.

The clique profile of a triangle free graph without singleton vertices, e.g., trees and bipartite graphs, is $\langle 2, 2, \ldots, 2 \rangle$. In a clique profile of a planar graph, $k_1 \leq 4$ because a planar graph does not contain a clique of size 5. If a graph can be colored with c colors then $k_1 \leq c$ because a clique of size larger than c cannot be colored with c colors. In the clique profile of a graph with maximum degree Δ there is no clique number greater than $\Delta + 1$.

Observe that the only way to realize $k_i = 1$ is when vertex i is a singleton vertex.

The clique profile admits the following complete characterization: A k-specification vector $K = \langle k_1 \geq k_2 \geq \cdots \geq k_n \rangle$ is a clique profile if and only if $k_1 = k_2 = \cdots = k_{k_1}$.

The "only if" part is straightforward since if vertex 1 is a member of a clique of size k_1 then there are at least $k_1 - 1$ additional vertices whose clique number is k_1. For the if part, we show a realization with an interval graph.

Associate the open interval $I_i = (s_i, f_i) = (i - k_i, i)$ of length k_i with vertex i. The resulting graph G contains an edge (i, j) if and only if the intervals I_i and I_j overlap. See example in Fig. 2. We need to show that $\mathcal{CLIQUE}(G) = K$.

By definition, $f_1 < f_2 < \cdots < f_n$. Also $s_1 < s_2 < \cdots < s_n$ since $k_1 \geq k_2 \geq \cdots \geq k_n$. As a result, if the unit open interval $(j, j+1)$ is contained in I_i, then at most k_i other intervals contain $(j, j+1)$. Consequently, for $1 \leq i \leq n$, the clique number of vertex i is at most k_i. Finally, observe that the first k_1 intervals intersect at the unit open interval $(0, 1)$ and that for any $j > k_1$, the k_j intervals $I_{j-k_j+1}, \ldots, I_{k_j}$ intersect at the unit interval $(j - k_j, j - k_j + 1)$. Therefore, for $1 \leq i \leq n$, the clique number of vertex i is at least k_i.

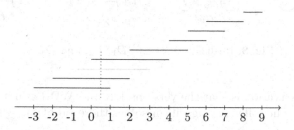

Fig. 2. A realization of $\langle 4, 4, 4, 4, 3, 2, 2, 2, 1 \rangle$. The dotted line represents the clique of the first $k_1 = 4$ intervals.

3.2 The Distance Profile

We next discuss several types of profiles representing distances. Generally, in a *distance profile* $\mathcal{DIST}(G) = \langle D_1, \ldots, D_n \rangle$ of a graph G, the profile D_i for every $1 \leq i \leq n$ is itself an n-entry vector, $D_i = \langle D_{i,1}, \ldots, D_{i,n} \rangle$, where each entry $D_{i,j}$ is a non-negative integer or ∞, for every $1 \leq j \leq n$, representing the distance between i and j in G (defined to be infinity when i and j reside in two disconnected components of G). Alternatively, the profile can be thought of as an $n \times n$ matrix D.

Given a matrix D, we need to decide whether it is a distance profile, namely, if there is an n-vertex unweighted undirected graph $G = (V, E)$ over $V = \{1, \ldots, n\}$ that realizes it, i.e., such that $dist(i, j, G) = D_{i,j}$ for every $1 \leq i, j \leq n$. We refer to this problem as the *distance realization (DR)* problem. We also consider the variant WDR of this question, where the realizing graph G is allowed to be a weighted graph. Two more variants we consider, named DR* and WDR*, permit some of the entries in the matrix D to be left unspecified. That is, we allow entries $D_{i,j} = *$, in which case $dist(i, j, G)$ may assume any value.

Note that D must be symmetric, as otherwise no realization is possible. Hence it suffices to look at the upper triangular part of D.

Example: Consider the (unweighted, fully specified) DR problem for $n = 3$ vertices. Consider the following five input matrices.

$$D_0 = \begin{pmatrix} 0 & \infty & \infty \\ - & 0 & \infty \\ - & - & 0 \end{pmatrix}, \quad D_1 = \begin{pmatrix} 0 & \infty & 1 \\ - & 0 & \infty \\ - & - & 0 \end{pmatrix}, \quad D_2 = \begin{pmatrix} 0 & 1 & 2 \\ - & 0 & 1 \\ - & - & 0 \end{pmatrix}, \quad D_3 = \begin{pmatrix} 0 & 1 & 1 \\ - & 0 & 1 \\ - & - & 0 \end{pmatrix}, \quad D_4 = \begin{pmatrix} 0 & 1 & 1 \\ - & 0 & 3 \\ - & - & 0 \end{pmatrix}.$$

Then the first four matrices can be realized, respectively, by the empty graph G_0, the graph G_1 consisting of the single edge $(1,3)$, the path graph $G_2 = (1,2,3)$, and the complete graph G_3 on $\{1,2,3\}$ (see Fig. 3). The last distance matrix, D_4, is unrealizable.

Fig. 3. Realizations of D_0, D_1, D_2, and D_3.

Our first observation is that the versions DR and WDR admit a polynomial time algorithm. The following algorithm solves these problem.

1. Initially set $V \leftarrow \{1, \ldots, n\}$ and $E \leftarrow \emptyset$.
2. For each $1 \leq i < j \leq n$, add an edge (i,j) to G of weight $D_{i,j}$. (In DR version of the problem we add an edge only if $D_{i,j} = 1$.)
3. Calculate the distance matrix of the resulting graph G, and check if it identical to D. If not identical, then return "Impossible", else return G.

In the partially specified weighted problem WDR*, the same algorithm applies, except that (i) while adding edge to G, all pairs (i,j) such that $D_{i,j} = *$ can be ignored, and (ii) we compare the distance matrix of the resulting graph G with D only at those index-pairs (i,j) where $D_{i,j} \neq *$.

Finally, we claim that the remaining version of the problem, namely, the unweighted partially specified version DR*, is NP-complete. The problem is clearly in NP. One can show that it is complete for NP by a reduction from the coloring problem.

3.3   Realizations by Vertex-Weighted Graphs

A class of more involved realization problems concerns settings where the sought graphs are vertex-weighted. Let us define two example profiles in this setting.

Example 5. **Max vertex-weighted neighbor:** The *maximum vertex-weighted neighbor* profile $\mathcal{MVWN}$, studied in [3], is defined as follows. For a simple undirected weighted graph $(G, \bar{\omega})$, where $G = (V, E)$, and a vector $\bar{\omega} = (\omega_1, \ldots, \omega_n)$ of positive integers, $\mathcal{MVWN}(G, \bar{\omega}) = \langle \varphi_1, \ldots, \varphi_n \rangle$, where $\varphi_i = \max_{j \in \Gamma(i)} \omega_j$. A φ-specification vector $\Phi = \langle \varphi_1, \ldots, \varphi_n \rangle$ of n positive integers is a φ-profile if there exists a realizing weighted graph $(G, \bar{\omega})$ such that $\mathcal{MVWN}(G, \bar{\omega}) = \Phi$.

The profile realizability problem for maximum vertex-weighted neighbor profiles was given a necessary and sufficient condition for realizability in [3].

Example 6. **Vertex-weighted neighborhood sum:** The *vertex-weighted neighborhood sum* profile, the main problem studied in [3], is defined as follows. For a weighted graph $(G, \bar{\omega})$ as above, $\mathcal{VWNS}(G, \bar{\omega}) = \langle \varphi_1, \ldots, \varphi_n \rangle$, where $\varphi_i = \sum_{j \in \Gamma(i)} \omega_j$. A φ-specification vector $\Phi = \langle \varphi_1, \ldots, \varphi_n \rangle$ of n positive integers is a φ-profile if there exists a realizing weighted graph $(G, \bar{\omega})$ such that $\mathcal{VWNS}(G, \bar{\omega}) = \Phi$.

The profile realizability problem for vertex-weighted neighborhood sum profiles was given necessary and sufficient conditions for even n, as well as for odd $n \leq 5$, but the conditions established for general (odd) n are not tight (although they are almost tight), so the problem is still open.

4 Extensions, Generalizations and Future Work

This paper focused mostly on illustrating possible questions rather than providing answers. Yet clearly, many additional directions for future study present themselves. Let us conclude by mentioning some of these.

1. Questions similar to those discussed in this paper can be raised in other contexts, such as directed graphs, multigraphs, edge-weighted graphs, hypergraphs, and more.
2. Similarly, one may explore such questions where the realizing graph must belong to some special graph class, such as connected graphs, trees and forests, bipartite graphs, planar graphs, and so on.
3. Profiles may be defined on the basis of the graph labels. For example, assuming the graph is labeled, let P_f be the property that the vertex i has an even number of neighbors whose label is greater than i. This yields the *large neighbors parity* profile. The specification vector $F = (1, 1, 1)$ is a realizable P_f-profile, as demonstrated by the 3-vertex path graph $(2 - 1 - 3)$ (or alternatively, by the 3-vertex graph G' composed of three singleton vertices). The complementary specification vector $F' = (0, 0, 0)$ is not realizable, since there is no 3-vertex graph where all vertices have odd degrees (as implied, e.g., by the known fact that in every graph, the number of odd degree vertices must be even).
4. Interesting profiles arise by combining two simple profiles into a more compound one. For example, one may consider the combined profile $\mathcal{DEG} \bigwedge \mathcal{H} = \langle (d_1, h_1), \ldots, (d_n, h_n) \rangle$, obtained from the degree profile $\mathcal{DEG} = \langle d_1, \ldots, d_n \rangle$ and some happiness profile $\mathcal{H} = \langle h_1, \ldots, h_n \rangle$. The input profile specifies, for every i, both the degree d_i and a "happiness bit" h_i, and a realizing graph should satisfy both.
5. The question of establishing necessary and sufficient conditions for the existence of *unique* realizations for various profiles promises to yield interesting challenges for future study.
6. In addition to the profile realizability, profile realization and approximate realization problems discussed so far, one may consider also the following questions:

- **Profile enumeration:** How many different realizable f-profiles exist?
- **Optimizing realization:** For a given realizable f-specification vector F, and assuming *costs* on graphs, construct an optimal-cost realizing graph G (namely, one whose f-profile is $\mathcal{F}(G) = F$ and whose cost is minimum).

7. Several other measures for happiness have been considered in the literature. Some examples are:
   - $d_i \geq \max_{j \in \Gamma(i)} d_j$ ("largest in the neighborhood").
   - d_i is greater than the degrees of half the neighbors ("above the median").
   - d_i is greater than the degrees of at least K neighbors.
8. An interesting question concerns distributed solutions for the realizability and realization problems, in a setting where vertices are aware only of their own portion of the profile (e.g., in the congested clique model).

Acknowledgments. We are grateful to Orr Fischer and Andrzej Pelc for helpful discussions and suggestions.

References

1. Aigner, M., Triesch, E.: Realizability and uniqueness in graphs. Discret. Math. **136**, 3–20 (1994)
2. Bar-Noy, A., Choudhary, K., Peleg, D., Rawitz, D.: Graph realizations for max- and min-neighborhood degree profiles. Unpublished manuscript (2018)
3. Bar-Noy, A., Peleg, D., Rawitz, D.: Vertex-weighted realizations of graphs. Unpublished manuscript (2017)
4. Blitzstein, J.K., Diaconis, P.: A sequential importance sampling algorithm for generating random graphs with prescribed degrees. Internet Math. **6**(4), 489–522 (2011)
5. Choudum, S.A.: A simple proof of the Erdös-Gallai theorem on graph sequences. Bull. Aust. Math. Soc. **33**(1), 67–70 (1991)
6. Cloteaux, B.: Fast sequential creation of random realizations of degree sequences. Internet Math. **12**(3), 205–219 (2016)
7. Eom, Y.-H., Jo, H.-H.: Generalized friendship paradox in complex networks. CoRR arxiv: abs/1401.1458 (2014)
8. Erdös, P., Gallai, T.: Graphs with prescribed degrees of vertices [hungarian]. Mat. Lapok **11**, 264–274 (1960)
9. Feld, S.L.: Why your friends have more friends than you do. Amer. J. Soc. **96**, 1464V–1477 (1991)
10. Hakimi, S.L.: On realizability of a set of integers as degrees of the vertices of a linear graph-I. SIAM J. Appl. Math. **10**(3), 496–506 (1962)
11. Havel, V.: A remark on the existence of finite graphs [in Czech]. Casopis Pest. Mat. **80**, 477–480 (1955)
12. Kelly, P.J.: A congruence theorem for trees. Pacific J. Math. **7**, 961–968 (1957)
13. Mihail, M., Vishnoi, N.: On generating graphs with prescribed degree sequences for complex network modeling applications. In: 3rd Workshop on Approximation and Randomization Algorithms in Communication Networks (2002)
14. Mossel, E., Ross, N.: Shotgun assembly of labeled graphs. CoRR arxiv: abs/1504.07682 (2015)

15. O'Neil, P.V.: Ulam's conjecture and graph reconstructions. Amer. Math. Monthly **77**, 35–43 (1970)
16. Sierksma, G., Hoogeveen, H.: Seven criteria for integer sequences being graphic. J. Graph Theory **15**(2), 223–231 (1991)
17. Tripathi, A., Tyagi, H.: A simple criterion on degree sequences of graphs. Discret. Appl. Math. **156**(18), 3513–3517 (2008)
18. Ulam, S.M.: A Collection of Mathematical Problems. Wiley, New York (1960)
19. Wang, D.L., Kleitman, D.J.: On the existence of n-connected graphs with prescribed degrees ($n > 2$). Networks **3**, 225–239 (1973)
20. Wormald, N.C.: Models of random regular graphs. Surv. Comb. **267**, 239–298 (1999)

A Self-Stabilizing Algorithm for Maximal Matching in Link-Register Model

Johanne Cohen[1], George Manoussakis[2(✉)], Laurence Pilard[3], and Devan Sohier[3]

[1] LRI-CNRS, Université Paris-Sud, Université Paris Saclay, Orsay, France
johanne.cohen@lri.fr
[2] Ben-Gurion University of the Negev, Beer-Sheva, Israel
gomanous@gmail.com
[3] LI-PaRAD, Université Versailles-St. Quentin,
Université Paris Saclay, Versailles, France
{laurence.pilard,devan.sohier}@uvsq.fr

Abstract. This paper presents a new distributed self-stabilizing algorithm solving the maximal matching problem under the fair distributed daemon. This is the first maximal matching algorithm in the link-register model under read/write atomicity. This work is composed of two parts. As we cannot establish a move complexity analysis under the fair distributed daemon, we first design an algorithm $\mathcal{A}_1$ under the unfair distributed daemon dealing with some relaxed constraints on the communication model. Second, we adapt $\mathcal{A}_1$ so that it can handle the fair distributed daemon, leading to the $\mathcal{A}_2$ algorithm. We prove that algorithm $\mathcal{A}_1$ stabilizes in $O(m\Delta)$ moves and algorithm $\mathcal{A}_2$ in $O(m\Delta)$ rounds, with Δ the maximum degree and m the number of edges.

1 Introduction

The matching problem consists in building disjoint pairs of adjacent nodes. The matching is maximal if no new pair can be built. This problem has a wide range of applications in networking and parallel computing, such as the implementation of load balancing [2,10]. We deal with the possible occurence of faults using the paradigm of self-stabilization [8]. In this context, there are two main *daemon* types, the *sequential* and *distributed* one. A daemon is said *fair* if every eligible process is eventually scheduled for execution or *unfair* if it only guarantees global progress. It is well-known that one cannot design a self-stabilizing distributed algorithm for a non-trivial task in the link-register model under read/write atomicity, converging under the unfair distributed daemon. In this paper we use the strongest possible daemon under this setting: the fair distributed one.

A network that uses *locally shared registers* can be modeled by a graph where nodes represent processors, and an edge joins two nodes if and only if the corresponding processors communicate directly. Two variants are defined by specifying whether the registers are single-writer/multi-reader and located at the nodes

© Springer Nature Switzerland AG 2018
Z. Lotker and B. Patt-Shamir (Eds.): SIROCCO 2018, LNCS 11085, pp. 14–19, 2018.
https://doi.org/10.1007/978-3-030-01325-7_2

(the *state model*), or single-writer/single-reader and located on the edges (the *link-register model*). Several kinds of atomicities exist in both models. In the *composite atomicity* model, a node can read in all its neighbors registers, and write in its own in one atomic step. In the *read/write atomicity* model, a node can perform either a single read operation or a single write operation in one register in one atomic step.

Thus, communication models are classified following two criteria: the register type and the atomicity. In the literature, the combination of these two criteria only leads to three distinct models [14]. No work is based on the link-register model with composite atomicity since by definition, the composite atomicity only makes sense with the state model. Figure 1 presents the different communication models; references only concern matching works.

	State Model	Link-Register Model
Composite atomicity	*Composite State Model* [1, 6, 18, 23, 28]	None
Read/Write atomicity	*Atomic State Model* [5]	*Atomic Register Model* **Our work**

Fig. 1. Three main communication models.

In this paper, we present two algorithms. Both of them are self-stabilizing and solve the maximal matching problem under the distributed daemon. The first one allows nodes to read in their own registers (so that they are not atomic registers). This allows us to design a self-stabilizing matching algorithm $\mathcal{A}_1$ under the distributed unfair daemon. We prove a $O(m\Delta)$ step complexity for this algorithm.

The second algorithm we present, $\mathcal{A}_2$, uses the classical atomic register model and converges under the unfair distributed daemon. We build $\mathcal{A}_2$ from $\mathcal{A}_1$ by removing from $\mathcal{A}_1$ all reading actions of the out-registers of a node and including some other actions to make the algorithm correct.

2   State of the Art

Various self-stabilizing algorithms for computing maximal matching have been designed in the composite state model (anonymous network [1] or not [23], weighted or unweighted, see [12] for a survey). For an unweighted graph, Hsu and Huang [15] gave the first self-stabilizing algorithm, and proved a bound of $O(n^3)$ on the number of moves under a sequential daemon, later improved by Hedetniemi *et al.* [13]. Manne *et al.* [19] gave a self-stabilizing algorithm that converges in $O(m)$ moves under a distributed unfair daemon. Cohen *et al.* [6] extended this result, and proposed a randomized self-stabilizing algorithm for computing a maximal matching in an anonymous network. The complexity is $O(n^2)$ moves with high probability, under the unfair distributed daemon.

Chattopadhyay *et al.* [5] presents this solution under the atomic state model and in a general anonymous network, under the fair distributed daemon, and with linear round complexity. Their algorithm assumes a different model than ours. Moreover, they assume that nodes know an upper bound on the system size while our algorithm does not.

It is possible to design transformers from the composite atomicity to the read/write atomicity. One approach is to implement a local mutual exclusion among the neighboring nodes. For example, a node u can execute an action of algorithm $\mathcal{A}$ only when is has the critical section access [3]. Moreover, a solution to the dining philosopher problem can also be considered as a solution to the local mutual exclusion problem, see [4, 7, 16, 21, 22]. Another solution using timestamp is presented in [20]. Finally, alternators are also a solution to the local mutual exclusion problem [11, 18]. On the other side, less is known when considering transformations from the state model to the link-register model. As far as we know, Higham and Johnen [14] present the only transformer from the atomic state model to the atomic register model under some additional conditions. However no time complexity is given.

Dolev [9] presents a transformer from the composite state model to the atomic register model under the fair sequential daemon. Together with the Manne *et al.* algorithm [19], this yields a self-stabilizing algorithm for a maximal matching construction in the atomic register model. But this transformer has exponential round complexity at worst. Thus, our solution assumes a stronger daemon and stabilizes more quickly.

Another approach would be to use communication primitives giving some nice properties on read and write atomic actions and leading to the simulation of the composite state model [17]. However, these primitives cannot be trivially used as a base for a transformer.

3   Model

The system consists in a set of nodes V and a set $E \subset V \times V$ of links with $n = |V|$ and $m = |E|$. The set of neighbors of a node u is noted $N(u) = \{v \in V/(u,v) \in E\}$; a node in $N(u)$ is said adjacent to u.

All nodes have the same local variables; if *var* is a variable, var_u denotes the instance of this variable on node u. Node u is the only node allowed to read or to write in var_u. Each node u has a unique identifier id_u; for the sake of simplicity, we do not distinguish between u and id_u. For every adjacent node v of u, there exists a shared register r_{uv} in which u is the only node allowed to write, and that v can read. Here, we consider two register models. The first is the classical link-register model that we use in algorithm $\mathcal{A}_2$ (Sect. 5). In this model, the only node allowed to read in r_{uv} is v. The second model, called *strong-link-register* model, allows u and v to read in r_{uv}. This model is used in algorithm $\mathcal{A}_1$ (Sect. 4).

The moment when a node u writes in register r_{uv} is the time from which the written value r_{uv} is available to v. Thus, the writing is analogous to a message reception by v in a message-passing model. In algorithm $\mathcal{A}_1$ which uses the strong-link-register model, a node u reads in its register r_{uv} in all guards. This allows to check the writing register of a node has reached its correct value. This can be paralleled with an acknowledgement. Observe that this cannot be done in $\mathcal{A}_2$ (Sect. 5) that uses the classical link-register model.

4 Algorithm $\mathcal{A}_1$ - Under the Unfair Daemon

Algorithm $\mathcal{A}_1$ builds a maximal matching under the strong-link-register model and the unfair distributed daemon. At worst, the algorithm has to take $O(m\Delta)$ moves before reaching a maximal matching.

Algorithm description. Each node u has two local variables. Variable $p_u \in N(u) \cup \{null\}$ is the identifier of the node u points to: nodes u and v are said to *be married* to each other if and only if $p_u = v$ (u points to v in the following) and $p_v = u$. We also use a variable m_u indicating the progress of u's marriage: $m_u \in \{0, 1, 2, 3\}$. Also, each node u has a four bit register r_{uv} for each of its neighbors v. The first two bits $r_{uv}.p$ can take the value Idle if u points to $null$, You if it points to v, and Other if it points to a node different from v. The last two bits $r_{uv}.m$ can be 0, 1, 2 or 3, and indicate the progress of u's marriage. In particular, a configuration solves the maximal matching if it is such that $\forall u, (p_u \neq null \Rightarrow p_{p_u} = u) \wedge (p_u = null \Rightarrow \forall v \in N(u), p_v \neq null)$.

Predicates and functions:

$Correct\_register\_value(u, a) \equiv$ if $p_u = null$ then return $(Idle, 0)$
                          else if $p_u = a$ then return (You, m_u)
                          else return $(Other, m_u)$

$PRabandonment(u) \equiv [p_u \neq null \wedge (r_{p_u u}.p \neq You \wedge (u > p_u \vee m_u \neq 0)) \vee (r_{p_u u} = (Other, 3)) \wedge u < p_u)]$

$PRreset(u) \equiv (p_u \neq null) \wedge (r_{p_u u}.p = You) \wedge ($
       $(|m_u - r_{p_u u}.m| \geq 2)$
       $\vee (m_u = 0 \wedge r_{p_u u}.m = 1 \wedge u > p_u) \vee (m_u = 1 \wedge r_{p_u u}.m = 0 \wedge u < p_u)$
       $\vee (m_u = 1 \wedge r_{p_u u}.m = 2 \wedge u < p_u) \vee (m_u = 2 \wedge r_{p_u u}.m = 1 \wedge u > p_u)$
       $\vee (m_u = 3 \wedge r_{p_u u}.m = 2 \wedge u > p_u) \vee (m_u = 2 \wedge r_{p_u u}.m = 3 \wedge u < p_u))$

Rules for each node u:

$\forall a \in N(u),$ **Write(a)**   :: $r_{ua} \neq Correct\_register\_value(u, a) \rightarrow r_{ua} := Correct\_register\_value(u, a)$

         Seduction(a) :: $p_u = null \wedge r_{ua} = Correct\_register\_value(u, a)$
                              $\wedge r_{au} = (Idle, 0) \wedge (u < a) \rightarrow (p_u, m_u) := (a, 0)$

         Marriage(a) :: $p_u = null \wedge r_{ua} = Correct\_register\_value(u, a)$
                              $\wedge r_{au} = (You, 0) \wedge (u > a) \rightarrow (p_u, m_u) := (a, 0)$

Increase :: $p_u \neq null \wedge r_{up_u} = Correct\_register\_value(u, p_u) \wedge (r_{p_u u}.p = You) \wedge ($
         $(m_u = 0) \wedge [(u < p_u \wedge r_{p_u u}.m = 1) \vee (u > p_u \wedge r_{p_u u}.m = 0)]$
         $\vee (m_u = 1) \wedge [(u < p_u \wedge r_{p_u u}.m = 1) \vee (u > p_u \wedge r_{p_u u}.m = 2)]$
         $\vee (m_u = 2) \wedge [(u < p_u \wedge r_{p_u u}.m = 2) \vee (u > p_u \wedge r_{p_u u}.m = 3)])$
         $\rightarrow m_u := m_u + 1$

Reset :: $p_u \neq null \wedge r_{up_u} = Correct\_register\_value(u, p_u) \wedge (PRabandonment(u)$
         $\vee PRreset(u)) \rightarrow (p_u, m_u) := (null, 0)$

5 Algorithm $\mathcal{A}_2$ - Under the Atomic Register Model and the Fair Daemon

In $\mathcal{A}_2$, we juste add a *ToTrue* rule and a local variable $write_u$. $write_u$ is a boolean array indexed by the neighbors of u. This array, together with the $ToTrue$ rule, is used to cyclically update all out-registers of u. Then, algorithm $\mathcal{A}_2$ mimics the behavior of $\mathcal{A}_1$ in a setting in which it is not allowed to read its output registers.

Rules for each node u:

$\forall a \in N(u)$, **Write(a)**   :: $write_u[a] \to write_u[a] := false$; $r_{ua} := Correct\_register\_value(u,a)$

   Seduction(a) :: $p_u = null \wedge \neg write_u[a] \wedge r_{au} = (Idle, 0) \wedge (u < a)$
      $\to write_u[a] := true$; $(p_u, m_u) := (a, 0)$

   Marriage(a) :: $p_u = null \wedge \neg write_u[a] \wedge r_{au} = (You, 0) \wedge (u > a)$
      $\to write_u[a] := true$; $(p_u, m_u) := (a, 0)$

Increase :: $p_u \neq null \wedge \neg write_u[p_u] \wedge (r_{p_u u}.p = You) \wedge$
   $(\quad (m_u = 0) \wedge [\,(u < p_u \wedge r_{p_u u}.m = 1) \vee (u > p_u \wedge r_{p_u u}.m = 0)\,]$
   $\vee (m_u = 1) \wedge [\,(u < p_u \wedge r_{p_u u}.m = 1) \vee (u > p_u \wedge r_{p_u u}.m = 2)\,]$
   $\vee (m_u = 2) \wedge [\,(u < p_u \wedge r_{p_u u}.m = 2) \vee (u > p_u \wedge r_{p_u u}.m = 3)\,]\,)$
      $\to write_u[p_u] := true$; $m_u := m_u + 1$

Reset :: $p_u \neq null \wedge \neg write_u[p_u] \wedge (PRabandonment(u) \vee PRreset(u))$
      $\to \forall a \in N(u) : write_u[a] := true$; $(p_u, m_u) := (null, 0)$

ToTrue :: $(\forall a \in N(u) : \neg write_u[a])$
   $\wedge [\,\forall a \in N(u) : \neg(p_u = null \wedge r_{au} = (Idle, 0) \wedge (u < a))$
      $\wedge \neg(p_u = null \wedge r_{au} = (You, 0) \wedge (u > a))\,]$
   $\wedge \neg[\, p_u \neq null \wedge (r_{p_u u}.p = You) \wedge ($
      $(m_u = 0) \wedge [\,(u < p_u \wedge r_{p_u u}.m = 1) \vee (u > p_u \wedge r_{p_u u}.m = 0)\,]$
      $\vee (m_u = 1) \wedge [\,(u < p_u \wedge r_{p_u u}.m = 1) \vee (u > p_u \wedge r_{p_u u}.m = 2)\,]$
      $\vee (m_u = 2) \wedge [\,(u < p_u \wedge r_{p_u u}.m = 2) \vee (u > p_u \wedge r_{p_u u}.m = 3)\,] \,) \,]$
   $\wedge \neg(p_u \neq null \wedge (PRabandonment(u) \vee PRreset(u)))$
      $\to \forall a \in N(u) : write_u[a] := true$

References

1. Asada, Y., Inoue, M.: An efficient silent self-stabilizing algorithm for 1-maximal matching in anonymous networks. In: Rahman, M.S., Tomita, E. (eds.) WALCOM 2015. LNCS, vol. 8973, pp. 187–198. Springer, Cham (2015). https://doi.org/10.1007/978-3-319-15612-5_17
2. Berenbrink, P., Friedetzky, T., Martin, R.A.: On the stability of dynamic diffusion load balancing. Algorithmica **50**(3), 329–350 (2008)
3. Boulinier, C., Petit, F., Villain, V.: When graph theory helps self-stabilization. In: PODC, pp. 150–159. ACM (2004)
4. Cantarell, S., Datta, A.K., Petit, F.: Self-stabilizing atomicity refinement allowing neighborhood concurrency. In: Huang, S.-T., Herman, T. (eds.) SSS 2003. LNCS, vol. 2704, pp. 102–112. Springer, Heidelberg (2003). https://doi.org/10.1007/3-540-45032-7_8
5. Chattopadhyay, S., Higham, L., Seyffarth, K.: Dynamic and self-stabilizing distributed matching. In: PODC, pp. 290–297. ACM (2002)
6. Cohen, J., Lefevre, J., Maâmra, K., Pilard, L., Sohier, D.: A self-stabilizing algorithm for maximal matching in anonymous networks. PPL **26**(04), 1–17 (2016)
7. Danturi, P., Nesterenko, M., Tixeuil, S.: Self-stabilizing philosophers with generic conflicts. In: Datta, A.K., Gradinariu, M. (eds.) SSS 2006. LNCS, vol. 4280, pp. 214–230. Springer, Heidelberg (2006). https://doi.org/10.1007/978-3-540-49823-0_15

8. Dijkstra, E.W.: Self-stabilizing systems in spite of distributed control. Commun. ACM **17**(11), 643–644 (1974)
9. Dolev, S.: Self-Stabilization. MIT Press (2000)
10. Ghosh, B., Muthukrishnan, S.: Dynamic load balancing by random matchings. J. Comput. Syst. Sci. **53**(3), 357–370 (1996)
11. Gouda, M.G., Haddix, F.F.: The alternator. Distrib. Comput. **20**(1), 21–28 (2007)
12. Guellati, N., Kheddouci, H.: A survey on self-stabilizing algorithms for independence, domination, coloring, and matching in graphs. J. Parallel Distrib. Comput. **70**(4), 406–415 (2010)
13. Hedetniemi, S.T., Pokrass Jacobs, D., Srimani, P.K.: Maximal matching stabilizes in time O(m). Inf. Process. Lett. **80**(5), 221–223 (2001)
14. Higham, L., Johnen, C.: Relationships between communication models in networks using atomic registers. In: IPDPS, Proceedings, pp. 25–29 (2006)
15. Hsu, S.-C., Huang, S.-T.: A self-stabilizing algorithm for maximal matching. Inf. Process. Lett. **43**(2), 77–81 (1992)
16. Huang, S.-T.: The fuzzy philosophers. In: Rolim, J. (ed.) IPDPS 2000. LNCS, vol. 1800, pp. 130–136. Springer, Heidelberg (2000). https://doi.org/10.1007/3-540-45591-4_16
17. Johnen, C., Lavallee, I., Lavault, C.: Reliable self-stabilizing communication for quasi rendezvous. arXiv preprint arXiv:1005.5630 (2010)
18. Kulkarni, S.S., Bolen, C., Oleszkiewicz, J., Robinson, A.: Alternators in read/write atomicity. Inf. Process. Lett. **93**(5), 207–215 (2005)
19. Manne, F., Mjelde, M., Pilard, L., Tixeuil, S.: A new self-stabilizing maximal matching algorithm. Theor. Comput. Sci. (TCS) **410**(14), 1336–1345 (2009)
20. Mizuno, M., Nesterenko, M.: A transformation of self-stabilizing serial model programs for asynchronous parallel computing environments. Inf. Process. Lett. **66**(6), 285–290 (1998)
21. Nesterenko, M., Arora, A.: Dining philosophers that tolerate malicious crashes. In: Proceedings 22nd International Conference on Distributed Computing Systems, pp. 191–198 (2002)
22. Nesterenko, M., Arora, A.: Stabilization-preserving atomicity refinement. J. Parallel Distrib. Comput. **62**(5), 766–791 (2002)
23. Turau, V., Hauck, B.: A new analysis of a self-stabilizing maximum weight matching algorithm with approximation ratio 2. Theor. Comput. Sci. (TCS) **412**(40), 5527–5540 (2011)

Message-Efficient Self-stabilizing Transformer Using Snap-Stabilizing Quiescence Detection

Anaïs Durand[(✉)] and Shay Kutten

Technion - Israel Institute of Technology, Haifa, Israel
danais@technion.ac.il, kutten@ie.technion.ac.il

Abstract. By presenting a message-efficient snap-stabilizing quiescence detection algorithm, we also facilitate a transformer that converts non self-stabilizing algorithms into self-stabilizing ones. We propose a message-efficient snap-stabilizing ongoing quiescence detection algorithm. (Notice that by definition it is also self-stabilizing and can detect termination.) This algorithm works for diffusing computations. We are not aware of any other self-stabilizing or snap-stabilizing ongoing quiescence or termination detection algorithm.

Keywords: Fault-tolerance · Snap-stabilization · Quiescence
Termination · Diffusing computations

1 Introduction

Self-stabilization [11] is a property of distributed systems that withstand transient faults. After transient faults set it into an arbitrary state, a self-stabilizing system recovers in finite time a correct behavior. Multiple transformers that transform non self-stabilizing algorithms $\mathcal{A}$ into self-stabilizing ones [2, 4–6] works roughly as follows. First, $\mathcal{A}$ is executed. When $\mathcal{A}$ terminates, a local checking algorithm is executed (called "local detection" algorithm [2] or local *verifier* of a *Proof Labeling Scheme* [19]). This verifier detects an illegal state if and only if a fault occurred. A self-stabilizing *reset* algorithm, *e.g.*, [3], is then executed to bring all the nodes to an initial state that is legal for $\mathcal{A}$. The cycle is then started again, *i.e.*, $\mathcal{A}$ is executed, termination detected, *etc.* Note that a proof labeling scheme has to be designed especially for $\mathcal{A}$, and some change to $\mathcal{A}$ may be needed in order to generate the specific proof labeling scheme.

The above transformers assume a *synchronous* network in order to know that $\mathcal{A}$ terminated and the verifier could be activated to verify the output (otherwise, the verifier would signal a fault since the output is not yet computed). We do not want this assumption. Alternatively [18], such transformers use a self-stabilizing synchronizer [3,7]. This is a very message intensive function. It uses $\Omega(m)$ messages per round (where m is the number of edges). For example, if $\mathcal{A}$'s time complexity is $\Omega(n)$, its self-stabilizing version (using such a transformer),

© Springer Nature Switzerland AG 2018
Z. Lotker and B. Patt-Shamir (Eds.): SIROCCO 2018, LNCS 11085, pp. 20–24, 2018.
https://doi.org/10.1007/978-3-030-01325-7_3

would need $\Omega(nm)$ messages till stabilization. An earlier transformer uses even more messages [17]. (It assumed a self stabilizing leader election, which was then provided by [2]). The snap-stabilizing quiescence detection algorithm presented here is a much more message-efficient termination detection method in place of the self-stabilizing synchronizer, at least for diffusing computations [12] (e.g., DFS, BFS, token circulation).

Still, one needs yet another component for the transformer. Indeed, since $\mathcal{A}$ is not self-stabilizing, if a fault occurs, $\mathcal{A}$ may never terminates. The missing component is one that enforces termination. Here we use a very simple enforcer: assume that a node sends at most some x messages executing $\mathcal{A}$ when there·is no fault. To implement the enforcer, each node just refuses to send more than x messages. It turns out that even under these constraints (diffusing computations and the simple enforcer) the resulting transformer sends less messages than the traditional ones for various algorithms.

Quiescence detection. A distributed system reaches *quiescence* [9, 21] when no messages are in the communication links and a local indicator of stability holds at every process. Termination and deadlock are two examples of quiescence properties. Detecting quiescence is fundamental. When a deadlock is detected, some measures can be taken such as initiating a reset. Detecting the termination of a task allows the system to use its computed result or issue another operation. In particular, a distributed application is often composed of several modules where one must wait for the termination of a module before starting the next one. It is considered easier to design a task that eventually terminates and combine it with a termination detection protocol, see [15].

The quiescence detection problem and its sub-problems have been extensively studied in distributed computing since the seminal works of Dijkstra and Scholten [12] and Francez [14] on termination detection. One can distinguish two main kinds of quiescence detection algorithms. *Ongoing detection* algorithms must monitor the execution since its beginning and eventually detects quiescence when it is reached, e.g., [12]. *Immediate detection* algorithms answers whether the system has reached quiescence by now or not, e.g., [14]. Ongoing quiescence detection is needed for the transformer, and for most other applications. Ongoing detection can be designed using an immediate detection algorithm by repeatedly executing the detection algorithm until it actually detects quiescence, however it might be highly inefficient.

Cournier *et al.* [10] explain how to design a snap-stabilizing[1] immediate termination detection algorithm using their *Propagation of Information and Feedback (PIF)* algorithm in the locally shared memory model. This does not seem applicable for the message efficient transformer - not only this is not an ongoing detection, the memory requirement is large since the whole state of the system must be locally computed and stored (this can also increase the message complexity in the $\mathcal{CONGEST}$ model).

---

[1] *Snap-stabilization* [8] is a variant of self-stabilization that ensures immediate recovery after transient faults. Notice that a snap-stabilizing algorithm is also self-stabilizing.

Contributions. We propose the first self-stabilizing and snap-stabilizing ongoing quiescence detection algorithm $\mathcal{Q}$ for *diffusing computations.*[2] Using $\mathcal{Q}$, we also implement a message-efficient self-stabilizing transformer.

$\mathcal{Q}$ requires $O(\Delta \log n)$ bits per process, where Δ is the maximum degree. If the execution is *k-synchronous*[3] [16], its cost in messages depends on t_{ab}, m_A, t_A, n, m and k, where t_{ab} is the number of rounds needed to empty all the initial messages out of the channels and reach stabilization of the alternating bit protocol of Afek and Brown [1]. The message (resp. round) complexity of $\mathcal{A}$, the monitored algorithm, is denoted m_A (resp. t_A). The additional cost of $\mathcal{Q}$ is $O(t_{ab} + m_A + n)$ rounds.

2   Quiescence Detection Algorithm $\mathcal{Q}$

We assume the $\mathcal{CONGEST}$ model [20] with FIFO channels of message capacity one (see [1,4] to enforce this).

A *(global) quiescent* property is defined by a *local quiescent-indicator quiet*(p) at each process p such that: (a) while $quiet(p)$ holds, p does not send messages and, as long as p does not receives a message, $quiet(p)$ continues to hold; (b) the channels are empty and $quiet(p)$ holds at every process p if and only if quiescence is reached.

In the context of snap-stabilization (see [8]), a quiescence detection algorithm can start from an arbitrary configuration that leads processes to signal quiescence even if quiescence is not actually reached. In particular, some message can initially be in some channel (p, q) while neither p or q are aware of it until q receives it. Thus, processes have two output signals: $SignalQ()$ and $SignalE()$. A process calls $SignalQ()$ when it detects (global) quiescence. $SignalE()$ is called when an error is detected, *i.e.*, the execution did not start from a *clean configuration*. For example, in a clean configuration of our algorithm $\mathcal{Q}$ it is required, among other things, that channels are empty and an execution of $\mathcal{A}$ starting from this configuration is actually a diffusing computation.

Definition 1. $\mathcal{Q}$ is a *snap-stabilizing ongoing quiescence detection* algorithm if, for every execution Γ where $\mathcal{Q}$ monitors algorithm $\mathcal{A}$ since the beginning of its execution:

- *Eventual Detection:* If the execution of $\mathcal{A}$ reaches quiescence, a process eventually calls $SignalQ()$ or $SignalE()$.
- *Soundness:* If $SignalQ()$ is called, either the execution of $\mathcal{A}$ actually reached quiescence or the initial configuration of $\mathcal{Q}$ was not clean.
- *Relevance:* If the execution of $\mathcal{A}$ satisfies $\mathcal{E}$ and the initial configuration of $\mathcal{Q}$ is clean, no process ever calls $SignalE()$.

---

[2] In a *diffusing computation*, a unique process, the *initiator*, can spontaneously send a message to one or more of its neighbors and only once [12]. After receiving their first message, the other processes can freely send messages to their neighbors.

[3] In a k-synchronous execution, the difference of speed between any two processes is at most k.

The relevance property prevents a trivial and useless detection algorithm where a process calls $SignalE()$ all the time. Notice that there is no hypothesis on $\mathcal{A}$, *i.e.*, we do not require $\mathcal{A}$ to be self-stabilizing or even to compute a correct result.

Overview of the Algorithm. $\mathcal{A}$ and $\mathcal{Q}$ are composed using a fair composition [13]. To avoid confusion, we call *packets* the messages of $\mathcal{A}$. The idea of $\mathcal{Q}$ adapts the algorithm of Dijkstra and Scholten [12] to the snap-stabilizing context using local checking [2]. To monitor $\mathcal{A}$ and detect quiescence, $\mathcal{Q}$ builds the tree of the execution. The initiator of the diffusing computation is the root. When a process that is not in the tree receives a packet m, it joins the tree by choosing the sender of m as parent. When a process p has no children and $quiet(p)$ holds, p leaves the tree. $SignalQ()$ is called when the initiator has no children and its local quietness-indicator holds.

To ensure that quiescence is not signaled when some messages are traveling, $\mathcal{Q}$ uses acknowledgments to wait until messages are received before taking any action of leaving the tree. In [12], counters are used to keep track of how many messages have not been acknowledged yet. In a stabilizing context, maintaining counters is not easy. Thus, $\mathcal{Q}$ sends and receives packets of $\mathcal{A}$ using a self-stabilizing *alternating bit protocol* [1]. Simple proof labeling schemes [19] are used in various parts of the algorithm to make sure it performs correctly. (Those schemes are somewhat generalized in the sense that they are used to verify properties of the algorithm while the algorithm is still running.) See the full version of the paper.

Acknowledgement. This research was carried with a partial support of the Israel Ministry of Science and Technology.

References

1. Afek, Y., Brown, G.M.: Self-stabilization over unreliable communication media. Distrib. Comput. **7**(1), 27–34 (1993)
2. Afek, Y., Kutten, S., Yung, M.: The local detection paradigm and its application to self-stabilization. Theor. Comput. Sci. **186**(1–2), 199–229 (1997)
3. Awerbuch, B., Kutten, S., Mansour, Y., Patt-Shamir, B., Varghese, G.: Time optimal self-stabilizing synchronization. In: STOC 1993, pp. 652–661 (1993)
4. Awerbuch, B., Patt-Shamir, B., Varghese, G.: Self-stabilization by local checking and correction (extended abstract). In: FOCS 1991, pp. 268–277 (1991)
5. Awerbuch, B., Patt-Shamir, B., Varghese, G., Dolev, S.: Self-stabilization by local checking and global reset. In: Tel, G., Vitányi, P. (eds.) WDAG 1994. LNCS, vol. 857, pp. 326–339. Springer, Heidelberg (1994). https://doi.org/10.1007/BFb0020443
6. Awerbuch, B., Varghese, G.: Distributed program checking: a paradigm for building self-stabilizing distributed protocols. In: FOCS 1991, pp. 258–267 (1991)
7. Boulinier, C., Petit, F., Villain, V.: When graph theory helps self-stabilization. PODC **2004**, 150–159 (2004)

8. Bui, A., Datta, A.K., Petit, F., Villain, V.: State-optimal snap-stabilizing PIF in tree networks. In: WSS 1999, pp. 78–85 (1999)
9. Chandy, K.M., Misra, J.: An example of stepwise refinement of distributed programs: quiescence detection. ACM TOPLAS 8(3), 326–343 (1986)
10. Cournier, A., Datta, A.K., Devismes, S., Petit, F., Villain, V.: The expressive power of snap-stabilization. Theor. Comput. Sci. 626, 40–66 (2016)
11. Dijkstra, E.W.: Self-stabilizing systems in spite of distributed control. Commun. ACM 17(11), 643–644 (1974)
12. Dijkstra, E.W., Scholten, C.S.: Termination detection for diffusing computations. Inf. Process. Lett. 11(1), 1–4 (1980)
13. Dolev, S.: Self-Stabilization. MIT Press, Cambridge (2000)
14. Francez, N.: Distributed termination. ACM TOPLAS 2(1), 42–55 (1980)
15. Francez, N., Rodeh, M., Sintzoff, M.: Distributed termination with interval assertions. In: Díaz, J., Ramos, I. (eds.) ICFPC 1981. LNCS, vol. 107, pp. 280–291. Springer, Heidelberg (1981). https://doi.org/10.1007/3-540-10699-5_105
16. Hendler, D., Kutten, S.: Bounded-wait combining: constructing robust and high-throughput shared objects. Distrib. Comput. 21(6), 405–431 (2009)
17. Katz, S., Perry, K.J.: Self-stabilizing extensions for message-passing systems. Distrib. Comput. 7(1), 17–26 (1993)
18. Korman, A., Kutten, S., Masuzawa, T.: Fast and compact self-stabilizing verification, computation, and fault detection of an MST. In: PODC 2011, pp. 311–320 (2011)
19. Korman, A., Kutten, S., Peleg, D.: Proof labeling schemes. Distrib. Comput. 22(4), 215–233 (2010)
20. Peleg, D.: Distributed Computing: A Locality-sensitive Approach. Society for Industrial and Applied Mathematics (2000)
21. Shavit, N., Francez, N.: A new approach to detection of locally indicative stability. In: Kott, L. (ed.) ICALP 1986. LNCS, vol. 226, pp. 344–358. Springer, Heidelberg (1986). https://doi.org/10.1007/3-540-16761-7_84

Constant-Space Self-stabilizing Token Distribution in Trees

Yuichi Sudo[1][(✉)], Ajoy K. Datta[2], Lawrence L. Larmore[2],
and Toshimitsu Masuzawa[1]

[1] Osaka University, 1-5, Yamadaoka, Suita, Osaka, Japan
y-sudou@ist.osaka-u.ac.jp
[2] University of Nevada, Las Vegas, 4505 S Maryland Pkwy, Las Vegas, NV, USA

1 Introduction

The token distribution problem was originally defined by Peleg and Upfal in their seminal paper [4]. Consider a network of n processes and n tokens. Initially, the tokens are arbitrarily distributed among processes but with up to a maximum of l tokens in any process. The problem is to uniformly distribute the tokens such that every process ends up with exactly one token. We generalize this problem as follows: the goal is to distribute nk tokens such that every process holds k tokens where k is any given number. We present a self-stabilizing algorithm that solves this generalized problem. As we deal with self-stabilizing systems, the network (tree in this paper) can start in an arbitrary configuration where the total number of tokens in the network may not be exactly equal to nk. Each process holds an arbitrary number, from zero to l, of tokens in an initial configuration. Thus, we assume that only the root process can push/pull tokens to/from the *external store* as needed.

We present three silent and self-stabilizing token distribution algorithms for rooted tree networks in this paper. The performances of the algorithms are summarized in Table 1. First, we present a self-stabilizing token distribution algorithm *Base*. This algorithm has the optimal convergence time, $O(nl)$ (asynchronous) rounds. However, *Base* may have a large number of redundant token moves; $\Theta(nh\epsilon)$ *redundant* (or unnecessary) token moves happen in the worst case where $\epsilon = \min(k, l-k)$ where h is the height of the tree network. Next, we combine the algorithm *Base* with a synchronizer or PIF waves to reduce redundant token moves, which results in *SyncTokenDist* or *PIFTokenDist*, respectively. Algorithm *SyncTokenDist* reduces the number of redundant token moves to $O(nh)$ without any additional costs while *PIFTokenDist* drastically reduces the number of redundant token moves to the asymptotically optimal value, $O(n)$, at the expense of increasing convergence time from $O(nl)$ to $O(nhl)$ in terms of rounds. Work space complexities, i.e., the amount of memory to store information except for tokens, of all the algorithms are constant both per process and per link register.

This work was partially supported by Japan Science and Technology Agency (JST) SICORP.

Z. Lotker and B. Patt-Shamir (Eds.): SIROCCO 2018, LNCS 11085, pp. 25–29, 2018.
https://doi.org/10.1007/978-3-030-01325-7_4

Table 1. Token distribution algorithms for rooted trees. ($\epsilon = \min(k, l - k)$)

	Conv. time	#Red. token moves	Work space (Process)	Work space (Link)
Base	$O(nl)$ rounds	$\Theta(nh\epsilon)$	0	$O(1)$
SyncTokenDist	$O(nl)$ rounds	$O(nh)$	$O(1)$	$O(1)$
PIFTokenDist	$O(nhl)$ rounds	$O(n)$	$O(1)$	$O(1)$
Lowerbounds	$\Omega(nl)$ rounds	$\Omega(n)$	-	-

2   Preliminaries

We consider a tree network $T = (V, E)$ where V is the set of n processes and E is the set of $n-1$ links. The tree network is rooted, that is, there exists a designated process $v_{\text{root}} \in V$, and every process v other than v_{root} knows its parent $p(v)$. We denote the set of v's children by $C(v)$. We define $N(v) = C(v) \cup \{p(v)\}$. Each link $\{u, v\} \in E$ has two *link registers* or just *registers* $r_{u,v}$ and $r_{v,u}$. We call $r_{u,v}$ (resp. $r_{v,u}$) an output register (resp. an input register) of u. Process u can read from the both registers and can write only to output register $r_{u,v}$.

A process v holds at most l tokens at a time, each of which is a bit sequence of length b. These tokens are stored in a dedicated memory space of the process, called *token store*. We denote the token store of v by $v.\texttt{tokenStore}$ and the number of the tokens in it by $|v.\texttt{tokenStore}|$. We use a link register to send and receive a token between processes. Each register $r_{u,v}$ contains at most one token in a dedicated variable $r_{u,v}.\texttt{token}$. The root process v_{root} can access another token store called *the external token store*, in which an infinite number of tokens exist. The root v_{root} can reduce the total number of tokens in the tree by pushing a token into the external store and can increase it by pulling a token from the external store.

Given $k \leq l$, our goal is to reach a configuration where every process holds exactly k tokens in a self-stabilizing fashion. All tokens must not disappear from the network except in the case that root v_r pushes them to the external store. A process must not create a new token. A new token appears only when the root pulls it from the external store.

We evaluate token distribution algorithms with three metrics—time complexity, space complexity, and the number of token moves. We measure the time complexity in terms of (asynchronous) *rounds*. We measure the space complexity as the *work space complexity* in each process and in each register of an algorithm. The work space complexity in each process (resp. in each register) is the bit length to represent all variables on the process (resp. in the register) except for $\texttt{tokenStore}$ (resp. $\texttt{token}$). We evaluate the number of token moves as follows. Generally, a token is transferred from a process u to a process v in the following two steps: (i) u moves the token from $u.\texttt{tokenStore}$ to $r_{u,v}.\texttt{token}$; (ii) v moves the token from $r_{u,v}.\texttt{token}$ to $v.\texttt{tokenStore}$. In this paper, we regard the above two steps together as one token move and consider the number of token moves as the number of the occurrences of the former steps. We are interested in *the number of redundant token moves*. Let $\tau(v)$ be the number of tokens in input registers of process v. Then, we define $\Delta(v) = \sum_{u \in T_v} d(u)$ where $d(u) = |u.\texttt{tokenStore}| + \tau(u) - k$ and T_v is the sub-tree consisting of all the descendants of v (including v itself). Intuitively, $\Delta(v)$ is the number of tokens that v must send to $p(v)$ to achieve the token distribution if $\Delta(v) \geq 0$; Otherwise, $p(v)$ must send $-\Delta(v)$ tokens to v. We define the number of redundant token moves in an execution as the total number of token moves in the execution minus $\sum_{v \in V} |\Delta(v)|$ of the initial configuration of the execution.

3 Algorithms

3.1 Algorithm *Base*

We use common notation $\text{sgn}(x)$ for real number x, that is, $\text{sgn}(x) = 1$, $\text{sgn}(x) = 0$, and $\text{sgn}(x) = -1$ if $x > 0$, $x = 0$, and $x < 0$, respectively.

The basic idea of *Base* is simple. Each process v always tries to estimate $\text{sgn}(\Delta(v))$, that is, tries to find whether $\Delta(v)$ is positive, negative, or just zero. Then, process v other than v_{root} reports its estimation to its parent $p(v)$ using a shared variable $r_{v,p(v)}.\texttt{est}$. When its estimation is negative, $p(v)$ sends a token to v if $p(v)$ holds a token and $r_{p(v),v}.\texttt{token}$ is empty. When the estimation is positive, v sends a token to its parent $p(v)$ if v holds a token and $r_{v,p(v)}.\texttt{token}$ is empty. Root v_{root} always pulls a new token from the external store to increase $\Delta(v_{\text{root}})$ when its estimation is negative, and pushes a token to the external store to decrease $\Delta(v_{\text{root}})$ when the estimation is positive. If all processes v correctly estimate $\text{sgn}(\Delta(v))$, each of them eventually holds k tokens. After that, no process sends a token.

Thus, estimating $\text{sgn}(\Delta(v))$ is the key of algorithm *Base*. Each process v estimates $\text{sgn}(\Delta(v))$ as follows ($Est(v)$ is the estimation):

$$Est(v) = \begin{cases} 1 & (d(v) > 0 \land \forall u \in C(v) : r_{u,v}.\texttt{est} \in \{1, 0^+, 0\}) \\ 0^+ & (d(v) = 0 \land \forall u \in C(v) : r_{u,v}.\texttt{est} \in \{1, 0^+, 0\} \land \exists w \in C(v) : r_{u,v}.\texttt{est} \in \{0^+, 1\}) \\ 0 & (d(v) = 0 \land \forall u \in C(v) : r_{u,v}.\texttt{est} = 0) \\ 0^- & (d(v) = 0 \land \forall u \in C(v) : r_{u,v}.\texttt{est} \in \{-1, 0^-, 0\} \land \exists w \in C(v) : r_{u,v}.\texttt{est} \in \{0^-, -1\}) \\ -1 & (d(v) < 0 \land \forall u \in C(v) : r_{u,v}.\texttt{est} \in \{-1, 0^-, 0\}) \\ \bot & (\text{otherwise}), \end{cases}$$

where the candidate values 1, 0^+, 0, 0^-, -1, and $\bot$ of $Est(v)$ represent that the estimation is positive, "never negative", zero, "never positive", negative, and "unsure", respectively. A process sends a token to its parent only when its estimation is 1, and it send a token to its child only when the estimation of the child is -1. Our detailed analysis proves that this simple constant-space algorithm shows the performance listed in Table 1.

3.2 Algorithm *SyncTokenDist*

The key idea of *SyncTokenDist* is simple. It is guaranteed that every process v has correct estimation in variable est within $2h$ asynchronous rounds and no redundant token moves happen thereafter. However, some processes can send or receive many tokens in the first $2h$ asynchronous rounds, which makes $\Omega(nh\epsilon)$ redundant token moves in total in the worst case. Algorithm *SyncTokenDist* simulates an execution of *Base* with a simplified version of the $\mathbb{Z}_3$ synchronizer [3], which loosely synchronizes an execution of *Base* so that the following property holds;

> For any integer x, if a process executes the procedure of *Base* at least $x + 2$ times, then every neighboring process of the process must execute the procedure of *Base* at least x times.

Thus, every process v can execute the procedure of *Base* at most $O(h)$ times until all agents have correct estimation, after which no redundant token moves happen.

3.3 Algorithm *PIFTokenDist*

Algorithm *PIFTokenDist* uses Propagation and Information with Feedback (PIF) scheme [1] to reduce the number of redundant token moves. For our purpose, we use a simplified version of PIF. The pseudo code is shown in Algorithm 1. Each process v has a local variable v.wave $\in \{0, 1, 2\}$, a shared variable $r_{v,u}$.wave $\in \{0, 1, 2\}$ for all $u \in N(v)$, and all the variables of *Base*. Process v always copies the latest value of v.wave to $r_{v,u}$.wave for all $u \in N(v)$ (Line 4). An execution of *PIFTokenDist* repeats the cycle of three waves — the 0-wave, the 1-wave, and the 2-wave. Once v_{root}.wave $= 0$, the zero value is propagated from v_{root} to leaves (Line 1, the 0-value). In parallel, each process v changes v.wave from 0 to 1 after verifying that all its children already have the zero value in variable wave (Line 2, the 1-wave). When the 1-wave reaches a leaf, the wave bounces back to the root, changing the wave-value of processes from 1 to 2 (Line 3, the 2-wave). When the 2-wave reaches the root, it resets v_{root}.wave to 0, thus the next cycle begins. A process v executes the procedure of *Base* every time it receives the 2-wave, that is, every time it changes v.wave from 1 to 2 (Line 3).

Algorithm 1 *PIFTokenDist*

[Actions of process v]

1: v.wave $\leftarrow 0$ if $(v = v_{\text{root}} \land v$.wave $= 2) \lor (v \neq v_{\text{root}} \land r_{p(v),v}$.wave $= 0)$
2: v.wave $\leftarrow 1$ if $(v$.wave $= 0) \land (v = v_{\text{root}} \lor r_{p(v),v}$.wave $= 1) \land \forall u \in C(v) : r_{u,v}$.wave $= 0)$
3: v.wave $\leftarrow 2$ and execute the procedure of *Base* if $(v$.wave $= 1) \land (\forall u \in C(v) : r_{u,v}$.wave $= 2)$
4: $r_{v,u}$.wave $\leftarrow v$.wave for all $u \in N(v)$

The *PIFTokenDist* shown in Algorithm 1 is not silent, but it can get the silence property with slight modification such that the root begins the 0-wave at Line 1 only when it detects that the simulated algorithm (*Base*) is not terminated. This modification is easily implemented by using the enabled-signal-propagation technique presented in [2].

References

1. Bui, A., Datta, A.K., Petit, F., Villain, V.: Snap-stabilization and PIF in tree networks. Distrib. Comput. **20**(1), 3–19 (2007)
2. Datta, A.K., Larmore, L.L., Masuzawa, T., Sudo, Y.: A self-stabilizing minimal k-grouping algorithm. In: Proceedings of the 18th International Conference on Distributed Computing and Networking, pp. 3:1–3:10. ACM (2017)
3. Datta, A.K., Larmore, L.L., Masuzawa, T.: Constant space self-stabilizing center finding in anonymous tree networks. In: Proceedings of the International Conference on Distributed Computing and Networking, pp. 38:1–38:10 (2015)
4. Peleg, D., Upfal, E.: The token distribution problem. SIAM J. Comput. **18**(2), 229–243 (1989)

Distributed Counting Along Lossy Paths Without Feedback

Vitalii Demianiuk[1,2(✉)], Sergey Gorinsky[1], Sergey Nikolenko[2,3],
and Kirill Kogan[1]

[1] IMDEA Networks Institute, Leganés, Spain
{vitalii.demianiuk,sergey.gorinsky,kirill.kogan}@imdea.org
[2] Steklov Institute of Mathematics at St. Petersburg, Saint Petersburg, Russia
sergey@logic.pdmi.ras.ru
[3] Neuromation OU, 10111 Tallinn, Estonia

Abstract. Network devices need packet counters for a variety of applications. For a large number of concurrent flows, on-chip memories can be too small to support a separate counter per flow. While a single network element might struggle to implement flow accounting on its own, in this work we study alternatives leveraging underutilized resources elsewhere in the network and implement flow accounting on multiple network devices. This paper takes the first step towards understanding the design principles for robust network-wide accounting with lossy unidirectional channels without feedback.

1 Background and Problem Settings

Scalability Chalenges. Per-flow packet counting in network devices is a crucial functionality in network operation, management, and accounting [1–5]. When a packet arrives to a network device that needs to perform per-flow counting, fast-path processing in the device determines the flow associated with the packet and increments the flow counter. Packet counting is traditionally done in a single network device, but it becomes prohibitively expensive—if at all feasible—to maintain per-flow counters as the number of flows and link speeds grow. Proposed solutions either sacrifice counting accuracy or adopt complex memory architectures. Our paper explores an alternative of network-wide packet counting.

Horizontal vs. Vertical Counter Split. To count packets in a flow, network devices involved in distributed accounting have to lie on the flow's path; at the very least, each flow traverses two switches, its source and destination. Assuming reliable communication, a flow counter can be allocated in any one of the network elements in its path; we call this representation a *horizontal split*. In this work, we relax the constraints on interconnecting links as much as possible, assuming

---

This work was partially supported by a grant from the Cisco University Research Program Fund, an advised fund of Silicon Valley Community Foundation and by the Regional Government of Madrid on Cloud4BigData grant S2013/ICE-2894.

Z. Lotker and B. Patt-Shamir (Eds.): SIROCCO 2018, LNCS 11085, pp. 30–33, 2018.
https://doi.org/10.1007/978-3-030-01325-7_5

an unreliable unidirectional communication without feedback. Then horizontal split becomes infeasible since packets can be dropped before they ever reach the counter, so some part of the counter should be allocated on the source network element. We split the counter into two chunks for source and destination switches; we call this a *vertical split*. Since any given switch stores only a fraction of the counter, it needs less memory to support counting the same number of flows. With our relaxed assumptions on interconnecting links, it is crucial to make distributed execution robust to packet reordering and loss; moreover, we also assume that each packet is allowed to carry only a few bits, which can significantly complicate the operation of distributed counters.

Problem Statement. An asynchronous network delivers a flow f of packets from source switch S to destination switch D, where p_i is an i-th packet of f at S, $i = 0, \ldots, |f| - 1$. Our goal is to compute $|f|$, i.e., number of packets received by ingress switch S from a flow f. Switches S, D maintain partial counter states of at most n bits. The proposed distributed counter representations should be able to exactly reconstruct the counter value after a flow terminates; during a flow's lifetime, counter values returned by queries should not decrease in time and cannot exceed the actual counter value. We study the problem of correctly executing counters under space constraints despite potential packet reordering and loss. We assume that a packet can be prepended by at most t bits.

2   Proposed Method

Splitting a counter between source and destination switches is a ubiquitous model since it does not make any assumptions about routing. Robustness of the distributed accounting to packet reordering and loss certainly has its fundamental limits, e.g., the loss of *all* packets in the network disables stateful communication. To characterize the limits of achievable robustness, we represent delivery disruptions with two parameters:

- *reordering parameter R* is the maximal extent of packet reordering, i.e., the destination switch can receive packet p_j before packet p_i only if $j \leq i + R$;
- *loss parameter L* is the length of a maximal interval of consecutive losses, i.e., the destination switch receives at least one packet from any range $p_i, \ldots, p_{i+L}$.

To overcome delivery disruptions, both S and D use t bits of the n-bit counter chunk as *synch bits* to synchronize the two counter chunks. These t synch bits are the most significant bits in S's chunk c_1, least significant in D's chunk c_2, and middle bits in flow f's merged two-chunk counter c, which counts up to 2^{2n-t}. Upon receiving a packet p, switch S records synch bits from its counter c_1 into packet header $h[p]$ and increments the n-bit counter. When p arrives to D, the latter computes the difference between packet header $h[p]$ and the t synch bits in counter c_2. If this difference is between 1 and 2^{t-1}, switch D adds it to c_2. Upon the completion of flow f, the controller managing accounting network infrastructure collects the c_1 and c_2 values from S and D to obtain $|f|$, i.e., the total number of flow f's packets received by switch S: the controller sets

---

Algorithm 1: Two-switch counting.

---

procedure SOURCEUPDATE(p)
  $h[p] = c_1 \gg (n - t)$
  $c_1 = (c_1 + 1) \bmod 2^n$
end procedure

function DIFFERENCE(a,b,t)
  $\delta := (a + 2^t - b \bmod 2^t) \bmod 2^t$
  return δ
end function

procedure DESTINATIONUPDATE(p)
  diff := **Difference**($h[p], c_2, t$)
  If $1 \leq$ diff $\leq 2^{t-1}$ **then**
    $c_2 := c_2 +$ diff
end procedure

procedure TOTALCOUNT(c_1, c_2)
  $c := c_2 \ll (n - t)$
  $c := c +$ **Difference**(c_1, c, n)
end procedure

---

the n most significant bits of counter c to c_2 and then adds to c the difference between c_1 and the n least significant bits of c. Algorithm 1 consists of the SOURCEUPDATE, DESTINATIONUPDATE, and TOTALCOUNT procedures described above.

Theorem 1. *Algorithm 1 correctly counts up to 2^{2n-t} packets under the following conditions:*

$$L + R < 2^{n-1} \quad and \quad R \leq 2^{n-1} - 2^{n-t}. \tag{1}$$

Proof. To prove correctness, we have to show that each update of c_2 by switch D is correct, i.e., c_2 becomes equal to $i \gg (n - t)$ after D receives any packet that updates c_2. We prove it by induction. When D receives its first packet, the packet's index is at most $L + R$ (this can happen if the first L packets of the flow are lost, and packet p_{L+R} arrives to D first, before $p_L, \ldots, p_{L+R-1}$). For this packet $h[p] \leq (L + R) \gg (n - t) \leq 2^{t-1}$, therefore, it correctly updates c_2.

For the induction step, suppose $c_2 = i \gg (n - t)$ after D processes packet p_i updating c_2 and consider the next arrival of packet p_j to D that updates c_2. Figure 1 partitions the packet sequence at switch S into groups of 2^{n-t} consecutive packets that have the same synch bits. Let I and J denote the groups of p_i and p_j respectively. For $I < J \leq I + \lceil \frac{L+R+1}{2^{n-t}} \rceil$, Algorithm 1 correctly updates c_2 due to $\lceil \frac{L+R+1}{2^{n-t}} \rceil \leq 2^{t-1}$, i.e., the difference between packet header $h[p_j]$ and c_2 is at most 2^{t-1}. Since at least one packet from the considered sequence of groups arrives to D after p_i, no packet from a group later than $I + 2^t$ arrives first due to R. By definition of j, packet p_j does not belong to groups $I + 2^{t-1} + 1$ through $I + 2^t$ because D does not update c_2 for a packet of these groups. $J \leq I$ is impossible since conditions (1) imply that $J \geq I - \lceil \frac{R}{2^{n-t}} \rceil \geq I - \frac{2^{n-1} - 2^{n-t}}{2^{n-t}} > I - 2^{t-1}$, the difference between $h[p_j]$ and c_2's synch bits is either 0 or greater than 2^{t-1}, and Algorithm 1 appropriately does not change c_2. This establishes correctness for each update of c_2.

When flow f ends, c_1 contains the n least significant bits of $|f|$, and the index of the last packet that updates c_2 differs from $|f|$ by at most $L + R + 1$ packets. Since $L + R + 1 < 2^n$, Algorithm 1 accounts for all subsequent missing packets when TOTALCOUNT increases $c_2 \ll (n - t)$ by making its n least significant bits equal to c_1. Thus, counter c correctly computes $|f|$.

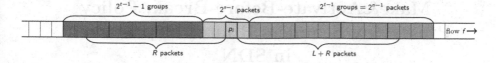

Fig. 1. Packet sequence in the proof of Theorem 1.

Parameter t represents a tradeoff between increasing R and decreasing L and $|f|$. Without packet reordering, when $R = 0$, Algorithm 1 correctly counts packets with loss of up to $L < 2^{n-1}$ consecutive packets. When $R > 2^{n-1}$, Algorithm 1 is never guaranteed to work correctly. Since c_2 never decreases, and $c_2 \ll (n-t)$ never exceeds the number of packets that have arrived to switch S, $c_2 \ll (n-t)$ can be used as a real-time lower bound on the number of packets arrived to S.

Algorithm 1 has attractive robustness in practical settings. For example, when a counter chunk contains 12 bits and uses 2 of them as synch bits, Algorithm 1 correctly counts up to 2^{22} packets despite the loss of up to 1023 consecutive packets and reordering stretch up to 1024 packets. Doubling the number of synch bits from 2 to 4 increases the tolerated reordering stretch to 1792 packets, reducing loss tolerance to 255 consecutive packets and decreasing supported flow size to 2^{20} packets.

3   Conclusion

In this work, we have studied distributed counter implementation under packet reordering and loss. The basic idea of our design is to exploit the state overlap between two communicating switches to maintain correctness of distributed counter state under network noise.

References

1. Lu, Y., Montanari, A., Prabhakar, B., Dharmapurikar, S., Kabbani, A.: Counter braids: a novel counter architecture for per-flow measurement. In: SIGMETRICS, pp. 121–132 (2008)
2. Ramabhadran, S., Varghese, G.: Efficient implementation of a statistics counter architecture. In: SIGMETRICS, pp. 261–271 (2003)
3. Shah, D., Iyer, S., Prabhakar, B., McKeown, N.: Analysis of a statistics counter architecture. In: HOTI, pp. 107–111 (2001)
4. Wang, N., Ho, K.H., Pavlou, G., Howarth, M.P.: An overview of routing optimization for internet traffic engineering. IEEE Commun. Surv. Tutorials **10**(1–4), 36–56 (2008)
5. Zhao, Q., Xu, J.J., Liu, Z.: Design of a novel statistics counter architecture with optimal space and time efficiency. In: SIGMETRICS/Performance, pp. 323–334 (2006)

Make&Activate-Before-Break: Policy Preserving Seamless Routes Replacement in SDN

Yefim Dinitz[1], Shlomi Dolev[1], and Daniel Khankin[1,2(✉)]

[1] Ben-Gurion University of the Negev, Beersheba, Israel
{dinitz,dolev}@cs.bgu.ac.il,danielkh@post.bgu.ac.il
[2] Shamoon College of Engineering (SCE), Beersheba, Israel
daniehe@ac.sce.ac.il

Keywords: Software-Defined Networking · SDN · MABB · RRV
Routes update · Seamless routes update · Network policy · Network
function · Dependence graph

Software-Defined Networking (SDN) allows decoupling of the control plane from the data plane. With this separation, the network switches become simple forwarding devices, while the control logic is implemented in a logically centralized controller [5]. Such separation allows frequent modifications to the routing rules, simplified policy enforcement, and flexible network updates. In SDN, devices do not react to network events such as topological changes, failures, or modifications to table entries. In turn, the controller manages the network update for every such event by orchestrating the switches (see, e.g., [4]).

The routes replacement problem is important in networking and plays a central role in SDN. One line of its study began with the work [2], where the single route replacement problem was considered. There, executing a sequence of sub-routes replacements was suggested for solving it. The new sub-routes are launched in order from the end of the new route to its beginning. For verifying the readiness of a new sub-route N' for flow F, going from switch r_1 to switch r_2, the following idea was proposed in [2]. The controller sends a special packet from itself to r_1 via N' to r_2 and then back to itself. The arrival of that packet at the controller is an evidence that sub-route N' is ready. After that, the controller reroutes F to N' and proceeds to replace the next sub-route. As a result, a way of seamless route updating was declared. However, the suggested idea remained raw, since no network protocol was suggested in [2].

In further research [3], the study was continued for the problem of several routes replacements in SDN, taking into account possible congestion on links. This problem is known to be NP-hard [1]. In [3], finding a schedule of sub-routes replacements was suggested for solving it. No a priori restriction on the order of the new sub-route launchings is posed. The dependence graph model was suggested there for detecting legitimate sub-route replacements at any stage of the replacement process. That model was proposed to establish foundations for

Z. Lotker and B. Patt-Shamir (Eds.): SIROCCO 2018, LNCS 11085, pp. 34–37, 2018.
https://doi.org/10.1007/978-3-030-01325-7_6

AI solutions to the problem. In [3], the verification idea of [2] was implemented in the form of a (high level) network protocol, which supports the seamless processing of the sequence of flow packets in order.

In the first part of this research, the extended setting of the routes updates problem including preserving network policies is studied. A network policy requires packets belonging to some flow to pass in order through a certain network function (NF). A migration of NF to another place (switch) requires moving it with its state, if any, and preserving passing the flow packets through it in order.

In one of the considered problem settings, NF migrations are allowed. We observe that the dependence graph model of [3] suits the extended problem. We develop network protocols for a sub-route replacement for all cases of NF migration. We suggest delivering the NF state to the new location encapsulated in a message. As an alternative, we suggest a duplication of packets (see [2] for details on such update method) based migration, so that a copy of a duplicated packet, followed by the recently updated state, arrive at the current and new NFs. For this, we assume that each of the NFs is wrapped into an active function that handles the arrival of duplicated packets in a way that preserves the state consistency of the NFs, until the migration process is certainly finished.

In the other problem setting, migrations are forbidden, which defines a new type of deadlock. (A deadlock during the replacement process is the situation in which no new sub-route launching is possible.) We reveal the concrete obstacles to be eliminated in order to avoid such deadlocks. Then, we enrich the formal dependence graph model by adding special explicit dependences, while preserving its equivalence. Such explicit dependences are essential for solving the problem using AI methods.

In the second part of this work, we develop in detail a proper implementation of the sub-route replacement considered in [2,3] on a high level. The OpenFlow communication interface standard was developed for SDN [9]. It is known that OpenFlow provides no way to verify when the controller commands are actually executed by switches. This gap substantially restricts SDN abilities; using half-made routes might result in routing cycles, forwarding inconsistencies, and network functions losing their states. This situation is widely known; see, e.g., [12, Sect. 2.2] (also [6,11]). The state-of-art even includes SOS style suggestions, such as waiting a constant time before a new configuration is installed [10] and the time-triggered updates method [8], which was proposed in [7] to become an SDN standard for use in network updates. Summarizing, no reliable verification method currently exists in SDN.

We implement the verification idea of [2] up to a network protocol based on the OpenFlow standard. We suggest to call that implementation *Route Readiness Verifier* (RRV) tool for SDN, and stress that it is suited for general use in SDN. We believe that using RRV and its variants would enable closing most of, if not the entire, above-mentioned gap in the OpenFlow standard.

Notably, RRV supports the *Make&Activate-Before-Break* (MABB) approach used in [2,3] as a general one for rerouting in SDN. Summarizing, for the routes replacement problem, MABB suggests: (a) activating a new route part only after it is completely ready for use, and (b) correct stitching of the previously sent flow on the replaced route part with its continuation on the replacing route part.

The sequence of steps in the suggested RRV tool is as follows. We assume that the new sub-route N' of flow F goes from switch r_1 to switch r_2. We denote by O' the part of the current route of F from r_1 to r_2, and by C the controller.

- Instructing the switches on N', except for r_1, on forwarding packets of F along N'.
- Sending a special tagged packet p_1 of F (that is marked artificially as belonging to F) from C via N' to itself (that is, sending it to r_1, with the special rule to return it from r_2 to C). Afterwards, C is waiting for the return of p_1 from r_2. Maybe several such rounds would be needed, recurrently sending p_1 up to success.
- Instructing r_2 to pause packets of F arriving along N'. For verifying the execution of this instruction, C sends a tagged packet p_2 of F to the switch before r_2 of N', with the special rule to return it from r_2 to C. C waits for the message from r_2 on suspension of p_2. As above, maybe several such rounds would be needed.
- Instructing r_1 to forward packets of F to the next switch on N'. For verifying the execution of this instruction, C sends a tagged packet p_3 of F to r_1. Its tag prescribes returning p_3 to C at both switches following r_1 at N' and O'. C waits for the return of p_3 from the next switch of N'. As above, maybe several such rounds would be needed.
- Sending a tagged packet p_4 of F from C via O' to itself, and waiting for the return of p_4 from r_2. Getting p_4 at C ensures that no packets are left traversing O', because of the FIFO order of processing the packets in F.
- Instructing r_2 to cancel pausing F, thus releasing all packets of F sent along N' and previously suspended by r_2.
- Instructing the switches along O', except for r_1 and r_2, to remove the rule for forwarding packets of F. In order to verify that, the controller sends tagged packet p_x to each switch r_x along O', excluding r_1 and r_2, with a special instruction to return each p_x back to C from r_2. Each switch r_x should return p_x back to C if the forwarding rule of F has already been removed. The controller C repeats this process for each switch r_x that did not return p_x back.

Acknowledgments. This research was (partially) funded by the Office of the Israel Innovation Authority of the Israel Ministry of Economy under Neptune - the Israeli Consortium for Network Programming, generic research project, and by the Lynne and William Frankel Center for Computer Science.

References

1. Amiri, S.A., Dudycz, S., Schmid, S., Wiederrecht, S.: Congestion-Free Rerouting of Flows on DAGs. [cs, math], November 2016. arXiv: 1611.09296
2. Delaet, S., Dolev, S., Khankin, D., Tzur-David, S., Godinger, T.: Seamless SDN route updates. In: 2015 IEEE 14th International Symposium on Network Computing and Applications, pp. 120–125, September 2015. https://doi.org/10.1109/NCA.2015.24
3. Dinitz, Y., Dolev, S., Khankin, D.: Dependence graph and master switch for seamless dependent routes replacement in SDN (extended abstract). In: 2017 IEEE 16th International Symposium on Network Computing and Applications (NCA), pp. 1–7, October 2017. https://doi.org/10.1109/NCA.2017.8171386
4. Foerster, K.T., Schmid, S., Vissicchio, S.: A Survey of Consistent Network Updates (2016)
5. Kreutz, D., et al.: Software-defined networking: a comprehensive survey. Proc. IEEE **103**(1), 14–76 (2015)
6. Kuzniar, M., Canini, M., Kostic, D.: OFTEN testing OpenFlow networks. In: 2012 European Workshop on Software Defined Networking, pp. 54–60, October 2012. https://doi.org/10.1109/EWSDN.2012.21
7. Mizrahi, T., Moses, Y.: Time4: Time for SDN. IEEE Trans. Netw. Serv. Manag. **13**(3), 433–446 (2016). https://doi.org/10.1109/TNSM.2016.2599640
8. Mizrahi, T., Moses, Y.: Time-based updates in software defined networks. In: Proceedings of the Second ACM SIGCOMM Workshop on Hot Topics in Software Defined Networking, HotSDN 2013, pp. 163–164. ACM, New York, NY, USA (2013). https://doi.org/10.1145/2491185.2491214
9. ONF: OpenFlow Switch Specification Ver 1.5.1. Open Networking Foundation (2015). https://www.opennetworking.org/software-defined-standards/specifications/
10. Reitblatt, M., Foster, N., Rexford, J., Schlesinger, C., Walker, D.: Abstractions for network update. SIGCOMM Comput. Commun. Rev. **42**(4), 323–334 (2012). https://doi.org/10.1145/2377677.2377748
11. Talayco, D.: [openflow-discuss] Question about barrier messages, March 2010. https://mailman.stanford.edu/pipermail/openflow-discuss/2010-March/000820.html
12. Zhang, P., Li, H., Hu, C., Hu, L., Xiong, L., Wang, R., Zhang, Y.: Mind the gap: monitoring the control-data plane consistency in software defined networks. In: Proceedings of the 12th International on Conference on Emerging Networking EXperiments and Technologies, CoNEXT 2016, pp. 19–33. ACM, New York, NY, USA (2016). https://doi.org/10.1145/2999572.2999605

Brief Announcement: Fast Approximate Counting and Leader Election in Populations

Othon Michail[1]($\boxtimes$), Paul G. Spirakis[1,2]($\boxtimes$), and Michail Theofilatos[1]($\boxtimes$)

[1] Department of Computer Science, University of Liverpool, Liverpool, UK
{Othon.Michail,P.Spirakis,Michail.Theofilatos}@liverpool.ac.uk
[2] Computer Engineering and Informatics Department,
University of Patras, Patras, Greece

Keywords: Population protocol · Epidemic · Leader election
Counting · Approximate counting · Polylogarithmic time protocol

1 Introduction

Population protocols [2] are networks that consist of very weak computational entities (also called *nodes* or *agents*), regarding their individual capabilities and it has been shown that are able to perform complex computational tasks when they work collectively. *Leader Election* is the process of designating a single agent as the coordinator of some task distributed among several nodes. The nodes communicate among themselves in order to decide which of them will get into the *leader* state, starting from the same initial state q. An algorithm A solves the leader election problem if eventually the states of agents are divided into *leader* and *follower*, a unique leader remains elected and a follower can never become a leader. A randomized algorithm R solves the leader election problem if eventually only one leader remains in the system w.h.p.. *Counting* is the problem where nodes must determine the size n of the population. We call *Approximate Counting* the problem in which nodes must determine an estimation $\hat{n}$ of the population size, where $\frac{\hat{n}}{a} < n < \hat{n}$. We call a the estimation parameter. Consider the setting in which an agent is in an initial state a, the rest $n - 1$ agents are in state b and the only existing transition is $(a, b) \rightarrow (a, a)$. This is the *one-way epidemic* process and it can be shown that the expected time to convergence under the uniform random scheduler is $\Theta(n \log n)$ (e.g., [3]), thus *parallel time* $\Theta(\log n)$.

2 Related Work

The framework of population protocols was first introduced by Angluin et al. [2] in order to model the interactions in networks between small resource-limited

All authors were supported by the EEE/CS initiative NeST. The last author was also supported by the Leverhulme Research Centre for Functional Materials Design.

Z. Lotker and B. Patt-Shamir (Eds.): SIROCCO 2018, LNCS 11085, pp. 38–42, 2018.
https://doi.org/10.1007/978-3-030-01325-7_7

mobile agents. There are many solutions to the problem of leader election, such as in networks with nodes having distinct labels or anonymous networks [1,4,5]. In a recent work, Gasieniec and Stachowiak [5] designed a space optimal ($O(\log \log n)$ states) leader election protocol, which stabilizes in $O(\log^2 n)$ parallel time. They use the concept of phase clocks (introduced in [3] for population protocols), which is a synchronization and coordination tool in distributed computing. Regarding the counting problem, in a recent work, Michail [6] proposed a terminating protocol in which a pre-elected leader equipped with two n-counters computes an approximate count between $n/2$ and n in $O(n \log n)$ parallel time w.h.p..

3 Contribution

In this work we employ the use of simple epidemics in order to provide efficient solutions to approximate counting and also to leader election in populations. Our model is that of population protocols. Our goal for both problems is to get polylogarithmic parallel time and to use small memory per agent. *(a)* We start by providing a protocol which provides an upper bound $\hat{n}$ of the size n of the population, where $\hat{n}$ is at most n^a for some $a > 1$. This protocol assumes the existence of a unique leader in the population. The runtime of the protocol until stabilization is $\Theta(\log n)$ parallel time. Each node except the unique leader uses only a constant number of states. However, the leader is required to use $\Theta(\log^2 n)$ states. *(b)* We then look into the problem of electing a leader. We assume an approximate knowledge of the size of the population and provide a protocol (parameterized by the size m of a counter for drawing local random numbers) that elects a unique leader w.h.p. in $O(\frac{\log^2 n}{\log m})$ parallel time, with number of states $O(\max\{m, \log n\})$ per node. By adjusting the parameter m between a constant and n, we obtain a leader election protocol whose time and space can be smoothly traded off between $O(\log^2 n)$ to $O(\log n)$ parallel time and $O(\log n)$ to $O(n)$ states.

4 The Model

In this work, the system consists of a population V of n distributed and anonymous (i.e., do not have unique IDs) agents, that are capable to perform local computations. Each of them is executing as a deterministic state machine from a finite set of states Q according to a transition function $\delta : Q \times Q \to Q \times Q$. Their interaction is based on the probabilistic (uniform random) scheduler, which picks in every discrete step a random edge from the complete graph G on n vertices. When two agents interact, they mutually access their local states, updating them according to the transition function δ. The transition function is a part of the population protocol which all nodes store and execute locally. *The time is measured as the number of steps until stabilization, divided by n (parallel time).*

5   Fast Counting with a Unique Leader

Our probabilistic algorithm for solving the approximate counting problem requires a unique leader who is responsible to give an estimation on the number of nodes. There is initially a unique leader l and all other nodes are in state q. The leader l stores two counters in its local memory, initially both set to 0, and after the first interaction it starts an epidemic by turning a q node into an a node. Whenever a q node interacts with an a node, its state becomes a. Whenever the leader l interacts with a q node, the value of the counter c_q is increased by one and whenever l interacts with an a node, c_a is increased by one. The termination condition is $c_q = c_a$ and then the leader holds a constant-factor approximation of $\log n$. Chernoff bounds then imply that repeating this protocol a constant number of times suffices to obtain $n/2 \leq n_e \leq 2n$ w.h.p..

Analysis

THEOREM 1. *Our Approximate Counting protocol obtains a constant-factor approximation of* $\log n$ *in* $O(\log n)$ *parallel time w.h.p..*

PROOF. We divide the process into two phases, with the first phase starting when the unique leader initiates the spreading of an epidemic, and the second phase starting when half of the agents become infected. During the first phase, c_q reaches $O(\log n)$, while c_a is increased by a small constant number w.h.p.. This means that our protocol does not terminate w.h.p. until more than half of the population has been infected. During the second phase, when the infected agents are in the majority, c_q is increased by a small constant number, while c_a eventually catches up the first counter. The termination condition ($c_q = c_a$) is satisfied and the leader obtains a constant-factor approximation of $\log n$. Finally, our protocol terminates after $\Theta(\log n)$ parallel time w.h.p.. After half of the population has been infected, it holds that $|c_q - c_a| = \Theta(\log n)$. When the a nodes are in the majority, this difference reaches zero after $\Theta(\log n)$ leader interactions. Thus, the total parallel time to termination is $\Theta(\log n)$.

6   Leader Election with Approximate Knowledge of n

We assume that the nodes know *an upper bound on the population size* n^b, *where* n *is the number of nodes and* b *is any big constant number*. All nodes store three variables; the round e, a random number r and a counter c and they are able to compute random numbers within a predefined range $[1, m]$ (m is *the maximum number that the nodes can generate*). We define two types of states; the leaders (l) and the followers (f). Initially, all nodes are in state l, indicating that they are all potential leaders. The protocol operates in rounds and in every round, the leaders compete with each other trying to survive (i.e., do not become followers). During the first interaction of two l nodes, one of them becomes follower, a random number in $[1, m]$ is being generated, the leader enters the first round and the follower copies the tuple (r, e) from the leader to its local memory. The followers are only being used for information spreading purposes among the potential leaders and they cannot become leaders again.

Information Spreading. All leaders try to spread their tuple (r, e) throughout the population, but w.h.p. all of them except one eventually become followers. We say that a node x wins during an interaction with node y if: (a) $e_x > e_y$ or (b) if $(e_x = e_y)$, $r_x > r_y$. One or more leaders L are in the *dominant state* if their tuple (r_1, e_1) wins every other tuple in the population. Then, the tuple (r_1, e_1) is being spread as an epidemic throughout the population, independently of the other leaders' tuples (all leaders or followers with the tuple (r_1, e_1) always win their competitors). We also call leaders L the *dominant leaders*.

Transition to Next Round. After the first interaction, a leader l enters the first round. As long as a leader survives (i.e., does not become a follower), in every interaction it increases it's counter c by one. When c reaches $b \log n$, where n^b is the upper bound on n, it resets it and round e is increased by one. Finally, the followers can never increase their round or generate random numbers.

Stabilization. Our protocol stabilizes, as the whole population will eventually reach in a final configuration of states. To achieve this, when the round of a leader l reaches $\lceil \frac{2b \log n - \log(b \log^2 n)}{\log m} \rceil$, l stops increasing its round e, unless it interacts with another leader. This rule guarantees the stabilization of our protocol.

Analysis. The protocol proceeds by monotonously reducing the set of possible leaders, until only one candidate for a leader remains. There are initially $k_0 = n$ leaders in the population (round $e = 0$) and between successive rounds, the number of the dominant leaders is given by $k_e = \frac{n}{m^e}$.

THEOREM 2. *Our Leader Election protocol elects a unique leader in* $O(\frac{\log^2 n}{\log m})$ *parallel time w.h.p..*

PROOF. During a round e, the dominant tuple spreads throughout the population in $\Theta(\log n)$ parallel time. No leader can enter to the next round if their epidemic has not been spread throughout the whole population before, thus, for $m = b \log n$ the overall parallel time is $O(\frac{\log^2 n}{\log \log n})$. Finally, during an execution of the protocol, at least one leader will always exist in the population (i.e., a unique leader can never become follower) and a follower can never become leader again. The rule which says that leaders stop increasing their rounds if $e >= \frac{2b \log n - \log (b \log^2 n)}{\log m}$, unless they interact with another leader, implies that the population stabilizes in $O(\frac{\log^2 n}{\log m})$ parallel time w.h.p. and when this happens, there will exist only one leader in the population and eventually, our protocol always elects a unique leader.

References

1. Alistarh, D., Gelashvili, R.: Polylogarithmic-time leader election in population protocols. In: Halldórsson, M.M., Iwama, K., Kobayashi, N., Speckmann, B. (eds.) ICALP 2015. LNCS, vol. 9135, pp. 479–491. Springer, Heidelberg (2015). https://doi.org/10.1007/978-3-662-47666-6_38
2. Angluin, D., Aspnes, J., Diamadi, Z., Fischer, M.J., Peralta, R.: Computation in networks of passively mobile finite-state sensors. Distrib. Comput. 18(4), 235–253 (2006)
3. Angluin, D., Aspnes, J., Eisenstat, D.: Fast computation by population protocols with a leader. Distrib. Comput. 21(3), 183–199 (2008)
4. Fischer, M., Jiang, H.: Self-stabilizing Leader election in networks of finite-state anonymous agents. In: Shvartsman, M.M.A.A. (ed.) OPODIS 2006. LNCS, vol. 4305, pp. 395–409. Springer, Heidelberg (2006). https://doi.org/10.1007/11945529_28
5. Gasieniec, L., Stachowiak, G.: Fast space optimal leader election in population protocols. In: SODA 2018: ACM-SIAM Symposium on Discrete Algorithms, pp. 265–266 (2018)
6. Michail, O.: Terminating distributed construction of shapes and patterns in a fair solution of automata. In: Proceedings of the 2015 ACM Symposium on Principles of Distributed Computing, pp. 37–46 (2015)

One-Max Constant-Probability Networks: Results and Future Work

Mark Korenblit[✉]

Holon Institute of Technology, Holon, Israel
korenblit@hit.ac.il

Abstract. In a number of our works we present and use the tree-like network models, so called *one-max constant-probability models* characterized by the following newly studied principles: (i) each new vertex may be connected to at most one existing vertex; (ii) any connection event is realized with the same probability p due to external factors; (iii) the probability Π that a new vertex will be connected to vertex i depends not directly on its degree d_i but on the place of d_i in the sorted list of vertex degrees. In this announcement we describe features and applications of these models and discuss possible ways of their generalization.

According to the well-known Barabási-Albert random graph model [1], *scale-free networks* are characterized by two main mechanisms: continuous growth and preferential attachment. That is, (a) the networks expand continuously by addition of new vertices, and (b) there is a higher probability that a new vertex will be linked to a vertex already having many connections (high-degree vertex). Vertex degrees in a scale-free network are distributed by a power law. Most vertices have only a few connections while there are a few highly connected hubs. Vertices of a scale-free network are the elements of any system and its edges represent the interaction between them. Irrespective of the nature, many complex systems may be simulated using scale-free networks.

The Barabási-Albert model is described as follows. Starting with a small number m_0 of vertices, at every time step we add a new vertex with $m \leq m_0$ edges that link the new vertex to m different vertices already present in the system. To incorporate preferential attachment, we assume that the probability Π that a new vertex will be connected to vertex i depends on the degree d_i of that vertex.

The mechanism of preferential attachment is assumed to be linear in the model, i.e., $\Pi(d_i)$ is proportional to d_i [1]. However, as noted in the same work, in general relationship between $\Pi(d_i)$ and d_i could have an arbitrary form and, therefore, different types of preferential attachment may be considered. For instance, [2] presents the two-levels network model using a preference function that takes into consideration a vertex's degree and degrees of vertices connected to this vertex. Weighted scale-free networks [3] are created by attachment of new vertices to ends of preferentially chosen weighted edges and by updating the weights of these edges.

© Springer Nature Switzerland AG 2018
Z. Lotker and B. Patt-Shamir (Eds.): SIROCCO 2018, LNCS 11085, pp. 43–47, 2018.
https://doi.org/10.1007/978-3-030-01325-7_8

In the special case, when in every step a new vertex is connected to only one of the old vertices ($m = 1$) we have a tree-like network model. This model has a number of applications. Specifically, it may serve for modeling pyramidal structures based on the principle "success breeds success" and for simulation of stock markets. Tree-like network models may also be convenient for simulation of scale-free networks with m_0 close to 1 in which m does not exceed 1 in most of steps. Decentralized and centralized networks of hubs to which a new node (e.g., personal computer or server, Web page or media file) may always be added, present examples of such networks.

In [5] we introduced a number of tree-like network models which are not exactly scale-free, so called *one-max constant-probability (CP) models*. These models are characterized by the following features: (i) each new vertex may be connected to at most one old vertex; (ii) any connection event is realized with the same probability p due to external factors; iii) the probability Π that a new vertex will be connected to vertex i depends not directly on its degree d_i but on the place of d_i in the sorted list of vertex degrees.

The proposed network model is rather realistic because in real life the choice of an object may be determined not by an absolute characteristic of the object but by a relative status of this object among other objects. The status itself depends, in its turn, on the objects' characteristics. Besides, this model explicitly defines the order of priorities in the search of appropriate connection and, therefore, it allows not just to analyze the topology of networks, but also to examine the network dynamics step-by-step. As noted in [4], one of disadvantages of commonly used techniques for the random generation of graphs is their lack of bias with respect to history of the evolution of the graph. Our model introduces an explicit dependency of the graph's topology on its previous evolution.

In accordance to one of the models presented in [5], so called *constant-probability ordered model (CPOM)*, the list of existing vertices is kept sorted in decreasing order of their degrees so that the vertex with a maximum degree is in the top of the list. The list is scanned from the top and a new vertex is connected to the first vertex v which "is allowed to be connected by the probability p". The degree of vertex v is incremented by 1 and this vertex is moved toward the top of the list to find a proper new place for it. The new vertex's degree is assigned to 1 and this vertex is inserted into the list above vertices with degrees 0 (*isolated vertices*) if it has been connected to any vertex.

CPOM is characterized by the following phenomenon for low p. Some vertices which come first may remain isolated since while a network is not large, a new vertex may rather connect to no existing vertices and find oneself at the bottom of the list. Next later vertices will find more vertices in the network and the probability of their connecting to one of existing vertices will be higher. At that, they will be linked with a higher probability to vertices with larger degrees and their degrees after connection will be 1. Therefore, as the size of the network increases, the chance of vertices with zero degrees "to be found" by new vertices decreases. Figure 1 illustrates the simulation results for $p = 0.1$. A network after

100 time steps (Fig. 1(a)) and the same network after 1000 time steps (Fig. 1(b)) have the same six isolated vertices with order numbers 1, 5, 11, 15, 23, 27.

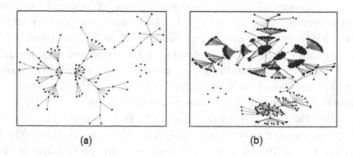

(a)   (b)

Fig. 1. The phenomenon of first isolated vertices for CPOM.

That is, given an n-vertex network based on CPOM, the expected number I_n of isolated vertices in the network reaches saturation for large n, i.e., $\lim_{n\to\infty} I_n = I(p)$, where $I(p)$ is a function of p.

Another model called *constant-probability ordered non-0 model* (*CPOM-N0*) is designed to neutralize the negative effect described above, when some vertices which come first may remain isolated. A new vertex connected to one of existing vertices is not inserted above isolated vertices and remains at the bottom of the list. Thus old vertices with zero degrees will not be at the bottom and the list will be sorted only concerning degrees exceeding 1. Simulations indicate that in n-vertex networks based on CPOM-N0, isolated vertices disappear for large n, i.e., $\lim_{n\to\infty} I_n = 0$.

Some laws of network evolution discovered in [5] have been applied in the algorithms for struggle with malicious networks [6]. In accordance to the proposed approach, a wrong network development is forced using these laws and by means of short-term information distortions. Specifically, artificial decentralization of a network is caused by periodical swapping of the vertices which are in the top of the list of vertex degrees (network's centers) with random vertices in the list that prevents occurrence of high-degree vertices in the malicious network.

As shown in [5], an isolated vertex is a start vertex of a new autonomous part in the network. Therefore, the number of *connected components* (collections of connected vertices which have no connections to one another) in the network is equal to the number of vertices which were isolated some time. It is proved that in an n-vertex one-max CP network, the expected number of connected components tends to $\frac{1}{p}$ with increase of n. In order to create the network consisting of more than $\frac{1}{p}$ connected components which appear not only in the first steps, we proposed the mechanism of appearance of artificial isolated vertices that leads to artificial fragmentation of the malicious network [6].

The proposed model opens the way for future research of the network topology evolution. One possible generalization of the one-max CP network is the *m-max CP network* ($m \geq 1$) in which every new vertex may be connected to at most m old vertices (non-tree-like networks). The probability of connecting the new vertex to none of old vertices does not depend on m and, therefore, m does not influence on the expected number of appearances of isolated vertices. However, increase of m increases the chance of isolated vertices "to be found" by new vertices and, ultimately, leads do decrease of the number of isolated vertices. Moreover, the new vertex may be linked to vertices belonging to different autonomous parts of the network and thus to connect them into one component. Therefore, the number of connected components in a large m-max CP network that grows naturally is expected to be less than $\frac{1}{p}$. Our intent is to study structures and behaviors of these networks.

Another direction of the future research is to move from a constant-probability model to a *degree-dependent-probability model* (DDPM) in which connection of a new vertex to vertex i is realized with probability p_i depending on its degree d_i. Thus the probability $\Pi(d_i)$ that a new vertex will be connected to vertex i depends both on the absolute value of d_i and on the place of d_i in the sorted list of vertex degrees. Our preliminary studies of the one-max DDPM have yielded results significantly different from findings described above. The probability of connecting a new vertex to a "grabbed the lead" vertex is set after first time steps to an approximately constant value while probabilities of its connecting to vertices with less degrees decrease in the course of the network's growth. That is, the measure of concentration around the maximum-degree vertex is higher than for CP models and networks based on the one-max DDPM are more centralized. Notice that isolated vertices do not stop to appear on the network growth, even for rather high probability that is set for the maximum-degree vertex. That is, a new vertex that has not been connected to the "leader", has a small chance to link to any other vertex and "slides down the list".

The most general network model which is the *m-max degree-dependent-probability model* will form our future research agenda.

Acknowledgments. The author thanks Ilya Levin, Eugene Levner and Vadim Talis for their contribution to the studies.

References

1. Barabási, A.-L., Albert, R.: Emergence of scaling in random networks. Science **286**(5439), 509–512 (1999)
2. Dangalchev, C.: Generation models for scale-free networks. Phisica A **338**, 659–671 (2004)
3. Dorogovtsev, S.N., Mendes, J.F.F.: Minimal models of weighted scale-free networks, arXiv.org, Cornell University Library, http://arxiv.org/abs/cond-mat/0408343 (2004)
4. Gustedt, J.: Generalized attachment models for the genesis of graphs with high clustering coefficient. In: Fortunato, S., Mangioni, G., Menezes, R., Nicosia, V. (eds.) Complex Networks. Studies in Computational Intelligence, vol. 207. Springer, Heidelberg (2009). https://doi.org/10.1007/978-3-642-01206-8_9

5. Korenblit, M., Talis, V., Levin, I.: One-max constant-probability models for complex networks. In: Contucci, P., Menezes, R., Omicini, A., Poncela-Casasnovas, J. (eds.) Complex Networks V. SCI, vol. 549, pp. 181–188. Springer, Cham (2014). https://doi.org/10.1007/978-3-319-05401-8_17
6. Korenblit, M.: A new approach to weakening and destruction of malicious internet networks. In: Pichardo-Lagunas, O., Miranda-Jiménez, S. (eds.) MICAI 2016. LNCS (LNAI), vol. 10062, pp. 460–469. Springer, Cham (2017). https://doi.org/10.1007/978-3-319-62428-0_37

Reaching Distributed Equilibrium
with Limited ID Space

Dor Bank$^{(\boxtimes)}$, Moshe Sulamy$^{(\boxtimes)}$, and Eyal Waserman

Tel-Aviv University, Tel Aviv, Israel
dorbank@gmail.com, moshesulamy@mail.tau.ac.il

Abstract. We examine the relation between the size of the *id* space and the number of rational agents in a network under which equilibrium in distributed algorithms is possible. When the number of agents in the network is not a-priori known, but the *id* space is limited, a single agent may duplicate to gain an advantage but each duplication involves a risk of being caught. Given an *id* space of size L, we provide a method of calculating the threshold, the minimal value t such that agents know that $n \geq t$, such that the algorithm is in equilibrium. We apply the method to Leader Election and Knowledge Sharing, and provide a constant-time approximation $t \approx \frac{L}{5}$ of the threshold for Leader Election.

Keywords: Rational agents · Game theory · Leader Election

1 Introduction

We consider the model of distributed game theory [1–5,8,9], in which the participants are rational agents, and may deviate from the algorithm when it increases their personal gain. The goal is to design distributed algorithms that are *in equilibrium*, that is, where no agent has an incentive to cheat.

Previous works [3–5,8,9] assumed that n, the number of agents in the network, is a-priori known to all agents. When n is not a-priori known, in some distributed algorithms an agent may cheat by duplicating itself (perform a Sybil Attack [7]) in order to gain an advantage. We consider the case where the *id* space is limited and any duplication involves a risk of detection, i.e., guessing an *id* that might already be taken by some other agent.

For the *id*-space $ID = \{1, 2, ..., L\}$, and when all agents a-priori know that n, the true number of agents in the network, distributes uniformly $n \sim U[t, L]$, what is the minimal threshold t we must provide the agents for the algorithm to reach equilibrium?

2 Model

The model is a standard synchronous message-passing model of a 2-vertex connected network of $n \geq 3$ nodes, each node representing an agent.

This research was supported by the Israel Science Foundation (grant 1386/11).

Z. Lotker and B. Patt-Shamir (Eds.): SIROCCO 2018, LNCS 11085, pp. 48–51, 2018.
https://doi.org/10.1007/978-3-030-01325-7_9

Each agent a-priori know its input (if any), its id, the id-space $\{1, 2, ..., L\}$ and the threshold $t \in \mathbb{N}$ s.t $3 \leq t \leq n \leq L$. We assume the prior over any unknown information is uniformly distributed over all possible values. We assume all agents start the protocol together. If not, we can use the Wake-Up building block [5] to relax this assumption.

Each rational agent $\mathcal{A}$ wants to maximize its utility function $u_\mathcal{A} : \mathcal{O} \to \mathbb{R}$ where $\mathcal{O}$ is the set of all possible outputs to the algorithm. A rational agent participates in the algorithm but may deviate from it if a deviation increases its *expected* utility, while assuming all other agents follow the protocol.

To differentiate from Byzantine faults, all utility functions must satisfy the *Solution Preference* [5] property, which ensures agents never prefer an outcome in which the algorithm fails over one in which it terminates correctly. An algorithm is said to be *in equilibrium* if no agent, at any point in the algorithm execution, can unilaterally increase its utility by deviating from the algorithm.

2.1 Duplication

Since n is not a-priori known to agents, an agent $\mathcal{A}$ can deviate by simulating m imaginary agents. Each duplicated agent must be assigned an id and duplication involves a risk of choosing an id that already exists, rendering it non-unique, and causing the algorithm failure. We assume m and the ids of all m duplicated agents must be chosen at round 0, before the algorithm starts.

2.2 Leader Election

Each agent $\mathcal{A}$ outputs $o_\mathcal{A} \in \{0, 1\}$, $o_\mathcal{A} = 1$ if $\mathcal{A}$ was elected leader, and $o_\mathcal{A} = 0$ otherwise. The set of legal output vectors is defined as: $O_L = \{\underline{o} \mid \exists \mathcal{A} : o_\mathcal{A} = 1, \quad \forall \mathcal{A}' \neq \mathcal{A} : o_{\mathcal{A}'} = 0\}$

We assume a *fair* leader election [3] where, at the beginning of the algorithm, each agent has an equal chance to be elected leader, and assume agents prefer to be elected leader.

2.3 Knowledge Sharing

In the problem (from [4], adapted from [5]), each agent $\mathcal{A}$ has a private input $i_\mathcal{A}$ and a function q, where q is identical at all agents. An output is *legal* if all agents output the same value. An output is *correct* if all agents output $q(I)$ where $I = \{i_1, ..., i_n\}$. The function q satisfies the Full Knowledge property [4,5], which states that when one or more input values are not known, any output in the range of q is *equally* possible. We assume that each agent $\mathcal{A}$ prefers a certain output value $p_\mathcal{A}$. Following [4], in this paper we only discuss Knowledge Sharing in ring graphs.

3   Solution Basis

Equation 1 defines the necessary condition for equilibrium in the distributed problem in the presence of rational agents:

$$\sum_{k=t}^{L} e_0(k) \geq \max_{m} \sum_{k=t}^{L-m} p_m(k) e_m(k) \tag{1}$$

Where $e_m(k)$ is the expected utility of an agent simulating m false duplicates, when k true agents participate in the network; $p_m(k)$ is the probability of successfully choosing m ids that are not yet taken, generally $p_m(k) = \frac{\binom{L-k}{m}}{\binom{L-1}{m}}$. We are interested in the minimal threshold t that satisfies Eq. 1, and it can be calculated in $O(L^3)$ running time, by trying all values for t.

3.1   Enhancements

Linear Threshold. For most algorithms there exists L_0 such that for any $L > L_0$, there exists a pivot value t_0 such that for any $t \geq t_0$ the algorithm is in equilibrium, and for any $t < t_0$ it is not in equilibrium. In such cases we can use binary search to improve the running time to $O(L^2 \log L)$.

Limited Duplications. For some algorithms there exists a specific duplication number m', such that if there exists m for which agent has an incentive to deviate, then it also has an incentive to deviate with m' duplications. For such algorithms we only need to examine a single duplication value, improving the running time to $O(L \log L)$.

For algorithms that satisfy both enhancements, the running time is improved to $O(L \log L)$.

4   Contributions

Here we summarize our contributions. Details and full proofs are provided in the full paper [6].

4.1   Leader Election

The Leader Election algorithm [3,5] satisfies both enhancements. Thus, the minimal threshold can be found in $O(L \log L)$ time.

Particularly, whenever an agent has an incentive to deviate by duplicating m agents, it also has an incentive to deviate by duplicating 1 agent. Thus, to check for equilibrium it suffices to check the case $m = 1$.

Furthermore, we prove a constant-time approximation of the Leader Election threshold that shows the minimal threshold t for equilibrium is in the range $0.2L < t < 0.21L$.

4.2   Knowledge Sharing

The Knowledge Sharing algorithm [4,5] (in a ring) satisfies only the "Linear Threshold" enhancement. Thus, the minimal threshold can be found in $O(L^2 \log L)$ time.

Acknowledgment. We would like to thank Yehuda Afek for helpful discussions and his course on Distributed Computing which has inspired this research, and to Sivan Schick for his contributions to this paper.

References

1. Abraham, I., Alvisi, L., Halpern, J.Y.: Distributed computing meets game theory: combining insights from two fields. SIGACT News **42**(2), 69–76 (2011). https://doi.org/10.1145/1998037.1998055
2. Abraham, I., Dolev, D., Gonen, R., Halpern, J.Y.: Distributed computing meets game theory: robust mechanisms for rational secret sharing and multiparty computation. In: PODC, pp. 53–62 (2006)
3. Abraham, I., Dolev, D., Halpern, J.Y.: Distributed protocols for leader election: a game-theoretic perspective. In: Afek, Y. (ed.) DISC 2013. LNCS, vol. 8205, pp. 61–75. Springer, Heidelberg (2013). https://doi.org/10.1007/978-3-642-41527-2_5
4. Afek, Y., Rafaeli, S., Sulamy, M.: Cheating by duplication: equilibrium requires global knowledge. ArXiv e-prints, November 2017
5. Afek, Y., Ginzberg, Y., Landau Feibish, S., Sulamy, M.: Distributed computing building blocks for rational agents. In: Proceedings of the 2014 ACM Symposium on Principles of Distributed Computing, PODC 2014, pp. 406–415. ACM, New York (2014). https://doi.org/10.1145/2611462.2611481
6. Bank, D., Sulamy, M., Waserman, E.: Reaching distributed equilibrium with limited ID space. ArXiv e-prints, April 2018
7. Douceur, J.R.: The sybil attack. In: Druschel, P., Kaashoek, F., Rowstron, A. (eds.) IPTPS 2002. LNCS, vol. 2429, pp. 251–260. Springer, Heidelberg (2002). https://doi.org/10.1007/3-540-45748-8_24
8. Halpern, J.Y., Vilaça, X.: Rational consensus: extended abstract. In: Proceedings of the 2016 ACM Symposium on Principles of Distributed Computing, PODC 2016, pp. 137–146. ACM, New York (2016). https://doi.org/10.1145/2933057.2933088
9. Yifrach, A., Mansour, Y.: Fair leader election for rational agents. In: PODC 2018 (2018)

Full Papers

Crash-Tolerant Consensus in Directed Graph Revisited (Extended Abstract)

Ashish Choudhury[1]([✉]), Gayathri Garimella[3], Arpita Patra[2], Divya Ravi[2], and Pratik Sarkar[4]

[1] International Institute of Information Technology Bangalore, Bengaluru, India
ashish.choudhury@iiitb.ac.in
[2] Department of Computer Science and Automation,
Indian Institute of Science, Bengaluru, India
{arpita,divya.ravi}@iisc.ernet.in
[3] Oregon State University, Corvallis, USA
AnnapurnaGayathri.Garimella@iiitb.org
[4] Boston University, Boston, USA
pratik.sarkar@iisc.ernet.in

Abstract. Fault-tolerant distributed consensus is a fundamental problem in secure distributed computing. In this work, we consider the problem of distributed consensus in directed graphs tolerating crash failures. Tseng and Vaidya (PODC'15) presented necessary and sufficient condition for the existence of consensus protocols in directed graphs. We improve the round and communication complexity of their protocol. Moreover, we prove that our protocol requires the optimal number of communication rounds, required by any protocol belonging to a restricted class of crash-tolerant consensus protocols in directed graphs.

Keywords: Directed graph · Consensus · Crash failure
Round complexity

1 Introduction

Fault-tolerant reliable consensus [7,8,12] is a fundamental problem in distributed computing. Informally, a consensus protocol allows a set of n mutually distrusting parties, each with some private input, to agree on a common output. This is ensured even in the presence of a *computationally unbounded* centralized adversary, who may corrupt any f out of the n parties and try to prevent the remaining parties from achieving consensus. Since its inception [12], the problem has been widely studied in the literature and several interesting results have been obtained regarding the possibility, feasibility and optimality of reliable consensus

---

P. Sarkar and G. Garimella—The research was conducted while these two authors were at the Indian Institute of Science and International Institute of Information Technology Bangalore, respectively.

Z. Lotker and B. Patt-Shamir (Eds.): SIROCCO 2018, LNCS 11085, pp. 55–71, 2018.
https://doi.org/10.1007/978-3-030-01325-7_10

(see [2,9,11,14] and their references for the exhaustive list of work done in this area). However, all these results are derived assuming the underlying communication network to be a *complete undirected graph*, where the parties are assumed to be directly connected with each other by pair-wise private and authentic channels. There are scenarios, where such undirected graphs may not be available. For example, in a typical wireless network, the communication links may support only uni-directional communication. In a radio network, a base station can communicate to the receiving nodes, but communication in the other direction may not be possible. Further, it may be the case that a node is connected to some other node "indirectly" via intermediate nodes. Thus in a practical network like the Internet it is hard to ensure that every user is directly connected with every other user by a dedicated channel. This scenario can be appropriately modelled by a more generic *incomplete directed graph*. We are interested in the consensus problem in such arbitrary directed graphs.

In a series of beautiful work [15–17], the possibility of consensus protocols in arbitrary directed graphs is studied, where necessary and sufficient conditions are presented for the existence of consensus protocols. Separate conditions are derived for the *fail-stop* and *Byzantine* adversary model. The fail-stop model is a weaker adversary model and assumes that the adversary can crash any f nodes during the execution of a protocol. The more stronger Byzantine adversary model assumes that the adversary has full control over the set of f nodes under its control, which can be forced to behave in any arbitrary fashion during the protocol execution. In this work, we revisit the crash-tolerant version of the consensus problem in arbitrary directed graphs; specifically we look into the round complexity of crash-tolerant consensus protocols in arbitrary directed graphs. We stress that even though the fail-stop model is a weaker adversary model, it is practically motivated. For instance, in a typical distributed system, the chances that some of the "components" of the system stop working are more compared to some of the components behaving erratically. More specifically, it is relatively simpler for an attacker to crash a system and make it stop working completely, compared to taking full control of it and make it behave in an erroneous fashion. Hence studying the round complexity of consensus protocols in arbitrary directed networks against fail-stop corruption is practically motivated.

Existing Results for Crash-tolerant Consensus in Directed Graphs: The necessary (and sufficient) condition for the existence of crash-tolerant consensus protocol in directed graphs is presented in [15] and this is not a straight-forward extension of the necessary condition for the existence of crash-tolerant consensus in undirected[1] graphs. Informally, in directed graphs the necessary condition demands that even if an arbitrary set of f nodes crashes, there should still exist a special node in the graph, called *source*, which should have a directed path to every other node in the remaining graph (see Sect. 2 for the formal definition of source node and other related terms). The authors in [15] proved the sufficiency

---

[1] In undirected graphs, $f + 1$ node connectivity is both and necessary and sufficient for the existence of crash-tolerant consensus.

of their necessity condition by presenting two consensus protocols, one for the binary and the other for the multi-valued case.[2]

The protocols of [15] are significantly different from the traditional consensus protocols developed for undirected graphs. Specially they belong to a special class of consensus protocols, based on "flooding". In more detail, the protocols consist of several "phases", each consisting of d rounds of "send-receive-update", where d is called the *crash-tolerant diameter* of a directed graph. Informally, d is the maximum distance of any node from a potential source in the graph. Thus any given potential source can propagate its value to all remaining nodes in a single phase within the d rounds of flooding. In a round every node (including the source) broadcasts its value to its neighbours. At the end of the round, each node "updates" its value, by locally applying an update function to the received values. In the subsequent round, nodes broadcast their updated value. Now, two types of update function applied are a min function for a *min phase* and a max function for a *max phase*. The min function requires nodes to update their value by taking the minimum of all the received values (including its own value) and symmetrically in the max function nodes update by taking the maximum of all the received values. In [15], it was also shown that the usage of two different types of update function is necessary to achieve consensus for *anonymous* directed graphs.

The binary consensus protocol of [15] requires $2f + 2$ alternate min-max phases, each with d rounds. The round complexity of the protocol is $(2f + 2) \cdot d$ rounds and the communication complexity is $\mathcal{O}(nfd)$ bits (the number of neighbours of a node is $\mathcal{O}(n)$). In [15] the authors claimed that their binary consensus protocol based on min-max strategy cannot be extended trivially to the multi-valued case. Hence they present a different multi-valued consensus protocol, which in essence runs their binary consensus protocol K times, when the inputs are in the set $\{0, \dots, K\}$. The idea is to run an instance of the min-max based binary consensus for each candidate $k \in \{0, \dots, K\}$ to verify if some source node has a value k and if so, then try to reach agreement on this value k. The protocol requires $(2f+2) \cdot d \cdot K$ rounds of communication and the communication complexity is $\mathcal{O}(nfdK \log K)$ bits since a node has to communicate $\log K$ bits to every neighbour in a round. Clearly the protocol has exponential round and communication complexity, as $K = 2^{\log K}$.

Our Motivation and Results: In this work, we revisit the consensus protocols of [15] based on min-max strategy. Our main motivation is to improve the round and communication complexity of their protocols because the number of communication rounds and the amount of communication done in each round are crucial resources in a distributed protocol. We consider the binary consensus protocol of [15] and observe that if instead of d, we allow $d + 1$ rounds of communication in each of the phases, then it is possible to achieve consensus with just $f + 2$ alternate min-max phases, thus making the round complexity $(f + 2)(d + 1)$. We then show an optimization of our protocol, where we allow *only* d rounds in the

---

[2] In the binary consensus problem, the inputs of each node is a binary value. On the other hand in the multi-valued case, the inputs belong to a publicly known domain.

first and the *last* phase, thus reducing the round complexity to $(f+2)(d+1)-2$. Interestingly, we show that our protocol works even for the multi-valued case, with *no* modifications what so ever. Hence, unlike [15], the round complexity of our multi-valued consensus protocol is *independent* of K. The communication complexity of our protocol is $\mathcal{O}(nfd\log K)$ bits and for significantly large values of K our protocol improves upon the round and communication complexity of the multi-valued consensus protocols of [15]. Moreover, we improve the number of rounds for the binary consensus, for every $f, d \geq 2$.

We also address the problem of lower bound on the minimum number of rounds required by any crash-tolerant consensus protocol in a directed graph, based on min-max strategy and derive three interesting lower bounds. We first consider the case, where only $f + 1$ min-max phases are allowed in the protocol and with *no restriction* on the number of communication rounds in each phase. We show that it is impossible to achieve crash-tolerant consensus within $f + 1$ phases. Next we consider min-max based consensus protocols with *at least d* rounds in each phase. For such protocols, we show that it is impossible to achieve consensus in general with $(f + 2)(d + 1) - 3$ rounds in total. This further shows that our min-max based protocol with $(f+2)(d+1) - 2$ rounds is *round optimal*. Finally we consider min-max based consensus protocols with *exactly d* rounds of communication in each phase. Note that the consensus protocols of [15] belong to this class. For several values of f and d, we show that the minimum number of phases required to achieve consensus in this case is $2f + 2$, thus showing that the binary consensus protocol of [15] has the *optimal* number of communication rounds.

All the above lower bounds are derived by presenting non-trivial directed graphs and corresponding adversary strategies, which ensure that consensus is not achieved till sufficient number of min-max phases are allowed in the underlying protocol. We stress that different graphs and adversary strategies are required to derive the lower bound for different cases. Even though the lower bounds are for a restricted class of protocols (namely the one based on min-max strategy), to the best of our knowledge, these are the first (non-trivial) lower bounds on the round complexity of consensus protocols in directed graphs. More importantly, the lower bounds establish that our protocol is the best in terms of the round complexity if one is interested to design consensus protocols based on min-max strategy. Hence to obtain further improvements in the round complexity, a different approach (other than the min-max based strategy) is required.

Informal Discussion on Our Protocol: Our starting point is the binary consensus protocol of [15] with $2f + 2$ phases, each with d rounds. The correctness of their protocol is based on the guaranteed occurrence of two *consecutive* crash-free phases, among the $2f + 2$ alternate min-max phases, within which consensus is shown to be achieved. We observe that if instead of d rounds, we allow $d + 1$ rounds in each phase then consensus can be achieved if we either have two consecutive crash-free phases or a crashed phase followed by a crash-free phase, provided *only one* node crashes during the crashed phase. The base of our observation is the following: if during the crashed phase the single node to be crashed is a non-source node, then it is equivalent to having two consecutive crash-free phases (with source node(s) being unaltered) and so consensus will be achieved within these two phases. On the other hand, if during the crashed phase the single node to be crashed is a source node, then at least one of new source nodes will be at a distance of *one* from the crashed source (this observation lies at the heart of our protocol). If the crashed source node sends its value to one of the new source node before crashing, there will be still d rounds left for this new source node in the crashed phase to further propagate the crashed source node's value in the remaining graph. So in essence, we still get the effect of two consecutive crash-free phases. We further show that with $f + 2$ alternate min-max phases, there always exist either two crash-free phases or a crashed phase with a single crash, followed by a crash-free phase, irrespective of the way adversary crashes the f nodes.

Moving from the binary case to the multi-valued case, we find that the above ideas are applicable even for the multi-valued case. For simplicity, we consider the case when there are two crash-free phases and without loss of generality, let these be a min phase followed by a max phase. Let λ^{min} be the least value among the source nodes at the beginning of crash-free min phase. If the non-source nodes have their value greater than or equal to λ^{min} at the beginning of this phase, then clearly consensus will be achieved at the end of this min phase itself; this is because each node will update their value to λ^{min} at the end of the min phase. On the other hand, if some *non-source* node has a value smaller than λ^{min} at the beginning of the crash-free min phase, then consensus will not be achieved in this phase. However, at the end of this min phase, the modified values of all the nodes (both source as well as non-source) is upper bounded by λ^{min}; moreover all the *source nodes* will have λ^{min} as their modified value. Hence in the next crash-free phase which is a max phase, the value λ^{min} of the source nodes will be the maximum value in the graph and hence consensus will be achieved at the end of the crash-free max phase[3]. The above argument also works for the case when there is a crashed phase followed by a crash-free phase, where it is guaranteed that exactly one node crashes during the crashed phase.

Related Work: In [5], possibility of *approximate* crash-tolerant consensus in dynamic directed graphs is studied; informally in an approximate consensus protocol, the fault-free nodes are supposed to produce outputs within a certain constant ϵ of each other, where $\epsilon > 0$. On contrary, we are interested in the *exact* consensus, where $\epsilon = 0$. As mentioned earlier, most of the literature on consensus considers a complete graph, where parties are connected by pair-wise reliable channels and where the graph is assumed to be *static*. However, there are few works which consider different variations of this model. For example, [3] considers *undirected* graphs and shows that all-pair reliable communication is not necessary to achieve consensus against *Byzantine* adversary, provided nodes can use some authentication mechanism. In [1], Byzantine consensus in *unknown*

---

[3] This argument also shows that the binary consensus protocol of [15] with $2f + 2$ alternate min-max phases will work for the multi-valued case as well, with no modifications; this is because there will be always two consecutive crash-free phases.

networks is considered, where the underlying network remains *fully connected*. [4,10] considers *fault-free, approximate* consensus protocols where there are no faults, but the underlying graph is partially connected and *dynamic*. In [13], the authors consider edge corruptions, where edges may get Byzantine corrupted, but nodes remain fault-free. All these variations of consensus is different from the setting considered in this paper and so these results are incomparable to ours. Hence we do not consider these works for further discussion.

2 Preliminaries, Definitions and Notations

We consider a distributed synchronous network modelled as a simple directed graph $\mathbf{G} = (\mathsf{V}, \mathsf{E})$ where V represents the set of n nodes $\{v_1, v_2, \ldots, v_n\}$ and E represents the set of directed edges between the nodes in V. The communication network is assumed to be *static*; i.e. edges and nodes are not allowed to be inserted or deleted dynamically. Node v_i can communicate to node v_j if and only if the directed edge $(v_i, v_j) \in \mathsf{E}$. Moreover we assume that each node can send messages to itself. For a node $v \in \mathsf{V}$, the set N_v^+ denotes the set of "outgoing neighbours" of v in $\mathbf{G}$. That is, $\mathsf{N}_v^+ = \{v_j | (v, v_j) \in \mathsf{E}\}$. Thus, v can "directly" send messages to the nodes in the set N_v^+. The set N_v^- denotes the set of "incoming neighbours" of v in $\mathbf{G}$. That is, $\mathsf{N}_v^- = \{v_j | (v_j, v) \in \mathsf{E}\}$. Thus, the nodes in N_v^- can "directly" send messages to the node v. The network is assumed to be synchronous where all the nodes are synchronised and there exists a known upper bound on message delay. Any protocol in such a network is assumed to proceed as a sequence of rounds, where in every round, each node sends messages to its outgoing neighbours, receives messages sent by its incoming neighbours in that round, followed by local computation. We assume a *computationally unbounded* adaptive adversary $\mathcal{A}$, which can corrupt any f nodes in $\mathbf{G}$ in a fail-stop fashion, where a corrupted node can crash at any point of time during the execution of a protocol; however till the node crashes, it *honestly* follows the instructions of the underlying protocol. We also assume that if a node crashes during a round, then an arbitrary subset of its outgoing messages for that round are delivered to the corresponding neighbours, as decided by $\mathcal{A}$. We next define the consensus problem.

Definition 1 (Multi-valued Crash-Tolerant Consensus [15]). *Let Π be a synchronous protocol for the n nodes in $\mathbf{G}$, where each node $v_i \in \mathsf{V}$ has an input $in_i \in \{0, \ldots, K\}$ and each party has an output $out_i \in \{0, \ldots, K\}$, where K is publicly known. Then Π is called a crash-tolerant consensus protocol tolerating $\mathcal{A}$ if the following holds:* (1) **Agreement***: All fault-free nodes should have the same output. That is, for every fault-free nodes $v_i, v_j \in \mathsf{V}$, the condition $out_i = out_j$ holds.* (2) **Validity***: the output at any fault-free node must be some node's input. That is $out_i \in \{in_1, \ldots, in_n\}$ should hold.* (3) **Termination***: every fault-free node eventually decides on an output.*

We next recall few definitions from [15].

Definition 2 (Reduced Graph [15]). *Given a directed graph $G = (V, E)$ and a subset $F \subset V$, the reduced graph induced by F is $G_F = (V_F, E_F)$, where $V_F = V - F$ and $E_F = E \setminus \{(v_i, v_j) | v_i \in F \text{ or } v_j \in F\}$.*

Definition 3 (Crash-tolerant Node Connectivity [15]). *A graph $G = (V, E)$ is said to satisfy k crash-tolerant node connectivity if for any $F \subset V$ with $|F| \leq k$, there is at least one node $s \in V \setminus F$ that has a directed path to all the nodes in the corresponding reduced graph G_F.*

Definition 4 (Source of a Reduced Graph [15]). *Let $G = (V, E)$ be a graph and let $F \subset V$, with G_F being the corresponding reduced graph. Then a node v_s in G_F is called the source of G_F if there exists a directed path from v_s to all the nodes in G_F.*

For a reduced graph G_F, we denote by S_{G_F} the set of source nodes. The necessity condition for the existence of crash-tolerant consensus in a directed graph is given in Theorem 1.

Theorem 1 (Necessary Condition for Crash-Tolerant Consensus [15]). *Crash-tolerant consensus tolerating $\mathcal{A}$ is possible in a directed graph G only if G has f crash-tolerant node connectivity.*

We end this section with the definition of crash-tolerant diameter d of a directed graph. Informally it denotes the maximum number of rounds over all possible reduced graphs induced by various subsets of size atmost f, within which the message of a potential source node can reach all the remaining nodes in a reduced graph.

Definition 5 (Crash-tolerant Diameter [15]). *A spanning tree in a directed graph $G = (V, E)$ is said to be a rooted spanning tree rooted at a node $v_r \in V$ if v_r has a directed path to all the nodes in V. Let tree(v_r, G) denote the set of rooted spanning trees, rooted at v_r. We define height(v_r, G) as the minimum height of all the trees $T \in tree(v_r, G)$. That is:*

$$height(v_r, G) = \min_{T \in tree(v_r, G)} (height \, of \, T).$$

The crash-tolerant diameter d is defined as follows:

$$d = \max_{F \subset V, |F| \leq f} \max_{v_s \in S_{G_F}} (height(v_s, G_F)).$$

Note that in a directed graph with n nodes, d is always upper bounded by n.

2.1 Some Properties of Graphs with f Crash-Tolerant Node Connectivity

In this section we state few properties of reduced graphs which will be used in the rest of the paper. In the rest of this section we consider an arbitrary directed graph $\mathbf{G} = (\mathsf{V}, \mathsf{E})$ which has f crash-tolerant node connectivity. Moreover we consider a scenario where during the execution of an arbitrary protocol, $\mathcal{A}$ crashes a subset of nodes $\mathsf{F} \subset \mathsf{V}$, where $|\mathsf{F}| \leq f$. The corresponding reduced graph is denoted as $\mathbf{G_F} = (\mathsf{V_F}, \mathsf{E_F})$. If $\mathcal{A}$ further crashes additional $|\mathsf{T}|$ nodes in $\mathbf{G_F}$, with $|\mathsf{F} \cup \mathsf{T}| \leq f$, then the corresponding reduced graph is denoted as $\mathbf{G_{F'}} = (\mathsf{V_{F'}}, \mathsf{E_{F'}})$, where $\mathsf{F'} = \mathsf{F} \cup \mathsf{T}$ denotes the set of nodes, crashed by $\mathcal{A}$ so far. We use the notation $\mathbf{G_F} \rightarrow \mathbf{G_{F'}}$ to denote the transition when the additional nodes in T get crashed. The set $\mathsf{S_{G_F}}$ and $\mathsf{S_{G_{F'}}}$ will denote the set of source nodes for $\mathbf{G_F}$ and $\mathbf{G_{F'}}$ respectively. Note that both $\mathsf{S_{G_F}}$ and $\mathsf{S_{G_{F'}}}$ will be non-empty, as we are assuming $\mathbf{G}$ to have f crash-tolerant node connectivity. Due to space constraints, the proofs of the following properties are available in [6].

The following proposition states that if a node has a directed path to a source node in a reduced graph then the node also is a source node of the reduced graph.

Proposition 1. *If a node $v_i \in \mathsf{V_F}$ has a directed path to any node $v_j \in \mathsf{S_{G_F}}$ in $\mathbf{G_F}$, then $v_i \in \mathsf{S_{G_F}}$.*

As an immediate corollary of the above we get the following:

Corollary 1. *Let $v_i, v_j \in \mathsf{S_{G_F}}$ with $v_i \neq v_j$. Then all intermediate nodes along any directed path[4] between v_i and v_j also belong to $\mathsf{S_{G_F}}$.*

We next claim that during the execution of a protocol, a non-source node in a reduced graph cannot become a source node in the next reduced graph, as long as there exists at least one source node in the old reduced graph that is not crashed by the adversary.

Claim. Consider an arbitrary $v_i \in \mathsf{V_F}$, such that $v_i \notin \mathsf{S_{G_F}}$. Moreover let $v_i \in \mathsf{V_{F'}}$ (i.e. node v_i is not crashed during the transition $\mathbf{G_F} \rightarrow \mathbf{G_{F'}}$). Let there exist at least one node, say $v_j \in \mathsf{S_{G_F}}$ that is not crashed by the adversary during the transition $\mathbf{G_F} \rightarrow \mathbf{G_{F'}}$ (i.e. $v_j \notin \mathsf{T}$). Then $v_i \notin \mathsf{S_{G_{F'}}}$.

Based on the above claim, we next claim that during the execution of a protocol, the source set remains intact in reduced graphs, unless some subset of nodes within the source set crashes.

Claim. If during the transition $\mathbf{G_F} \rightarrow \mathbf{G_{F'}}$ $\mathsf{T} \cap \mathsf{S_{G_F}} = \emptyset$, then $\mathsf{S_{G_F}} = \mathsf{S_{G_{F'}}}$.

Finally we claim that if a source node of $\mathbf{G_F}$ crashes during the transition $\mathbf{G_F} \rightarrow \mathbf{G_{F'}}$, then at least one of the outgoing neighbours of this crashed source node will be the source for the next reduced graph.

Claim. Let $\mathsf{T} = \{v_s\}$, where $v_s \in \mathsf{S_{G_F}}$ (i.e. during $\mathbf{G_F} \rightarrow \mathbf{G_{F'}}$ the only node to crash is v_s). Then $\mathsf{N}_{v_s}^+ \cap \mathsf{S_{G_{F'}}} \neq \emptyset$, where $\mathsf{N}_{v_s}^+$ denotes the set of outgoing neighbours of v_s in the reduced graph $\mathbf{G_F}$.

---

[4] Note that a directed path will exist from v_i to v_j and from v_j to v_i in $\mathbf{G_F}$ as both v_i and v_j are source nodes.

3   Multi-valued Consensus Protocol Based on Min-Max Strategy

Let $\mathbf{G} = (\mathsf{V}, \mathsf{E})$ be a directed graph where $|\mathsf{V}| = n$, such that $\mathbf{G}$ has f crash-tolerant node connectivity. We present a multi-valued crash-tolerant consensus protocol called MinMax (see Fig. 1) tolerating $\mathcal{A}$. Similar to the consensus protocols of [15], the algorithm is based on min-max strategy, consisting of $f + 2$ phases, with even numbered phases being a min phase while odd numbered phases being a max phase. Each phase further consists of $d + 1$ rounds, where d denotes the crash-tolerant diameter of $\mathbf{G}$.

Protocol MinMax($\mathbf{G}$)

The input of the ith node is in_i where $in_i \in [0, K]$. For $i = 1, \ldots, n$, each node v_i executes the following code:

- Repeat for phase $p = 1$ to $f + 2$:
  - If $p \bmod 2 = 0$ then repeat the following steps $d + 1$ times (**Min Phase**):
    - Send in_i to all the nodes in $\mathsf{N}_{v_i}^+$.
    - Receive values from the nodes in^a $\mathsf{N}_{v_i}^-$.
    - Set in_i to the minimum of all the values received.
  - Else repeat the following steps $d + 1$ times (**Max Phase**):
    - Send in_i to all the nodes in $\mathsf{N}_{v_i}^+$.
    - Receive values from the nodes in $\mathsf{N}_{v_i}^-$.
    - Set in_i to the maximum of all values received.
- Output $out_i := in_i$ and terminate.

---

^{a} We assume that each node can communicate to itself. Hence v_i also sends its value to itself and receives the same.

Fig. 1. Multi-valued crash-tolerant consensus based on min-max strategy

We now prove the properties of MinMax. We first claim that irrespective of the strategy followed by $\mathcal{A}$ during the execution of MinMax, there always exist either two consecutive crash-free phases or a crashed phase with a single crash, followed by a crash-free phase. Formally, let k_i denote the total number of crashes that occur during the ith phase of MinMax, where $k_1 + \ldots + k_{f+2} \leq f$ and each $k_i \in \{0, \ldots, f\}$. Then we have the following lemma.

Lemma 1. *Irrespective of the strategy followed by $\mathcal{A}$ during* MinMax, *there exists at least one subsequence $k_{i-1}k_i$ such that either $k_{i-1} = 0, k_i = 0$ or $k_{i-1} = 1, k_i = 0$, where $i \in \{2, \ldots f + 2\}$.*

Proof. We prove the lemma using strong induction, over the values of f.

1. Consider the base case where, $f = 1$ and the number of phases are $f + 2 = 3$. The set of all possible sequence $k_1k_2k_3$ is $\{000, 100, 010, 001\}$ and each of them has either the subsequence 00 or 10.

2. Assume the lemma is true for all f where $1 \leq f \leq t - 1$; i.e. either the subsequence 00 or 10 occurs among all possible sequence $k_1 k_2 \ldots k_{f+2}$, where $f \leq t - 1$.
3. Now consider $f = t$. We focus on the last term k_{t+2}, where there are two possible cases:
   - If $k_{t+2} \geq 1$ then it implies that at most $t - 1$ faults could occur in the first $t + 1$ phases. However by induction hypothesis it follows that irrespective of the adversary strategy, the subsequence 00 or 10 will occur among all possible sequence $k_1 k_2 \ldots k_{t+1}$ if at most $t - 1$ faults are allowed during $t + 1$ phases. This automatically implies that the subsequence 00 or 10 will occur among all possible sequence $k_1 k_2 \ldots k_{t+2}$.
   - If $k_{t+2} = 0$ then we further focus on the term k_{t+1}. If k_{t+1} has value 1 or 0 we meet our subsequence requirement. However if $k_{t+1} \geq 2$ then it implies that at most $t - 2$ faults could occur in the first t phases. However by induction hypothesis it follows that irrespective of the adversary strategy, the subsequence 00 or 10 will occur among all possible sequence $k_1 k_2 \ldots k_t$ if at most $t - 2$ faults are allowed during t phases. This automatically implies that the subsequence 00 or 10 will occur among all possible sequence $k_1 k_2 \ldots k_{t+2}$.

$\square$

We next claim that if there are two consecutive crash-free phases then protocol MinMax achieves consensus within those two phases.

Lemma 2. *Let G have f crash-tolerant node connectivity. If during the execution of* MinMax *there are two consecutive phases, say p_t and p_{t+1}, such that no crash occurs in any of these two phases then consensus is achieved by the end of phase p_{t+1}.*

Proof. Without loss of generality let p_t be a min phase and p_{t+1} be a max phase. Let F be the set of nodes that have crashed before phase p_t. So p_t and p_{t+1} are executed over the reduced graph $G_F = (V_F, E_F)$ with the set of source nodes S_{G_F}. Note that the source set does not change during these two phases since no crash occurs. Let λ^{min} denote the minimum value among the nodes in S_{G_F} at the beginning of p_t and let $S_{G_F}^{min}$ be the set of source nodes possessing the minimum value λ^{min}. Now there are two possible cases:

- *All the nodes in the set $V_F \setminus S_{G_F}^{min}$ have values which are greater that or equal to λ^{min} at the beginning of p_t*: In this case, at the end of p_t, all the nodes in V_F will have their value equal to λ^{min}. This is because p_t is a min phase and by the definition of crash-tolerant diameter, the value λ^{min} will propagate to each node in V_F within d rounds of communication. Hence in this case, consensus is achieved at the end of p_t.
- *At least one node in the set $V_F \setminus S_{G_F}$ has value less than λ^{min} at the beginning of p_t*: In this case, all the nodes in the source set S_{G_F} will have their value set to λ^{min} at the end of p_t. This is because no node in the source set S_{G_F} will

ever see a value smaller than λ^{min} being propagated during p_t, otherwise it contradicts the assumption that λ^{min} is the minimum value among the nodes in S_{G_F} at the beginning of p_t. Moreover all the nodes in the set $V_F \setminus S_{G_F}$ will have their values set to a value which is less than or equal to λ^{min} at the end of p_t. This is because p_t is a min phase and the minimum value propagated to any node in $V_F \setminus S_{G_F}$ within d rounds of communication will be either λ^{min} or a value less that it. This implies that at the beginning of the next phase p_{t+1}, the value λ^{min} will be the maximum value of any node. Since all the nodes in S_{G_F} have λ^{min} as their value during p_{t+1} and since p_{t+1} is a max phase, no node will ever see a value greater than λ^{min} being propagated during p_{t+1}. Moreover by the definition of d, the value λ^{min} will be propagated to every non-source node during p_{t+1}. Hence at the end of p_{t+1} all the nodes in G_F will set their values to λ^{min}, thus achieving consensus at the end of p_{t+1}.

□

We next show that consensus will be achieved even if there is crashed phase followed by a crash-free phase, provided only one node crashes during the crashed phase.

Lemma 3. *Let G have f crash-tolerant node connectivity. If during the execution of MinMax there are two consecutive phases, say p_t and p_{t+1}, such that only one node crashes during p_t and no node crashes during p_{t+1}, then consensus is achieved by the end of phase p_{t+1}.*

Proof. Without loss of generality let p_t be a min phase and p_{t+1} be a max phase. Let F denote the nodes that have crashed before the phase p_t and let $G_F = (V_F, E_F)$ be the reduced graph at the beginning of p_t. Let $G_F \rightarrow G_{F'}$ denote the transition, where a single node in G_F, say v_c, crashes during the phase p_t, resulting in the reduced graph $G_{F'}$. Let S_{G_F} and $S_{G_{F'}}$ denote the set of source nodes for the reduced graphs G_F and $G_{F'}$ respectively. If the crashed node $v_c \notin S_{G_F}$ then the proof of the lemma is exactly the same as Lemma 2, as in this case $S_{G_F} = S_{G_{F'}}$ (follows from Claim 2.1). So we next consider the case when $v_c \in S_{G_F}$. We have two further sub-cases:

- *If the node v_c crashes during the first round of p_t and without propagating its value to any node in $S_{G_{F'}}$:* in this case we can effectively ignore the effect of the initial source set S_{G_F}. Moreover, from the second round onward of p_t, each node in the source set $S_{G_{F'}}$ will get the required d rounds of communication to propagate their value to every other node in $G_{F'}$ during p_t. Furthermore, during p_{t+1}, the source set remains the same as $S_{G_{F'}}$. In essence, this is equivalent to as if p_t and p_{t+1} are executed over the reduced graph $G_{F'}$ with at least d rounds of communication, with the source set being $S_{G_{F'}}$ and with no crash occurring in these phases. So using exactly the same arguments as in Lemma 2, we can conclude that consensus will be achieved by the end of phase p_{t+1}.

- *If the node v_c crashes after propagating its value to at least one node in $S_{G_{F'}}$:*
  let λ denote the value of v_c at the beginning of p_t. We first note that there
  exists at least one source node, say $v_{s'} \in S_{G_{F'}}$, such that $v_{s'}$ receives λ from
  v_c at the end of the first round of p_t. This is because the node $v_{s'}$ will be
  an outgoing neighbour of the node v_c in G_F (this follows from Claim 2.1).
  Moreover, the node $v_{s'}$ can propagate λ to all the remaining nodes in $G_{F'}$
  during p_t. This follows from the definition of d and the fact that p_t has still
  d rounds of communication left.
  Let $\lambda_{G'}^{\min}$ denote the minimum value among the nodes in $S_{G_{F'}}$ at the beginning
  of p_t and let $\lambda^{\min} = \min(\lambda, \lambda_{G'}^{\min})$. We claim that at the end of p_t, all the nodes
  in $S_{G_{F'}}$ will have $\lambda^{\min}$ as their updated value. This is because no node in $S_{G_{F'}}$
  will ever see a value smaller than $\lambda^{\min}$ being propagated. Next we claim that
  all the non-source nodes in $G_{F'}$ will have their value updated to $\lambda^{\min}$ or a
  smaller value at the end of p_t. More specifically, if the non-source nodes in
  $G_{F'}$ have their value greater than $\lambda^{\min}$ at the beginning of phase p_t, then all
  these non-source nodes will set $\lambda^{\min}$ as their updated value at the end of p_t
  and consensus will be achieved at the end of p_t. This is because $\lambda^{\min}$ will be
  propagated to all these nodes. On the other hand, if some non-source node
  has a value smaller than $\lambda^{\min}$ at the beginning of p_t, then the node will set
  a value smaller than $\lambda^{\min}$ as its updated value at the end of p_t. In this case,
  at the beginning of p_{t+1}, the value $\lambda^{\min}$ will be maximum value of any node.
  Moreover, all the nodes in $S_{G_{F'}}$ will have $\lambda^{\min}$ as their value. Since p_{t+1} is a
  max phase and no crash occurs during p_{t+1}, it follows from the definition of
  d that $\lambda^{\min}$ will be propagated to all the nodes in $G_{F'}$ and every node will set
  $\lambda^{\min}$ as their updated value at the end of p_{t+1}, thus attaining consensus.   □

The following theorem follows from Lemmas 1–3 and the fact that every node
will terminate the protocol after $(f + 2)(d + 1)$ rounds. In every round, each
node has to send $\log |K|$ bits to all its outgoing neighbours and there are $\mathcal{O}(n)$
outgoing neighbours of every node; this proves the communication complexity.

Theorem 2. *Let $G = (V, E)$ be a directed graph with f crash-tolerant node
connectivity and crash-tolerant diameter d, where $|V| = n$. Then protocol* MinMax
*is a $(f + 2)(d + 1)$ round protocol for multi-valued consensus tolerating $\mathcal{A}$. The
protocol has communication complexity $\mathcal{O}(nfd\log |K|)$ bits.*

Further optimization in the round complexity of MinMax: In [6], we
present a modified version of MinMax called MinMax″, where we allow the first
phase and the last phase to have exactly d rounds; the remaining f phases still
consist of $d + 1$ rounds of communication. Hence the total round complexity is
$(f + 2)(d + 1) - 2$. We show that MinMax″ still achieves consensus. The idea is
as follows: we first consider a variation MinMax′ of MinMax, where only the first
phase is restricted to d rounds and show that MinMax′ still achieves consensus
within $f + 2$ phases. This is argued depending upon whether the first phase of
MinMax′ is crash-free or not. If it is crash-free, then the execution of MinMax′ is
"equivalent" to that of MinMax, where there are $d + 1$ rounds in the first phase
and where the first phase is crash-free. On the other hand, if the first phase of

MinMax$'$ is not crash-free, then in the remaining $f + 1$ phases (each of which has $d + 1$ rounds), at most $f - 1$ crashes can occur. Now using Lemma 1, we can say that in these $f + 1$ phases, either there will be at least two consecutive crash-free phases or a crashed phase with a single crash followed by a crash-free phase. So consensus will be achieved by the end of $(f + 2)th$ phase. We then consider protocol MinMax$''$, which is a variation of MinMax$'$ in that the last phase is now restricted to d rounds. Now again depending upon whether the last phase of MinMax$''$ is crash-free or not we show that consensus will be achieved by MinMax$''$.

4   Lower Bounds on Round Complexity of Consensus Protocols Based on Min-Max Strategy

In this section we consider crash-tolerant consensus protocols based on min-max strategy, consisting of alternate min and max phases and show few impossibility results regarding the minimum number of rounds required by consensus protocols. Based on these results, we conclude that our protocol MinMax requires *optimal* number of communication rounds. We assume protocols which have alternate min-max phases, where the first phase is a min phase (this is without loss of generality).

4.1   Impossibility of Consensus in $f + 1$ Phases (Irrespective of the Number of Rounds)

Consider the family of graphs with f-crash-tolerant node connectivity such that every graph $\mathsf{G} = (\mathsf{V}, \mathsf{E})$ has the following properties (see Fig. 2): $\mathsf{V} = \{v_1, v_2, ..., v_{f+3}\}$ and $d = 1$. The edge set $\mathsf{E} = \{(v_i, v_j)\}$, where $i < j$ and $1 \le i \le f + 2$. Clearly $|\mathsf{E}| = \frac{((f+3)^2 - (f+3))}{2} - 1$. The graph has two "sink" nodes v_{f+2}, v_{f+3}, which do not have any outgoing edges. Node v_1 has input 1, while $v_2, ..., v_{f+3}$ have input 0.

Let Π_{MinMax} be an arbitrary protocol for G, consisting of $f + 1$ alternate min-max phases, where the first phase is a min phase. Moreover, each phase has *at least* d rounds of communication[5]. We consider the following adversary strategy $\mathcal{A}_{\mathsf{MinMax}}$ by $\mathcal{A}$ during Π_{MinMax}: No crash occurs during phase p_1. Adversary crashes node v_i during phase p_{i+1} for $i = 1, ..., f$ in the following fashion: during the first round of

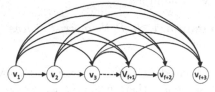

Fig. 2. The family of graphs in which it is impossible to achieve consensus in $f + 1$ phases.

---

[5] The number of rounds in each phase need to be finite so that Π_{MinMax} should terminate for each node.

p_{i+1}, the node v_i sends its value to nodes $v_{i+2}, \ldots, v_{f+3}$ and crashes. Hence, except v_{i+1}, all the neighbours of v_i receive v_i's value. The adversarial strategy ensures the following: if p_{i+1} is a min (resp. max) phase, then v_i will be the source node at the beginning of p_{i+1} with value 0 (resp. 1), while all the remaining nodes $v_{i+1}, \ldots, v_{f+3}$ will have value 1 (resp. 0). At the end of p_{i+1}, the node v_{i+1} will be the source node with value 1 (resp. 0), while all the remaining nodes $v_{i+2}, \ldots, v_{f+3}$ will have value 0 (resp. 1). Hence at the end of each min (resp. max) phase, all the nodes in the graph except the source will have value 0 (resp. 1). So at the end of Π_{MinMax}, the reduced graph will have nodes v_{f+1}, v_{f+2} and v_{f+3}, with v_{f+1} being the source and where v_{f+2} and v_{f+3} will have values, different from v_{f+1}. The formal proof is available in [6].

Theorem 3. *Consensus will not be achieved in* **G** *(Fig. 2) by* Π_{MinMax} *against the strategy* $\mathcal{A}_{\mathsf{MinMax}}$.

4.2 Impossibility of Consensus with $(f + 2)(d + 1) - 3$ Rounds in Total

Here we consider min-max protocols with $f + 2$ phases and $(f + 2)(d + 1) - 3$ rounds in total, with each phase having at least d rounds. We present a family of directed graphs and a corresponding adversarial strategy against which consensus will not be achieved at the end of $f + 2$ phases. This shows that the minimum number of rounds required is $(f + 2)(d + 1) - 2$, implying that our protocol MinMax″ (optimized variant of MinMax) is *round optimal*. Note that the adversarial strategy $\mathcal{A}_{\mathsf{MinMax}}$ and the graph of Fig. 2 cannot be used to derive the lower bound. This is because if we allow $f + 2$ phases then consensus will be achieved in the graph of Fig. 2 against $\mathcal{A}_{\mathsf{MinMax}}$. Hence we need to modify the graph and also the adversary strategy.

Let Π be an arbitrary min-max based protocol with $f + 2$ phases $p_1 \ldots, p_{f+2}$ and $(f + 2)(d + 1) - 3$ rounds, with each phase having at least d rounds. We first state few properties of Π, based on counting arguments, whose proofs are available in [6].

Lemma 4. *In* Π *there exist at least three phases consisting of exactly d rounds each.*

Based on Lemma 4, we next show that in Π, other than the first and the last phase, there exists at least one "intermediate phase" $p_\ell \in \{p_2, \ldots, p_{f+1}\}$ consisting of *exactly d* rounds; moreover if $p_\ell \neq p_{f+1}$ then it holds that the phase $p_{\ell+2}$ in Π has *at most $d + 1$* rounds[6]. More specifically, let $r_1, \ldots, r_{f+2}$

---

[6] Note that if $p_\ell = p_{f+1}$ then there is no phase $p_{\ell+2}$ in Π. If phase $p_{\ell+2}$ exists in Π then it will have either d or $d + 1$ rounds because in Π each phase has at least d rounds.

denote the number of rounds in phase $p_1, \ldots, p_{f+2}$ of Π respectively, where each $r_i \geq d$ and where $r_1 + \ldots + r_{f+2} = (f+2)(d+1) - 3$. Then we have the following lemma.

Lemma 5. *In protocol Π, one of the following holds:*

- *There exists a phase $p_\ell \in \{p_2, \ldots, p_f\}$, where $r_\ell = d$ and $r_{\ell+2} \leq d+1$.*
- *$r_{f+1} = d$.*

Now consider the directed graph **G** (see Fig. 3) with f crash-tolerant node connectivity and with crash-tolerant diameter d. We next specify an adversary strategy $\mathcal{A}_\Pi$ consisting of sub-strategies $\mathcal{A}_\Pi^1$ and $\mathcal{A}_\Pi^2$ against the protocol Π, executed over **G**:

- From p_1 to $p_{\ell-1}$, $\mathcal{A}_\Pi^1$ is followed, which is same as the strategy $\mathcal{A}_{\mathsf{MinMax}}$. Namely p_1 is crash-free. Then in $p_2, \ldots, p_{\ell-1}$, the source node crashes during the first round of the phase after sending its value to all its outgoing neighbours, except the next source in the $L1$ layer.
- Between p_ℓ and $p_{\ell+2}$, the sub-strategy $\mathcal{A}_\Pi^2$ is followed: at the beginning of phase p_ℓ, the node $v_{\ell-1}$ in the $L1$ layer will be the source node. During the first round of p_ℓ, the source $v_{\ell-1}$ crashes after sending its value to the next source v_ℓ. The next phase $p_{\ell+1}$ is a crash-free phase. During $p_{\ell+2}$, the source node v_ℓ crashes in one of the following two ways, depending upon whether $p_{\ell+2}$ has d or $d+1$ rounds:
  - $p_{\ell+2}$ *has d rounds*: here v_ℓ crashes after sending its value to all its neighbours (both in the $L1$ layer as well as the $P(1)$ layer), except the next source $v_{\ell+1}$.
  - $p_{\ell+2}$ *has $d+1$ rounds*: here v_ℓ crashes after sending its value only to its neighbours in the $L1$ layer (and not to the neighbours in $P(1)$ layer), except $v_{\ell+1}$.
- For the remaining phases $p_{\ell+3}, \ldots, p_{f+2}$, sub-strategy $\mathcal{A}_\Pi^1$ is followed, where in each phase, the source node in $L1$ layer crashes after sending its value to all its neighbour (in $L1$ layer and $P(1)$ layer), except the next source in the $L1$ layer.

Note that in the above description, we assumed that $p_\ell \neq p_{f+1}$, so that phase $p_{\ell+2}$ exists in Π. In case $p_\ell = p_{f+1}$, then $\mathcal{A}_\Pi^1$ is followed from p_1 to p_f, while $\mathcal{A}_\Pi^2$ is followed during p_{f+1} and p_{f+2}. Finally we state the following theorem.

Theorem 4. *Let Π be a min-max protocol wth $f+2$ phases with total $(f+2)(d+1) - 3$ rounds, where each phase has at least d rounds. Then consensus will not be achieved in **G** (Fig. 3) by Π against $\mathcal{A}_\Pi$.*

4.3   Impossibility of Consensus Based on Min-Max Strategy in $2f + 1$ Phases with D Rounds

Here we consider protocols based on min-max strategy, consisting of $2f +$ 1 alternate min and max phases, with each phase having d rounds of communication. For several values of f and d, we show that there exist graphs for which it is impossible to reach consensus within $2f+1$ phases. The adversary strategy in all these graphs will be the following: without loss of generality, we will assume that the first phase is a min phase. Hence there will be $f + 1$ min phases and f max phases. There will be *no* crash during the min phases and during each max phase, *one new* node

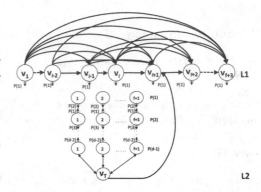

Fig. 3. Graphs in which it is impossible to achieve consensus in $f + 2$ phases with each round having at least d rounds and with total $(f + 2)(d + 1) - 3$ rounds.

will get crashed by the adversary. Moreover, the node to be crashed will be the *source* node of the reduced graph at the beginning of that phase. It will be always ensured that every reduced graph has *only one* source node. It will be ensured that no consensus is achieved during any of the min phases. This is achieved by ensuring that at the beginning of each min phase, the source has value 1 and there exists at least one non-source node with value 0. As a result, at the end of each min phase, the source will retain 1 as its value (as it will never see the value 0 during the min phase because there will be only one source), while there will be at least one node, which retains 0 as its value. During each of the f max phases, the adversary will crash the source node. The crashed node will crash during the first round of a max phase, after sending its value 1 *only* to the new source of the reduced graph. The new source will further get $d - 1$ rounds to propagate the value 1 that it received from the crashed source. However, it will be ensured that there is some node with value 0 which is d distance apart from the new source, such that it never sees the value 1 during the max phase and as a result, it retains its original value 0, thus preventing consensus being achieved during the crashed max phase.

Presenting a generalized graph which maintains the above properties for a general value of f and d is an extremely challenging task. In [6], we present graphs for several values of f and d, where consensus is achieved only at the end of $2f + 2$ phases.

Acknowledgments. We thank the anonymous referees of SIROCCO 2018 for their helpful comments. The work of the first two authors is financially supported by Infosys foundation. The third author would like to acknowledge the financial support by SERB Women Excellence Award from Science and Engineering Research Board of India and INSPIRE Faculty Fellowship from Department of Science & Technology, India

References

1. Alchieri, E.A.P., Bessani, A.N., da Silva Fraga, J., Greve, F.: Byzantine consensus with unknown participants. In: Baker, T.P., Bui, A., Tixeuil, S. (eds.) OPODIS 2008. LNCS, vol. 5401, pp. 22–40. Springer, Heidelberg (2008). https://doi.org/10.1007/978-3-540-92221-6_4
2. Attiya, H., Welch, J.: Distributed Computing: Fundamentals. Simulation and Advanced Topics, Wiley series on Parallel and Distributed Computing (2004)
3. Bansal, P., Gopal, P., Gupta, A., Srinathan, K., Vasishta, P.K.: Byzantine agreement using partial authentication. In: Peleg, D. (ed.) DISC 2011. LNCS, vol. 6950, pp. 389–403. Springer, Heidelberg (2011). https://doi.org/10.1007/978-3-642-24100-0_38
4. Bertsekas, D.P., Tsitsiklis, J.N.: Parallel and Distributed Computation: Numerical Methods. Athena Scientific, Optimization and Neural Computation Series (1997)
5. Charron-Bost, B., Függer, M., Nowak, T.: Approximate consensus in highly dynamic networks: the role of averaging algorithms. In: Halldórsson, M.M., Iwama, K., Kobayashi, N., Speckmann, B. (eds.) ICALP 2015. LNCS, vol. 9135, pp. 528–539. Springer, Heidelberg (2015). https://doi.org/10.1007/978-3-662-47666-6_42
6. Choudhury, A., Garimella, G., Patra, A., Ravi, D., Sarkar, P.: Crash-tolerant consensus in directed graph revisited. Cryptology ePrint Archive, Report 2018/436 (2018)
7. Dolev, D.: The Byzantine generals strike again. J. Algorithms **3**(1), 14–30 (1982)
8. Fischer, M.J., Lynch, N.A., Merritt, M.: Easy impossibility proofs for distributed consensus problems. In: PODC, pp. 59–70. ACM (1985)
9. Fitzi, M.: Generalized communication and security models in byzantine agreement. Ph.D. thesis, ETH Zurich (2002)
10. Jadbabaie, A., Lin, J., Morse, A.S.: Coordination of groups of mobile autonomous agents using nearest neighbor rules. IEEE Trans. Automat. Contr. **48**(6), 988–1001 (2003)
11. Lynch, N.A.: Distributed Algorithms. Morgan Kaufmann (1996)
12. Pease, M., Shostak, R.E., Lamport, L.: Reaching agreement in the presence of faults. JACM **27**(2), 228–234 (1980)
13. Schmid, U., Weiss, B., Keidar, I.: Impossibility results and lower bounds for consensus under link failures. SIAM J. Comput. **38**(5), 1912–1951 (2009)
14. Tseng, L.: Recent results on fault-tolerant consensus in message-passing networks. In: Suomela, J. (ed.) SIROCCO 2016. LNCS, vol. 9988, pp. 92–108. Springer, Cham (2016). https://doi.org/10.1007/978-3-319-48314-6_7
15. Tseng, L., Vaidya, N.H.: Crash-tolerant consensus in directed graphs. CoRR, abs/1412.8532, 2014. Conference version appeared as [16]
16. Tseng, L., Vaidya, N.H.: Fault-tolerant consensus in directed graphs. In: PODC, pp. 451–460. ACM (2015)
17. Tseng, L., Vaidya, N.H.: A note on fault-tolerant consensus in directed networks. SIGACT News **47**(3), 70–91 (2016)

A Distributed Algorithm for Finding Hamiltonian Cycles in Random Graphs in $O(\log n)$ Time

Volker Turau$^{(\boxtimes)}$ (iD)

Institute for Telematics, Hamburg University of Technology,
Am Schwarzenberg-Campus 3, 21073 Hamburg, Germany
turau@tuhh.de
https://www.ti5.tuhh.de/staff/turau/

Abstract. It is known for some time that a random graph $G(n,p)$ contains w.h.p. a Hamiltonian cycle if p is larger than the critical value $p_{crit} = (\log n + \log \log n + \omega_n)/n$. The determination of a concrete Hamiltonian cycle is even for values much larger than p_{crit} a nontrivial task. In this paper we consider random graphs $G(n,p)$ with p in $\tilde{\Omega}(1/\sqrt{n})$, where $\tilde{\Omega}$ hides poly-logarithmic factors in n. For this range of p we present a distributed algorithm $\mathcal{A}_{\mathsf{HC}}$ that finds w.h.p. a Hamiltonian cycle in $O(\log n)$ rounds. The algorithm works in the synchronous model and uses messages of size $O(\log n)$ and $O(\log n)$ memory per node.

Keywords: Distributed algorithm · Hamiltonian cycle
Random graph

1 Introduction

Surprisingly few distributed algorithms have been designed and analyzed for random graphs. To the best of our knowledge the only work dedicated to the analysis of distributed algorithms for random graphs is [5,16,17]. This is rather surprising considering the profound knowledge about the structure of random graphs available since decades [3,10]. While algorithms designed for general graphs obviously can be used for random graphs the specific structure of random graphs often allows to prove asymptotic bounds that are far better. In the classical Erdős and Rényi model for random graphs a graph $G(n,p)$ is an undirected graph with n nodes where each edge independently exists with probability p [7]. The complexity of algorithms for random graphs often depends on p, e.g., Krzywdziński et al. [16] proposed a distributed algorithm that finds w.h.p. a coloring of $G(n,p)$ with $18np$ colors in $O(\ln \ln p^{-1})$ rounds.

In this work we focus on finding Hamiltonian cycles in random graphs. The decision problem, whether a graph contains a Hamiltonian cycle, is NP-complete.

This work is supported by the Deutsche Forschungsgemeinschaft (DFG) under grant DFG TU 221/6-2.

© Springer Nature Switzerland AG 2018
Z. Lotker and B. Patt-Shamir (Eds.): SIROCCO 2018, LNCS 11085, pp. 72–87, 2018.
https://doi.org/10.1007/978-3-030-01325-7_11

It is a non-local graph problem, i.e., it is required to always consider the entire graph in order to solve the problem. It is impossible to solve it in the *local neighborhoods*. For this reason there is almost no work on distributed algorithms for finding Hamiltonian cycles in general graphs. On the other hand it is well known that $G(n, p)$ contains w.h.p. a Hamiltonian cycle, provided $p \geq p_{crit} = (\log n + \log \log n + \omega(n))/n$, where $\omega(n)$ satisfies $\lim_{n \to \infty} \omega(n) = \infty$ [3, Th. 8.9]. There is a large body of work on sequential algorithms for computing w.h.p. a Hamiltonian cycle in a random graph (e.g. [1, 4, 21–23]).

We are only aware of two distributed algorithms for computing Hamiltonian cycles in random graphs. The algorithm by Levy et al. [17] outputs w.h.p. a Hamiltonian cycle provided $p = \omega(\sqrt{\log n}/n^{1/4})$. It works in synchronous distributed systems, terminates in linear worst-case number of rounds, requires $O(n^{3/4+\epsilon})$ rounds on expectation, and uses $O(n)$ space per node. The algorithm of Chatterjee et al. [5] works for $p \geq c \log n/n^{\delta}$ and has a run time of $\tilde{O}(n^{\delta})$.

The search for a distributed algorithm for a Hamiltonian cycle is motivated by the usage of virtual rings for routing in wireless networks [19, 25]. A virtual ring is a directed closed path involving each node of the graph, possibly several times. Virtual rings enable routing with constant space routing tables, messages are simply forwarded along the ring. The downside is that they may incur a linear path stretch. To attenuate this, distributed algorithms for finding *short* virtual rings have been proposed [12, 25]. Hamiltonian cycles are the *shortest* possible virtual rings and therefore of great interest. Short virtual rings are also of interest for all token circulation techniques as discussed in [8]. Kim et al. discuss the application of random Hamiltonian cycles for peer-to-peer streaming [13].

This paper uses the synchronous $\mathcal{CONGEST}$ model, i.e., each message contains at most $O(\log n)$ bits. Furthermore, each node has only $O(\log n)$ bits of local memory. Without these two assumptions there is a very simple solution provided the nodes have unique identifiers. First a BFS-tree rooted in a node v_0 is constructed. Then the adjacency list of each node is convergecasted to v_0 which applies a sequential algorithm to compute w.h.p. a Hamiltonian path (see Sec. 1.1). The result is broadcasted into the graph and thus each node knows its neighbor in the Hamiltonian cycle. This can be achieved in $O(diam(G))$ rounds. Note that if $p = \omega_n \log n/n$ then w.h.p. $diam(G(n, p)) = O(\log n/\log np)$ [6, 10]. In particular for p in $\tilde{\Omega}(1/\sqrt{n})$ w.h.p. the diameter of $G(n, p)$ is constant [2].

For the stated restrictions on message size and local storage we propose an algorithm that terminates in a logarithmic number of rounds, this is a significant improvement over previous work [5, 17]. Our contribution is the distributed algorithm $\mathcal{A}_{HC}$, its properties can be summarized as follows.

Theorem 1. *Let $G(n, p)$ with $p \geq (\log n)^{3/2}/\sqrt{n}$ be a random graph. Algorithm $\mathcal{A}_{HC}$ computes in the synchronous model w.h.p. a Hamiltonian cycle for G using messages of size $O(\log n)$. $\mathcal{A}_{HC}$ terminates in $O(\log n)$ rounds and uses $O(\log n)$ memory per node.*

1.1 Related Work

Pósa showed already in 1976 that almost all random graphs with $cn \log n$ edges possess a Hamiltonian cycle [21]. Later Komlós et al. determined the precise threshold p_{crit} for the existence of a Hamiltonian cycle in a random graph [14]. A sequential deterministic algorithm that works w.h.p. at this threshold requiring $O(n^{3+o(1)})$ time is due to Bollobás et al. [4]. For larger values of p or restrictions on the minimal node degree, more efficient algorithms are known [1,11].

The above cited algorithms were all designed for the sequential computing model. Some exact algorithms for finding Hamiltonian cycles in $G(n, p)$ on parallel computers have been proposed [9]. The first operates in the EREW-PRAM model and uses $O(n \log n)$ processors and $O(\log^2 n)$ time, while the second one uses $O(n \log^2 n)$ processors and $O((\log \log n)^2)$ time in the P-RAM model. MacKenzie and Stout proposed an algorithm for CRCW-PRAM machines that operates in $O(\log^* n)$ expected time and requires $n/\log^* n$ processors [18].

There are several approaches to construct a Hamiltonian cycle. The approach used by Levy et al. at least goes back to the work of MacKenzie and Stout [18]. They initially construct a small cycle with $\Theta(\sqrt{n})$ nodes. As many as possible of the remaining nodes are assorted in parallel into $\sqrt{n}$ vertex-disjoint paths. During the final phase, each path and each non-covered vertex is patched into the initial cycle. The second approach is used in the proofs to establish the critical value p_{crit} (e.g., [15,21]) and all derived sequential algorithms (e.g., [4]). Initially a preferably long path is constructed, e.g., using a depth first search algorithm [11]. This path is extended as long as the node at the head of the path has a neighbor that is not yet on the path. Then the path is *rotated* until it can be extended again. A rotation of the path cuts off a subpath beginning at the head, reverses the order of the subpath's nodes, and reattaches the subpath again. The procedure stops when no sequence of rotations leads to an extendable path. The algorithm in [5] follows this approach.

2 Computational Model and Assumptions

This work employs the synchronous $\mathcal{CONGEST}$ model of the *distributed message passing model* [20], i.e., each message contains at most $O(\log n)$ bits. Furthermore, each node has only $O(\log n)$ bits of local memory. The communication network is represented by an undirected graph $G = (V, E)$, where V is a set of n processors (nodes) and E represents the set of m bidirectional communication links (edges) between them. Each node carries a unique identifier. Communication between nodes is performed in synchronous rounds using messages exchanged over the links. Upon reception of a message, a node performs local computations and possibly sends messages to its neighbors. These operations are assumed to take negligible time.

The prerequisite of Algorithm $\mathcal{A}_{\mathsf{HC}}$ is a distinguished node v_0 which is the starting point of the Hamiltonian cycle and acts as a coordinator in the final phases of $\mathcal{A}_{\mathsf{HC}}$. The results proved in this work hold *with high probability* (w.h.p.) which means with probability tending to 1 as $n \to \infty$.

3 Informal Description of Algorithm $\mathcal{A}_{HC}$

Algorithm $\mathcal{A}_{HC}$ operates in sequential phases, each of them succeeds w.h.p. The first two phases last $O(\log n)$ rounds. Each subsequent phase requires a constant number of rounds only. Phase 0 lasts $3(3 \log n - 1)$ rounds and constructs a path P of length $3 \log n$ starting in v_0. In the next $3 \log n$ rounds Phase 1 closes P into a cycle C of length at most $4 \log n$. The following $16 \log(n)$ phases are called the middle phases. In each of those phases the number of nodes in C is increased. The increase is by a constant factor until C has $n/7$ nodes. Afterwards, the increase declines roughly linearly until C has $n - 3 \log n$ nodes. In each middle phase the algorithm tries to concurrently integrate as many nodes into C as possible. This is achieved by replacing edges (v, w) of C by two edges (v, x) and (x, w), where x is a node outside of C. At the end of the middle phases w.h.p. C has more than $n - 3 \log n$ nodes.

The integration of the remaining $3 \log n$ nodes requires a more sophisticated algorithm. This is done in the final phases. The idea is to remove two edges – not necessarily adjacent – from C and insert three new edges. This requires to reverse the edges of a particular segment of C of arbitrary length. Thus, this is no longer a local operation. Furthermore, segments may overlap and hence, the integration of several nodes can only be performed sequentially. Thus, this task requires coordination. Node v_0 takes over the role of a coordinator.

At the beginning of each final phase all nodes outside C that can be integrated report this to v_0, which in turn selects one of these nodes to perform this step. For this purpose a tree routing structure is set up, so that each node can reach v_0 w.h.p. in 3 hops. In order for the nodes of the segment to perform the reordering concurrently, the nodes of C are numbered in an increasing order (not necessarily consecutively) beginning with v_0. The assignment of numbers is embedded into the preceding phases with no additional overhead. The numbering is also maintained in the final $3 \log n$ integration steps. In order to accomplish the integration in a constant number of rounds – i.e., independent of the length of the segment – node v_0 floods the numbers of the terminal nodes of the segment to be reversed into the network. Upon receiving this information, each node can determine if it belongs to the segment to be reversed and can recompute its number to maintain the ordering. Note that this routing structure requires only $O(\log n)$ memory per node. Each of the $3 \log n$ final phases lasts a constant number of rounds.

Algorithm $\mathcal{A}_{HC}$ stops when either C is a Hamiltonian cycle or no more nodes can be integrated into C. The first event occurs w.h.p.

4 Formal Description of Algorithm $\mathcal{A}_{HC}$

Algorithm $\mathcal{A}_{HC}$ operates in synchronous rounds. By counting the rounds a node is always aware in which round and therefore also in which phase it is. Each phase lasts a known fixed number of rounds. If the work is completed earlier, the network is idle for the remaining rounds. This requires each node to know n.

Algorithm $\mathcal{A}_{\mathsf{HC}}$ gradually builds an oriented cycle C starting with node v_0. The cycle is maintained as a doubly linked list to support insertions. The orientation of C is administered with the help of variable $next$ – initially $null$ – which stores the identifier of the next node on the cycle in clockwise order. In the following each phase is described in detail.

4.1   Pre-processing

The algorithm is started by node v_0 which executes algorithm FLOOD [20] to construct a BFS tree. By Corollary 8 (i) of [2] the diameter of G is w.h.p. at most 3. Thus, in 3 rounds a BFS tree rooted in v_0 is constructed (Lemma 5.3.1, [20]). After a further 6 rounds each node is aware of n the number of nodes in the network. This allows to run each phase for the stated number of rounds.

4.2   Phase 0

In phase 0 an oriented path P starting in v_0 of length $3 \log n$ is constructed. Phase 0 lasts $3(3 \log n - 1)$ rounds. Initially $P = \{v_0\}$ and $v_0.next = v_0$. The following steps are repeated $3 \log n - 1$ times.

1. The final node v of P sends an invitation message to all neighbors. All neighbors not on P (i.e., nodes with $next = null$) respond to v.
2. If v does not receive any response the algorithm halts. Otherwise v randomly selects among the nodes that have responded a node w, sets $v.next := w$, informs w that it is the new final node, and instructs w to continue with phase 0. This message includes the id of node v_0, i.e., at any point in time all nodes of P know v_0.

4.3   Phase 1

In phase 1 path P is extended into an oriented cycle C of length at most $4 \log n$. The following steps are repeated at most $\log n$ times. Phase 1 lasts $3 \log n$ rounds.

1. The final node v of P sends an invitation message containing the id of node v_0 to all neighbors. All neighbors not on P respond to v. The response includes the information whether the recipient is connected to v_0.
2. If v does not receive any response the algorithm halts. If at least one responding node is connected to v_0, then v randomly selects such a node w, sets $v.next = w$, and informs w to close the cycle C, i.e., to set $w.next = v_0$. Otherwise v randomly selects a responding node w to extend P as in phase 0 and instructs w to repeat phase 1.
3. If after $\log n$ repetitions P is not a cycle then the algorithm halts otherwise the middle phases start.

4.4   Middle Phases

While in the first two phases actions were executed sequentially, in the middle phases many nodes are integrated concurrently. In each of the subsequent phases the following steps are performed (see Fig. 1). Each of the $16 \log n$ middle phases is performed in three rounds.

1. Each node w on C broadcasts its own id and the id of its predecessor on C using message I_1.
2. If a node v outside C receives a message I_1 from a node w such that the predecessor of w on C is a neighbor of v, it inserts w into the set C_v.
3. Each node v outside C with $C_v \neq \emptyset$ randomly selects a node w from C_v and sends an invitation message I_2 to the predecessor of w on C.
4. Each node $w \in C$ that received an invitation I_2 randomly selects a node v from which it received an invitation, sets $w.next = v$, and informs v with acceptance message I_3 to set its variable $next$ to the old successor w' of w. In other words the edge (w, w') is replaced by the edges (w, v) and (v, w').

It is unnecessary to store C_v. Using reservoir sampling this step can be implemented with $O(\log n)$ storage. Individual extensions do not interfere with each other. Each node outside C gets in the last round of a middle phase at most one request for extension and for each edge of C at most one request is sent.

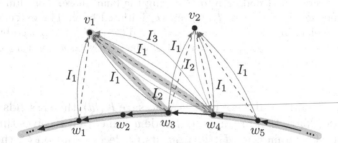

Fig. 1. The integration of nodes during a middle phase: Nodes w_i sent a message I_1 to all nodes outside C (red arrows). Nodes v_1 and v_2 sent a message I_2 back to w_4; v_2 might have also selected w_4 and sent I_3 to w_5. Node w_4 selected v_1 and sent back message I_3. Edge (w_4, w_3) is replaced by the edges (w_4, v_1) and (v_1, w_3). The extended cycle is depicted by the blue ribbon. (Color figure online)

4.5   Final Phases

After the completion of the middle phases the cycle C has w.h.p. at least $n - 3 \log n$ nodes. At that point the expected number of nodes $v \in V \setminus C$ that send an invitation I_2 becomes too low to complete the cycle. Therefore, the integration of the remaining nodes requires a more complex integration procedure as depicted in Fig. 2. The procedure of the final phases is as follows. Each node $v \in V \setminus C$ with identifier id sends a message $I_1(id)$ to each of its neighbors. A node $w_1 \in C$ that receives a message $I_1(id)$ sends a message $I_2(id)$ to its neighbor w_2 on C in clockwise order. If w_2 also received a message $I_1(id)$ (with the same id), then nodes w_1, w_2 and the initiating node v with identifier id form a triangle. Then v can be directly integrated into C as done in the middle phases. In this case w_1 asks v to initiate the integration step. A node on C receives w.h.p. at most 15 messages I_1 with different identifiers. Each node will aggregate all these identifiers into one message I_2. The same argument is applied to messages of type I_3. Thus, a final phase can be implemented with messages of size $O(\log n)$.

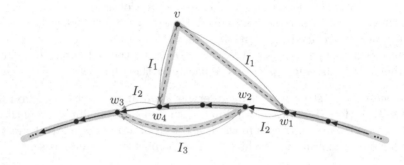

Fig. 2. The integration of node v into C during the final phase. The thin red arrows indicate the flow of the messages I_1, I_2, and I_3 initiated by v. The extended cycle is depicted by the blue ribbon. The edges (w_1, w_2) and (w_4, w_3) are replaced by the edges (w_1, v), (v, w_4), and (w_2, w_3). The order of the edges between w_4 and w_2 is reversed. (Color figure online)

Otherwise, if node w_2 did not receive a message $I_1(id)$, then it sends a message $I_3(id)$ to all neighbors that are on C. If a node w_3 on C that receives this message $I_3(id)$ also received a message $I_2(id)$ from its predecessor w_4 on C, then node v can be integrated into C as shown in Fig. 2. This is achieved by replacing edges (w_1, w_2) and (w_4, w_3) from C by edges (w_1, v), (v, w_4), and (w_2, w_3). Also, the edges on the segment from w_2 to w_4 must be traversed in opposite order, note that the number of nodes between w_2 and w_4 is not bounded. A naive explicit reversing of the order of the edges on the middle segment may require more than $O(\log n)$ rounds. Thus, we propose a different approach.

Apart from the reversal of the edges in the middle segment this integration can be implemented within five rounds. Node w_3 informs v about this integration possibility, this notification also includes the identifiers of nodes w_4 and w_2. The participating nodes w_4, w_2 and w_1 are also informed. The approach to invert the middle segment in a constant number of rounds is explained below.

Unfortunately there is another issue. While each node outside C can be integrated individually, these integration steps cannot be executed concurrently. A problem arises if the segments, which are inverted (e.g. from w_2 to w_4), overlap.

This can result in separate cycles as shown in Fig. 3. Even if the integration of the remaining nodes is performed sequentially, a problem appears if the reversal of the middle segment is not made explicit. In this case nodes receiving message I_1 may not have a consistent view with respect to the clockwise order of C.

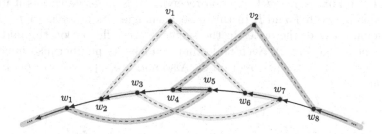

Fig. 3. The depicted scenario shows that the integration of two nodes with overlapping segments cannot be performed concurrently as this would lead to two cycles (shown in green and blue). If v_2 would be integrated first, then v_1 can no longer be integrated, since the predecessor of w_6 is then w_5 which is not connected to w_3. (Color figure online)

The solution to the problem of interfering concurrent integrations is to serialize all integration steps. For this purpose node v_0 acts as a coordinator. In each of the final phases each node v outside C first checks if can be integrated using the above described sequence of messages I_1 to I_3. If this is the case then v randomly selects one of these possibilities and informs v_0. This message includes information about the four nodes on C that characterize the integration (see below for details). Node v_0 selects among all offers a single node v and informs it. Upon receiving the integration order, a node v initialize the integration which is completed after fives rounds. Then the integration of the next node can start.

The solution for the second problem – the reversal of the segment – is based on an ascending numbering of the nodes. Such a numbering can easily be established in the first and middle phases. During phases 0 and 1 the nodes are numbered as follows: Node v_0 has number 0. In clockwise order the nodes have numbers $n^{14}, 2n^{14}, 3n^{14}, \ldots, \beta n^{14}$ for some integer constant $\beta \leq \lceil 4 \log n \rceil$. Thus, the difference between two consecutive nodes is n^{14}. During the middle phases when a node v is integrated into C between two nodes with numbers $f < l$ the integrated node gets the number $\lceil (f+l)/2 \rceil$. This is an integer strictly between f and l as long as $|f - l| \geq 2$. If a node is integrated between v_0 and the node with the highest number y, the new number is $y + \lceil (\beta+1)n/2 \rceil$. It is straightforward to verify that all numbers are different and are ascending along the cycle beginning with v_0. The choice of the initial numbers guarantees that the difference of the numbers of two consecutive nodes is always at least 2.

In case a node v is integrated during the final phase it gets the number $\lceil (n_1 + n_2)/2 \rceil$ as if it would be inserted between w_1 and w_2 with numbers n_1 and n_2 (see Fig. 2). The numbers of the nodes between w_2 and w_4 need to be updated such that overall the numbers are ascending. When a node can be integrated it includes in the notification message to v_0 the numbers of the end nodes of the segment that would be reversed if this node is integrated, i.e., the numbers of w_2 and w_4 (referred to as f and l in the following). Afterwards, when v_0 informs the selected node it distributes a message to all nodes in the network that also includes the numbers f and l. A node receiving this message checks if its own number x is between f and l. In this case it changes its number to $f + l - x$. Thus, the numbers of the nodes in the segment are reflected on the mid point of the segment (see Fig. 4). Each node that changes its number also updates it next pointer to the other neighbor on C. Also nodes v, w_1, and w_2 update their next pointer.

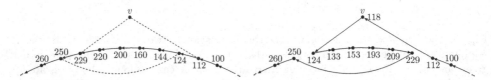

Fig. 4. Node v is to be integrated into C. The nodes w_1 and w_2 have the numbers $f = 124$ and $l = 229$. Node v will receive number $\lceil (112 + 124)/2 \rceil = 118$. Upon receiving the message form node v_0, nodes with a number between 124 and 229 change their numbers. The left sides shows the old numbers and the right side the new numbers.

This procedure results in a cycle including v with a numbering that is consistent with the orientation. Thus, when the integration phase of the next node starts, cycle C is in a consistent state. To carry out this phase a short route from each node to v_0 and vice versa is needed. This is provided by the BFS tree constructed in the pre-processing phase: Each node reaches v_0 in at most 3 hops. Thus, each final phase lasts 11 rounds.

5   Analysis of Algorithm $\mathcal{A}_{\mathsf{HC}}$

This section proves the correctness and analyzes the complexity of the individual phases and proves the main theorem. Proofs not contained in this paper can be found in [24]. First, we prove that $\mathcal{A}_{\mathsf{HC}}$ produces the numbering that guarantees that the final phases work correctly. Afterwards, the individual phases are analyzed. Some of the results are proved for values of p less than $(\log n)^{3/2}/\sqrt{n}$ to make them more general.

Lemma 1. *At the end of each phase each node has a different number and the numbers are ascending beginning with number 0 for node v_0 in clockwise order.*

Proof. After phase 1 starting with node v_0 the nodes have the numbers $n^{14}, 2n^{14}, 3n^{14}, \ldots, \beta n^{14}$, i.e., the difference between the numbers of two neighboring nodes on C is n^{14}. A node v that is inserted between two nodes with integral numbers

x and y in middle phase gets the number $\lceil (x+y)/2 \rceil$. Let $x < y$. If $x + y$ is even then $|x - \lceil (x+y)/2 \rceil| = |y - \lceil (x+y)/2 \rceil| = d/2$. If $x + y$ is odd then $|x - \lceil (x+y)/2 \rceil| = (d+1)/2$ and $|y - \lceil (x+y)/2 \rceil| = (d-1)/2$. This yields that the distance d between two consecutive numbers is approximately at most cut in half, i.e., the smaller part is at least $(d-1)/2$. After i middle phases the distance between to numbers is at least

$$d/2^i - (1 - 1/2^i) \tag{1}$$

Since there are $16 \log n$ middle phases the distance between two consecutive numbers is $n^{14}/2^{16 \log n} - (1 - 1/2^{16 \log n}) > 2^{(3 \log n)+1}$. This implies that after the middle phases the numbering of the nodes satisfies the stated condition.

Let v be a node that is inserted in a final phase into C. Assume that the smallest distance between the numbers of two consecutive nodes on C is at least 2. Consider Fig. 2 for reference. Let f (resp. l) the number of w_1 (resp. w_3) at the beginning of the corresponding final phase. Denote the nodes between w_2 and w_4 by $w'_1, \ldots, w'_k$ with $w_2 = w'_1$ and $w_4 = w'_k$. Furthermore, let $n'_1, \ldots, n'_k$ be the numbers of these nodes. Thus, $f < n'_1 < \ldots < n'_k < l$. The order of these nodes on C at the end of the phase is $w_1, v, w'_k, \ldots, w'_1, w_3$. Denote by n_i the new number of node w'_i, i.e., $n_i = n'_1 + n'_k - n'_i$. Thus, we need to prove

$$f < \lceil (f + n'_1)/2 \rceil < n_k < n_{k-1} < \ldots < n_1 < l.$$

Since $n'_1 > f + 1$ it follows $f < \lceil (f + n'_1)/2 \rceil$ and since $n_k = n'_1 + n'_k - n'_k = n'_1 > f + 1$ it follows $\lceil (f + n'_1)/2 \rceil < n_k$. Furthermore, $n'_i < n'_{i+1}$ implies $n_{i+1} = n'_1 + n'_k - n'_{i+1} < n'_1 + n'_k - n'_i = n_i$. Finally, $n_1 = n'_1 + n'_k - n'_1 = n'_k < l$.

As shown above at the end of the middle phases $d > 2^{(3 \log n)+1}$. Hence, after the last of the $3 \log n$ final phases we have $d > 1$ by Eq. (1). Thus, the numbers of all nodes are different and ascending.   □

The challenge in proving properties of iterative algorithms on random graphs is to organize the proof such that one only slowly uncovers the random choices in the input graph while constructing the desired structure, e.g., a Hamiltonian cycle. This is done in order to cleanly preserve the needed randomness and independence of events that establish the correctness proof. The *coupling technique* is well know to solve this problem ([10], p. 5). For $\gamma \in \mathbb{N}$ let $\hat{p} = 1 - (1 - p)^{1/\gamma \log n}$. Then $p = 1 - (1 - \hat{p})^{\gamma \log n}$. Thus $G(n, p)$ is equal to the union of $\gamma \log n$ independent copies of $G(n, \hat{p})$. For $p = (\log n)^{3/2}/\sqrt{n}$ we have

$$\left(1 - \frac{\sqrt{\log n}}{\gamma \sqrt{n}}\right)^{\gamma \log n} = e^{\frac{(\log n)^{3/2}}{\sqrt{n}} \log\left(1 - \frac{\sqrt{\log n}}{\gamma \sqrt{n}}\right) \frac{\gamma \sqrt{n}}{\sqrt{\log n}}} \geq e^{-\frac{(\log n)^{3/2}}{\sqrt{n}}} \geq 1 - \frac{(\log n)^{3/2}}{\sqrt{n}}$$

hence $\hat{p} \geq \sqrt{\log n}/\gamma \sqrt{n}$ and thus,

$$\bigcup_{i=1}^{\gamma \log n} G(n, \sqrt{\log n}/\gamma \sqrt{n}) \subseteq G(n, p).$$

We superimpose $\gamma \log n$ independent copies of $G(n, \sqrt{\log n}/\gamma\sqrt{n})$ and replace any double edge which may appear by a single one. In the following proof in each phase we will uncover a new copy of $G(n, \sqrt{\log n}/\gamma\sqrt{n})$. There will be $21 \log n$ phases, thus $\gamma = 21$. We set $q = \sqrt{\log n}/\gamma\sqrt{n}$ for the rest of this paper. All but the final phases also work for values of p slightly smaller than $(\log n)^{3/2}/\sqrt{n}$ and thus smaller values of q (i.e., $q = 1/\gamma\sqrt{n}$ for $p = \log n/\sqrt{n}$). This is reflected in the following proofs.

Let G^i be the union of i independent copies of $G(n, q)$. In phase i the constructed cycle C consists of edges belonging to G^i. The subsequent proofs use the following fact: The probability that any two nodes of V are connected with an edge from $G^{i+1} \setminus G^i$ is q. Thus, in each phase a new copy of $G(n, q)$ is revealed. In each phase we consider the nodes outside C. For each such node we consider *unused* edges incident to it, each of those exist with probability q independent of the choice of C, because C consist of edges of other copies of $G(n, q)$. Some of these unused edges may also exist in G^i, but that does not matter.

5.1   Phases 0 and 1

Phase 0 sequentially builds a path P by randomly choosing a node to extend P. Even for $p = \log n/n$ this allows to build paths of length $\Omega(\sqrt{n})$ in time proportional to the length of P. Since we aim at a runtime of $O(\log n)$ the following lemma suffices to prove that w.h.p. phase 0 terminates successfully.

Lemma 2. *If* $q \geq \log n/\gamma n$ *phase 0 completes w.h.p. after* $3 \log n$ *rounds with a path of length* $3 \log n$.

Phase 1 sequentially tries to extend P into a cycle C in at most $3 \log n$ rounds.

Lemma 3. *If* $q \geq 1/\gamma\sqrt{n}$ *phase 1 finds w.h.p. in* $3 \log n$ *rounds a cycle with at most* $4 \log n$ *nodes.*

5.2   Middle Phases

The middle phases contribute the bulk of nodes towards a Hamiltonian cycle. In each phase the number of nodes is increased by a constant factor w.h.p. by concurrently testing all edges in C for an extension. In the following we prove a lower bound for the number of nodes that are integrated w.h.p. into C in a middle phase. This will be done in two steps. First we state a lower bound for the number of nodes $v \in V \setminus C$ that send an invitation I_2. Based on this bound we prove a lower bound for the number of nodes that received an acceptance message I_3. Note that each node $v \in V \setminus C$ that receives an acceptance message I_3 is integrated into C and each $v \in V \setminus C$ receives at most one I_3 message.

Let $c = |C|$ and $v \in V \setminus C$. The event that an edge e of C together with v forms a triangle has probability q^2. Unfortunately these events are not independent in case the edges have a node in common. To have a lower bound for the probability that v is connected to at least one pair of consecutive nodes on C we consider only every second edge on C. Denote the edges of C by $e_0, \ldots, e_{c-1}$ with $e_i = (v^i, u^i)$.

Let $\pi_{v,i}$ be the event that node v forms a triangle with edge e_{2i} such that the edges (v, v^i) and (v, u^i) belong to newly uncovered copy of $G(n, q)$. For fixed v the events $\pi_{v,i}$ are independent and each occurs with probability q^2. Let π_v be the event that for node $v \in V \backslash C$ at least one of the events $\pi_{v,0}, \pi_{v,2}, \pi_{v,4}, \ldots, \pi_{v,c}$ occurs. Clearly the events π_v are independent and each occurs with probability $1 - (1 - q^2)^{c/2}$.

For $v \in V \backslash C$ let X_v be a random variable that is 1 if event π_v occurs. The variables $X_{v_1}, \ldots, X_{v_{n-c}}$ are independent Bernoulli-distributed random variables. Define a random variable X as $X = \sum_{v \in V \backslash C} X_v$. Then we have

$$E[X] = (n - c)(1 - (1 - q^2)^{c/2}). \tag{2}$$

Obviously X is a lower bound for the number of nodes of $V \backslash C$ that are connected to at least one pair of consecutive nodes on C, i.e., the number of nodes $v \in V \backslash C$ that sent an invitation I_2. Next let Y be a random variable denoting the number of nodes of $V \backslash C$ that receive an acceptance message I_3 provided that $X = x$ nodes sent an invitation I_2. We compute the conditional expected value $E[Y|X = x]$. The computation of Y can be reduced to the urns and balls model: The number of balls is x and the number of bins is c. Each ball is thrown randomly in any of the c bins. Note that the probability that a node v in C is connected to a node w in $V \backslash C$ is independent of v and w at least q. Thus, Y is equal to the number of nonempty bins and hence

$$E[Y|X = x] = c(1 - (1 - 1/c)^x). \tag{3}$$

Note that for a given value of x variable Y is the number of nodes inserted into C in one phase. Y/c is the ratio of the number of newly inserted nodes to the number of nodes in C. The next subsections give a lower bound for Y/c that holds w.h.p. We distinguish the cases $x \geq n/7$ and $x < n/7$. The reason is that the variance of X behaves differently in these two ranges: For $x < n/7$ the variance is rather large, whereas for $x \geq n/7$ the variance tends to 0. In both cases we first compute a lower bound for X and then derive a lower bound for Y/c with respect to the bound for X.

Instead of using $q = \sqrt{\log n}/\gamma \sqrt{n}$ the analysis of the middle phases is done for the smaller value $q = 1/\sqrt{n}$. This saves us from using the constant γ and simplifies the exposition of the proofs.

5.3   The Case $c < n/7$

Next we prove that while $c < n/7$ in each middle phase the number of nodes in C is increased by a factor of $2 - e^{-1/3}$ and that after $3 \log n$ phases the bound $n/7$ is exceeded.

Lemma 4. *Let $3 \log n < c < n/7$. Then there exists $d > 0$ such that $X > c/3$ with probability $1 - 1/n^d$.*

Lemma 5. *Let $\beta = 0.92$ and $3 \log n < c < n/7$. Then there exist $d > 0$ such that $\frac{Y}{c} \geq \beta \left(1 - \frac{1}{e^{1/3}}\right)$ with probability $1 - 1/n^d$.*

Proof. From equation (3) it follows

$$E[Y|X \geq c/3] \geq c \left(1 - \left(1 - \frac{1}{c}\right)^{c/3}\right).$$

Let $\delta^2 = 3\alpha \log n/c$ with $\alpha = (1 - \beta)^2$. Then $\delta^2 < 1$ and

$$e^{-E[Y|X\geq c/2]\delta^2/2} \leq e^{-3\alpha \log n \left(1-(1-1/c)^{c/3}\right)/2} = \left(\frac{1}{n}\right)^{3\alpha(1-(1-1/c)^{c/3})/2}.$$

The Chernoff bound implies that

$$Y|(X \geq c/3) > (1-\delta)E[Y|X \geq c/3] \geq \left(1 - \sqrt{\frac{3\alpha \log n}{c}}\right) c \left(1 - \left(1 - \frac{1}{c}\right)^{c/3}\right)$$

with probability $1 - 1/n^{3\alpha(1-(1-1/c)^{c/3})/2}$. Hence, by Lemma 4 there exists $d > 0$ such that

$$Y \geq \left(1 - \sqrt{\frac{3\alpha \log n}{c}}\right) c \left(1 - \left(1 - \frac{1}{c}\right)^{c/3}\right)$$

with probability $1 - 1/n^d$. This gives for any $c \geq 3\log n$

$$\frac{Y}{c} = \left(1 - \sqrt{\frac{3\alpha \log n}{c}}\right)\left(1 - \left(1 - \frac{1}{c}\right)^{c/3}\right) \geq \beta \left(1 - \frac{1}{e^{1/3}}\right).$$

$\square$

Lemma 6. *Let C be a cycle with at least $3\log n$ nodes. Then after at most $3\log n$ phases C has w.h.p. at least $n/7$ nodes.*

Proof. Lemma 5 yields that while the circle has less than $n/7$ nodes w.h.p. in i phases the number of nodes in C grows from c to $(1 + \beta(1 - \frac{1}{e^{1/3}}))^i c$, i.e., in three phases to $(1 + 0.92(1 - \frac{1}{e^{1/3}}))^3 c > 2c$, i.e., it doubles at least every three phases. Hence, starting with $c = 3\log n$, after at i phases C has at least $2^{i/3}3\log n$ nodes. Note that $2^{i_0/3}3\log n \geq n/7$ for $i_0 = 3\log(n/(21\log n))/\log 2$. Since $3\log n \geq i_0$, the union bound implies that after at most $3\log n$ phases w.h.p. the circle has at least $n/7$ nodes. $\square$

5.4   The Case $c \geq n/7$

Next we show that the size of C is still growing by a constant factor, but the factor is decreasing in each phase. This allows to infer that after $13\log n$ phases w.h.p. C has at least $n - 3\log n$ nodes.

Lemma 7. *Let $c = \xi n$ with $1/7 \leq \xi < 1 - 3(\log n)/n$. Then there exists $d > 0$ such that with probability $1 - 1/n^d$*

$$X > \left(1 - \sqrt{\frac{3\log n}{n(1-\xi)}}\right) c(1/\xi - 1)(1 - (1-q^2)^{c/2}).$$

Note that this Lemma proves that w.h.p. in each phase there exists at least one node that can be used to extend the cycle as long as $c < n - 3\log n$ holds.

Lemma 8. *Let $c = \xi n$ with $1/7 \leq \xi < 1 - 3(\log n)/n$. Then there exists $d > 0$ such that $\frac{Y}{c} \geq \left(1 - \sqrt{\frac{3\log n}{n(1-\xi)}}\right)\left(1 - e^{(1-1/\xi)(1-e^{-\xi/2})}\right)$ with probability $1 - 1/n^d$.*

Lemma 9. *Let $p \geq \log n/\sqrt{n}$ and C be a cycle with at least $n/7$ nodes. Then after $13\log n$ phases C has w.h.p. at least $n - 3\log n$ nodes.*

5.5 Final Phases

After the middle phases w.h.p. there are at most $3\log n$ nodes outside C. The following lemma proves the correctness of the final phases.

Lemma 10. *If $p \geq (\log n)^{3/2}/\sqrt{n}$ the final $3\log n$ phases integrate w.h.p. all remaining nodes into C.*

6 Proof of Theorem 1

The pre-processing phase lasts 9 rounds. By Lemmas 2 and 3 phases 0 and 1 terminate after $O(\log n)$ rounds w.h.p. with a cycle with at most $4\log n$ nodes. Each middle phase lasts a constant number of rounds. According to Lemma 6 after at $3\log n$ middle phases the cycle C has w.h.p. $n/7$ nodes and by Lemma 9 after another $13\log n$ middle phases w.h.p. $n-3\log n$ nodes. Then in $3\log n$ final phases, each lasting a constant number of rounds, C is w.h.p. a Hamiltonian cycle by Lemma 10. This leads to the total time complexity of $O(\log n)$ rounds. The statements about message size and memory per node are evident from the description of $\mathcal{A}_{\mathsf{HC}}$.

7 Conclusion

This paper presented an efficient distributed algorithm to compute in $O(\log n)$ rounds w.h.p. a Hamiltonian cycle for a random graph $G(n, p)$ provided $p \geq (\log n)^{3/2}/\sqrt{n}$. This constitutes a large improvement over the state of the art with respect to $p = c\log n/n^\delta$ $(0 < \delta \leq 1)$ and run time $\tilde{O}(n^\delta)$. It is well known that $G(n, p)$ contains w.h.p. a Hamiltonian cycle, provided $p \geq p_{crit}$. There is a large gap between $(\log n)^{3/2}/\sqrt{n}$ and p_{crit}. It appears that by maxing out the arguments of this paper it is possible to prove Theorem 1 for $p = \sqrt{\log n/n}$. All but the final phases already work for $p \geq \log n/\sqrt{n}$. We suspect that finding a distributed $O(\log n)$ round algorithm for $p \in o(1/\sqrt{n})$ is a hard task.

Acknowledgments. The author is grateful to the reviewers' valuable comments that helped to improve the paper.

References

1. Angluin, D., Valiant, L.: Fast probabilistic algorithms for hamiltonian circuits and matchings. J. Comput. Syst. Sci. **18**(2), 155–193 (1979)
2. Bollobás, B.: The diameter of random graphs. Trans. Am. Math. Soc. **267**(1), 41–52 (1981)
3. Bollobás, B.: Random Graphs, 2nd edn. Cambridge University Press, Cambridge (2001)
4. Bollobás, B., Fenner, T.I., Frieze, A.M.: An algorithm for finding hamilton paths and cycles in random graphs. Combinatorica **7**(4), 327–341 (1987)
5. Chatterjee, S., Fathi, R., Pandurangan, G., Dinh Pham, N.: Fast and efficient distributed computation of Hamiltonian cycles in random graphs. In: 38th IEEE International Conference on Distributed Computing Systems, ICDCS 2018, Vienna, Austria, 2–6 July 2018, pp. 764–774 (2018). https://doi.org/10.1109/ICDCS.2018.00079
6. Chung, F., Lu, L.: The diameter of sparse random graphs. Adv. Appl. Math. **26**(4), 257–279 (2001)
7. Erdős, P., Rényi, A.: On random graphs I. Publ. Math. (Debr.) **6**, 290–297 (1959)
8. Franceschelli, M., Giua, A., Seatzu, C.: Quantized consensus in hamiltonian graphs. Automatica **47**(11), 2495–2503 (2011)
9. Frieze, A.: Parallel algorithms for finding hamilton cycles in random graphs. Inf. Process. Lett. **25**(2), 111–117 (1987)
10. Frieze, A., Karoński, M.: Introduction to Random Graphs. Cambridge University Press, Cambridge (2015)
11. Frieze, A.M., Haber, S.: An almost linear time algorithm for finding hamilton cycles in sparse random graphs with minimum degree at least three. Random Struct. Algorithms **47**(1), 73–98 (2015)
12. Hélary, J., Raynal, M.: Depth-first traversal and virtual ring construction in distributed systems. Research Report RR-0704, INRIA Rennes (1987)
13. Kim, J., Srikant, R.: Peer-to-peer streaming over dynamic random hamilton cycles. In: 2012 Infernational Theory & Applications Workshop, pp. 415–419, February 2012
14. Komlós, J., Szemerédi, E.: Limit distribution for the existence of hamiltonian cycles in a random graph. Discret. Math. **43**(1), 55–63 (1983)
15. Krivelevich, M., Panagiotou, K., Penrose, M., McDiarmid, C.: Random Graphs, Geometry and Asymptotic Structure'. London Mathematical Society Student Texts (84). Cambridge University Press, Cambridge (2016)
16. Krzywdziński, K., Rybarczyk, K.: Distributed algorithms for random graphs. Theor. Comput. Sci. **605**, 95–105 (2015)
17. Levy, E., Louchard, G., Petit, J.: A distributed algorithm to find hamiltonian cycles in $\mathcal{G}(n,p)$ random graphs. In: López-Ortiz, A., Hamel, A.M. (eds.) CAAN 2004. LNCS, vol. 3405, pp. 63–74. Springer, Heidelberg (2005). https://doi.org/10.1007/11527954_7
18. MacKenzie, P.D., Stout, Q. F.: Optimal parallel construction of hamiltonian cycles and spanning trees in random graphs. In: Proceedings of the Fifth Annual ACM Symposium on Parallel Algorithms & Architectures, pp. 224–229, New York (1993)
19. Malkhi, D., Sen, S., Talwar, K., Werneck, R.F., Wieder, U.: Virtual ring routing trends. In: Keidar, I. (ed.) DISC 2009. LNCS, vol. 5805, pp. 392–406. Springer, Heidelberg (2009). https://doi.org/10.1007/978-3-642-04355-0_42

20. Peleg, D.: Distributed computing: a locality-sensitive approach. In: Monographs on Discrete Mathematics and Applications. Society for Industrial and Applied Mathematics, Philadelphia, PA, USA (2000)
21. Pósa, L.: Hamiltonian circuits in random graphs. Discret. Math. **14**(4), 359–364 (1976)
22. Shamir, E.: How many random edges make a graph hamiltonian? Combinatorica **3**(1), 123–131 (1983)
23. Thomason, A.: A simple linear expected time algorithm for finding a hamilton path. Discret. Math. **75**(1), 373–379 (1989)
24. Turau, V.: A Distributed Algorithm for Finding Hamiltonian Cycles in Random Graphs in $O(\log n)$ Time. (2018). arXiv preprint arXiv:1805.06728
25. Turau, V., Siegemund, G.: Scalable routing for topic-based publish/subscribe systems under fluctuations. In: Proceedings of the 37th International Conference on Distributed Computing Systems (2017)

Simple and Local Independent Set Approximation

Ravi B. Boppana[1], Magnús M. Halldórsson[2], and Dror Rawitz[3(✉)]

[1] Department of Mathematics, MIT, Cambridge, USA
rboppana@mit.edu
[2] School of Computer Science, Reykjavik University, Reykjavik, Iceland
mmh@ru.is
[3] Faculty of Engineering, Bar-Ilan University, Ramat Gan, Israel
dror.rawitz@biu.ac.il

Abstract. We bound the performance guarantees that follow from Turán-like bounds for unweighted and weighted independent sets in bounded-degree graphs. In particular, a randomized approach of Boppana forms a simple 1-round distributed algorithm, as well as a streaming and preemptive online algorithm. We show it gives a tight $(\Delta + 1)/2$-approximation in unweighted graphs of maximum degree Δ, which is best possible for 1-round distributed algorithms. For weighted graphs, it gives only a $(\Delta + 1)$-approximation, but a simple modification results in an asymptotic expected $0.529(\Delta + 1)$-approximation. This compares with a recent, more complex Δ-approximation [6], which holds deterministically.

1 Introduction

Independent sets are among the most fundamental graph structures. A classic result of Turán [25] says that every graph $G = (V, E)$ contains an independent set of size at least

$$\textsc{Turán}(G) \doteq \frac{n}{\bar{d} + 1},$$

where $n = |V|$ is the number of vertices and $\bar{d} = 2m/n$ is the average degree, where $m = |E|$. Turán's bound is tight for regular graphs, but for non-regular graphs an improved bound was given independently by Caro [10] and Wei [26]:

$$\alpha(G) \geq \textsc{CaroWei}(G) \doteq \sum_{v \in V} \frac{1}{d(v) + 1}, \tag{1}$$

where $\alpha(G)$ is the cardinality of a maximum independent set in G and $d(v)$ is the degree of vertex $v \in V$.

M. M. Halldórsson—Supported by grants nos. 152679-05 and 174484-05 from the Icelandic Research Fund.
D. Rawitz—Supported by the Israel Science Foundation (grant no. 497/14).

Z. Lotker and B. Patt-Shamir (Eds.): SIROCCO 2018, LNCS 11085, pp. 88–101, 2018.
https://doi.org/10.1007/978-3-030-01325-7_12

There are numerous proofs of the Caro-Wei bound, some involving simple greedy algorithms. Arguably the simplest argument known is a probabilistic one:

> *Uniformly randomly permute the vertices, and output the set of vertices that precede all their neighbors in the permutation.* $\qquad$ (2)

Each node v precedes its neighbors with probability $1/(d(v)+1)$, so by linearity of expectation the expected size of the output matches exactly $\textsc{CaroWei}(G)$. This argument, which first appeared in the book of Alon and Spencer [3], is due to Boppana [9]. It clearly leads to a very simple local decision rule once the permutation is selected.

An alternative formulation of the algorithm is practical in certain contexts.

> *Each vertex v picks a random real number x_v from $[0, 1]$. The vertex joins the independent set if its random number is larger than that of its neighbors.* $\qquad$ (3)

It suffices to select the numbers with precision $1/n^3$, for which collisions are very unlikely.

This leads to a fully *1-local* algorithm, in which each node decides whether to join the independent set after a single round of communication with its neighbors. The same $O(\log n)$ bits a node transmits go to all of its neighbors, which matches the Broadcast-CONGEST model of distributed algorithms. Furthermore, it is asynchronous. This is just about the simplest distributed algorithm one could hope for.

The simplicity of the approach also allows for other applications. The basic algorithm works well with edge streams, storing only the permutation and the current solution as a bit-vector. The storage can be reduced with an ϵ-min-wise permutation, at a small cost in performance. This can also be viewed as a preemptive online algorithm, where edges can cause nodes to be kicked out of the solution but never reenter.

1.1 Our Contribution

The main purpose of this essay is to analyze the performance guarantees of Boppana's algorithm on graphs of maximum degree Δ. We show that it achieves a tight $(\Delta + 1)/2$-approximation, which then also gives a bound on the fidelity of the Caro-Wei bound. In terms of the average degree $\bar{d}$, the performance is at most $(\bar{d} + 2)/1.657$. We also show that the Turán bound has strictly (but only slightly) worse performance than the Caro-Wei bound for bounded-degree graphs, or $(\Delta + 1)/2 + 1/(8\Delta)$.

We then address the case of weighted graphs, and find that unchanged Boppana's algorithm gives only a $(\Delta + 1)$-approximation. However, a slight modification yields an improved approximation which asymptotically approaches $0.529(\Delta + 1)$.

1.2   Related Work

Turán [25] showed that $\alpha(G) \geq$ Turán(G). Caro [10] and Wei [26] independently showed (in unpublished technical reports) that $\alpha(G) \geq$ CaroWei(G). The bound can also be seen to follow from an earlier work of Erdős [16], who showed that the bound is tight only for disjoint collections of cliques. Observe that CaroWei$(G) \geq$ Turán(G), for every graph G.

The min-degree greedy algorithm iteratively adds a minimum-degree node to the graph, removes it and its neighbors and repeats. It achieves the Caro-Wei bound [26] (see also [16]). Griggs [18] (see also Chvátal and McDiarmid [13]) showed that the max-degree greedy algorithm also attains the Caro-Wei bound, where the algorithm iteratively removes the vertex of maximum degree until the graph is an independent set. Sakai et al. [23] analyzed three greedy algorithms for weighted independent sets and showed them to achieve certain absolute bounds as well as a $(\Delta + 1)$-approximation.

The best sequential approximation known is $\tilde{O}(\Delta/\log^2 \Delta)$,[1] by Bansal et al. [5], which uses semi-definite programming. This matches the inapproximability result known, up to doubly-logarithmic factors, that holds assuming the Unique Games Conjecture [4]. The problem is known to be NP-hard to approximate within an $O(\Delta/\log^4 \Delta)$ factor [12]. For small values of Δ, a $(\Delta + 3)/5$-approximation [7] is achievable combinatorially, but requires extensive local search. As for simple greedy algorithms, it was shown in [20] that the performance guarantee of the min-degree greedy algorithm is $(\Delta+2)/3$, and also pointed out that the max-degree algorithm attains no better than a $(\Delta + 1)/2$ ratio.

Most works on distributed algorithms have focused on finding maximal independent sets, rather than optimizing their size. Boppana's algorithm corresponds to the first of $O(\log n)$ rounds of Luby's maximal independent set algorithm (see also Alon et al. [2]). As for approximations, $n^{\Theta(1/k)}$-approximation is achievable and best possible for local algorithms running in k rounds [8], where the upper bound assumes both unlimited bandwidth and computation. Bar-Yehuda et al. [6] gave a Δ-approximation algorithm for weighted independent sets using the local ratio technique that runs in time $O(\text{MIS} \cdot \log W)$ rounds in the CONGEST model, where MIS is the number of rounds needed to compute a maximal independent set and W is the ratio between the largest and smallest edge weight. We improve this approximation ratio by nearly a factor of 2 using only a single round, but at the price of obtaining a bound only on expected performance. Ghaffary, Kuhn, and Maus [17] gave a $(1 + \epsilon)$-approximation algorithm that requires a poly-logarithmic number of rounds in the LOCAL model.

Alon [1] gave nearly tight bounds for testing independence properties; his lower bound carries over to distributed algorithms, as we shall see in Sect. 2.4. For matchings, which correspond to independent sets in line graphs, Kuhn et al. [22] showed that achieving any constant factor approximation requires $\Omega(\max(\log \Delta/\log \log \Delta, \sqrt{\log n/\log \log n}))$ rounds. Censor-Hillel, Khoury, and

---

[1] $\tilde{O}(\cdot)$ suppresses $\log \log n$ factors.

Paz [11] presented a nearly quadratic lower bound on the number of rounds for solving maximum independent set exactly in the CONGEST model.

Halldórsson and Konrad [21] examined how well the Caro-Wei bound performs in different subclasses of graphs. They also gave a randomized one-round distributed algorithm where nodes broadcast only a single bit that yields an independent set of expected size at least $0.24 \cdot \text{CAROWEI}(G)$ on every graph G. This is provably the least requirement for an effective distributed algorithm, as without degree information, the bounds are polynomially worse.

Streaming algorithms (including Boppana's) achieving Turán-like bounds in graphs and hypergraphs were considered in [19], and streaming algorithms for approximating $\text{CAROWEI}(G)$ were given recently by Cormode et al. [14].

Motivated by a packet forwarding application, Emek et al. [15] considered the online set packing problem that corresponds to maintaining strong independent sets of large weight in hypergraphs under edge additions. We give a tight bound on their method for the special case of graphs.

2 Performance of Caro-Wei-Turán Bounds

We examine here how well the Caro-Wei and the Turán bounds perform on (unweighted) bounded-degree and sparse graphs.

Let OPT be an optimal independent set of size $\alpha = \alpha(G)$ and let $V' = V \setminus \text{OPT}$. We say that a bound $B(G)$ has a performance ratio $f(\Delta)$ if, for all graphs G with $\Delta(G) = \Delta$ it holds that

$$\alpha(G) \geq B(G) \geq \frac{\alpha(G)}{f(\Delta)}.$$

2.1 Caro-Wei in Bounded-Degree Graphs

Theorem 1. CAROWEI *has performance ratio* $(\Delta + 1)/2$.

Proof. Let G be a graph. Let O_i, for $i = 1, 2, \ldots, \Delta$, denote the number of vertices in OPT of degree i. Our approach is to separate the contributions of the different O_is to the Caro-Wei bound. The nodes of high degree have a smaller direct contribution, but also have an indirect contribution in forcing more nodes to be in V'.

Let m_{OPT} be the number of edges with an endpoint in OPT. Each such edge has the other endpoint in V', whereas nodes in V' are incident on at most Δ edges. Thus,

$$\sum_{i=1}^{\Delta} i \cdot O_i = m_{\text{OPT}} \leq \Delta |V'|. \tag{4}$$

We then obtain

$$
\begin{aligned}
\text{CAROWEI}(G) &= \sum_{v \in V} \frac{1}{d(v)+1} \\
&= \sum_{i=1}^{\Delta} O_i \cdot \frac{1}{i+1} + \sum_{v \in V'} \frac{1}{d(v)+1} \\
&\geq \sum_{i=1}^{\Delta} O_i \cdot \frac{1}{i+1} + |V'| \frac{1}{\Delta+1} \\
&\geq \frac{1}{\Delta+1} \sum_{i=1}^{\Delta} O_i \left(\frac{\Delta+1}{i+1} + \frac{i}{\Delta} \right) \qquad \text{(Applying (4))} \\
&= \frac{1}{\Delta+1} \sum_{i=1}^{\Delta} O_i \left(2 + \frac{\Delta-i}{i+1} - \frac{\Delta-i}{\Delta} \right) \\
&\geq \frac{1}{\Delta+1} \sum_{i=1}^{\Delta} O_i \cdot 2 \\
&= \frac{2}{\Delta+1} \alpha(G),
\end{aligned}
$$

obtaining the approximation upper bound claimed. Observe that the bound is tight only if the graph is regular.

To see that the ratio attained is no better than $(\Delta + 1)/2$, observe that in any regular graph, Boppana's algorithm achieves a solution of exactly $n/(\Delta+1)$, while in bipartite regular graphs the optimal solution has size $n/2$.   □

Remark. Selkow [24] generalized the Caro-Wei bound by extending Boppana's algorithm to two rounds. Namely, it adds also the nodes with no neighbor ordered earlier among those that did not get removed in the first round. For regular graphs, however, his bound reduces to the Caro-Wei bound, and thus does not attain a better performance ratio, given our lower bound construction.

2.2   Caro-Wei in Sparse Graphs

We now analyze the performance of the Caro-Wei bound in terms of the average degree $\overline{d} = 2m/n$. We shall use a certain application of the Cauchy-Schwarz inequality, which we state more generally in hindsight of its application in the following section.

Lemma 1. *If $x_1, x_2, \ldots, x_N$ and $w_1, w_2, \ldots, w_N$ are positive reals, then*

$$
\sum_{i=1}^{N} \frac{w_i^2}{x_i} \geq \frac{\left(\sum_{i=1}^{N} w_i \right)^2}{\sum_{i=1}^{N} x_i}.
$$

Proof. The Cauchy-Schwarz inequality implies that for $u_1, u_2, \ldots, u_N$ and $v_1, v_2, \ldots, v_N$,

$$\left(\sum_{i=1}^{N} u_i v_i \right)^2 \leq \left(\sum_{i=1}^{N} u_i^2 \right) \left(\sum_{i=1}^{N} v_i^2 \right).$$

The claim now follows using $u_i = \sqrt{x_i}$ and $v_i = w_i/\sqrt{x_i}$. $\qquad\square$

Note that applying Lemma 1 with $w_v = 1$ and $x_v = d(v) + 1$ yields that

$$\text{CaroWei}(G) = \sum_{v \in V} \frac{1}{d(v) + 1} \geq \frac{n^2}{\sum_v (d(v) + 1)} = \frac{n}{\bar{d} + 1} = \text{Turán}(G).$$

Theorem 2. CaroWei *has performance ratio at most* $(\bar{d} + 2)/1.657$.

Proof. Let OPT be an optimal independent set of size $\alpha = \alpha(G)$ and let $V' = V \setminus \text{OPT}$. Observe that when $|V'| = n - \alpha \geq \alpha$, the Turán bound gives $n/(\bar{d}+1) \geq \alpha \cdot 2/(\bar{d}+1)$, for a performance ratio of at most $(\bar{d}+1)/2$. We assume therefore that $\alpha \geq \frac{1}{2}n$.

Our approach is to first apply Lemma 1 separately on the parts of $\text{CaroWei}(G)$ corresponding to OPT and V'. We then show that the worst case occurs when all edges cross from OPT to V', indeed when the graph is bipartite with regular sides. Optimizing over the possible sizes of the sides then yields a tight upper and lower bounds.

Let m_{OPT} denote the number of edges with an endpoint in OPT, $m_{V'}$ the number of edges with both endpoints in V' and $m = m_{\text{OPT}} + m_{V'}$ be the total number of edges. Observe that $\sum_{v \in \text{OPT}} d(v) = m_{\text{OPT}}$ while $\sum_{v \in V'} d(v) = m_{\text{OPT}} + 2m_{V'}$.

Lemma 1 (with $w_v = 1$ and $x_v = d(v) + 1$) applied to OPT and V' separately yields that

$$\text{CaroWei}(G) = \sum_{v \in \text{OPT}} \frac{1}{d(v) + 1} + \sum_{v \in V'} \frac{1}{d(v) + 1}$$

$$\geq \frac{\alpha^2}{m_{\text{OPT}} + \alpha} + \frac{(n - \alpha)^2}{m_{\text{OPT}} + 2m_{V'} + (n - \alpha)},$$

Denoting $t = m_{\text{OPT}}/m$, we get that

$$\text{CaroWei}(G) \geq \frac{\alpha^2}{t \cdot m + \alpha} + \frac{(n - \alpha)^2}{(2 - t)m + n - \alpha}. \tag{5}$$

Considered as a function f of t, the r.h.s. of (5) has derivative

$$\frac{df}{dt} = -m \frac{\alpha^2}{(tm + \alpha)^2} + m \frac{(n - \alpha)^2}{((2 - t)m + n - \alpha)^2}.$$

Since we assume $\alpha \geq n/2$, it holds that $\alpha^2(m+n-\alpha)^2 \geq (n-\alpha)^2(m+\alpha)^2$, and thus $df/dt \leq 0$ for all $t \in [0,1]$. Hence, denoting $\tau = \alpha/n$, we obtain that

$$\text{CAROWEI}(G) \geq \frac{\alpha^2}{m+\alpha} + \frac{(n-\alpha)^2}{m+n-\alpha} = \alpha\left(\frac{\tau}{\bar{d}/2+\tau} + \frac{(1-\tau)^2/\tau}{\bar{d}/2+1-\tau}\right). \quad (6)$$

The expression in the parenthesis then upper bounds the reciprocal of the performance guarantee of CAROWEI.

To see that (6) is tightest possible, consider bipartite graphs G with regular sides. Let τ be such that τn is the size of the larger side and q is the degree of those vertices. Then the number of edges is $m = q \cdot \tau n$, average degree is $\bar{d} = 2m/n = 2q\tau$, and the degree of the nodes on the other side is $m/((1-\tau)n) = \bar{d}/(2(1-\tau))$. Clearly $\alpha(G) = \tau n$, while the Caro-Wei bound gives

$$\text{CAROWEI}(G) = \frac{\tau n}{\bar{d}/(2\tau)+1} + \frac{(1-\tau)n}{\bar{d}/(2(1-\tau))+1}$$

$$= \alpha(G)\left(\frac{1}{\bar{d}/(2\tau)+1} + \frac{(1-\tau)/\tau}{\bar{d}/(2(1-\tau))+1}\right),$$

which matches (6).

If we round up the lower order terms in the denominator of (6), we obtain a simpler expression for the asymptotic performance with $\bar{d}$:

$$\text{CAROWEI}(G) \geq \alpha(G)\left(\frac{\tau + (1-\tau)^2/\tau}{\bar{d}/2+1}\right),$$

which is minimized when $\tau = 1/\sqrt{2}$, for a performance ratio at most $(\bar{d}+2)/(4(\sqrt{2}-1)) \leq (\bar{d}+2)/1.657$.   □

2.3 Turán Bound

Recall Turán's theorem that

$$\alpha(G) \geq \text{TURÁN}(G) = \frac{n}{\bar{d}+1} = \frac{n^2}{2m+n}.$$

We find that the guarantee of the Turán bound is strictly weaker than that of Caro-Wei, yet asymptotically equivalent.

Theorem 3. TURÁN *has performance ratio* $\dfrac{(2\Delta+1)^2}{8\Delta} = \dfrac{\Delta+1}{2} + \dfrac{1}{8\Delta}$.

Proof. Because $\text{OPT} = V \setminus V'$ is independent, each of the m edges of G is incident to at least one vertex in V'. Conversely, each vertex in V' is incident to at most Δ edges. So by counting edges, we get

$$m \leq \Delta|V'| = \Delta(n-\alpha).$$

Therefore

$$2m + n \leq 2\Delta(n - \alpha) + n = (2\Delta + 1)n - 2\Delta\alpha.$$

Multiplying by $8\Delta\alpha$ and using the inequality $4xy \leq (x + y)^2$ gives

$$8\Delta\alpha(2m + n) \leq 4(2\Delta\alpha)[(2\Delta + 1)n - 2\Delta\alpha] \leq [(2\Delta + 1)n]^2.$$

Dividing both sides by $8\Delta(2m + n)$ gives

$$\alpha \leq \frac{(2\Delta + 1)^2}{8\Delta} \cdot \frac{n^2}{2m + n} = \frac{(2\Delta + 1)^2}{8\Delta} \text{TURÁN}(G).$$

The argument above shows that the performance ratio of Turán's bound is at most $\frac{(2\Delta+1)^2}{8\Delta}$. This performance ratio is tight as a function of Δ. To see why, given $\Delta > 0$, let A, B, and C be disjoint sets of size $2\Delta - 1$, $2\Delta - 1$, and 2, respectively. Let G be any Δ-regular bipartite graph with parts A and B, together with two isolated vertices in C. We can check that $n = 4\Delta$, $m = (2\Delta - 1)\Delta$, $\text{TURÁN}(G) = \frac{8\Delta}{2\Delta+1}$, and $\alpha(G) = 2\Delta + 1$. So the performance ratio of Turán's bound on this graph is indeed $\frac{(2\Delta+1)^2}{8\Delta}$. $\qquad\square$

2.4 Limitations of Distributed Algorithms

We may assume that we are equipped with unique labels from a universe of N labels, where $N \geq \Delta \cdot n$. The nodes have knowledge of n, Δ and N, and have unlimited bandwidth and computational ability. The nodes have distinct ports for communication with their neighbors, but do not initially know their labels.

Our result for Boppana's algorithm is optimal for 1-round algorithms. Observe that the lower bounds below hold also for randomized algorithms.

Theorem 4. *Every 1-round distributed algorithm has performance ratio at least* $(\Delta + 1)/2$, *even on unweighted regular graphs.*

Proof. In a single round, each node can only learn the labels of their neighbors and their random bits.

Consider the graph $G_1 = K_{\Delta+1}$, and G_2, which is any Δ-regular bipartite graph. Distributions over neighborhoods are identical. Hence, no 1-round algorithm can distinguish between these graphs.

All nodes will join the independent set with the same probability, averaged over all possible labelings, since they share the same views. This probability can be at most $1/(\Delta + 1)$, as otherwise the algorithm would produce incorrect answers on $K_{\Delta+1}$. The size of the solution is then at most $n/(\Delta + 1)$, while on every Δ-regular bipartite graphs, the optimal solution contains $n/2$ nodes. $\qquad\square$

It is not clear if better results can be obtained when using more rounds. A weaker lower bound holds even for nearly logarithmic number of rounds.

Theorem 5. *There are positive constants c_1 and c_2 such that the following holds: Every $c_1 \log_\Delta n$-round distributed algorithm has performance ratio at least $c_2 \Delta / \log \Delta$.*

Proof. Alon [1] constructs a Δ-regular graph G_1 of girth $\Omega(\log n / \log \Delta)$ with independence number $O(n/\Delta \cdot \log \Delta)$, and notes that it is well known that there exists a bipartite Δ-regular graph G_2 of girth $\Omega(\log n / \log \Delta)$. The distributions over the k-neighborhoods of these graphs are identical, for $k = O(\log n / \log \Delta)$. Hence, no k-round distributed algorithm can distinguish between the two.   □

3   Approximations for Weighted Graphs

In the weighted setting, each node v is assigned a positive integral weight $w(v)$ and the objective is to find an independent set I maximizing the total weight $\sum_{v \in I} w(v)$. For a set $X \subseteq V$, denote $w(X) = \sum_{x \in X} w(x)$.

Boppana's algorithm can be applied unchanged to weighted graphs, producing a solution B of expected weight

$$\mathbb{E}[w(B)] = \sum_{v \in B} w(v) \cdot \frac{1}{d(v) + 1},$$

by linearity of expectation. This immediately implies that $\mathbb{E}[w(B)] \geq w(V)/(\Delta+1)$, for a performance ratio at most $\Delta+1$. To see that this is also the best possible bound, consider the complete bipartite graphs $K_{N,N}$, where the nodes on one side have weight 1 and on the other side weight Q, for a parameter $Q \geq \Delta^2$. The expected weight of the algorithm solution is $(N+NQ)/(\Delta+1)$, while the optimal solution is of weight NQ. The performance ratio is then $(\Delta+1)/(1+1/Q)$, which goes to $\Delta + 1$ as Q gets large.

We therefore turn our attention to modifications that take the weights into account.

3.1   Modified Algorithm

We consider now a variation, MAX, previously considered in an online setting in [15].

Each node v picks a random real number x_v uniformly from $[0, 1]$. It broadcasts the values x_v and w_v to its neighbors, who compute from it $r_v = x_v^{1/w_v}$. As before, each node u joins the solution if its value r_u is the highest among its neighbors.

The only difference is the computation of r_v, which now depends on the weight w_v. Again the algorithm runs in a single round of Broadcast-CONGEST, with correctness following as before. The algorithm was previously shown in [15] to attain a Δ-approximation.

We obtain a tight bound, which does not have a nice closed expression.

Theorem 6. *The performance ratio $\rho(\Delta)$ of* MAX, *as a function of Δ, is given by*

$$\frac{1}{\rho(\Delta)} = \min_{x \in [0,1]} \left(\frac{x^2}{\Delta + x} + \frac{1}{x\Delta + 1} \right).$$

We prove Theorem 6 in the following subsection.

If we focus on the asymptotics as Δ gets large, we can ignore the additive terms in the denominators, obtaining that the performance ratio approaches

$$\frac{1}{\rho(\Delta)} \xrightarrow[\Delta \to \infty]{} \min_{x \in [0,1]} \left(\frac{x^2}{\Delta + 1} + \frac{1}{x(\Delta + 1)} \right) = \frac{\min_{x \in [0,1]} \left(x^2 + 1/x \right)}{\Delta + 1} \doteq \frac{1}{\tilde{\rho}(\Delta)}.$$

The derivative is of $g(x) = x^2 + 1/x$ is $\frac{dg(x)}{dx} = 2x - 1/x^2$. Hence, $\tilde{\rho}(\Delta)$ is maximized when $x = 2^{-1/3}$ for a ratio of

$$\tilde{\rho}(\Delta) = \frac{2^{2/3}}{3}(\Delta + 1) \sim (\Delta + 1)/1.89 \sim 0.529(\Delta + 1).$$

Theorem 7. *The asymptotic performance ratio of* MAX *is $\frac{2^{2/3}}{3}(\Delta + 1) \sim 0.529(\Delta + 1)$.*

Figure 1 shows $\rho(\Delta)/(\Delta + 1)$ as a function of Δ.

For $\Delta = 2$, we find that $1/\rho \sim 0.593$, or $\rho \sim 1.657 \sim 0.562(\Delta + 1)$, which is about 6% larger than $0.529(\Delta + 1)$, but 20% smaller than Δ. For $\Delta = 1$, the algorithm can be made optimal by preferring nodes with higher weight than their sole neighbor.

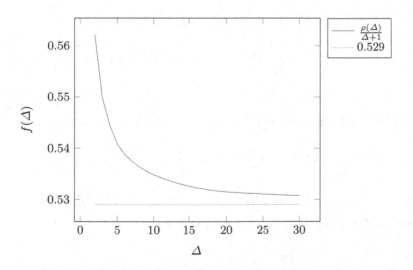

Fig. 1. Bounds on performance ratio, for small values of Δ.

3.2   Analysis

The key property of the MAX rule that leads to improved approximation is that the probability that a node is selected is now proportional to the fraction of its weight within its closed neighborhood (consisting of itself and its neighbors). We then obtain a bound in terms of weights of sets of nodes – the optimal solution and the remaining nodes – using the Cauchy-Schwarz inequality. We safely upper bound the degree of each node by Δ, but the main effort then is to show that the worst case occurs when the graph is bipartite with equal sides. This leads to matching upper and lower bounds.

Let $N(v)$ denote the set of neighbors of vertex v and $N[v] = \{v\} \cup N(v)$ its closed neighborhood. Let MAX also refer to the set of nodes selected by MAX.

The key property of the MAX rule is that the probability that a node is selected is now proportional to the fraction of its weight within its closed neighborhood. We provide a proof for the next lemma for completeness.

Lemma 2 ([15]). *For each vertex $v \in V$, we have that* $\mathbb{P}[v \in \text{MAX}] = \dfrac{w(v)}{w(N[v])}$.

Proof. Let $r_{\max} = \max\{r_u : u \in N(v)\}$. By independence of the random choices we have, for $\alpha \in [0,1]$, that

$$\mathbb{P}[r_{\max} < \alpha] = \prod_{u \in N(v)} \mathbb{P}[r_u < \alpha] = \prod_{u \in N(v)} \mathbb{P}[x_u < \alpha^{w(u)}] = \alpha^{\sum_{u \in N(v)} w(u)} = \alpha^{w(N(v))}.$$

It follows that $r_{\max}$ has distribution $D_{w(N(v))}$, where the distribution D_z has density $f_z(\alpha) = z\alpha^{z-1}$, for $\alpha \in [0,1]$. Hence,

$$\mathbb{P}[r_v > r_{\max}] = \int_0^1 \mathbb{P}[r_{\max} < \alpha] \cdot f_{r_v}(\alpha)d\alpha$$

$$= \int_0^1 \alpha^{w(N(v))} \cdot w(v)\alpha^{w(v)-1}d\alpha = \frac{w(v)}{w(N[v])},$$

as required.   □

Note that by Lemma 2 and linearity of expectation, we have that

$$\mathbb{E}[w(S \cap \text{MAX})] = \sum_{v \in S} \mathbb{P}[v \in \text{MAX}] \cdot w(v) = \sum_{v \in S} \frac{w(v)^2}{w(N[v])}, \tag{7}$$

for any subset $S \subseteq V$. Applying Lemma 1 (with $x_v = w(N[v])$) gives:

Lemma 3. *For any subset $S \subseteq V$ we have that*

$$\mathbb{E}[w(S \cap \text{MAX})] \geq \frac{w(S)^2}{\sum_{v \in S} w(N[v])}.$$

Applying Lemma 3 with $S = V$ gives an absolute lower bound on the solution size.

Lemma 4. $\mathbb{E}[w(\text{MAX})] \geq \dfrac{w(V)^2}{\sum_{v \in V} w(N[v])} = \dfrac{w(V)^2}{\sum_{v \in V}(d(v)+1)w(v)} \geq \dfrac{w(V)}{\Delta + 1}.$

We need the following lemma when showing that worst case occurs for bipartite graphs.

Lemma 5. *Let $a > b > 0$ and let $Z - Y \geq X > 0$. Then*

$$\min_{t \in [0,1]} \left\{ \frac{a}{Y + tX} + \frac{b}{Z + (1-t)X} \right\} = \frac{a}{Y + X} + \frac{b}{Z}.$$

Proof. Let $f(t) = \frac{a}{Y+tX} + \frac{b}{Z+(1-t)X}$. We have that

$$\frac{df(t)}{dt} = -\frac{aX}{(Y+tX)^2} + \frac{bX}{(Z+(1-t)X)^2},$$

which is negative for any $t \in [0,1]$, since $a > b$ and $Y + tX \leq Z + (1-t)X$. □

Now we are ready to prove Theorem 6.

Proof (of Theorem 6). Let OPT be an optimal solution, and define $V' \doteq V \setminus \text{OPT}$, and $\beta \doteq w(V')/w(\text{OPT})$. When $\beta \geq 1$, Lemma 4 implies that the performance ratio is at most $(\Delta + 1)/2$. We therefore focus on the case where $\beta < 1$.

We first apply Lemma 3 separately on OPT and on V', obtaining:

$$w(\text{MAX}) = w(\text{MAX} \cap \text{OPT})) + w(\text{MAX} \cap V')$$

$$\geq \frac{w(\text{OPT})^2}{\sum_{v \in \text{OPT}} w(N[v])} + \frac{w(V')^2}{\sum_{v \in V'} w(N[v])}. \tag{8}$$

Let $W = \sum_{v \in V'} w(v) \cdot |N(v) \cap \text{OPT}| = \sum_{v \in \text{OPT}} w(N(v))$ be the weighted degree of the nodes of V' into OPT, which can be viewed as the total of the weights of neighborhoods of nodes in OPT. Thus,

$$\sum_{v \in \text{OPT}} w(N[v]) = w(\text{OPT}) + \sum_{v \in V'} w(v)|N(v) \cap \text{OPT}| = w(\text{OPT}) + W. \tag{9}$$

and

$$\sum_{v \in V'} w(N[v]) = w(V') + \sum_{v \in V'} w(N(v))$$

$$= w(V') + \sum_{v \in \text{OPT}} w(v) \cdot |N(v) \cap V'| + \sum_{v \in V'} w(v) \cdot |N(v) \cap V'|$$

$$\leq w(V') + \Delta w(\text{OPT}) + \sum_{v \in V'} w(v) \cdot (\Delta - |N(v) \cap \text{OPT}|)$$

$$= w(V') + \Delta w(\text{OPT}) + \Delta w(V') - W. \tag{10}$$

Applying (9) and (10) to (8) gives

$$w(\text{MAX}) \geq \frac{w(\text{OPT})^2}{w(\text{OPT}) + W} + \frac{w(V')^2}{w(V') + \Delta w(\text{OPT}) + \Delta w(V') - W}.$$

Since $\beta < 1$ and $W \leq \Delta w(V')$ we can use Lemma 5 with $a = w(\text{OPT})^2$, $b = w(V')^2$, $Y = w(\text{OPT})$, $Z = w(V') + \Delta w(\text{OPT})$, $X = \Delta w(V')$, and $t = W/X$. Hence,

$$w(\text{MAX}) \geq \frac{w(\text{OPT})^2}{w(\text{OPT}) + \Delta w(V')} + \frac{w(V')^2}{w(V') + \Delta w(\text{OPT})}$$

$$= w(\text{OPT}) \cdot \left(\frac{1}{1 + \Delta\beta} + \frac{\beta^2}{\beta + \Delta} \right). \tag{11}$$

The upper bound of the theorem therefore follows.

To see that bound (11) is tight, consider any Δ-regular bipartite graph $G = (V, E)$ with V partitioned into two sets L and R, where $|L| = |R|$. Set the weight of nodes in L and in R as 1 and β, respectively, for some $\beta \leq 1$. Clearly, the weight of the optimal solution is $w(\text{OPT}) = |L|$. Observe that

$$w(\text{MAX}) = |L| \cdot \frac{1}{1 + \Delta\beta} + |R|\beta \cdot \frac{\beta}{\beta + \Delta} = w(\text{OPT}) \cdot \left(\frac{1}{1 + \beta\Delta} + \frac{\beta^2}{\beta + \Delta} \right),$$

matching (11). □

Remark. Sakai et al. [23] considered the following greedy algorithm (named GWMIN2): add the vertex v maximizing $w(v)/w(N[v])$ to the solution, remove its closed neighborhood, and recurse on the remaining graph. They derived a $(\Delta + 1)$-approximation upper bound but not a matching lower bound. Since their algorithm attains the bound (7) (see [23]), our analysis implies that it also attains the bound of Theorem 6.

4   Conclusion

It is surprising that the best distributed approximations known of independent sets are obtained by the simplest algorithm. Repeating the algorithm on the remaining graph will certainly give a better solution – the challenge is to quantify the improvement.

References

1. Alon, N.: On constant time approximation of parameters of bounded degree graphs. In: Goldreich, O. (ed.) Property Testing - Current Research and Surveys. LNCS, vol. 6390, pp. 234–239. Springer, Heidelberg (2010). https://doi.org/10.1007/978-3-642-16367-8_14
2. Alon, N., Babai, L., Itai, A.: A fast and simple randomized parallel algorithm for the maximal independent set problem. J. Algorithms **7**(4), 567–583 (1986)
3. Alon, N., Spencer, J.H.: The Probabilistic Method, 4th edn. Wiley, Hoboken (2016)
4. Austrin, P., Khot, S., Safra, M.: Inapproximability of vertex cover and independent set in bounded degree graphs. In: 24th IEEE CCC, pp. 74–80 (2009)
5. Bansal, N., Gupta, A., Guruganesh, G.: On the Lovász theta function for independent sets in sparse graphs. In: 47th ACM STOC, pp. 193–200 (2015)

6. Bar-Yehuda, R., Censor-Hillel, K., Ghaffari, M., Schwartzman, G.: Distributed approximation of maximum independent set and maximum matching. In: PODC, pp. 165–174 (2017)
7. Berman, P., Fujito, T.: On approximation properties of the independent set problem for low degree graphs. Theor. Comput. Syst. **32**(2), 115–132 (1999)
8. Bodlaender, M.H., Halldórsson, M.M., Konrad, C., Kuhn, F.: Brief announcement: local independent set approximation. In: PODC, pp. 377–378. ACM (2016)
9. Boppana, R.B.: Personal communication to Joel Spencer (1987)
10. Caro, Y.: New results on the independence number. Technical report, Tel Aviv Univ. (1979)
11. Censor-Hillel, K., Khoury, S., Paz, A.: Quadratic and near-quadratic lower bounds for the CONGEST model. In: 31st International Symposium on Distributed Computing, pp. 10:1–10:16 (2017)
12. Chan, S.O.: Approximation resistance from pairwise-independent subgroups. J. ACM **63**(3), 27 (2016)
13. Chvátal, V., McDiarmid, C.: Small transversals in hypergraphs. Combinatorica **12**(1), 19–26 (1992)
14. Cormode, G., Dark, J., Konrad, C.: Independent set size approximation in graph streams. Technical report arXiv:1702.08299, CoRR (2017)
15. Emek, Y., Halldórsson, M.M., Mansour, Y., Patt-Shamir, B., Radhakrishnan, J., Rawitz, D.: Online set packing. SIAM J. Comput. **41**(4), 728–746 (2012)
16. Erdős, P.: On the graph theorem of Turán. Mat. Lapok **21**, 249–251 (1970). (in Hungarian)
17. Ghaffari, M., Kuhn, F., Maus, Y.: On the complexity of local distributed graph problems. In: 49th Annual ACM SIGACT Symposium on Theory of Computing, pp. 784–797 (2017)
18. Griggs, J.R.: Lower bounds on the independence number in terms of the degrees. J. Combin. Theor. B **34**, 22–39 (1983)
19. Halldórsson, B.V., Halldórsson, M.M., Losievskaja, E., Szegedy, M.: Streaming algorithms for independent sets in sparse hypergraphs. Algorithmica **76**, 490–501 (2016)
20. Halldórsson, M., Radhakrishnan, J.: Greed is good: approximating independent sets in sparse and bounded-degree graphs. Algorithmica **18**(1), 145–163 (1997)
21. Halldórsson, M.M., Konrad, C.: Computing large independent sets in a single round. Distrib. Comput. **31**(1), 69–82 (2018)
22. Kuhn, F., Moscibroda, T., Wattenhofer, R.: Local computation: lower and upper bounds. J. ACM **63**(2), 17:1–17:44 (2016)
23. Sakai, S., Togasaki, M., Yamazaki, K.: A note on greedy algorithms for the maximum weighted independent set problem. Discrete Appl. Math. **126**(2–3), 313–322 (2003)
24. Selkow, S.M.: A probabilistic lower bound on the independence number of graphs. Discrete Math. **132**(1–3), 363–365 (1994)
25. Turán, P.: On an extremal problem in graph theory. Mat. Fiz. Lapok **48**, 436–452 (1941). (in Hungarian)
26. Wei, V.: A lower bound on the stability number of a simple graph. Technical report, Bell Laboratories (1981)

On the Strongest Message Adversary for Consensus in Directed Dynamic Networks

Ulrich Schmid, Manfred Schwarz$^{(\boxtimes)}$, and Kyrill Winkler

Embedded Computing Systems Group, TU Wien, Vienna, Austria
{s,mschwarz,kwinkler}@ecs.tuwien.ac.at

Abstract. Inspired by the successful chase for the weakest failure detector in asynchronous message passing systems with crash failures and surprising relations to synchronous directed dynamic networks with message adversaries established by Raynal and Stainer [PODC'13], we introduce the concept of message adversary simulations and use it for defining a notion for strongest message adversary for solving distributed computing problems like consensus and k-set agreement. We prove that every message adversary that admits all graph sequences consisting of perpetual star graphs and is strong enough for solving multi-valued consensus is a strongest one. We elaborate on seemingly paradoxical consequences of our results, which also shed some light on the fundamental difference between crash-prone asynchronous systems with failure detectors and synchronous dynamic networks with message adversaries.

Keywords: Dynamic networks · Strongest message adversary
Failure detectors · Consensus

1 Introduction

Synchronous distributed systems consisting of a possibly unknown number n of processes that never fail but where a *message adversary* (MA) [1] controls the ability to communicate is a well-established model for dynamic networks [24]. Runs are determined by sequences of communication graphs $\mathcal{G}^1, \mathcal{G}^2, \ldots$ here, where a directed edge (p, q) is in $\mathcal{G}^r$ iff the message adversary does not suppress the message sent by p to q in round r; we will use the notation $p \to_r q$ to concisely express this. A message adversary can be identified by the set of (infinite) graph sequences it may generate, which are called *admissible* graph sequences. Research has provided various possibility and impossibility results for agreement problems, in particular, (deterministic) consensus, in this setting, both in undirected [25] and directed dynamic networks [7–9,32,33]. Albeit these

This work has been supported by the Austrian Science Fund FWF under the projects ADynNet (P28182) and RiSE/SHiNE (S11405).

Z. Lotker and B. Patt-Shamir (Eds.): SIROCCO 2018, LNCS 11085, pp. 102–120, 2018.
https://doi.org/10.1007/978-3-030-01325-7_13

results enclose the impossibility/possibility border of consensus quite tightly, no "strongest" message adversary for consensus is known yet.

Raynal and Stainer [29] established an interesting relation between synchronous systems (abbreviated $\mathcal{SMP}$) with message adversaries and asynchronous message-passing systems (abbreviated $\mathcal{AMP}$) with process crashes augmented by failure detectors [12]. Among other results, they showed that $\mathcal{AMP}$ in conjunction with the weakest failure detector (Ω, Σ) for consensus with an arbitrary number of crashes [17] can be simulated in $\mathcal{SMP}$ with the message adversary (SOURCE, QUORUM) (and *vice versa*). This message adversary guarantees communication graphs that are all rooted[1] and where, for every pair of processes p, q and every pair of rounds r, r', there is some process ℓ such that $\ell \rightarrow_r p$ and $\ell \rightarrow_{r'} q$. Additionally, there is a process p and some round r_0 such that for all processes q and all rounds $r > r_0$ we have $p \rightarrow_r q$. Consequently, every (Ω, Σ)-based consensus algorithms for $\mathcal{AMP}$ can be employed atop of this simulation in $\mathcal{SMP}$ with (SOURCE, QUORUM).

However, the question arises whether failure detector-based consensus algorithms on top of failure detector implementations can indeed compete with specifically designed consensus algorithms for $\mathcal{SMP}$ for some given MA. In particular, is it always possible to implement the weakest failure detector (Ω, Σ) atop of $\mathcal{SMP}$ with a message adversary that admits a consensus algorithm? Conversely, given that (Ω, Σ) is a weakest failure detector for consensus, is it somehow possible to simulate $\mathcal{SMP}$ with a strong(est) message adversary for consensus atop of $\mathcal{AMP}$ augmented with (Ω, Σ)? Interestingly, it follows from our results in [9] that neither is possible: Σ cannot be implemented in $\mathcal{SMP}$ with some message adversary $\mathrm{VSSC}_{D,E}(\infty)$ that admits a consensus algorithm, and the latter cannot be implemented in $\mathcal{AMP}$ equipped with (Ω, Σ) either. Essentially, the reason is that all properties achievable in $\mathcal{AMP}$ with failure detectors are inherently time-free, i.e., of eventual-type, whereas $\mathcal{SMP}$ with message adversaries facilitates time-dependent properties: The latter are sometimes too short-lived to guarantee eventual properties, however, and, conversely, cannot be extracted from eventual properties either.

In this paper, we avoid the detour via failure detectors and introduce the simple concept of *message adversary simulations* in $\mathcal{SMP}$, using the HO model [15] as a basis. Inspired by the definition of a weakest failure detector, we also define a notion for a *strongest message adversary*: A strongest MA S for a given problem P is such that (i) it admits a solution algorithm for the problem P in $\mathcal{SMP}$ equipped with S, and (ii) every MA A that admits a solution algorithm for P in $\mathcal{SMP}$ with A allows to simulate $\mathcal{SMP}$ equipped with S.

Using MA simulations, we prove that the message adversary STAR (which generates an infinite sequence of identical star graphs $\mathcal{G}, \mathcal{G}, \dots$) is a strongest message adversary for multi-valued consensus. Moreover, we show that every message adversary A that satisfies STAR $\subseteq A$ and allows to solve multi-valued consensus is also a strongest message adversary. It hence turns out that both (SOURCE, QUORUM) and $\mathrm{VSSC}_{D,E}(\infty)$ are strongest message

---

[1] A graph is rooted if it has a rooted directed spanning tree.

adversaries, even though neither $(SOURCE, QUORUM) \subseteq VSSC_{D,E}(\infty)$ nor $(SOURCE, QUORUM) \supseteq VSSC_{D,E}(\infty)$ holds. Moreover, there are interesting and sometimes apparently paradoxical consequences resulting from our findings, like the one that $\mathcal{AMP}$ with (Ω, Σ) allows to simulate $\mathcal{SMP}$ with the strongest message adversary $(SOURCE, QUORUM)$, which in turn can be transformed into $\mathcal{SMP}$ with the strongest message adversary $VSSC_{D,E}(\infty)$, which in turn does not allow to simulate $\mathcal{AMP}$ with (Ω, Σ)!

The remainder of our paper is organized as follows: After a short account of related work in Sect. 2, we define our synchronous message passing models $\mathcal{SMP}$ with message adversaries in Sect. 3. Asynchronous message passing models $\mathcal{AMP}$ with failure detectors are introduced in Sect. 4, along with a collection of failure detector-related definitions and results. In Sect. 5, we introduce a simple simulation relation between message adversaries and prove that the message adversary STAR is strongest for solving consensus in $\mathcal{SMP}$. We also show that this result generalizes to a fairly large class of strongest message adversaries. In Sect. 6, we discuss the consequences arising from the fact that STAR is a strongest message adversary. Some conclusions in Sect. 7 complete our paper.

2   Related Work

There is a huge amount of work on relations between different distributed computing models. The implementability of failure detectors in timing-based models of computation has already been addressed in Chandra and Toueg's seminal work [12], and received quite some attention in the chase for the weakest system model for implementing Ω [2–5,19,21,26,27].

There are also several papers that establish more abstract relations between various synchronous models and asynchronous models with failure detectors. For example, Charron-Bost, Guerraoui and Schiper showed that the partially synchronous model [18] with $\Phi = 1$ and $\Delta \geq 1$ is not equivalent to the asynchronous model with perfect failure detectors in terms of problem solvability. Rajsbaum, Raynal and Travers showed in [28] that failure detectors do not increase the solution power of the iterated immediate snapshot model over asynchronous read/write shared memory. A more general relation between various eventually synchronous models and asynchronous models with stabilizing failure detectors, which also considers efficiency of the algorithmic transformations, has been established by Biely et. al. in [6]: Among other results, it establishes that Ω is essentially equivalent to models with an eventually timely source, as well as to eventual lock-step rounds. [14,22] shed some light on limitations of the failure detector abstraction in timing-based models of computation, which are relevant in our context.

Research on consensus in synchronous message passing systems subject to communication failures dates back at least to the seminal paper [30] by Santoro and Widmayer; generalizations have been provided in [10,13,15,16,31]. The term message adversary was coined by Afek and Gafni in [1]. Whereas the message adversaries in all the work above are oblivious, in the sense that they may choose the graph for a round arbitrarily from a fixed set of graphs, [7] and some follow-up work [8,32] allows arbitrary sequences of communication graphs (that can also model stabilizing behavior, for example). In [29], Raynal and Stainer related stabilizing message adversaries to asynchronous systems with failure detectors, which actually stimulated our interest in the problem addressed in this paper. As our first step, we showed in [9] that Σ cannot be implemented in $\mathcal{SMP}$ with message adversary $\mathrm{VSSC}_{D,E}(\infty)$ that admits a consensus algorithm, and the latter cannot be implemented in $\mathcal{AMP}$ equipped with (Ω, Σ) either. In this paper, we will show that these results also hold for a simpler message adversary.

Researchers have also developed several "round-by-round" frameworks, which allow to relate models of computation with different degrees of synchrony and failures. Examples are round-by-round fault detectors by Gafni [20], the GIRAF framework by Keidar and Shraer [23], and the HO model by Charron-Bost and Schiper [15].

3   The Model $\mathcal{SMP}$

The model for $\mathcal{SMP}$ used in this paper will be based on the HO model introduced in [15], which provides all the features needed for defining our MA simulations. It consists of a non-empty set $\Pi = \{p_1, \ldots, p_n\}$ of n processes with unique ids, and a set of messages M, which includes a null placeholder indicating the empty message. Each process $p \in \Pi$ consists of the following components: a set of states denoted by $states_p$, a subset $init_p$ of initial states, for each positive integer $r \in \mathbb{N}*$, called round number, a message sending function S_p^r mapping states $p \times \Pi$ to a unique (possibly null) message m_p, and a state-transition function T_p^r mapping $states_p$ and partial vectors (indexed by Π) z_p of elements of M to $states_p$. The collection of the pairs of message sending function and state-transition function of the processes for every round $r > 0$ is called an algorithm on Π.

Computations in the HO model are composed of infinitely many rounds, which are communication-closed layers in the sense that any message sent in a round can be received only at that round. In each round r, process p first applies S_p^r to the current state $\mathbf{s}_p^{r-1} \in states_p$, emits the messages to be sent to each process, and then, for a subset $HO(p, r)$ of Π (indicating the processes which p hears of), applies T_p^r to its current state and the partial vector of incoming messages whose support is $HO(p, r)$ to compute $\mathbf{s}_p^r$.

A communication predicate P is defined to be a predicate over heard-of collections, that is a Boolean function over the collections of subsets of Π indexed by $\Pi \times \mathbb{N}*$:

$$P : (2^\Pi)^{\Pi \times \mathbb{N}*} \Rightarrow \{true, false\}$$

Rather than directly using communication predicates for describing our message adversaries, however, we will exploit the fact that we can easily derive $\mathcal{G}^r$, given $HO(p, r)$ for all $p \in \Pi$. Consequently, we will usually stick to the admissible graph sequences of a given MA, and make the trivial assumption that they are translated to the according communication predicate.

In order to describe information propagation in a sequence $\mathcal{G}_r, \mathcal{G}_{r+1} \ldots, \mathcal{G}_{r+\ell}$ of communication graphs, the notion of edges in the compound graph $\mathcal{G}_r \circ \mathcal{G}_{r+1} \circ \cdots \circ \mathcal{G}_{r+\ell}$ becomes useful: Given two graphs $\mathcal{G} = \langle V, E \rangle$, $\mathcal{G}' = \langle V, E' \rangle$ with the same vertex-set V, the *compound graph* $\mathcal{G} \circ \mathcal{G}' := \langle V, E'' \rangle$ where $p \to q \in E''$ if and only if for some $p' \in V : p \to p' \in E$ and $p' \to q \in E'$. By $\mathbf{s}_p^r \rightsquigarrow \mathbf{s}_q^{r'}$, we express the fact that (the state $\mathbf{s}_p^r$ of) p at the end of round r *influences* (the state $\mathbf{s}_q^{r'}$ of) q at the end of round r', which obviously requires a chain of messages from p to q. Consequently, $\mathbf{s}_p^r \rightsquigarrow \mathbf{s}_q^{r'}$ if and only if $p \to q$ in $\mathcal{G}_r \circ \cdots \circ \mathcal{G}_{r'}$.

Following [29], the abbreviation used for such models in the sequel is $\mathcal{SMP}_n[adv : MA]$, where n is the number of processes and MA is the name assigned to a set of admissible graph sequences. For example, $\mathcal{SMP}_n[adv :$ SOURCE, QUORUM] denotes the synchronous model with message adversary (SOURCE, QUORUM) defined below.

We will restrict our attention to the (uniform[2]) consensus problem in this paper, which is defined as follows: Every process $p \in \Pi$ has an input value $x_p \in V$ from some arbitrary domain V (we use the term multi-valued consensus when it is important to stress that V is not restricted) and a decision value y_p, initially undefined $y_p = \bot$. Uniform consensus requires every process p that does not crash before it decides to irrevocably assign a value from V to y_p according to the following properties:

(V) Validity: y_p has to be equal to one of the x_q's.
(A) Agreement: $y_p = y_q$ for every pair of processes $p, q \in \Pi$ that decide.
(T) Termination: y_p has to be assigned a value in finite time at every process p that does not crash in the run.

A generalization/relaxation of consensus is k-set agreement [11], which allows at most k different decision values system-wide; 1-set agreement is equivalent to consensus.

We say that a problem like consensus is *impossible* under some model $\mathcal{SMP}_n[adv : MA]$, if there is no deterministic algorithm that solves the problem for every admissible communication graph sequence of MA. For example, every problem that requires at least some communication among the processes is impossible under the *unrestricted* message adversary ∞, as the sequence $\mathcal{G}, \mathcal{G}, \ldots$ where $\mathcal{G}$ does not contain even a single edge is also admissible here.

We will now specify four message adversaries, which are primarily used in this paper. The first two, Definitions 1 and 2, are simplified versions of MAs introduced in [9], where we strengthened the properties as much as possible, albeit in a way that neither sacrificed their sufficiency for solving consensus resp. k-set agreement nor their insufficiency for implementing certain failure detectors.

Definition 1 (Message adversary VSSC(∞)**).** The message adversary VSSC(∞) is the set of all sequences of communication graphs $(\mathcal{G}^r)_{r>0}$, where the following holds:

---

[2] Note that we will also study consensus in asynchronous systems with crash failures later on. In $\mathcal{SMP}$, no process ever crashes.

(i) For every round $r > 0$, $\exists p \in \Pi, \forall q \in \Pi$: there exists a (directed) path from p to q in every $\mathcal{G}^r$.

(ii) There exists a round $r > 0$ such that $\forall r' > r : S_r = S_{r'}$, where the set S_r is such that $p \in S_r$ if $\forall q \in \Pi$: there exists a path from p to q in $\mathcal{G}^r$.

The property that the set S consists of the same vertices for some duration or even, as in (ii) above, forever is called *vertex stability* [9].

Definition 2 (*k*-set message adversary $\mathrm{VSSC}_k(\infty)$). The message adversary $\mathrm{VSSC}_k(\infty)$ is the set of all sequences of communication graphs $(\mathcal{G}^r)_{r>0}$, where

(i) there is some $k > 0$ and $P_1 \cup \cdots \cup P_k = \Pi$, such that, in every $\mathcal{G}_r$, $r > 0$, every P_i is an isolated, weakly connected component,

(ii) $\mathrm{VSSC}(\infty)$ holds independently in every partition P_i.

Since it turns out that $\mathrm{VSSC}(\infty) \subset \mathrm{VSSC}_{D,E}(d)$ for $D = E = n - 1, d = \infty$, where $\mathrm{VSSC}_{D,E}(d)$ is a message adversary introduced in [9] that allows to solve consensus, it follows that consensus can also be solved in $\mathcal{SMP}_n[adv : \mathrm{VSSC}(\infty)]$ (by using the consensus algorithm for $\mathrm{VSSC}_{D,E}(d)$), and that k-set agreement can be solved in $\mathcal{SMP}_n[adv : \mathrm{VSSC}_k(\infty)]$ (by the same algorithm). In order to prove that $\mathrm{VSSC}(\infty)$ resp. $\mathrm{VSSC}_k(\infty)$ allow to solve consensus resp. k-set agreement, we need to restate some basic notation (Definitions 3–7) from [9].

Definition 3 (**Source Component**). A *source component* $S \neq \emptyset$ of a graph $\mathcal{G}$ is the set of vertices of a *strongly connected component* in $\mathcal{G}$ that has no incoming edges from other components, formally $\forall p \in S, \forall q \in \mathcal{G}: q \to p \in \mathcal{G} \Rightarrow q \in S$.

Note that every weakly connected directed simple graph $\mathcal{G}$ has at least one source component. If a graph $\mathcal{G}$ contains only one source component S, it is called a root component.

Vertex-stable source/root components are source components that remain the same for multiple rounds in a given graph sequence, albeit their actual interconnection topology may vary.

Definition 4 (**Vertex-Stable Source Component**). Given a graph sequence $(\mathcal{G}^r)_{r>0}$, we say that the consecutive sub-sequence of communication graphs $\mathcal{G}^r$ for $r \in I = [a, b]$, $b \geq a$, contains an *I-vertex-stable source component* S, if, for $r \in I$, every $\mathcal{G}^r$ contains S as a source component.

We abbreviate I-vertex-stable source component as I-VSSC, and write $|I|$-VSSC if only the length of I matters. Note carefully that we assume $|I| = b - a + 1$ here, since $I = [a, b]$ ranges from the *beginning* of round a to the *end* of round b.

One can show that a certain amount of information propagation is guaranteed in any strongly connected component C that is vertex-stable, i.e., whose vertex set remains the same, for a given number of rounds:

Lemma 1. *Let $C \subseteq \Pi$ with $|C| > 1$, let $a \in \mathbb{N}$ and let C form a SCC of $\mathcal{G}^r$ for all $r \in [a + 1, a + |C| - 1]$. Then, $\forall p, q \in C$, it holds that $\mathsf{s}_p^a \rightsquigarrow \mathsf{s}_q^{a+|C|-1}$.*

Corollary 1 follows immediately from Lemma 1 and the fact that, by definition, VSSCs are strongly connected components.

Corollary 1. *For every I-vertex-stable source component S with $|S| > 1$ and $I = [a, b]$, it holds that $\forall p, q \in S$, $\forall x, y \in I$: $y \geq x + |S| - 2 \Rightarrow s_p^{x-1} \rightsquigarrow s_q^y$.*

In order to also model message adversaries that guarantee faster information propagation, Definition 5 introduces a system parameter $D \leq n - 1$, called the *dynamic source diameter*.

Definition 5 (D-bounded I-VSSC). A I-VSSC S is D-bounded with dynamic source diameter D, if $\forall p, q \in S$, $\forall r, r' \in I$: $r' \geq r + D - 1 \Rightarrow s_p^{r-1} \rightsquigarrow s_q^{r'}$.

Analogous considerations apply for the *dynamic network depth* $E \leq n - 1$ in communication graphs $\mathcal{G}^r$ with a single source component.

Definition 6 (E-influencing I-VSSC). A I-VSSC S is E-influencing with dynamic network depth E, if $\forall p \in S$, $\forall q \in \Pi$, $\forall r, r' \in I$: $r' \geq r + E - 1 \Rightarrow s_p^{r-1} \rightsquigarrow s_q^{r'}$.

We can now specify the message adversary $\text{VSSC}_{D,E}(d)$ introduced in [7,9], which allows to solve consensus for a sufficiently large d (in particular, for $d = \infty$):

Definition 7 (Consensus message adversary $\text{VSSC}_{D,E}(d)$). For $d > 0$, the message adversary $\text{VSSC}_{D,E}(d)$ is the set of all sequences of communication graphs $(\mathcal{G}^r)_{r>0}$, where

(i) for every round r, $\mathcal{G}^r$ contains exactly one source component,
(ii) all vertex-stable source components occurring in any $(\mathcal{G}^r)_{r>0}$ are D-bounded and E-influencing
(iii) for each $(\mathcal{G}^r)_{r>0}$, there exists some $r_{ST} > 0$ and an interval of rounds $J = [r_{ST}, r_{ST}+d-1]$ with a D-bounded and E-influencing J-vertex-stable source component.

We conclude this paragraphs by showing that $\text{VSSC}(\infty) \subset \text{VSSC}_{D,E}(d)$ for $D = E = n - 1$. This follows immediately from setting, in addition to $D = E = n - 1$, $d = \infty$ and the following facts:

(1) (i) demands in both adversaries that every graph is rooted.
(2) (ii) is trivially fulfilled by any rooted graph sequence if $D = E = n - 1$, as shown by Corollary 1.
(3) If $d = \infty$, (iii) demands in both cases that the single source component eventually consists of the same vertices forever.

Thus, $\text{VSSC}(\infty) = \text{VSSC}_{n-1,n-1}(\infty)$ holds. Solvability for $\text{VSSC}_k(\infty)$ follows from the fact that the solution algorithm for $\text{VSSC}(\infty)$ can be run in every partition: as consensus is solved in every partition, k-set agreement is guaranteed for the whole system.

The next two message adversaries have been introduced in [29]: SOURCE requires that, eventually, there is a round after which some process successfully sends a message to every process in the system:

Definition 8 (Message adversary SOURCE). The message adversary SOURCE is the set of all sequences of communication graphs $(\mathcal{G}^r)_{r>0}$, where $\exists p \in \Pi : \exists r_0 \geq 1 : \forall r \geq r_0 : \forall q \in \Pi : p \rightarrow_r q$.

QUORUM requires that every two processes $p, q \in \Pi$ hear from some common process $\ell \in \Pi$, in every pair of rounds r_p, r_q. Moreover, it requires that the set of processes that appear strongly correct (formally introduced in Definition 14) is non-empty:

Definition 9 (Message adversary QUORUM). The message adversary QUORUM is the set of all sequences of communication graphs $(\mathcal{G}^r)_{r>0}$, where $\forall p, q \in \Pi : \forall r_p, r_q : (\{\ell : \ell \rightarrow_{r_p} p\} \cap \{\ell : \ell \rightarrow_{r_q} q\} \neq 0)]$, and the set of strongly correct processes is not empty.

$\mathcal{SMP}_n[adv : \text{SOURCE}, \text{QUORUM}]$ and $\mathcal{SMP}_n[adv : \text{VSSC}(\infty)]$ allow a solution algorithm for consensus according to [29] and [9], respectively.

4 Failure Detectors in Asynchronous Systems

In asynchronous systems ($\mathcal{AMP}$), processes may take steps at any time, and the time between two steps must be finite and larger than 0. Given some initial configuration $C^0 = (s_1^0, \ldots, s_n^0)$ consisting of the initial states of all processes, a run (also called execution) in $\mathcal{AMP}$ is a sequence of infinitely many steps of every process starting from C^0, where every message sent must eventually be received and processed in finite time (except when failures occur, see below). Note that the end-to-end delay of a single message, from the time it is broadcast to the time it is received and processed, can be different for different recipients. Conceptually, we assume a (non-observable) clock with domain $\mathcal{T} = \{0, 1, 2, \ldots\}$ and require all computing steps in a run to occur synchronized with this clock.

A convenient way to characterize consensus and k-set solvability in distributed systems where processes are (usually) subject to crash failures are (weakest) *failure detectors* [12]. Again, we restate the appropriate notations from [9].

Definition 10 (Process crashes). We say that process p_i crashes at time $t \geq 0$, if it stops executing its computing step at time t (possibly leaving it incomplete) and does not execute further steps at time $t' > t$.

A failure detector [12] is an oracle that can be queried by any process in any computing step. Formally, a failure detector $\mathcal{D}$ with range $\mathcal{R}$ maps each failure pattern F to a non-empty set of histories with range $\mathcal{R}$, where a history H with range $\mathcal{R}$ is a function $H : \Pi \times \mathcal{T} \rightarrow \mathcal{R}$. The failure pattern is a function $F : \mathcal{T} \rightarrow 2^{\Pi}$ that maps each $t \in \mathcal{T}$ to the processes that have crashed by t. The set of all possible failure patterns is called the environment. Finally, $\mathcal{D}(F)$ denotes the set of possible failure detector histories permitted by $\mathcal{D}$ for the failure pattern F.

Definition 11. $\mathcal{AMP}_{n,x}[fd : FD]$ denotes the asynchronous message passing model consisting of n processes, at most x of which may crash in a run, augmented by failure detectors FD.

Two important failure detectors for consensus in $\mathcal{AMP}$ are Σ and Ω, as their combination (Ω, Σ) is known to be a weakest failure detector in the wait-free environment [17].

Definition 12. The eventual leader failure detector Ω has range Π. For each failure pattern F, for every history $H \in \Omega(F)$, there is a time $t \in \mathcal{T}$ and a correct process s.t. for every process p, $H(p,t) = q$.

Definition 13. The quorum failure detector Σ has range 2^{Π}. For each failure pattern F, for every $H \in \Sigma(F)$, two properties hold: (1) for every $t, t' \in \mathcal{T}$ and $p, q \in \Pi$ we have $H(p,t) \cap H(q,t') \neq \emptyset$ and (2) there is a time $t \in \mathcal{T}$ s.t. for every process p, $H(p,t) \subseteq \Pi \setminus \bigcup_{t \in \mathcal{T}} F(t)$.

In order to relate such failure detector models to our message adversaries, we use the simple observation that the externally visible effect of a process crash can be expressed in our setting: Since correct processes in asynchronous message passing systems perform an infinite number of steps, we can assume that they send an infinite number of (possibly empty) messages that are eventually received by all correct processes. As in [29], we hence assume that the correct (= noncrashing) processes in the simulated $\mathcal{AMP}$ are the *strongly correct processes*. Informally, a strongly correct process is able to disseminate its state to all other processes infinitely often.

Definition 14 (Faulty and strongly correct processes). Given an infinite sequence of communication graphs σ, process p is *faulty* in a run with σ if there is a round r s.t., for some process q, for all $r' > r$: $\mathbf{s}_p^r \nrightarrow \mathbf{s}_q^{r'}$.

Let $\mathcal{C}(\sigma) = \left\{ p \in \Pi \mid \forall q \in \Pi, \forall r \in \mathbb{N}, \exists r' > r : \mathbf{s}_p^r \rightsquigarrow \mathbf{s}_q^{r'} \right\}$ denote the strongly correct (= non-faulty) processes in any run with σ.

If a given process influences just one strongly correct process infinitely often, it would transitively influence all processes in the system, hence would also be strongly correct. Therefore, in order not to be strongly correct, a faulty process must not influence *any* strongly correct process infinitely often. We can hence define failure patterns as follows:

Definition 15 (Failure Pattern). The failure pattern associated with communication graph sequence σ is a function $F_\sigma : \mathbb{N} \to 2^{\Pi}$ s.t. $p \in F_\sigma(r)$ if, and only if, for all processes $q \in \mathcal{C}(\sigma)$, for all $r' > r$: $\mathbf{s}_p^r \nrightarrow \mathbf{s}_q^{r'}$.

Note that $F_\sigma(r) \subseteq F_\sigma(r+1)$ as required.

The following lemmas have originally been proven for the message adversary $\mathrm{VSSC}_{D,E}(d)$ in [9]. Fortunately, these proofs translate almost literally to the weaker message adversaries given in Definitions 1 and 2.

Lemma 2. $\mathcal{SMP}_n[adv : \text{VSSC}(\infty)]$ *does not allow to implement* $\mathcal{AMP}_{n,n-1}$ $[fd : \Sigma]$.

Proof. [9, Proof of Lemma 28]: We will prove our lemma for $n = 2$ for simplicity, as it is straightforward to generalize the proof for arbitrary n. Suppose that, for all rounds r and any processes p, some algorithm $\mathcal{A}$ computes $out(p,r)$ s.t. for any admissible failure pattern F, $out \in \Sigma(F)$. Consider the graph sequence $\sigma = (p \to q)_{r \geq 1}$. Clearly, the failure pattern associated with σ is $F_\sigma(r) = \{q\}$. Hence, in the run ε starting from some initial configuration C^0 with sequence σ, there is some round r' s.t. $out(p,r) = \{p\}$ for any $r > r'$ by Definition 13. Let $\sigma' = (p \to q)_{r=1}^{r'}(p \leftarrow q)_{r>r'}$. By similar arguments as above, in the run ε' that starts from C^0 with sequence σ', there is a round r'' such that $out(q,r) = \{q\}$ for any $r > r''$. Finally, for $\sigma'' = (p \to q)_{r=1}^{r'}(p \leftarrow q)_{r=r'+1}^{r''}(p \leftrightarrow q)_{r>r''}$, let ε'' denote the run starting from C^0 with graph sequence σ''. Until round r', $\varepsilon'' \sim_p \varepsilon$, hence, as shown above, $out(p,r') = \{p\}$ in ε''. Similarly, until round r'', $\varepsilon'' \sim_q \varepsilon'$ and hence $out(q,r'') = \{q\}$ in ε''. Clearly, $\sigma, \sigma', \sigma'' \in \text{VSSC}(\infty)$ and $F_{\sigma''}(r) = \{\}$, that is, no process is faulty in σ''. However, in ε'', $out(p,r') \cap out(q,r'') = \emptyset$, a contradiction to Definition 13. $\square$

We continue with the definitions of generalized failure detectors for the k-set agreement setting in crash-prone asynchronous message passing systems.

Definition 16. The range of the failure detector Ω_k is all k-subsets of 2^Π. For each failure pattern F, for every history $H \in \Omega_k(F)$, there $\exists LD = \{q_1, \ldots, q_k\} \in 2^\Pi$ and $t \in \mathcal{T}$ such that $LD \cap C \neq \emptyset$ and for all $t' \geq t, p \in C : H(p,t') = LD$.

Definition 17. The failure detector Σ_k has range 2^Π. For each failure pattern F, for every $H \in \Sigma_k(F)$, two properties must hold: (1) for every $t,t' \in \mathcal{T}$ and $S \in \Pi$ with $|S| = k+1$, $\exists p, q \in S : H(p,t) \cap H(q,t') \neq \emptyset$, (2) there is a time $t \in \mathcal{T}$ s.t. for every process p, for every $t' \geq t$: $H(p,t') \subseteq C$.

k-set agreement in our lock-step round model with link failures allows non-temporary partitioning, which in turn makes it impossible to use the definition of crashed and correct processes from the previous section: In a partitioned system, every process p has at least one process q such that $\forall r' > r : \mathbf{s}_p^r \rightsquigarrow \mathbf{s}_q^{r'}$, but no p usually reaches all $q \in \Pi$ here. Definition 10 hence implies that there is no correct process in this setting. Hence, we employ the following generalized definition:

Definition 18. Given a infinite graph sequence σ, let a *minimal source set* S in σ be a set of processes with the property that $\forall q \in \Pi, \forall r > 0$ there exists $p \in S, r' > r$ such that $\mathbf{s}_p^r \rightsquigarrow \mathbf{s}_q^{r'}$. The set of *weakly correct* processes $\mathcal{WC}(\sigma)$ of a sequence σ is the union of all minimal source sets S in σ.

This definition is a quite natural extension of correct processes in a model, which allows perpetual partitioning of the system. Based on this definition of weakly correct processes, it is possible to generalize some of our consensus-related results (obtained for Σ and Ω). First, we show that Σ_k cannot be implemented, since $\text{VSSC}_k(\infty)$ allows the system to partition into k isolated components.

Lemma 3. $\mathcal{AMP}_{n,n-1}[fd : \Sigma_k]$ *cannot be implemented in* $\mathcal{SMP}_n[adv : \mathrm{VSSC}_k(\infty)]$.

Proof. [9, Proof of Lemma 30]: For $k = 1$, we can rely on Lemma Lemma 2, as every $\sigma \in \mathrm{VSSC}(\infty)$ is also admissible in $\mathrm{VSSC}_k(\infty)$. Hence, $\Sigma_1 = \Sigma$ cannot be implemented in $\mathrm{VSSC}_k(\infty)$.

The impossibility can be expanded to $k > 1$ by choosing some σ that (i) perpetually partitions the system into k components $\widetilde{P} = \{P_1, \ldots, P_k\}$ that each have a single source component and consist of the same processes throughout the run, and (ii) demands eventually a vertex stable source component in every partition forever. Pick an arbitrary partition $P \in \widetilde{P}$. If $|P| > 1$, such a sequence does not allow to implement Σ in P (e.g., the message adversary could emulate the graph sequence used in Lemma Lemma 2 in P). We hence know that $\exists p, p' \in P$ and $\exists r, r'$ such that $out(p, r) \cap out(p', r') = \emptyset$. Furthermore, and irrespective of $|P|$, as for every $p \in P$, it is indistinguishable whether any $q \in \widetilde{P} \setminus P$ is faulty in σ or not, p has to assume that every process $q \in \widetilde{P} \setminus P$ is faulty. Hence, for every $p \in P$, we must eventually have $out(p, r_i) \subseteq P$ for some sufficiently large r_i.

We now construct a set S of $k+1$ processes that violates Definition Definition 17: fix some $P \in \widetilde{P}$ with $|P| > 1$ and add the two processes $p, p' \in P$, as described above, to S. For every partition $P_j \in \widetilde{P} \setminus P$, add one process p_i from P_j to S. Since there exist r, r' such that $out(p, r) \cap out(p', r') = \emptyset$, and $\forall P_j \in \widetilde{P} \setminus P, \forall p \in P_j, \exists r_i : out(p_i, r_i) \subseteq P_i$ and, by the construction of S, we have that $\forall p, q \in S, \exists r_i, r_j$ such that $out(p, r_i) \cap out(q, r_j) = \emptyset$. This set S clearly violates Definition Definition 17, as required. □

Lemma 4. *For* $k > 1$, $\mathcal{AMP}_{n,n-1}[fd : \Omega_k]$ *cannot be implemented in* $\mathcal{SMP}_n[adv : \mathrm{VSSC}_k(\infty)]$.

Proof. [9, Proof of Lemma 31]: We show the claim for $k = 2$ and $n = 3$, as it is straightforward to derive the general case from this. So suppose that some algorithm implements Ω_k under this message adversary. The following graph sequences (a)–(e) are all admissible sequences under $\mathrm{VSSC}_k(\infty)$ (we assume that nodes not depicted are isolated):

(a) $(p_3 \leftarrow p_1 \rightarrow p_2)_{r>0}$
(b) $(p_3 \leftarrow p_2 \rightarrow p_1)_{r>0}$
(c) $(p_2 \leftarrow p_3 \rightarrow p_1)_{r>0}$
(d) $(p_1 \rightarrow p_2)_{r>0}$
(e) $(p_1 \rightarrow p_3)_{r>0}$

Let $\varepsilon_a, \ldots, \varepsilon_e$ be the runs resulting from the above sequences applied to the same initial configuration. By Definitions 16 and 18, LD has to include p_1 in ε_a, p_2 in ε_b, and p_3 in ε_c. By Definition 16, in ε_d, because $\varepsilon_a \sim_{p_1} \varepsilon_d$ and $\varepsilon_c \sim_{p_3} \varepsilon_d$ in all rounds, for some $t > 0$, for all $t' > t$, $out(p_1, t') = \{p_1, p_3\}$. A similar argument shows that in ε_e, for some $t > 0$, for all $t' > t$, $out(p_1, t') = \{p_1, p_2\}$, because $\varepsilon_a \sim_{p_1} \varepsilon_e$ and $\varepsilon_b \sim_{p_2} \varepsilon_e$. The indistinguishability $\varepsilon_d \sim_{p_1} \varepsilon_e$ provides the required contradiction, as for some $t > 0$, for all $t' > t$, $out(p_1, t')$ should be the same in ε_d and ε_e.   □

Lemma 3 may come as a surprise, since the proof of the necessity of Σ_k for k-set agreement (hence the necessity of $\Sigma = \Sigma_1$ for consensus) developed by Raynal et. al. [11] only relies on the availability of a correct k-set agreement algorithm. However, their reduction proof works only in $\mathcal{AMP}_{n,n-1}$, i.e., crash-prone asynchronous message passing systems: It relies crucially on the fact that there cannot be a safety violation (i.e., a decision on a value that eventually leads to a violation of k-agreement) in any finite prefix of a run. This is not the case in the simulation running atop of $\mathcal{SMP}_{n,0}[adv : VSSC(\infty)]$, however, as we cannot ensure the crash failure semantics of faulty processes (that is needed for ensuring safety in arbitrary prefixes) here. Hence, we cannot apply their result (or adapt their proof) in our setting.

From these negative results, we conclude that, given $\mathcal{SMP}$ with some message adversary, looking out for simulations of $\mathcal{AMP}_{n,n-1}[fd : (\Sigma, \Omega)]$ in order to be able to run standard failure detector-based consensus algorithms is not a viable alternative to the development of a tailored consensus algorithm, and is hence also no substitute for the chase for strongest message adversaries in $\mathcal{SMP}$. We hence need a different way for approaching the latter, which will be presented in the following section.

5   Message Adversary Simulations and the Strongest Message Adversary for Consensus

The lemmas in the previous section showed that $\mathcal{AMP}_{n,n-1}[fd : \Sigma, \Omega]$ cannot be simulated atop of $\mathcal{SMP}_n[adv : MA]$ with some message adversary MA that allows to solve consensus. Even though failure detectors cannot hence be used directly to find a strongest message adversary, the concept of comparing models with different restrictions in terms of their computational power is nevertheless attractive. This idea was already used in [15] to structure communication predicates in the HO model, albeit the "general translations" introduced for this purpose suffered from the fact that one would need to solve repeated consensus. In sharp contrast, the message adversary simulations introduced below only need a single instance of consensus.

Our equivalent of a failure detector simulation is a *message adversary simulation* of MA M atop of M', using a suitable simulation algorithm A running in $\mathcal{SMP}_n[adv : M']$ that emulates $\mathcal{SMP}_n[adv : M]$. Note that A may also depend on the algorithm $\mathcal{A}$ that is to be run in $\mathcal{SMP}_n[adv : M]$ here. If such a simulation exists, for every $\mathcal{A}$, then M' and M have the same computational power, i.e., M' allows a solution for every problem where M allows a solution. We will now describe the details of our MA simulation, using the HO model as a basis.

Consider the HO model corresponding to $\mathcal{SMP}_n[adv : M']$, and let A be a still to-be-defined algorithm that maintains a variable $NewHO_p \subseteq \Pi$ at every process p. For some positive integer k, let the macro-round $\rho \geq 1$ for process

p be the sequence of the k consecutive rounds $r_1 = k(\rho - 1) + 1, \ldots, r_k = k\rho$. Note that $k = k(p, \rho)$ may be different for different (receiver) processes p and macro rounds ρ here. We say that A *emulates* (macro-)rounds $\rho \in \{1, 2, \ldots\}$ of $\mathcal{SMP}_n[adv : M]$, if, in any run of the latter, the value of $NewHO_p^{(\rho)}$ computed at the end of macro-round ρ satisfies:

(E1) $q \in NewHO_p^{(\rho)}$ iff $\mathbf{s}_q^{r_1-1} \rightsquigarrow \mathbf{s}_p^{r_k}$, i.e., if there exist an integer l in $\{1, \ldots, k\}$, a chain of $l + 1$ processes $p_0, p_1, \ldots, p_l$ from $p_0 = q$ to $p_l = p$, and a subsequence of l increasing round numbers $r_1, \ldots, r_l$ in macro-round ρ such that, for any index $i, 1 \leq i \leq l$, we have $p_{i-1} \in HO(p_i, r_i)$.

(E2) The collection $NewHO_p^{(\rho)}$ for all $p \in \Pi, \rho > 0$ satisfies M.

Clearly, the purpose of (E1) and (E2) is to guarantee well-defined and correct emulations, respectively.

Implementing the above emulation, i.e., the emulation algorithm A, is trivial: Let $m_{q \to p}^r$ represent the message sent by q to p in round r in $\mathcal{SMP}_n[adv : M']$, and $m_{q \to p}^{(\rho)}$ the message sent in macro-round ρ in the simulated $\mathcal{SMP}_n[adv : M]$. A just piggy-backs $m_{q \to p}^{(\rho)}$ on message $m_{q \to p}^j$, for every $(\rho - 1)k + 1 \leq j \leq \rho k$, and delivers $m_{q \to p}^{(\rho)}$ in $z_p^{(\rho)}$ in macro-round ρ, along with maintaining $NewHO_p^{(\rho)}$ in accordance with (E1). Unfortunately, however, this emulation is too restrictive for our purpose.

Our next step will hence be to define a more abstract *simulation* of $\mathcal{SMP}_n[adv : M]$, by relaxing (E1) in a way that still guarantees well-defined simulations. We recall that, by definition, $q \in HO(p, r)$ iff $m_{q \to p}^r \in z_p^r$, and that $m_{q \to p}^r = S_q^r(\mathbf{s}_q^{r-1}, p)$. Now consider the following relaxed variant of (E1), where we replace the requirement of p having *received* the message $m_{q \to p}^r$ by the requirement of q having attempted to *send* $m_{q \to p}^r$:

(E1') $q \in NewHO_p^{(\rho)}$, iff there exists at least one $j, (\rho - 1)k + 1 \leq j \leq \rho k$, for which p has acquired local knowledge of m' with $m' = S_q^j(\mathbf{s}_q^{j-1}, p)$.

We say that A *simulates* (macro-)rounds $\rho > 0$ of $\mathcal{SMP}_n[adv : M]$, if, in any run of the latter, the value of $NewHO_p^{(\rho)}$ computed at the end of macro-round ρ satisfies (E1') and (E2). At the first glance, (E1') appears to be equivalent to (E1), as it has the same outcome in the case where a chain of messages from q to p as specified in (E1) exists. However, the essential difference is played out in the case where such a chain does not exist: Sometimes, it may be possible for the simulation algorithm A at process p to locally simulate the execution of A at process q, and hence to locally compute m' without actual communication!

Using this type of message adversary simulations, in conjunction with the fact that every communication predicate can be viewed as a message adversary, we will prove in Lemma 5 below that consensus solvability and the ability to simulate the communication predicate SP_UNIF introduced in [15] are equivalent.

Definition 19. Let SP_UNIF be the communication predicate where for all $p, q, r : HO(p, r) = HO(q, r)$.

Lemma 5. *The following assertions are equivalent:*

(1) For any set of initial values V, there is an algorithm A that solves consensus in $\mathcal{SMP}_n[adv : M']$.

(2) M' allows to simulate $\mathcal{SMP}_n[adv : SP\_UNIF]$ in the execution of every algorithm $\mathcal{A}$.

Proof. The direction (2) $\to$ (1) follows from the fact that [15] provided a (trivial) algorithm that solves multi-valued consensus. We can hence plug-in this algorithm in (2) to obtain a consensus algorithm in $\mathcal{SMP}_n[adv : M']$.

To show the direction (1) $\to$ (2), let A be an algorithm that solves multi-valued consensus in $\mathcal{SMP}_n[adv : M']$, and consider an arbitrary algorithm $\mathcal{A}$ to be executed in $\mathcal{SMP}_n[adv : SP\_UNIF]$. We design an algorithm B based on A and $\mathcal{A}$, which allows to simulate $\mathcal{SMP}_n[adv : SP\_UNIF]$ in the execution of $\mathcal{A}$. Note that B executes only one instance of the consensus algorithm A throughout its execution.

To simulate the first macro-round $\rho = 1$, B first executes A on every process until consensus is solved. More specifically, p starts A with the local input value $x_p = state_p^{(0)}$, where $state_p^{(0)}$ denotes algorithm $\mathcal{A}$'s initial state. Let v be the common decision value, and $v.id = \ell$ for $v = state_\ell^{(0)}$. When A terminates at process p, B sets $NewHO_p^{(1)} := \{v.id\}$. By validity, v is indeed the initial state $state_\ell^{(0)}$ of some process $\ell \in \Pi$, and by agreement, $NewHO_p^{(1)} = NewHO_q^{(1)}$ for every $p, q \in \Pi$.

Now, assuming inductively that every process p knows $state_\ell^{(\rho-1)}$ and $state_p^{(\rho-1)}$ (as well as $\mathcal{A}$), B at p can also locally compute the message $m_{\ell \to \ell}^{(\rho)} = S_\ell^{(\rho)}(state_\ell^{(\rho-1)}, \ell)$ and $m_{\ell \to p}^{(\rho)} = S_\ell^{(\rho)}(state_\ell^{(\rho-1)}, p)$ sent by ℓ in macro round ρ. Moreover, B sets the message vector $z_\ell^{(\rho)}$ of the messages "received" by the simulated algorithm $\mathcal{A}$ for process ℓ to $z_\ell^{(\rho)} = \{m_{\ell \to \ell}^{(\rho)}\}$ and $z_p^{(\rho)} = \{m_{\ell \to p}^{(\rho)}\}$, from where it can locally compute $state_\ell^{(\rho)} = T_\ell^{(\rho)}(state_\ell^{(\rho-1)}, z_\ell^{(\rho)})$ and $state_p^{(\rho)} = T_p^{(\rho)}(state_p^{(\rho-1)}, z_p^{(\rho)})$. Finally, p sets $NewHO_p^{(\rho)} = \{\ell\}$ accordingly.

By construction, (E1') clearly holds. Moreover, since agreement secures $NewHO_p^{(1)} = NewHO_q^{(1)}$, which in turn leads to $NewHO_p^{(\rho)} = NewHO_q^{(\rho)}$ for every $\rho \geq 1$ due to the identical local computations at p and q, B indeed simulates $\mathcal{SMP}_n[adv : SP\_UNIF]$, which confirms also (E2). $\square$

With these preparations, we will now define and discuss our notion of a *strongest message adversary*:

Definition 20 (Strongest message adversary). A message adversary M is a strongest message adversary for some problem $\mathcal{P}$, if $\mathcal{P}$ is solvable in $\mathcal{SMP}_n[adv : M]$ and if, on top of every $\mathcal{SMP}_n[adv : M']$ for which $\mathcal{P}$ is solvable, we can (algorithmically) simulate some feasible execution of any algorithm $\mathcal{A}$ that solves $\mathcal{P}$ in $\mathcal{SMP}_n[adv : M]$.

A property that follows directly from Definition 20 is:

Corollary 2. *If a strongest message adversary for multi-valued consensus allows to solve some problem $\mathcal{P}$, it holds that every message adversary that allows to solve multi-valued consensus also allows to solve $\mathcal{P}$.*

By Lemma 5, SP_UNIF is a strongest message adversary for multi-valued consensus. Even more, the simulation algorithm A used in the proof of Lemma 5 actually simulates the message adversary $STAR \subset SP\_UNIF$, where there is some $p \in \Pi$ such that $HO(q, r) = p$ for every $r \geq 0$ and every $q \in \Pi$. Since multi-valued consensus is trivially solvable under $STAR$, this reveals that $STAR$ is also a strongest message adversary for multi-valued consensus.

Since every other message adversary that contains $STAR$ and allows to solve multi-valued consensus is also a strongest message adversary by definition, we finally obtain the following Corollary 3:

Corollary 3 (Class of strongest message adversaries for consensus). *Let $STAR$ be the message adversary that consists of all sequences of any perpetually repeated star graph. Every message adversary that includes $STAR$ and allows to solve multi-valued consensus, is a strongest message adversary for multi-valued consensus.*

Examples for such message adversaries are (SOURCE, QUORUM), VSSC(∞) introduced in Sect. 3, and SP_UNIF.

Interestingly, the findings above can be easily be adapted for k-set agreement as well: The same simulation algorithm B as used in the proof of Lemma 5 can be used to simulate any perpetually repeated graph that consists of k star graphs, atop of a message adversary M' that allows to solve k-set agreement: As any k-set agreement algorithm A guarantees at most k different decision values, B indeed allows to simulate any perpetually repeated graph that consists of at most k star graphs, with the k decisions as the centers. Hence:

Corollary 4 (Class of strongest message adversaries for k-set agreement). *Every message adversary that contains all sequences of any perpetually repeated graph that consists of k star graphs and allows to solve k-set agreement, is a strongest message adversary for k-set agreement.*

Examples for strongest message adversaries for k-set agreement are $VSSC_k(\infty)$ and the message adversary $VSSC_{D,H}(n, \infty) + MAJINF(k)$ introduced in [9].

6   Consequences of Our Results

The results of Sect. 4, in particular, Lemma 2, reveal the following facts:

(i) Since VSSC(∞) does not allow to implement Σ, we cannot hope to run (Σ, Ω)-based consensus algorithms on top if it.

(ii) The message adversary (SOURCE, QUORUM) considered in [29], VSSC(∞) and SP_UINF are all incomparable in terms of graph sequence inclusion, even though they all belong to the class of strongest message adversaries.

(ii) There are message adversaries like the one introduced in [32], which (unlike VSSC(∞)) do not even guarantee a single strongly correct process in some runs. Implementing Σ subject to Definition 13 atop of such message adversaries is trivially impossible, as its specification becomes void.

On the other hand, the results of Sect. 5, in particular, Lemma 5, reveals that it is possible to simulate the message adversary SP_UNIF atop of any message adversary (hence also VSSC(∞)) that allows to solve multi-valued consensus. However, it is trivial to simulate (Σ, Ω) in $\mathcal{AMP}$ in $\mathcal{SMP}_n[adv : SP\_UNIF]$: Initially, process p outputs p as the leader and Π as the quorum. At the end of round 1, both the leader and the quorum is set to $NewHO(p, r)$. Therefore, we seem to have arrived at a contradiction of Lemma 2!

This seemingly paradoxical result is traceable to the fact that the set $NewHO(p, r)$ provided by the simulation of SP_UNIF need *not* contain a strongly correct process! Indeed, recall that the infinite repetition of $\mathcal{G}$ can also be achieved by letting every process p in the system *locally* simulate the behavior of some ℓ's algorithm. This is possible, since p knows both ℓ's deterministic algorithm and its initial state, from the star graph G in round 1.

Hence, it finally turns out that the impossibility of implementing Σ established in Lemma 2 depends crucially on the assumption to consider strongly correct processes as correct in the simulated $\mathcal{AMP}$. In principle, it might be possible to implement Σ (and also Ω) atop of any message adversary that allows to solve consensus if a weaker alternative of Definition 10 of correct processes in $\mathcal{AMP}$ was used: For ℓ, it would essentially be sufficient if it managed to disseminate its initial state to all processes in the system once. Quite obviously, though, such a definition of a correct process would severely affect the semantics of failure detectors and hence the wealth of known results.

In addition, the principal ability to simulate (Σ, Ω) atop of the simulated system $\mathcal{SMP}_n[adv : SP\_UNIF]$ is not very useful in practice, as it hinges on the availability of a consensus algorithm for the bottom-level message adversary M'. Consequently, this possibility does not open up a viable alternative to the development of consensus algorithms tailored to specific message adversaries like the ones introduced in [32,33].

Overall, it turns out that strongest message adversaries according to Definition 20 do not have much discriminating power, as essentially all message adversaries known to us that allow to solve consensus are strongest according to Corollary 3. Finding a better definition of a strongest message adversary is a topic of future research. Note, however, that naive ideas like one that (i) admits a solution algorithm and (ii) is maximal w.r.t. its set of admissible graph sequences it may generate do not easily work out: Given that the latter set is usually uncountable, as admissible graph sequences are infinite, it is not clear whether (ii) is well-defined in general.

7 Conclusions

We defined message adversary simulations as a means for defining a notion of a strongest message adversary for consensus in synchronous directed dynamic networks. It turned out that every message adversary that allows to solve consensus and admits all sequences consisting of perpetual star graphs is a strongest one. We elaborate on some seemingly paradoxical consequences of our results and their relation to asynchronous systems with failure detectors.

Acknowledgments. This work has been supported by the Austrian Science Fund FWF under the projects ADynNet (P28182) and RiSE/SHiNE (S11405).

References

1. Afek, Y., Gafni, E.: Asynchrony from synchrony. In: Frey, D., Raynal, M., Sarkar, S., Shyamasundar, R.K., Sinha, P. (eds.) ICDCN 2013. LNCS, vol. 7730, pp. 225–239. Springer, Heidelberg (2013). https://doi.org/10.1007/978-3-642-35668-1_16
2. Aguilera, M.K., Delporte-Gallet, C., Fauconnier, H., Toueg, S.: On implementing Omega with weak reliability and synchrony assumptions. In: Proceeding of the 22nd Annual ACM Symposium on Principles of Distributed Computing (PODC 2003), pp. 306–314. ACM Press, New York (2003)
3. Aguilera, M.K., Delporte-Gallet, C., Fauconnier, H., Toueg, S.: Stable leader election. In: Welch, J. (ed.) DISC 2001. LNCS, vol. 2180, pp. 108–122. Springer, Heidelberg (2001). https://doi.org/10.1007/3-540-45414-4_8
4. Aguilera, M.K., Delporte-Gallet, C., Fauconnier, H., Toueg, S.: Communication-efficient leader election and consensus with limited link synchrony. In: PODC 2004, pp. 328–337. ACM Press, St. John's, Newfoundland (2004)
5. Anceaume, E., Fernández, A., Mostéfaoui, A., Neiger, G., Raynal, M.: A necessary and sufficient condition for transforming limited accuracy failure detectors. J. Comp. Sys. Sci. **68**(1), 123–133 (2004)
6. Biely, M., Hutle, M., Penso, L.D., Widder, J.: Relating stabilizing timing assumptions to stabilizing failure detectors regarding solvability and efficiency. In: Masuzawa, T., Tixeuil, S. (eds.) SSS 2007. LNCS, vol. 4838, pp. 4–20. Springer, Heidelberg (2007). https://doi.org/10.1007/978-3-540-76627-8_4
7. Biely, M., Robinson, P., Schmid, U.: Agreement in directed dynamic networks. In: Even, G., Halldórsson, M.M. (eds.) SIROCCO 2012. LNCS, vol. 7355, pp. 73–84. Springer, Heidelberg (2012). https://doi.org/10.1007/978-3-642-31104-8_7
8. Biely, M., Robinson, P., Schmid, U., Schwarz, M., Winkler, K.: Gracefully degrading consensus and k-set agreement in directed dynamic networks. In: Bouajjani, A., Fauconnier, H. (eds.) NETYS 2015. LNCS, vol. 9466, pp. 109–124. Springer, Cham (2015). https://doi.org/10.1007/978-3-319-26850-7_8
9. Biely, M., Robinson, P., Schmid, U., Schwarz, M., Winkler, K.: Gracefully degrading consensus and k-set agreement in directed dynamic networks. Theor. Comput. Sci. **726**, 41–77 (2018). https://doi.org/10.1016/j.tcs.2018.02.019. http://www.sciencedirect.com/science/article/pii/S0304397518301166
10. Biely, M., Schmid, U., Weiss, B.: Synchronous consensus under hybrid process and link failures. Theor. Comput. Sci. **412**(40), 5602–5630 (2011). https://doi.org/10.1016/j.tcs.2010.09.032. http://www.sciencedirect.com/science/article/pii/S0304397510005359

11. Bonnet, F., Raynal, M.: On the road to the weakest failure detector for k-set agreement in message-passing systems. Theor. Comput. Sci. **412**(33), 4273–4284 (2011)
12. Chandra, T.D., Toueg, S.: Unreliable failure detectors for reliable distributed systems. J. ACM **43**(2), 225–267 (1996). http://www.cs.cornell.edu/home/sam/ FDpapers/CT96-JACM.ps
13. Charron-Bost, B., Függer, M., Nowak, T.: Approximate consensus in highly dynamic networks: the role of averaging algorithms. In: Halldórsson, M.M., Iwama, K., Kobayashi, N., Speckmann, B. (eds.) ICALP 2015. LNCS, vol. 9135, pp. 528–539. Springer, Heidelberg (2015). https://doi.org/10.1007/978-3-662-47666-6_42
14. Charron-Bost, B., Hutle, M., Widder, J.: In search of lost time. Inf. Process. Lett. **110**(21), 928–933 (2010)
15. Charron-Bost, B., Schiper, A.: The Heard-Of model: computing in distributed systems with benign faults. Distrib. Comput. **22**(1), 49–71 (2009)
16. Coulouma, É., Godard, E., Peters, J.G.: A characterization of oblivious message adversaries for which consensus is solvable. Theor. Comput. Sci. **584**, 80–90 (2015). http://dx.doi.org/10.1016/j.tcs.2015.01.024
17. Delporte-Gallet, C., Fauconnier, H., Guerraoui, R., Hadzilacos, V., Kouznetsov, P., Toueg, S.: The weakest failure detectors to solve certain fundamental problems in distributed computing. In: PODC 2004, pp. 338–346. ACM Press (2004)
18. Dwork, C., Lynch, N., Stockmeyer, L.: Consensus in the presence of partial synchrony. J. ACM **35**(2), 288–323 (1988)
19. Fernández, A., Raynal, M.: From an asynchronous intermittent rotating star to an eventual leader. IEEE Trans. Parallel Distrib. Syst. **21**(9), 1290–1303 (2010)
20. Gafni, E.: Round-by-round fault detectors (extended abstract): unifying synchrony and asynchrony. In: Proceedings of the Seventeenth Annual ACM Symposium on Principles of Distributed Computing. pp. 143–152. ACM Press, Puerto Vallarta (1998)
21. Hutle, M., Malkhi, D., Schmid, U., Zhou, L.: Chasing the weakest system model for implementing omega and consensus. IEEE Trans. Dependable Secur. Comput. **6**(4), 269–281 (2009). http://www.vmars.tuwien.ac.at/documents/ extern/1803/paper.pdf
22. Jayanti, P., Toueg, S.: Every problem has a weakest failure detector. In: PODC 2008, pp. 75–84. ACM, New York (2008)
23. Keidar, I., Shraer, A.: Timeliness, failure detectors, and consensus performance. In: PODC 2006, pp. 169–178. ACM Press, New York (2006)
24. Kuhn, F., Oshman, R.: Dynamic networks: models and algorithms. SIGACT News **42**(1), 82–96 (2011)
25. Kuhn, F., Oshman, R., Moses, Y.: Coordinated consensus in dynamic networks. In: PODC 2011, ACM (2011)
26. Malkhi, D., Oprea, F., Zhou, L.: Ω Meets Paxos: leader election and stability without eventual timely links. In: Fraigniaud, P. (ed.) DISC 2005. LNCS, vol. 3724, pp. 199–213. Springer, Heidelberg (2005). https://doi.org/10.1007/11561927_16
27. Mostéfaoui, A., Raynal, M.: Solving consensus using chandra-toueg's unreliable failure detectors: a general quorum-based approach. In: Jayanti, P. (ed.) DISC 1999. LNCS, vol. 1693, pp. 49–63. Springer, Heidelberg (1999). https://doi.org/10. 1007/3-540-48169-9_4
28. Rajsbaum, S., Raynal, M., Travers, C.: An impossibility about failure detectors in the iterated immediate snapshot model. Inf. Process. Lett. **108**(3), 160–164 (2008). https://doi.org/10.1016/j.ipl.2008.05.001

29. Raynal, M., Stainer, J.: Synchrony weakened by message adversaries vs asynchrony restricted by failure detectors. In: PODC 2013, pp. 166–175 (2013)
30. Santoro, N., Widmayer, P.: Time is not a healer. In: Monien, B., Cori, R. (eds.) STACS 1989. LNCS, vol. 349, pp. 304–313. Springer, Heidelberg (1989). https://doi.org/10.1007/BFb0028994
31. Schmid, U., Weiss, B., Keidar, I.: Impossibility results and lower bounds for consensus under link failures. SIAM J. Comput. **38**(5), 1912–1951 (2009). http://www.vmars.tuwien.ac.at/documents/extern/2554/paper.pdf
32. Schwarz, M., Winkler, K., Schmid, U.: Fast consensus under eventually stabilizing message adversaries. In: ICDCN 2016, pp. 7:1–7:10. ACM, New York (2016). http://doi.acm.org/10.1145/2833312.2833323
33. Winkler, K., Schwarz, M., Schmid, U.: Consensus in directed dynamic networks with short-lived stability. CoRR abs/1602.05852 (2016). http://arxiv.org/abs/1602.05852

Symmetric Rendezvous with Advice: How to Rendezvous in a Disk

Konstantinos Georgiou[✉], Jay Griffiths, and Yuval Yakubov

Department of Mathematics, Ryerson University,
350 Victoria St., Toronto, ON M5B 2K3, Canada
{konstantinos,jay.griffiths,yyakubov}@ryerson.ca

Abstract. In the classic Symmetric Rendezvous problem on a Line (SRL), two robots at known distance 2 but unknown direction execute the same randomized algorithm trying to minimize the expected rendezvous time. A long standing conjecture is that the best possible rendezvous time is 4.25 with known upper and lower bounds being very close to that value. We introduce and study a geometric variation of SRL that we call Symmetric Rendezvous in a Disk (SRD) where two robots at distance 2 have a common reference point at distance ρ. We show that even when ρ is not too small, the two robots can meet in expected time that is less than 4.25. Part of our contribution is that we demonstrate how to adjust known, even simple and provably non-optimal, algorithms for SRL, effectively improving their performance in the presence of a reference point. Special to our algorithms for SRD is that, unlike in SRL, for every fixed ρ the worst case distance traveled, i.e. energy that is used, in our algorithms is finite. In particular, we show that the energy of our algorithms is $O\left(\rho^2\right)$, while we also explore time-energy tradeoffs, concluding that one may be efficient both with respect to time and energy, with only a minor compromise on the optimal termination time.

1 Introduction

In a rendezvous game two players reside at unknown locations in a given domain and they wish to minimize the (expected) meeting (rendezvous) time. Various rendezvous problems have been studied intensively, with applications in computer science and real-world modeling, such as the search for a mate problem in which species with a low spatial density try to find suitable partners [9]. Rendezvous problems can be classified as *asymmetric*, in which each agent may use a different strategy, or *symmetric*, in which each agent follows the same algorithm; moreover, strategies can be classified as *mixed*, incorporating randomness, or *pure* which are deterministic.

A full version of this work is posted on the Computing Research Repository [28].

K. Georgiou—Research supported in part by NSERC Discovery Grant.

J. Griffiths—Research supported in part by NSERC Undergraduate Student Research Award.

Y. Yakubov—Research supported in part by the FoS Undergraduate Research Program, Ryerson University.

Z. Lotker and B. Patt-Shamir (Eds.): SIROCCO 2018, LNCS 11085, pp. 121–133, 2018.
https://doi.org/10.1007/978-3-030-01325-7_14

In this paper, we discuss *symmetric rendezvous with advice*. Two speed-1 robots (mobile agents) start at known distance but at unknown locations and they are trying to meet (rendezvous). At any time, robots have the option to meet at a known immobile reference point that is initially placed ρ away from both agents. The goal is to design mixed strategies so as to minimize the expected rendezvous time, i.e. the expected value of the first time that robots meet. After scaling, our problem can be equivalently described as a Symmetric Rendezvous problem in a unit Disk (SRD), where mobile agents lie at the perimeter of disk at known arc distance 2α, having the option to always meet at the origin.

SRD is a geometric variation of the well-studied Symmetric Rendezvous problem on a Line (SRL) where no reference point is available, and for which a long-standing conjecture stipulates that it can be solved in expected time 4.25. Critical differences between the two problems is that in SRD (a) the rendezvous can always be realized deterministically, (b) the performance can be much better than the distance from the reference point ρ and better than the conjectured 4.25 even for not too small values of ρ and (c) the worst case rendezvous time can be bounded in ρ even when one tries to minimize the expected rendezvous time. The latter is an important property, since if the two agents are vehicles with limited fuel, our strategies can be used to guarantee rendezvous before the fuel runs out.

1.1   Related Work

The rendezvous problem is a special type of a search game where two or more agents (robots) attempt to occupy the same location at the same time in a domain. Search games and rendezvous have a long history; see [4,9] for a thorough introduction to the area, and [3] for a not so recent survey. The challenge of the task (search or rendezvous) is induced by limitations related to communication, coordination, synchronization, mobility, visibility, or other types of resources, whereas examples of rendezvous domains include networks, discrete nodes and geometric environments. Notably, each of the aforementioned specifications, along with combinations of them, have given rise to a long list of publications, a short representative list of which we discuss below.

The rendezvous problem was first proposed informally by Alpern [1] in 1976, and received attention due to the seminal works of Anderson and Weber [12] for discrete domains and of Alpern [2] for continuous domains. Our work is a direct generalization of the special and so-called Symmetric Rendezvous Search Problem on a Line (SRL) proposed by Alpern [2] in 1995. In that problem, two blind agents are at known distance 2 on a line, and they can perform the same synchronized randomized algorithm (with no shared randomness). The original algorithm of Alpern [2] had performance (expected rendezvous time) 5, which was later improved to 4.5678 [13], then to 4.4182 [15], then to 4.3931 [36], and finally to the best performance known of 4.2574 [29] by Han et al. Similarly, a series of proven lower bounds [8,36] have lead to the currently best value known of 4.1520 [29].

A number of variations of SRL have been exhaustively studied, and below we mention just a few. The symmetric rendezvous problem with unknown initial distance or with partial information about it has been considered in [16,17]. A number of different topologies have been considered including labeled network [10], labeled line [18], ring [27,32] (see survey monograph [31]), torus [30], planar lattice [5], and high dimensional host spaces [7]. We note here that the topology we consider in this work follows a long list studies of relevant search/rendezvous-type problems in the disk. The rendezvous problem with faulty components has been studied in [24,25]. Asynchronous strategies have been explored in [34,35]. Studied variations of robots capabilities include sense of direction [6,14], memory [20], visibility [22], speed [26], power consumption [11] and location awareness [19]. Interesting variations of communication models between agents have been studied in [23] (whiteboards), [21] (tokens), [31] (mobile tokens), and [34] (look-compute-move model). Finally, [33] is a comprehensive survey in deterministic rendezvous in networks.

1.2 Formal Definitions, Notation and Terminology

Problem Definition. In the Symmetric Rendezvous problem in a Disk (SRD) two agents (robots) are initially placed on the plane at known distance from each other but at unknown location. A common reference point O is at known distance and known location to both robots. The robots can move at speed 1 anywhere on the plane, and they detect each other only if they are at the same location, i.e. when the meet. Given that robots run the same (randomized) and synchronized algorithm, the goal is to design trajectory movements so as to minimize the (expected) meeting, also known rendezvous, time.

The natural way to model SRD is to have robots start on the perimeter of disk, where its center serves as the common reference point. We adopt two equivalent parameterizations of the problem that arise by either normalizing robots' initial distance or the radius of the disk. In SRD^ρ the disk has radius ρ, and the robots have Euclidean distance 2, while in SRD_α robots start on the perimeter of a unit disk at arc distance 2α.

As we explain below, SRD^ρ is the natural extension of the well-studied rendezvous on a line problem, while SRD_α is convenient for analyzing the performance of trajectory movements. We will use both perspectives of the problem interchangeably. Clearly, the initial Euclidean distance of the two robots in SRD_α is $2\sin(\alpha)$. Hence, after scaling the instance by $1/\sin(\alpha)$, the initial distance of the robots becomes 2, and the reference point (the origin) is at distance $\rho = 1/\sin(\alpha)$. Therefore, SRD^ρ and SRD_α are equivalent under transformation $\alpha = \arcsin(1/\rho)$. Moreover, we will silently assume that $0 < \alpha < \pi/4$ as otherwise SRD_α is degenerate, or that $\rho > \sqrt{2}$ for SRD^ρ.

The Related Rendezvous on a Line Problem. In the well-studied Rendezvous problem on a Line (SRL), two robots, with the same specifications as in SRD are placed at known distance 2, but at unknown locations on the line.

The objective is again to minimize the (expected) rendezvous time. Note that SRL is exactly the same as SRD$^\infty$.

Natural randomized algorithms for solving SRL are so-called k-*Markovian Strategies*, i.e. random processes that iterate indefinitely, so that in every iteration each robot follows a partial trajectory of total length k (or k times more than the original distance of the agents). The simplest 2-*Markovian Strategy* achieves expected rendezvous time 7: each robot with probability $1/2$ moves distance 1 to the left and then to the right, back to its original position (and robot follows the symmetric trajectory to the right with the complementary probability). Note that robots meet with probability $1/4$ after time 1, and otherwise they repeat the experiment after moving distance 2. If f denotes the expected meeting time, then clearly $f = \frac{1}{4} + \frac{3}{4}(2 + f)$ from which we obtain $f = 7$.

An elegant refinement was proposed by Alpern [2] and achieves expected rendezvous time 5. In this 3-*Markovian Strategy* each robot with probability $1/2$ moves distance 1 to the left, then to the right back to its original position and then further right at distance (and robot follows the symmetric trajectory to the right with the complementary probability). Robots meet with probability $1/4$ after time 1, and with probability $1/4$ after time 3, otherwise the repeat the same process. If f denotes the expected meeting time, then $f = \frac{1}{4} + \frac{1}{4}3 + \frac{1}{2}(3 + f)$ from which we obtain $f = 5$. Interestingly, this is also the best possible 3-Markovian strategy.

Alpern's algorithm above is a distance-preserving algorithm, that is, after each iteration robots either meet or they preserve their original distance (but not their original locations). After a series of improvements, this idea was fruitfully generalized to k-*Markovian Strategies* by Han et al. [29] giving the best known rendezvous time 4.2574 (for $k = 15$). Notably, the best lower bound know is 4.1520 [29], which has resulted into the believable conjecture that 4.25 is the best rendezvous time possible.

Measures of Efficiency. SRD and SRL can be viewed as online problems, where robots attempt to solve the problem only with partial input information. The natural measure of efficiency of any proposed *online algorithm* is the so-called competitive ratio, defined as the ratio between the (expected) online algorithm performance over the best possible performance achievable by an *offline* algorithm that knows the input. With this terminology in mind, it is immediate that Alpern's Algorithm [2] for SRL is 5-competitive, while the conjecture above stipulates that 4.25 is the best possible competitive ratio for the problem.

Using the terminology above, the best offline algorithm can solve SRD$^\rho$ in time 1, and SRD$_\alpha$ in time $\sin(\alpha)$, hence for our competitive analysis we will always scale the expected performance of our randomized algorithms accordingly. As a result, the competitive ratio of our algorithms will be described by functions of ρ and α for SRD$^\rho$ and SRD$_\alpha$, respectively, that are at least 1 for all values of the parameters.

Our main goal will be to beat the psychological threshold of 4.25 for SRD$^\rho$, even for not too small values of ρ, demonstrating this way both the usefulness of a reference point and the effectiveness of our algorithms. In order to quantify this more explicitly, we introduce one more alternative measure of efficiency: an algorithm for SRD$^\rho$ will be called δ-*effective*, if δ is the largest value of ρ for which the expected rendezvous time is no more than 4.25. If such ρ does not exist, i.e. if the algorithm has expected rendezvous time at least 4.25 for all $\rho > \sqrt{2}$, then we call the algorithm 0-effective. To conclude, apart from calculating the competitive ratio of our algorithms for SRD$^\rho$, we will complementarily comment also on the effectiveness, with the understanding that the higher their value is, the better the algorithm is. Note that the naive algorithm that simply has robots go to the reference point is ρ-competitive and 4.25-effective.

Finally, we also consider the worst case performance of our algorithms that we call *energy*. Formally, the energy of a rendezvous algorithm is defined as the supremum of the time by when the rendezvous is realized with probability 1. Note that any algorithm for SRL is bound to have infinite energy, whereas we show in this paper a family of algorithms for SRD that have bounded energy.

1.3   Our Results

Techniques Outline. Our main contribution is the exploration of 3-Markovian strategies for SRD. In particular, we adjust Alpern's optimal 3-Markovian algorithm [2] so as to take advantage of the reference point. Similar to the algorithm for SRL, our algorithm uses infinitely many random bits. In each random step, robots attempt to meet twice. If the rendezvous is not realized, then the projection of their trajectory to the perimeter of the original disk has length 3, however agents reside in a smaller disk but still at the same arc-distance. Then, robots repeat the process, so that, overall, the distances of the possible meeting points to the origin are strictly decreasing, i.e. the disk is sequentially shrinking. The trajectories of the robots are determined by *two critical angles*, that determine the distance of the possible meeting points to the origin, i.e. how much the disk are shrunk.

If in each iteration, the disk is shrunk "a lot", then robots move much more than half their Euclidean distance in order to meet, however when they repeat the experiment, they are solving a simpler problem since they are at the same arc-distance but the reference point is closer. If, on the other hand, the new disk is comparable to the original one, then robots attempt to greedily rendezvous as fast as possible, however if the meeting is not realized, robots have to solve an identical rendezvous problem (and such a strategy is bound to have a competitive ratio no better than 5, i.e. the ratio of the original SRL). Hence, the heart of the difficulty is to determine the two critical angles so that the instance that robots have to solve in each step shrinks by the right amount. Part of our contribution is that we demonstrate how to model the latter problem as a non-trivial non-linear optimization problem, which we also solve.

High Level Contributions. As it is typical in online algorithmic problems, the impossibility of achieving optimal solutions is due to the unknown input (in our case the exact location of the robots). Our work contributes toward the fundamental algorithmic question as to whether additional resources (partial information about the unknown input - in our case a reference point) could yield improved upper bounds. Not only we answer this question in the positive, and we quantify properly our findings, but our trajectories also demonstrate how a rendezvous can be realized in 2 dimensions, even though the detection visibility of the robots in one dimensional. Part of our contribution is to also demonstrate how to adjust known algorithms for SRL so as to solve SRD. In particular, our methods can be generalized and induce improved competitive ratio upper bounds when the starting rendezvous algorithm is some other k-Markovian trajectory, $k > 3$ (see [29]). However, each such adaptation requires the determination of more than two critical angles, and the induced non-linear optimization problems would be possible to solve only numerically, rather than analytically as we do in this work. At the end, our algorithms are simple, yet powerful enough to induce good performance for a wide range of SRD instances.

Discussion on Energy. We also consider the *worst case* rendezvous time for our algorithms that we deliberately call *energy*. In real-life applications, robots are bound to run only for limited time due to restricted resources (e.g. fuel). Assuming that the actual energy spent (fuel burnt) by a robot is proportional to it's operation time, we view the worst-case running time of our algorithms as the minimum *energy* required by the robots that ensures that the execution of the algorithm terminates successfully with probability 1. Note that in the original SRL problem, and for any feasible rendezvous strategy, there is a positive probability (though exponentially small) that the rendezvous is arbitrarily large. Given that mobile robots should have access to bounded *energy* (fuel), the probability that the rendezvous is never realized is positive. In contrast, we show that our algorithms for SRD require bounded energy, that there is a finite time by when the rendezvous is realized with probability 1. We show that this property holds true under mild conditions for our algorithms, and in particular it holds true for our algorithm that minimizes the expected rendezvous time. For the latter algorithm we show that the energy required in $\Theta\left(\rho^2\right)$. Finally, and somehow surprising, we also show that by compromising slightly on the expected termination time, the required energy becomes $\Theta\left(\rho\right)$.

Paper Organization. Section 2 is devoted to the optimization problem of minimizing the expected rendezvous time. First, in Sect. 2.1 we introduce some simple rendezvous algorithms that are mostly used as benchmark results for what will follow. Section 2.2 introduces the first non-trivial refinement, by providing a single random bit 1-Markovian algorithm. Our observations and results of that section are later used in Sect. 2.3, where we discuss general 3-Markovian strategies. Our main contribution is the determination of optimal critical angles, as well as of the induced competitive ratio, and induced effectiveness. We also provide the asymptotic behavior of the critical angles, as well as the convergence to competitive ratio 5, as the distance ρ of the reference point goes to infinity.

Then, in Sect. 3 we study the worst case rendezvous time induced by our most efficient algorithm for SRD. In particular, the main contribution of Sect. 3.1 is the asymptotic analysis of the worst case rendezvous time for our algorithm that is meant to minimize the expected rendezvous time, and is shown to be $\Theta\left(\rho^2\right)$. Motivated by this, we study in Sect. 3.2 time-energy tradeoffs. More specifically, we show that asymptotically in ρ, the expected termination time can stay optimal achieving improved but still $\Theta\left(\rho^2\right)$ energy, while only slightly suboptimal termination time allows for $\Theta(\rho)$ energy. Many of our proofs are omitted from this extended abstract due to space limitations. The interested reader may consult [28] for a full version of this paper.

2 Rendezvous Algorithms in a Disk

2.1 Some Immediate Benchmark Upper Bounds

First we establish some immediate positive results that can be used as benchmarks for rendezvous trajectories that we will present in subsequent sections. Recall that the naive "go-to-origin" algorithm is 4.25-effective.

The first attempt is to blindly implement the 4.2574-competitive algorithm of [29] for SRL. Indeed, given instance SRD_α, robots can be restricted to move on the perimeter of the disk. It is clear that the resulting algorithm has expected rendezvous time α, and hence competitive ratio $4.2574\frac{\alpha}{\sin(\alpha)}$ for SRD_α (note that $\frac{\alpha}{\sin(\alpha)} \geq 1$). However, one can slightly improve upon this by making robots move along chords instead. Indeed, the algorithm of [29] for SRL has the property that robots always move and attempt to meet at integral points, assuming that one of the robots starts from the origin of the real line. Now for problem SRD^ρ in the disk, and given any initial location of the robots, consider an infinite sequence of clockwise and of counterclockwise arcs of length 2, along with their corresponding chords of length $2\sin(1)$. Any integral movement of robots in the line can be simulated by movements on the chords by multiples of $\sin(1)$, while $\sin(1)$ is also the optimal offline solution. Therefore, we immediately obtain the following.

Theorem 1. *SRD$^\rho$ admits an online algorithm which is 4.2574-competitive and 0-effective.*

Next we show that Theorem 1 admits an easy refinement using a simple 3-Markovian process, which is a direct application of [2].

Theorem 2. *SRD$^\rho$ admits an online algorithm which is* $\left(\frac{7\rho^2+8\sqrt{\rho^2-1}\rho-3}{3\rho^2+1}\right)$-*competitive and 2.57-effective.*

2.2   Rendezvous with Minimal Randomness

Theorems 1 and 2 were obtained by algorithms that use infinitely many random bits. This section is devoted into showing that even with 1 random bit, we can perform better than the naive "go-to-origin" algorithm, as well as of the algorithms of Theorems 1 and 2, at least for certain values of α, ρ. This will also help as a warm-up for our later results.

Consider instance SRD_α and mobile agents at arc distance α. Each of them knows that their peer is α away either clockwise or counterclockwise, and consider the corresponding arcs. Notice that in both algorithms of Theorems 1 and 2 robots attempt to meet at the bisectors of the two arcs. Given a fixed angle β, each robot, and at each iteration chooses uniformly at random either the cw or the ccw direction, and moves in that direction with respect to the origin till the bisector is hit. We call this move a *random β-darting*. Notice that 0-darting corresponds to going to the origin, while the algorithm of Theorem 2 we have $\beta = \pi/2 - \alpha$. The main idea behind 1RB with parameter β is to choose the optimal $\beta \in [0, \pi/2 - \alpha)$ that minimizes the expected termination time.

Algorithm 1. 1RB$_\beta$

1: Do a random β-darting.
2: Go to origin (if peer is not already met).

Theorem 3. *The optimal 1RB$_\beta$ algorithm uses* $\overline{\beta} = \max\left(0, -\sin\left(1/\rho\right) + \arccos\left(\frac{3}{4}\right)\right)$ *in which case the algorithm is* $\frac{3\sqrt{\rho^2-1}+\sqrt{7}}{4}$*-competitive and 4.88813-effective.*

2.3   Improved Rendezvous with 3-Markovian Trajectories

In this section we generalize the algorithm of Sect. 2 in two ways; first we allow more random bits, and second, in every random trial, we allow robots trajectories two darting attempts (recall that Algorithm 1RB$_\beta$ allows for only one darting attempts. In the language of the established results for SRL we will adopt Alpern's 3-Markovian trajectory [2].

The main idea behind our new algorithms is as follows

At every random step, robots will reside at the perimeter of a disk, and they will be at constant arc distance α. As in 1RB$_\beta$, each robot is associated with two bisectors in which robot will make an attempt to meet her peer. A fixed angle β along with a random bit will determine the direction (cw or ccw) of the random β-darting that will bring the robot in one of the bisectors. Note that due to the symmetry imposed by the trajectory, a meeting is realized in this step with probability 1/4. If the rendezvous is not realized, the robot will attempt a deterministic γ-darting to the other bisector, and the meeting is realized in this step with probability 1/4 as well. If the rendezvous fails again, then the process repeats or robots go to the origin to meet. A process that involves k random bits

---

Algorithm 2. k-RB$_{\beta,\gamma}$

1: Repeat k times
2:       Do a random β-darting.
3:       Do a γ-darting in the opposite direction
4: Go to origin (if peer is not already met).

(and hence $2k$ possible meeting points) will be referred to as k-step 3-Markovian. Note that we allow $k = \infty$. The formal description of the algorithm is as follows.

Observe that the algorithm of Theorem 2 can be alternatively described as ∞-RB$_{\pi/2-\alpha/2,\pi/2-\alpha/2}$, while 1RB$_\beta$ is equivalent to 1-RB$_{\beta,0}$. Next we analyze k-RB$_{\beta,\gamma}$ for all values of k, β, γ. Our goal is to analyze the expected rendezvous time, denoted by $\mathcal{R}_k(\beta, \gamma)$. We adopt the language either of SRD$^\rho$ or of SRD$_\alpha$ depending on what is more convenient, in which case $\mathcal{R}_k(\beta, \gamma)$ will be either a function of ρ or of α. To make this more explicit in our notation, and in order to remove any ambiguity, we will be writing $\mathcal{R}_k^\rho(\beta, \gamma)$ and $\mathcal{R}_k^\alpha(\beta, \gamma)$ for the expected running time in SRD$^\rho$ and SRD$_\alpha$, respectively. Note that $\mathcal{R}_k^\rho(\beta, \gamma) = \frac{\mathcal{R}_k^\alpha(\beta,\gamma)}{\sin(\alpha)}$.

Theorem 4. *Consider problem SRD$^\rho$. If $\rho < \csc\left(\frac{1}{2}\cos^{-1}\left(\frac{2}{3}\right)\right) \approx 2.44949$, then the optimal 1-RB$_{\beta,\gamma}$ algorithm is obtained for $\overline{\gamma} = 0$, and the algorithm is identical to the optimal 1RB$_\beta$ algorithm (see Theorem 3).*

If $\rho \geq \csc\left(\frac{1}{2}\cos^{-1}\left(\frac{2}{3}\right)\right)$, then the optimal 1-RB$_{\beta,\gamma}$ is obtained for the following parameters

$$\overline{\gamma} = \cos^{-1}\left(\frac{2}{3}\right) - 2\sin^{-1}\left(\frac{1}{\rho}\right)$$

$$\overline{\beta} = \cos^{-1}\left(\frac{3}{4}\cos\left(\cos^{-1}\left(\frac{2}{3}\right) - 2\sin^{-1}\left(\frac{1}{\rho}\right)\right)\right) - \sin^{-1}\left(\frac{1}{\rho}\right)$$

For the optimal parameters, the algorithm has competitive ratio $\cos\left(\overline{\beta}\right)$ which equals

$$\frac{1}{2}\left(-\frac{\sqrt{5}}{\rho^2} + \sqrt{\rho^2 - 1} - \frac{2\sqrt{\rho^2-1}}{\rho^2} + 2\sqrt{1 - \frac{\left(\rho\left(\sqrt{5-\frac{5}{\rho^2}}+\rho\right)-2\right)^2}{4\rho^4} + \sqrt{5}}\right)$$

and it is 5.32366-effective.

We can now compute also the optimal parameters for ∞-RB$_{\beta,\gamma}$. Since the competitive ratio becomes a lengthy expression in ρ for SRD$^\rho$, we choose to only comment on the effectiveness of the resulting algorithm. The competitive ratio will be explicit from our calculations.

Theorem 5. *For all $\rho \geq 1/\sin(1/2) \approx 2.08583$, the optimal ∞-RB$_{\beta,\gamma}$ algorithm for SRD$^\rho$ uses parameters $\overline{\beta}, \overline{\gamma}$ satisfying equations*

$$\frac{3}{4}\cos(\overline{\gamma}) = \cos\left(\arcsin(1/\rho) + \overline{\beta}\right) \tag{1}$$

$$\frac{2}{3}\cos(\overline{\beta}) = \cos\left(2\arcsin(1/\rho) + \overline{\gamma}\right). \tag{2}$$

In particular, we have

$$\overline{\beta} := \arctan\left(\frac{-v + \sqrt{v^2 - (\frac{9}{4}\cos^2\alpha - 1)(\frac{5}{4} - v^2)}}{\frac{9}{4}\cos^2\alpha - 1}\right) \tag{3}$$

$$\overline{\gamma} := \arccos\left(\frac{4}{3}\cos\left(\alpha + \overline{\beta}\right)\right). \tag{4}$$

where $v := (2\cos\alpha - \cos 2\alpha)\csc 2\alpha$ *and* $\alpha = \arcsin(1/\rho)$. *Also for these values of* $\overline{\beta}, \overline{\gamma}$, *the algorithm is* 7.13678-*effective.*

We conclude this section by providing some asymptotic analysis for the optimal parameters $\overline{\beta}, \overline{\gamma}$ of Algorithm ∞-RB$_{\beta,\gamma}$ as $\rho \to \infty$. As expected, both $\overline{\beta}, \overline{\gamma}$ tend to $\pi/2$, as well as $\mathcal{R}^{\rho}_{\infty}\left(\overline{\beta}, \overline{\gamma}\right)$ tends to 5 (the competitive ratio of the SRL algorithm we are extending). This is what we make explicit with the next theorem, by also providing the rate of convergence.

Theorem 6. *For the optimal parameters* $\overline{\beta} = \overline{\beta}(\rho), \overline{\gamma} = \overline{\gamma}(\rho)$ *of Algorithm* ∞-*RB$_{\beta,\gamma}$, we have* $\lim_{\rho\to\infty} \rho^2(5 - \mathcal{R}^{\rho}_{\infty}\left(\overline{\beta}, \overline{\gamma}\right)) = 289/6$.

3   Energy-Efficient Rendezvous

3.1   Energy Analysis of Our Infinite-Step Rendezvous Algorithm

A unique feature of the SRD problem is that, unlike in SRL, the worst case rendezvous time can be finite. As before we distinguish whether we calculate the energy of ∞-RB$_{\beta,\gamma}$ in SRD$^{\rho}$ or in SRD$_{\alpha}$ by writing $\mathcal{E}^{\rho}_{\infty}(\beta,\gamma)$ and $\mathcal{E}^{\alpha}_{\infty}(\beta,\gamma)$, respectively.

Lemma 1. *The energy* $\mathcal{E}^{\alpha}_{\infty}(\beta,\gamma)$ *of* ∞-*RB$_{\beta,\gamma}$ for SRD$_{\alpha}$ is finite if and only if* $\sin(\beta)\sin(\gamma) < \sin(\alpha + \beta)\sin(2\alpha + \gamma)$. *Moreover*

$$\mathcal{E}^{\alpha}_{\infty}(\beta,\gamma) := \frac{\sin(\alpha)\csc(\alpha + \beta) + \sin(\beta)\csc(\alpha + \beta)\sin(2\alpha)\csc(2\alpha + \gamma)}{1 - \sin(\beta)\csc(\alpha + \beta)\sin(\gamma)\csc(2\alpha + \gamma)}. \tag{5}$$

Lemma 2. *For any fixed* ρ, *the energy* $\mathcal{E}^{\rho}_{\infty}\left(\overline{\beta}, \overline{\gamma}\right)$ *of the optimal* ∞-*RB$_{\overline{\beta},\overline{\gamma}}$ is finite.*

Using values $\overline{\beta}, \overline{\gamma}$ (see (3) and (4) of Theorem 5), and substituting in (5) of Lemma 1 we obtain an explicit function of α (or equivalently of $\rho = 1/\sin(\alpha)$) for $\mathcal{E}^{\alpha}_{\infty}\left(\overline{\beta}, \overline{\gamma}\right)$. Using MATHEMATICA we can observe graphically that $\mathcal{E}^{\rho}_{\infty}\left(\overline{\beta}, \overline{\gamma}\right)$ is strictly increasing (which is also expected), and that $\mathcal{E}^{\rho}_{\infty}\left(\overline{\beta}, \overline{\gamma}\right)/\rho^2$ is strictly decreasing in $\rho > 2$. However a formal proof is eluding us due to the complication of the formulas. Nevertheless, we can find the asymptotic behaviour of the energy as $\rho \to \infty$.

Theorem 7. *For the optimal parameters* $\overline{\beta} = \overline{\beta}(\rho), \overline{\gamma} = \overline{\gamma}(\rho)$ *of Algorithm* ∞-*RB$_{\beta,\gamma}$, we have* $\lim_{\rho\to\infty} \frac{\mathcal{E}^{\rho}_{\infty}(\overline{\beta},\overline{\gamma})}{\rho^2} = \frac{18}{79}$.

An immediate corollary of Theorem 7 is that $\mathcal{E}^{\rho}_{\infty}\left(\overline{\beta}, \overline{\gamma}\right) = \Theta(\rho^2)$. As long as the rendezvous between the two agents is not realized, both follow random-walk-like trajectories.

3.2   Expected Rendezvous Time - Energy Tradeoffs

In this section we attempt to understand how energy constraints can impact the performance of $\infty\text{-RB}_{\beta,\gamma}$. By Theorem 6 we know that the optimal $\infty\text{-RB}_{\overline{\beta},\overline{\gamma}}$ Algorithm induces competitive ratio 5, asymptotically in $\rho \to \infty$. By Theorem 7 we know that the same algorithm (with the same parameters) requires $\Theta\left(\rho^2\right)$ energy. In the other extreme, if the energy is less that ρ, then the problem admits no solution (and if the energy equals ρ, then the best rendezvous is attained when robots go directly to the reference point). Hence, we are motivated to study the problem of minimizing the expected rendezvous time in SRD^ρ given that agents' energy is between ρ and $\frac{18}{79}\rho^2$. Somehow surprisingly, we show below that for every $\epsilon > 0$ we can preserve a competitive ratio of 5 and energy no more than $\epsilon\rho^2 + o(\rho^2)$ or competitive ratio $5 + \epsilon$ and energy no more than $\frac{2}{\sqrt{\epsilon}}\rho + o(\rho)$, both asymptotically in ρ.

Theorem 8. *The following claims are true asymptotically for SRD^ρ as $\rho \to \infty$. For every $\epsilon > 0$, there exist β_1, γ_1 so that the competitive ratio of $\infty\text{-RB}_{\beta_1,\gamma_1}$ is 5, as well as $\mathcal{E}^\rho_\infty\left(\beta_1, \gamma_1\right)/\rho^2 \leq \epsilon$. Moreover, for every $\delta > 0$, there exist β_2, γ_2 so that the competitive ratio of $\infty\text{-RB}_{\beta_2,\gamma_2}$ is $5 + \delta$, as well as $\mathcal{E}^\rho_\infty\left(\beta_1, \gamma_1\right)/\rho \leq 2/\sqrt{\delta}$.*

4   Conclusion

We introduced and studied a new geometric variant of symmetric rendezvous that we call Symmetric Rendezvous in a Disk (SRD). Our main contribution pertains to the algorithmic reduction of known suboptimal algorithms for the classic Symmetric Rendezvous problem on a Line (SRL) to SRD. Since SRD can also be interpreted as a variant of SRL in which agents are equipped with additional advice, our results demonstrate how this advice can be beneficial to the expected rendezvous time, beating in some cases the conjectured best possible time for SRL. Special to SRD is also that, unlike in SRL, our algorithms induce bounded worst case (energy) performance. Motivated by this, we also studied energy-efficiency tradeoffs, and we showed that, somehow surprisingly, one can achieve rendezvous with limited energy (and with probability 1) by compromising only slightly on the expected rendezvous time.

Our techniques can be generalized for all known improved rendezvous protocols for SRL, however optimal reductions will be challenging to obtain. Nevertheless, it is interesting to investigate heuristic reductions, which we leave as an open research direction. Other interesting variants of our problem include the introduction of more agents, or relaxations of the notion of advice that we are using.

References

1. Alpern, S.: Hide and seek games. Seminar (1976)
2. Alpern, S.: The rendezvous search problem. SIAM J. Control Optim. **33**(3), 673–711 (1995)

3. Alpern, S.: Rendezvous search: a personal perspective. Oper. Res. **50**(5), 772–795 (2002)
4. Alpern, S.: Ten Open Problems in Rendezvous Search, pp. 223–230. Springer, New York (2013). https://doi.org/10.1007/978-1-4614-6825-7_14
5. Alpern, S., Baston, V.: Rendezvous on a planar lattice. Oper. Res. **53**(6), 996–1006 (2005)
6. Alpern, S., Baston, V.: A common notion of clockwise can help in planar rendezvous. Eur. J. Oper. Res. **175**(2), 688–706 (2006)
7. Alpern, S., Baston, V.: Rendezvous in higher dimensions. SIAM J. Control Optim. **44**(6), 2233–2252 (2006)
8. Alpern, S., Gal, S.: Rendezvous search on the line with distinguishable players. SIAM J. Control Optim. **33**(4), 1270–1276 (1995)
9. Alpern, S., Gal, S.: The theory of search games and rendezvous. In: International Series in Operations Research & Management Science, vol. 55. Springer, Heidelberg (2003). https://doi.org/10.1007/b100809
10. Steve Alpern and Wei Shi Lim: Rendezvous of three agents on the line. Naval Res. Logist. (NRL) **49**(3), 244–255 (2002)
11. Anaya, J., Chalopin, J., Czyzowicz, J., Labourel, A., Pelc, A., Vaxès, Y.: Collecting information by power-aware mobile agents. In: Aguilera, M.K. (ed.) DISC 2012. LNCS, vol. 7611, pp. 46–60. Springer, Heidelberg (2012). https://doi.org/10.1007/978-3-642-33651-5_4
12. Anderson, E.J., Weber, R.R.: The rendezvous problem on discrete locations. J. Appl. Probab. **27**(4), 839–851 (1990)
13. Anderson, E.J., Essegaier, S.: Rendezvous search on the line with indistinguishable players. SIAM J. Control Optim. **33**(6), 1637–1642 (1995)
14. Barrière, L., Flocchini, P., Fraigniaud, P., Santoro, N.: Rendezvous and election of mobile agents: impact of sense of direction. Theory Comput. Syst. **40**(2), 143–162 (2007)
15. Baston, V.J.: Two rendezvous search problems on the line. Naval Res. Logist. **46**, 335–340 (1999)
16. Baston, V., Gal, S.: Rendezvous on the line when the players' initial distance is given by an unknown probability distribution. SIAM J. Control Optim. **36**(6), 1880–1889 (1998)
17. Beveridge, A., Ozsoyeller, D., Isler, V.: Symmetric rendezvous on the line with an unknown initial distance. Technical report (2011)
18. Chester, E.J., Tütüncü, R.H.: Rendezvous search on the labeled line. Oper. Res. **52**(2), 330–334 (2004)
19. Collins, A., Czyzowicz, J., Gąsieniec, L., Kosowski, A., Martin, R.: Synchronous rendezvous for location-aware agents. In: Peleg, D. (ed.) DISC 2011. LNCS, vol. 6950, pp. 447–459. Springer, Heidelberg (2011). https://doi.org/10.1007/978-3-642-24100-0_42
20. Cooper, C., Frieze, A., Radzik, T.: Multiple random walks and interacting particle systems. In: Albers, S., Marchetti-Spaccamela, A., Matias, Y., Nikoletseas, S., Thomas, W. (eds.) ICALP 2009. LNCS, vol. 5556, pp. 399–410. Springer, Heidelberg (2009). https://doi.org/10.1007/978-3-642-02930-1_33
21. Czyzowicz, J., Dobrev, S., Kranakis, E., Krizanc, D.: The power of tokens: rendezvous and symmetry detection for two mobile agents in a ring. In: Geffert, V., Karhumäki, J., Bertoni, A., Preneel, B., Návrat, P., Bieliková, M. (eds.) SOFSEM 2008. LNCS, vol. 4910, pp. 234–246. Springer, Heidelberg (2008). https://doi.org/10.1007/978-3-540-77566-9_20

22. Czyzowicz, J., Pelc, A., Labourel, A.: How to meet asynchronously (almost) every-where. ACM Trans. Algorithms **8**(4), 37:1–37:14 (2012)
23. Das, S.: Distributed computing with mobile agents: solving rendezvous and related problems. Ph.D. thesis, University of Ottawa, Canada (2007)
24. Das, S.: Mobile agent rendezvous in a ring using faulty tokens. In: Rao, S., Chatterjee, M., Jayanti, P., Murthy, C.S.R., Saha, S.K. (eds.) ICDCN 2008. LNCS, vol. 4904, pp. 292–297. Springer, Heidelberg (2007). https://doi.org/10.1007/978-3-540-77444-0_29
25. Das, S., Luccio, F.L., Markou, E.: Mobile agents rendezvous in spite of a malicious agent. In: Bose, P., Gąsieniec, L.A., Römer, K., Wattenhofer, R. (eds.) ALGO-SENSORS 2015. LNCS, vol. 9536, pp. 211–224. Springer, Cham (2015). https://doi.org/10.1007/978-3-319-28472-9_16
26. Feinerman, O., Korman, A., Kutten, S., Rodeh, Y.: Fast Rendezvous on a cycle by agents with different speeds. In: Chatterjee, M., Cao, J., Kothapalli, K., Rajsbaum, S. (eds.) ICDCN 2014. LNCS, vol. 8314, pp. 1–13. Springer, Heidelberg (2014). https://doi.org/10.1007/978-3-642-45249-9_1
27. Flocchini, P., Kranakis, E., Krizanc, D., Santoro, N., Sawchuk, C.: Multiple mobile agent rendezvous in a ring. In: Farach-Colton, M. (ed.) LATIN 2004. LNCS, vol. 2976, pp. 599–608. Springer, Heidelberg (2004). https://doi.org/10.1007/978-3-540-24698-5_62
28. Georgiou, K., Griffiths, J., Yakubov, Y.: Symmetric rendezvous with advice: How to rendezvous in a disk. CoRR, abs/1805.03351 (2018)
29. Han, Q., Donglei, D., Vera, J., Zuluaga, L.F.: Improved bounds for the symmetric rendezvous value on the line. Oper. Res. **56**(3), 772–782 (2008)
30. Kranakis, E., Krizanc, D., Markou, E.: Mobile agent rendezvous in a synchronous torus. In: Correa, J.R., Hevia, A., Kiwi, M. (eds.) LATIN 2006. LNCS, vol. 3887, pp. 653–664. Springer, Heidelberg (2006). https://doi.org/10.1007/11682462_60
31. Kranakis, E., Krizanc, D., Markou, E.: The Mobile Agent Rendezvous Problem in the Ring. Synthesis Lectures on Distributed Computing Theory. Morgan & Claypool Publishers, San Rafael (2010)
32. Kranakis, E., Santoro, N., Sawchuk, C., Krizanc, D.: Mobile agent rendezvous in a ring. In: Distributed Computing Systems, pp. 592–599. IEEE (2003)
33. Pelc, A.: Deterministic rendezvous in networks: a comprehensive survey. Networks **59**(3), 331–347 (2012)
34. Prencipe, G.: Impossibility of gathering by a set of autonomous mobile robots. Theor. Comput. Sci **384**(2–3), 222–231 (2007)
35. Ta-Shma, A., Zwick, U.: Deterministic rendezvous, treasure hunts, and strongly universal exploration sequences. ACM Trans. Algorithms **10**(3), 12:1–12:15 (2014)
36. Patchrawat Patch Uthaisombut: Symmetric rendezvous search on the line using move patterns with different lengths. Working paper (2006)

Two Rounds Are Enough
for Reconstructing Any Graph (Class)
in the Congested Clique Model

Pedro Montealegre[1($\boxtimes$)], Sebastian Perez-Salazar[2], Ivan Rapaport[3],
and Ioan Todinca[4]

[1] Facultad de Ingeniería y Ciencias, Universidad Adolfo Ibáñez, Santiago, Chile
p.montealegre@edu.uai
[2] ISyE, Georgia Institute of Technology, Atlanta, USA
sperez@gatech.edu
[3] DIM-CMM (UMI 2807 CNRS), Universidad de Chile, Santiago, Chile
rapaport@dim.uchile.cl
[4] Université d'Orléans, INSA Centre Val de Loire, LIFO EA 4022, Orléans, France
ioan.todinca@univ-orleans.fr

Abstract. In this paper we study the *reconstruction problem* in the congested clique model. In the reconstruction problem nodes are asked to recover *all the edges* of the input graph G. Formally, given a class of graphs $\mathcal{G}$, the problem is defined as follows: if $G \notin \mathcal{G}$, then every node must reject; on the other hand, if $G \in \mathcal{G}$, then every node must end up knowing all the edges of G. It is not difficult to see that the cost Rb of *any algorithm* that solves this problem (even with public coins) is at least $\Omega(\log |\mathcal{G}_n|/n)$, where $\mathcal{G}_n$ is the subclass of all n-node labeled graphs in $\mathcal{G}$, R is the number of rounds and b is the bandwidth.

We prove here that the lower bound above is in fact tight and that it is possible to achieve it with only $R = 2$ rounds and private coins. More precisely, we exhibit (i) a one-round algorithm that achieves this bound for hereditary graph classes; and (ii) a two-round algorithm that achieves this bound for arbitrary graph classes. Later, we show that the bound $\Omega(\log |\mathcal{G}_n|/n)$ cannot be achieved in one round for arbitrary graph classes, and we give tight algorithms for that case.

From (i) we recover all known results concerning the reconstruction of graph classes in one round and bandwidth $\mathcal{O}(\log n)$: forests, planar graphs, cographs, etc. But we also get new one-round algorithms for other hereditary graph classes such as unit-disc graphs, interval graphs, etc. From (ii), we can conclude that *any problem* restricted to a class of graphs of size $2^{\mathcal{O}(n \log n)}$ can be solved in the congested clique model in two rounds, with bandwidth $\mathcal{O}(\log n)$. Moreover, our general two-round algorithm is valid for any set of labeled graphs, not only for graph classes.

Partially supported by CONICYT PIA/Apoyo a Centros Científicos y Tecnológicoss de Excelencia AFB 170001 (P.M. and I.R.), Fondecyt 1170021 (I.R.) and CONICYT + PAI + CONVOCATORIA NACIONAL SUBVENCIÓN A INSTALACIÓN EN LA ACADEMIA CONVOCATORIA AÑO 2017 + PAI77170068 (P.M.).

Z. Lotker and B. Patt-Shamir (Eds.): SIROCCO 2018, LNCS 11085, pp. 134–148, 2018.
https://doi.org/10.1007/978-3-030-01325-7_15

Keywords: Congested clique · Round complexity
Reconstruction problem · Graph classes · Hereditary graphs

1   Introduction

The *congested clique* model –a message-passing model of distributed computation where the underlying communication network is the complete graph [20]– is receiving increasingly more attention [4,8–11,14]. This model allows us to separate and understand the impact of congestion in distributed computing. The point is the following: if the communication network is a complete graph and the cost of local computation is ignored, then the only obstacle to perform any task is due to congestion alone. In other words, by isolating the effect of the bandwidth, we intend to understand it. Despite the theoretical motivation of the congested clique model, examples of distributed and parallel systems, where the efficiency depends heavily on the bandwidth and therefore might benefit from our results, are increasingly less exceptional [5,21,26]. For instance, in [12], the authors show that fast algorithms in the congested clique model can be translated into fast algorithms in the MapReduce model. Many theoretical models, aiming to bridging the gap between theory and previously mentioned softwares, have emerged [1,16,17] (These models are all very similar, but not completely identical, to the congested clique model).

The congested clique model is defined as follows. There are n nodes which are given distinct identities (IDs), that we assume for simplicity to be numbers between 1 and n. In this paper we consider the situation where the joint input to the nodes *is a graph* G. More precisely, each node v receives as input an n-bit boolean vector $x_v \in \{0,1\}^n$, which is the indicator function of its neighborhood in G. Note that the input graph G is an arbitrary n-node graph, *a subgraph of the communication network* K_n. Nodes execute an algorithm, communicating with each other in synchronous rounds and their goal is to compute some function f that depends on G. In every round, each of the n nodes may send up to $n-1$ different b-bit messages through each of its $n-1$ communication links. When an algorithm stops *every node must know* $f(G)$. We call $f(G)$ the *output* of the distributed algorithm. The parameter b is known as the *bandwidth* of the algorithm. We denote by R the *number of rounds*. The product Rb represents the total number of bits received by a node through one link, and we call it the *cost* of the algorithm.

An algorithm may be deterministic or randomized. We distinguish two sub-cases of randomized algorithms: the private-coin setting, where each node flips its own coin; and the public-coin setting, where the coin is shared between all nodes. An ε-error algorithm $\mathcal{A}$ that computes a function f is a randomized algorithm such that, for every input graph G, $\Pr(\mathcal{A}\,\text{outputs}\,f(G)) \geq 1 - \varepsilon$. In the case where $\varepsilon \to 0$ as $n \to \infty$, we say that $\mathcal{A}$ computes f with high probability (whp).

Function f defines the problem to be solved. A $0-1$ function corresponds to a decision problem (such as connectivity [11]). For other, more general types of problems, f should be defined, in fact, as a relation. This happens, for instance, when we want to construct a minimum spanning tree [9,14], a maximal independent set [10], a 3-ruling set [13], all-pairs shortest-paths [4], etc.

The most difficult problem one could attempt to solve is the *reconstruction problem*, where nodes are asked to reconstruct the input graph G. In fact, if at the end of the algorithm every node v has full knowledge of G, then it could answer *any question* concerning G. (This holds because in the congested clique model nodes have unbounded computational power).

In centralized, classical graph algorithms, a widely used approach to cope with NP-hardness is to restrict the class of graphs where the input G belongs. We are going to use an analogous approach here, in the congested clique model. But, as we are going to explain later, surprisingly, the complexity of the reconstruction problem *will only depend on the cardinality* of the subclass of n-node graphs in $\mathcal{G}$.

Formally, for any fixed set of graphs $\mathcal{G}$, we are going to introduce two problems. The first one, the *strong recognition problem* $\mathcal{G}$-STRONG-REC, is the following.

$\mathcal{G}$-STRONG-REC

| *Input:* | An arbitrary graph G |
| *Output:* | $\begin{cases} \text{all the edges of } G & \text{if } G \in \mathcal{G}; \\ \text{reject} & \text{otherwise.} \end{cases}$ |

Recall that the output is computed by *every node* of the network. In other words, every node of an algorithm that solves $\mathcal{G}$-STRONG-REC must end up knowing whether G belongs to $\mathcal{G}$; and, in the positive cases, every node also finishes knowing all the edges of G.

We also define a *weak recognition problem* $\mathcal{G}$-WEAK-REC. This is a promise problem, where the input graph G is promised to belong to $\mathcal{G}$. In other words, for graphs that do not belong to $\mathcal{G}$, the behavior of an algorithm that solves $\mathcal{G}$-WEAK-REC does not matter.

$\mathcal{G}$-WEAK-REC

| *Input:* | $G \in \mathcal{G}$ |
| *Output:* | all the edges of G |

For any positive integer n we define $\mathcal{G}_n$ as the set of n-node graphs in $\mathcal{G}$. There is an obvious lower bound for Rb, even for the weak reconstruction problem $\mathcal{G}$-WEAK-REC and even in the public-coin setting. In fact, $Rb = \Omega(\log |\mathcal{G}_n|/n)$. This can be easily seen if we note that, in the randomized case, there must be at least one outcome of the coin tosses for which the correct algorithm reconstructs the input graph in at least $(1 - \varepsilon)$ of the cases.

In this paper we are going to prove that this bound is essentially tight even with $R = 1$ (if $\mathcal{G}$ is an hereditary class of graphs) and $R = 2$ (in the general case).

1.1   Our Results

We start this paper by studying a very natural family of graph classes known as *hereditary*. A class $\mathcal{G}$ is hereditary if, for every graph $G \in \mathcal{G}$, every induced subgraph of G also belongs to $\mathcal{G}$. Many graph classes are hereditary: forests, planar graphs, bipartite graphs, k-colorable graphs, bounded tree-width graphs, d-degenerate graphs, etc. [3]. Moreover, any intersection class of graphs –such as interval graphs, chordal graphs, unit disc graphs, etc.– is also hereditary [3].

In Sect. 3 we give, for every hereditary class of graphs $\mathcal{G}$, a one-round private-coin randomized algorithm that solves $\mathcal{G}$-STRONG-REC with bandwidth $\mathcal{O}(\max_{k \in [n]} \log |\mathcal{G}_k|/k + \log n)$.

We emphasize that our algorithm runs in one round, and therefore it runs in the *broadcast congested clique*, a restricted version of the congested clique model where, in every round, the $n-1$ messages sent by a node must be the same. (This equivalence is a consequence of the requirement that *all nodes* must compute the output after one round). We also remark that for many hereditary graph classes, including all classes listed above, our algorithm is tight. Moreover, our result implies that $\mathcal{G}$-STRONG-REC can be solved in one round with bandwidth $\mathcal{O}(\log n)$ when $\mathcal{G}$ is the class of forests, planar graphs, interval graphs, unit-circle graphs, or any other hereditary graph class $\mathcal{G}$ such that $|\mathcal{G}_n| = 2^{\mathcal{O}(n \log n)}$.

In Sect. 4 we give a very general result, showing that two rounds are sufficient to solve $\mathcal{G}$-STRONG-REC in the congested clique model, for any set of graphs $\mathcal{G}$. More precisely, we provide a two-round deterministic algorithm that solves $\mathcal{G}$-WEAK-REC and a two-round private-coin randomized algorithm that solves $\mathcal{G}$-STRONG-REC whp. We also give a three-round deterministic algorithm solving $\mathcal{G}$-STRONG-REC. All algorithms run using bandwidth $\mathcal{O}(\log |\mathcal{G}_n|/n + \log n)$, so they are asymptotically optimal when $|\mathcal{G}_n| = 2^{\Omega(n \log n)}$.

Our result implies, in particular, that $\mathcal{G}$-STRONG-REC can be solved in two rounds with bandwidth $\mathcal{O}(\log n)$, when $\mathcal{G}$ is *any set* of graphs of size $2^{\mathcal{O}(n \log n)}$. The only property of the set of graphs $\mathcal{G}$ used by our algorithm is the cardinality of $\mathcal{G}_n$. Our algorithm does not require $\mathcal{G}$ to be closed under isomorphisms.

In Sect. 5 we revisit the one-round case. Our general algorithm can be adapted to run in one round (i.e., in the broadcast congested clique model) by allowing a larger bandwidth. We show that, for every set of graphs $\mathcal{G}$, there is a one-round deterministic algorithm that solves $\mathcal{G}$-WEAK-REC, and a one-round private-coin algorithm that solves $\mathcal{G}$-STRONG-REC whp, both of them using bandwidth $\mathcal{O}(\sqrt{\log |\mathcal{G}_n| \log n} + \log n)$. We finish Sect. 5 pointing out that these algorithms, with respect to the bandwidth, are tight.

1.2   Some Remarks

Lenzen's Algorithm. Lenzen's algorithm performs a load balancing procedure in the congested clique model [18]. Therefore, if the input graph is sparse, it solves the reconstruction problem very fast (by simply distributing all the edges among the nodes, and then broadcasting everything). For instance, if the input graph G contains $\mathcal{O}(n)$ edges, then it reconstruct G in a constant number of rounds with bandwidth $\mathcal{O}(\log n)$. Our result is much more general. We do not need the graphs to be sparse. We just need the class to be *small*. For example, the class of *interval graphs* contains very dense graphs (including the clique), but it is small, since it contains $2^{\mathcal{O}(n \log n)}$ different labeled graphs. In Sect. 4 we prove that, if the class $\mathcal{G}$ is such that $|\mathcal{G}_n| = 2^{\mathcal{O}(n \log n)}$, then there exists a three-round deterministic algorithm that reconstructs $\mathcal{G}$ using bandwidth $\mathcal{O}(\log n)$. Therefore, our three-round deterministic algorithm can be applied to sparse graphs, interval graphs, etc.

Broadcast Congested Clique. Consider the case where $\mathcal{G}$ is indeed sparse but we want to reconstruct it using the *broadcast* congested clique model (and therefore we can not use Lenzen's algorithm). Suppose, for instance, that the number of edges of graphs in $\mathcal{G}$ is $\mathcal{O}(n)$. The naive algorithm, where every node broadcasts its incident edges, may take $\Omega(n/b)$ rounds, because some nodes may have $\Omega(n)$ neighbors (recall that b is the bandwidth). In Sect. 5 we prove that, in the broadcast congested clique model, we can reconstruct any class of graphs $\mathcal{G}$ in one round using bandwidth $b = \mathcal{O}(\sqrt{\log |\mathcal{G}_n| \log n} + \log n)$. The class of graphs having $\mathcal{O}(n)$ edges satisfies that $\log |\mathcal{G}_n| = \mathcal{O}(n \log n)$. Hence, we can reconstruct it *in one round* using bandwidth $b = \mathcal{O}(\sqrt{n} \log n)$. This algorithm is much faster than the naive one, that would take, for the same bandwidth, $\Omega(\sqrt{n}/\log n)$ rounds.

Reconstruction Versus Recognition. The recognition problem is the classical decision problem, where we simply want to decide whether the input graph belongs to some class $\mathcal{G}$. It is clear that finding a formal proof showing some type of equivalence between the reconstruction and the recognition problems would yield a non-trivial lower bound on the recognition problem. However, in [6], the authors show that any non-trivial unconditional lower bound on a decision problem in the congested clique model would imply novel Boolean circuit complexity lower bounds. Nevertheless, proving lower bounds for explicit Boolean functions in the theory of circuit complexity has been an elusive goal for decades. Therefore, even though for some graph classes $\mathcal{G}$, *it seems that the only strategy to decide whether $G \in \mathcal{G}$ is to reconstruct G*, proving this is as difficult as proving fundamental conjectures in circuit complexity, a notoriously difficult challenge.

1.3   Techniques

The main techniques we use in this paper are *fingerprints* and *error correcting codes*. A fingerprint is a small representation of a large vector which statisfies that, if two vectors are different, then their fingerprints, whp, are also different [25]. We define in this paper the *fingerprint of a graph*, which is simply the collection of fingerprints of the rows of its adjacency matrix. Consider two graphs G and H defined on the same set of nodes. The fingerprints of these two graphs are different with a probability that grows exponentially with respect to the number of nodes having different neighborhoods in G and H. Therefore, roughly speaking, if $\mathcal{G}$ is a set of graphs where all graphs are *very different*, then each graph in $\mathcal{G}$ will have a different fingerprint.

What happens when G differs from H only in a few nodes? We have two different answers, depending on whether: (i) the graphs belong to some hereditary class of graphs $\mathcal{G}$; (ii) the graphs are arbitrary. In the first, hereditary case, we prove that, for any graph G, the number of graphs $H \in \mathcal{G}$ which are *close* to G (in terms of the number different rows in the corresponding adjacency matrices) is *small*. Therefore, the fingerprints will be different even for graphs which are close between themselves.

In the second, general case, we use, together with fingerprints, error-correcting codes. More precisely, we use *Reed-Solomon codes* [23]. The idea of these codes consists in mapping a vector into a slightly larger one, satisfying that the mapping of two different vectors differ in *many* coordinates. With this, we define *error-correcting-graphs* where, instead of vectors, we map any graph into a slightly larger one. The mapping of two different graphs will have *many* nodes with different neighborhoods. We show that the fingerprint of such mapping uniquely identifies the graphs in $\mathcal{G}$, for any $\mathcal{G}$. The advantage of our constructions is that it mainly depends on the neighborhoods of the nodes (i.e., rows of the adjacency matrix), and can be implemented efficiently in the congested clique model.

1.4 Related Work

All known results concerning the reconstruction of graphs obtained so far, have been obtained in the context of hereditary graph classes. For instance, let $\mathcal{G}$ be the class of *cograph*, that is, the class of graphs that do not contain the 4-node path as an induced subgraph. This class is obviously hereditary. In [15], the authors presented a one-round public-coin algorithm that solves $\mathcal{G}$-STRONG-REC with bandwidth $\mathcal{O}(\log n)$. Note that $|\mathcal{G}_n| = 2^{\Theta(n \log n)}$. The result we give in this paper is stronger, because our one-round algorithm needs the same bandwidth but uses private coins.

In [2,22] it is shown that, if $\mathcal{G}$ is the class of d-*degenerate* graphs, then there is a one-round deterministic algorithm that solves $\mathcal{G}$-STRONG-REC with bandwidth $\mathcal{O}(d \log n) = \mathcal{O}(\log n)$. A graph G is d-degenerate if one can remove from G a vertex r of degree at most d, and then proceed recursively on the resulting graph $G' = G - r$, until obtaining the empty graph. Note that planar graphs (or more generally, bounded genus graphs), bounded tree-width graphs, graphs without a fixed graph H as a minor, are all d-degenerate, for some constant $d > 0$. Since the class of d-degenerate graphs is hereditary and satisfies $|\mathcal{G}_n| = 2^{\Theta(n \log n)}$, it follows, from this paper, the existence of a one-round private-coin randomized algorithm that solves $\mathcal{G}$-STRONG-REC with bandwidth $\mathcal{O}(\log n)$. However, the result of [2] for this particular class is stronger, since their algorithm is deterministic.

Another example of reconstruction with one-round deterministic algorithms can be found in [6]. There, the authors consider the class of graphs defined by one forbidden subgraph H. They show that such classes can be reconstructed deterministically in one round with bandwidth $b = \mathcal{O}((ex(n, H) \log n)/n)$, where $ex(n, H)$ is the *Turán number* of H, defined as be the maximum number of edges in an n-node graph not containing an isomorphic copy of H as a subgraph. For example, if C_4 is the cycle of length 4, then $ex(n, C_4) = \mathcal{O}(n^{3/2})$. This implies that, if we define $\mathcal{G}$ as the class of graphs not containing C_4 as a subgraph, then there is a one-round *deterministic* algorithm that solves $\mathcal{G}$-STRONG-REC with bandwidth $\mathcal{O}(\sqrt{n}\log n)$.

2   Preliminaries

2.1   Some Graph Terminology

Two graphs G and H are *isomorphic* if there exists a bijection $\varphi : V(G) \to V(H)$ such that any pair of vertices u, v are adjacent in G if and only if $f(u)$ and $f(v)$ are adjacent in H. A *class of graphs* $\mathcal{G}$ is a set of graphs which is *closed under isomorphisms*, i.e., if G belongs to $\mathcal{G}$ and H is isomorphic to G, then H also belongs to $\mathcal{G}$. For a class of graphs $\mathcal{G}$ and $n > 0$, we call $\mathcal{G}_n$ the subclass of n-node graphs in $\mathcal{G}$. For a graph $G = (V, E)$ and $U \subseteq V$ we denote $G[U]$ the subgraph of G induced by U. More precisely, the vertex set of $G[U]$ is U and the edge set consists of all of the edges in E that have both endpoints in U. A class $\mathcal{G}$ is *hereditary* if it is closed under taking induced subgraphs, i.e., for every $G = (V, E) \in \mathcal{G}$ and every $U \subseteq V$, the induced subgraph $G[U] \in \mathcal{G}$.

For a graph $G = (\{v_1, \ldots, v_n\}, E)$, we call $A(G)$ its *adjacency matrix*, i.e., the $0 - 1$ square matrix of dimension n where $[A(G)]_{ij} = 1$ if and only if v_i is adjacent to v_j. Let M be a square matrix of dimension n, and let $i \in [n] = \{1, \ldots, n\}$. We call M_i the i-th row of M. Let N be another square matrix of dimension n. We denote by $d_r(M, N)$ the *row-distance* between M and N, that is, the number of rows that are different between M and N. In other words, $d_r(M, N) = |\{i \in [n] : M_i \neq N_i\}|$. For $k > 0$ and $G = (V, E)$, we denote by $D(G, k)$ the set of all graphs $H = (V, E')$ such that $d_r(A(G), A(H)) = k$.

2.2   Fingerprints

Let n be a positive integer and p be a prime number. In the following, we denote by $\mathbb{F}_p$ the finite field of size p and by $\mathbb{F}_p[X]$ the polynomial ring on $\mathbb{F}_p$. A polynomial $P \in \mathbb{F}_p[X]$ is an expression of the form $P(x) = \sum_{i=1}^{d} a_i x^{i-1}$, where $a_i \in \mathbb{F}_p$.

Let q be a prime number such that $q < n < p$. For each $a \in \mathbb{F}_q^n$, consider the polynomial $FP(a, \cdot) \in \mathbb{F}_p[X]$ defined as $FP(a, x) = \sum_{i \in [n]} a_i x^{i-1} \mod p$. (Note that we interpret the coordinates of a as elements of $\mathbb{F}_p$).

For $t \in \mathbb{F}_p$, we call $FP(a, t)$ the *fingerprint* of a and t. The following lemma is direct.

Lemma 1. [19] *Let n be a positive integer, p and q be two prime numbers such that $q < n < p$. Let $a, b \in (\mathbb{F}_q)^n$ such that $a \neq b$. Then, $|\{t \in \mathbb{F}_p : P(a, t) = P(b, t)\}| \leq n - 1$.*

We extend the definition of fingerprints to matrices. Let M be a square matrix of dimension n and coordinates in $\mathbb{F}_q$, and let T be an element of $(\mathbb{F}_p)^n$. We call $FP(M,T) \in (\mathbb{F}_p)^n$ the *fingerprint of M and T*, defined as $FP(M,T) = (FP(M_1,T_1),\ldots,FP(M_n,T_n))$, where M_i is the i-th row of M, for each $i \in [n]$. Moreover, for a graph of size n, and $T \in (\mathbb{F}_p)^n$ we call $FP(G,T)$ the fingerprint of $A(G)$ and T.

3   Reconstructing Hereditary Graph Classes

In this section we start giving the main result. Later we explain the consequence of this result on well-known hereditary graph classes.

Theorem 1. *Let $\mathcal{G}$ be an hereditary class of graphs. There exists a one-round private-coin algorithm that solves $\mathcal{G}$-STRONG-REC whp and bandwidth*

$$\mathcal{O}(\max_{k \in [n]}(\log(|\mathcal{G}_k|)/k) + \log n).$$

Proof. In the algorithm, nodes use a prime number p, whose value will be chosen later. The algorithm consists in: (1) Each node i picks t_i in $\mathbb{F}_p$ uniformly at random (using private coins), and computes $FP(x_i,t_i)$. (2) Each node communicates t_i and $FP(x_i,t_i)$. (3) Every node constructs $T = (t_1,\ldots t_n)$ and $FP(G,T) = (FP(x_1,t_1),\ldots,F(x_n,t_n))$ from the messages sent in the communication round. Finally: (4) Every node looks in $\mathcal{G}_n$ for a graph H such that $FP(H,T) = FP(G,T)$. If such graph H exists, the algorithm outputs H, otherwise it *rejects*.

Let T in $(\mathbb{Z}_p)^n$, picked uniformly at random. We aim to show that, for every G, if some $H \in \mathcal{G}_n$ satisfies $FP(H,T) = FP(G,T)$, then $G = H$ whp. First, note that

$$\Pr(\exists H \in \mathcal{G}_n \text{ s.t. } H \neq G \text{ and } FP(G,T) = FP(H,T))$$
$$\leq$$
$$\sum_{k \in [n]} \Pr(\exists H \in \mathcal{G}_n \cap D(G,k) \text{ s.t. } FP(G,T) = FP(H,T)).$$

Now suppose that $H \neq G$ and let $k > 0$ such that H belongs to $D(G,k) \cap \mathcal{G}_n$. From Lemma 1 we deduce that $Pr(FP(G,T) = FP(H,T)) \leq \left(\frac{n}{p}\right)^k$. It follows that

$$\Pr(\exists H \in \mathcal{G}_n \cap D(G,k) \text{ s.t. } FP(G,T) = FP(H,T)) \leq \left(\frac{n}{p}\right)^k \cdot |\mathcal{G}_n \cap D(G,k)|.$$

We now claim that $|\mathcal{G}_n \cap D(G,k)| \leq \binom{n}{k}|\mathcal{G}_k|$. Indeed, we can interpret a graph H in $D(G,k)$ as a graph built by picking k vertices $\{v_1,\ldots v_k\}$ of G and then adding or removing edges between those vertices. Since $\mathcal{G}$ is hereditary, the graph induced by $\{v_1, \ldots, v_k\}$ must belong to $\mathcal{G}_k$. Therefore, $|\mathcal{G}_n \cap D(G, k)| \leq \binom{n}{k} |\mathcal{G}_k|$. This claim implies:

$$\Pr(\exists H \in \mathcal{G}_n \cap D(G, k) \text{ s.t. } FP(G, T) = FP(H, T)) \leq \left(\frac{n}{p}\right)^k \cdot \left(\frac{ne}{k}\right)^k \cdot |\mathcal{G}_k|$$

$$\leq \left(\frac{n^2 \cdot e \cdot (|\mathcal{G}_k|)^{1/k}}{p}\right)^k.$$

Let $f : \mathbb{N} \to \mathbb{R}$ be defined as $f(n) = n \cdot \max_{k \in [n]} \frac{\log |\mathcal{G}_k|}{k}$. Note that this function is increasing, satisfies $f(n)/n \leq f(n+1)/(n+1)$, and $\log |\mathcal{G}_n| \leq f(n)$. Therefore, $(|\mathcal{G}_k|)^{1/k} \leq 2^{f(k)/k} \leq 2^{f(n)/n}$. We deduce:

$$\Pr(\exists H \in \mathcal{G}_n \text{ s.t. } H \neq G \text{ and } FP(G, T) = FP(H, T)) \leq \sum_{k \in [n]} \left(\frac{n^2 \cdot e \cdot 2^{f(n)/n}}{p}\right)^k.$$

We now fix p as the smallest prime number greater than $n^4 \cdot e \cdot 2^{f(n)/n}$, and we get that with probability at least $1 - 1/n$, either $G = H$ or $F(H, T) \neq F(G, T)$, for every $H \in \mathcal{G}_n$. Hence, the algorithm solves $\mathcal{G}$-STRONG-REC whp.

Note that the bandwidth required by node i in the algorithm equals the number of bits required to represent the pair $(t_i, F(x_i, t_i))$, which are two integers in $[p]$. Therefore, the bandwidth of the algorithm is

$$2\lceil \log p \rceil = \mathcal{O}(f(n)/n + \log n) = \mathcal{O}\left(\max_{k \in [n]}(\log(|\mathcal{G}_k|)/k) + \log n\right).$$

$\square$

We deduce the following corollary.

Corollary 1. *Let $\mathcal{G}$ be an hereditary class of graphs, and h be an increasing function such that $|\mathcal{G}_n| = 2^{\theta(nh(n))}$. Then, our private-coin algorithm solves $\mathcal{G}$-STRONG-REC whp, in one-round, with bandwidth $\Theta(\log |\mathcal{G}_n|/n + \log n)$. This matches the lower bound on the cost Rb (which must be satisfied even in the public coin setting).*

In [24], Scheinerman and Zito showed that hereditary graph classes have a very specific growing rate. They showed (Theorem 1 in [24]) that, for any hereditary class of graphs $\mathcal{G}$, one of the following behaviors must hold: $|\mathcal{G}_n| \in \{\mathcal{O}(1), n^{\Theta(1)}, 2^{\Theta(n)}, 2^{\Theta(n \log n)}, 2^{\omega(n \log n)}\}$. Corollary 1 implies that our algorithm is tight for any hereditary class of graphs such that $|\mathcal{G}_n| = 2^{\Theta(n \log n)}$.

4   Reconstructing Arbitrary Graph Classes

In this section we show that there exists a two-round private-coin algorithm in the congested clique model that solves $\mathcal{G}$-STRONG-REC whp and bandwidth $\mathcal{O}(\log |\mathcal{G}_n|/n + \log n)$. Our algorithm is based, roughly, on the same ideas used to reconstruct hereditary classes of graphs. But the problem we encounter is the following: while in the case of hereditary classes of graphs, we had for every graph G and $k > 0$, a bound on the number of graphs contained in $D(G, k) \cap \mathcal{G}_n$, this is not the case in an arbitrary family $\mathcal{G}$. Therefore, fingerprints alone are not enough to differentiate graphs. To cope with this obstacle, we use error correcting codes.

Definition 1. Let $0 \leq k \leq n$, and let q be the smallest prime number greater that $n + k$. An *error correcting code with parameters* (n, k) is a mapping $C : \{0, 1\}^n \to (\mathbb{F}_q)^{n+k}$, satisfying:

(1) For every $x \in \{0, 1\}^n$ and $i \in [n]$, $C(x)_i = x_i$.
(2) For each $x, y \in \{0, 1\}^n$, $x \neq y$ implies $|\{i \in [n + k] : C(x)_i \neq C(y)_i\}| \geq k$.

For sake of completeness, we give the construction of an error correcting code with parameters (n, k). For $x \in \{0, 1\}^n$, let P_x be the unique polynomial of degree (at most) n in $\mathbb{F}_q[X]$ satisfying $P_x(i) = x_i$ for each $i \in [n]$. The function C is then defined as $C(x) = (P_x(1), \ldots, P_x(n + k))$. This function satisfies properties (1) and (2). We now adapt the definition of error correcting codes to graphs.

Definition 2. For a graph G, we call $C(G)$ the square matrix of dimension $n + k$ with elements in $\mathbb{F}_q$ defined as follows.

- For each $i \in [n]$, the i-th row of $C(G)$ is $C(A(G)_i) \in (\mathbb{F}_q)^{n+k}$ (recall that $A(G)_i$ is the i-th row of the adjacency matrix of G).
- For each $i \in [k]$, the $(n + i)$-th row of $C(G)$ is the vector

$$(C(x_1)_{n+i}, \ldots, C(x_n)_{n+i}, \mathbf{0}) \in (\mathbb{F}_q)^{n+k},$$

where $\mathbf{0}$ is the zero-vector of $\mathbb{F}_q^d$, and $C(x)_j \in \mathbb{F}_q$ is the j-th element of $C(x)$.

We can represent $C(x)$ as a pair $(x, \tilde{x})$, where $\tilde{x}$ belongs to $(\mathbb{F}_q)^k$. Similarly, for a graph G, we can represent $C(G)$ as the symmetric matrix:

$$C(G) = \begin{bmatrix} A(G) & A(\tilde{G}) \\ A(\tilde{G})^T & 0 \end{bmatrix},$$

where $A(\tilde{G})$ is the matrix with rows $C(A(G)_i)_{n+1}, \ldots, C(A(G)_i)_{n+k}$, with $i \in [n]$.

Remark 1. Note that $d_r(C(G), C(H)) > k$, for every two different n-node graphs H and G. Indeed, if $G \neq H$, there exists $i \in [n]$ such that $A(G)_i$ is different than $A(H)_i$. Then, by definition of C, $|\{j \in [n + k] : C(A(G))_{i,j} \neq C(A(H))_{i,j}\}| > k$. This means that $d_r(C(G), C(H)) > k$, because $C(G)$ and $C(H)$ are symmetric matrices.

Lemma 2. *Let $\mathcal{G}$ be a set of graphs, C the error correcting code with parameters (n, k), and let p be the smallest prime number greater than $(n + k) \cdot |\mathcal{G}_n|^{2/k}$. Then, there exists $T \in (\mathbb{F}_p)^{n+k}$ depending only on $\mathcal{G}$, satisfying $FP(C(G), T) \neq FP(C(H), T)$ for all different $G, H \in \mathcal{G}_n$.*

Proof. From Remark 1, we know that $d_r(C(G), C(H)) > k$, for every two different n-node graphs H and G. Then, if we pick $T \in (\mathbb{F}_p)^{n+k}$ uniformly at random we have, from Lemma 1:

$$\Pr(FP(C(G), T) = FP(C(H), T)) < \left(\frac{n+k}{p}\right)^k.$$

Then, by the union bound

$$\Pr(\exists G, H \in \mathcal{G}_n \text{ s.t. } G \neq H \quad \text{and } FP(C(G), T) = FP(C(H), T))$$
$$< \left(\frac{n+k}{p}\right)^k \cdot |\mathcal{G}_n|^2 \leq 1.$$

The last inequality follows from the choice of p. Therefore, there must exist a $T \in (\mathbb{F}_p)^{n+k}$ such that $FP(C(G), T) \neq FP(C(H), T)$, for all different $G, H \in \mathcal{G}_n$. □

Theorem 2. *Let $\mathcal{G}$ be a set of graphs. The following holds:*

(1) There exists a two-round deterministic algorithm in the congested clique model that solves $\mathcal{G}$-WEAK-REC with bandwidth $\mathcal{O}(\log |\mathcal{G}_n|/n + \log n)$.

(2) There exists a three-round deterministic algorithm in the congested clique model that solves $\mathcal{G}$-STRONG-REC with bandwidth $\mathcal{O}(\log |\mathcal{G}_n|/n + \log n)$.

(3) There exists a two-round private-coin algorithm in the congested clique model that solves $\mathcal{G}$-STRONG-REC with bandwidth $\mathcal{O}(\log |\mathcal{G}_n|/n + \log n)$ whp.

Proof.

(1) Let p be the first prime greater than $2n \cdot |\mathcal{G}_n|^{2/n}$ (then $p \leq 4n \cdot |\mathcal{G}_n|^{2/n}$), and let q be the smallest prime number greater than $2n$. In the algorithm, node i first computes $C(x_i)$, where C is the error correcting code with parameters (n, n). Then, for each $j \in [n]$ node i communicates $C(x_i)_{j+n}$ to node j. This communication round requires bandwidth $\lceil \log q \rceil = \mathcal{O}(\log n)$. After the first communication round, node i knows $C(x_i)$ and $(C(x_1)_{i+n}, \ldots, C(x_n)_{i+n})$, i.e., it knows rows i and $i + n$ of matrix $C(G)$. Each node computes a vector $T \in (\mathbb{F}_p)^{2n}$ such that $FP(C(G), T) \neq FP(C(H), T)$, for all different $G, H \in \mathcal{G}_n$ (each node computes the same T). The existence of T is given by Lemma 2. Then, node i communicates (broadcasts) $FP(C(G)_i, T_i)$ and $FP(C(G)_{i+n}, T_{i+n})$. This communication round requires bandwidth $2\lceil \log p \rceil = \mathcal{O}((\log |\mathcal{G}_n|)/n + \log n)$. After the second communication round, each node knows $FP(C(G), T)$. Then, they locally compute the unique $H \in \mathcal{G}_n$ such that $FP(C(H), T) = FP(C(G), T)$. Since G belongs to $\mathcal{G}_n$, then necessarily $G = H$.

(2) Suppose now that we are solving $\mathcal{G}$-STRONG-REC. In this case G does not necessarily belong to $\mathcal{G}_n$. After receiving the fingerprints of $C(G)$, nodes look for a graph H in $\mathcal{G}_n$ that satisfies $FP(C(G), T) = FP(C(H), T)$. If such a graph exists, we call it a *candidate*. Otherwise, every node decides that G is not in $\mathcal{G}_n$, so they *reject*. Note that, if the candidate exists, then it is

unique, since $FP(C(H_1), T) \neq FP(C(H_2), T)$ for all different H_1, H_2 in $\mathcal{G}_n$. So, if the candidate H exists, each node i checks whether the neighborhood of vertex i on G and H are equal, and announces the answer in the third round (communicating one bit). If every node announces affirmatively, then they output $G = H$. Otherwise, it means that G is not in $\mathcal{G}_n$, so every node *rejects*.

(3) We now show that, if we allow the algorithm to be randomized, then we can spare the third round. Let $p' \in [n^2, 2n^2]$ be a prime number. In the second round, node i picks $S_i \in \mathbb{F}_p$, and it communicates, together with $FP(C(G)_i, T_i)$ and $FP(C(G)_{i+n}, T_{i+n})$, also $FP(x_i, S_i)$. After the second round of communication, if a candidate $H \in \mathcal{G}_n$ exists, each node computes $S = (S_1, \ldots, S_n)$, $FP(G, S) = (FP(x_1, S_1), \ldots, FP(x_n, S_n))$. If $FP(G, S) = FP(H, S)$, then nodes deduce that $G = H$. Otherwise, they deduce that $G \notin \mathcal{G}_n$ and *rejects*. Note that if G belongs to $\mathcal{G}_n$, then the algorithm always give the correct answer. Otherwise, it rejects whp. Indeed, if $G \notin \mathcal{G}_n$, then $H \neq G$, and from Lemma 1, $\Pr(FP(G, T) = FP(H, T)) \leq 1/n$.

$\square$

Our private-coin algorithm for $\mathcal{G}$-STRONG-REC has one-sided error. In fact, if the input graph belongs to $\mathcal{G}$, then our algorithm reconstructs it with probability 1. On the other hand, if G does not belong to $\mathcal{G}$, then our algorithm fails to discard the candidate with probability at most $1/n$.

5   Revisiting the One Round Case

In this section we revisit the one-round case (and therefore the broadcast congested clique model). But, instead of studying hereditary graph classes, we study arbitrary graph classes, and we show that for this general case we need a larger bandwith. Our results, in terms of the bandwidth, are tight.

Theorem 3. *Let $\mathcal{G}$ be a set of graphs. The following holds:*

(1) There exists a one-round deterministic algorithm in the congested clique model that solves $\mathcal{G}$-WEAK-REC with bandwidth $\mathcal{O}(\sqrt{\log |\mathcal{G}_n|} \log n + \log n)$.

(2) There exists a one-round private-coin algorithm in the congested clique model that solves $\mathcal{G}$-STRONG-REC with bandwidth $\mathcal{O}(\sqrt{\log |\mathcal{G}_n|} \log n + \log n)$ whp.

Proof. The algorithm in this case are very similar to the algorithms we provided in the proof of Theorem 2. Let k be a parameter whose value will be chosen at the end of the proof, and let C be the error-correcting-code with parameters (n, k). Let p be the smallest prime number greater than $2n \cdot |\mathcal{G}|^{2/k}$. Let $T \in (\mathbb{F}_p)^{n+k}$ be the vector given by Lemma 2, corresponding to $\mathcal{G}$. In the algorithm, every node i computes $C(x_i)$, and communicates $FP(C(x_i), T_i)$ together with $C(x_i)_{n+1}, \ldots, C(x_i)_{n+k} \in (\mathbb{F}_q)^k$, where q is the smallest prime greater than $k+n$. Note that the communication round requires bandwidth

$$\mathcal{O}(\log p + k \cdot \log(n+k)) = \mathcal{O}(\log |\mathcal{G}_n|/k + (k+1) \cdot \log n).$$

After the communication round, every node knows $FP(C(x_i), T_i)$, for all $i \in [n]$, and also knows the matrix $A(\widetilde{G})$. Therefore, every node can compute $FP(C(x_i), T_i)$, for all $i \in \{n+1, \ldots, n+k\}$, and, moreover, compute $FP(C(G), T)$.

From the construction of T, there is at most one graph $H \in \mathcal{G}_n$ such that $FP(C(G), T) = FP(C(H), T)$. Therefore, if G belongs to $\mathcal{G}$, every node can reconstruct it.

On the other hand, if we are solving $\mathcal{G}$-STRONG-REC, then we proceed as in the algorithm of Theorem 2, either testing whether $H = G$ in one more round, or sending a fingerprint of G to check with high probability if a candidate $H \in \mathcal{G}_n$ such that $FP(C(G), T) = FP(C(H), T)$ is indeed equal to G. This verification requires to send $\mathcal{O}(\log n)$ more bits, which fits in the asymptotic bound of the bandwidth. The optimal value of k, that is, the one which minimizes the expression $\mathcal{O}(\log |\mathcal{G}_n|/k + (k+1) \cdot \log n)$, is such that $k = \mathcal{O}\left(\sqrt{\frac{\log |\mathcal{G}_n|}{\log n}}\right)$. Therefore, the bandwidth is $\mathcal{O}(\sqrt{\log |\mathcal{G}_n| \log n} + \log n)$.   □

Now we are going to show that previous algorithms for solving $\mathcal{G}$-WEAK-REC and $\mathcal{G}$-STRONG-REC are in fact tight. For proving this, we are going to exhibit a class of graphs $\mathcal{G}$ satisfying $|\mathcal{G}_n| \leq 2^{\mathcal{O}(n)}$ such that every algorithm (deterministic or randomized) solving $\mathcal{G}$-WEAK-REC in the broadcast congested clique model has cost $Rb = \Omega(\sqrt{\log |\mathcal{G}_n|})$. This lower bound matches the upper *one-round* bound given in Theorem 3 (up to logarithmic factors).

Theorem 4. *There exists a class of graphs $\mathcal{G}^+$ satisfying $|\mathcal{G}_n^+| \leq 2^{\mathcal{O}(n)}$ such that, any ϵ-error public-coin algorithm in the broadcast congested clique model that solves $\mathcal{G}^+$-WEAK-REC, has cost $Rb = \Omega(\sqrt{n}) = \Omega(\sqrt{\log |\mathcal{G}_n^+|})$.*

Proof. Let $\mathcal{G}^+$ be the class of graphs defined as follows: G belongs to $\mathcal{G}_n^+$ if and only if G is the disjoint union of a graph H of $\lceil\sqrt{n}\rceil$ nodes and $n - |H|$ isolated nodes. Note that $|\mathcal{G}_n^+| = \binom{n}{\lceil\sqrt{n}\rceil} \cdot 2^{\binom{\lceil\sqrt{n}\rceil}{2}} \leq 2^{\mathcal{O}(n)}$. Indeed, there are $2^{\binom{\lceil\sqrt{n}\rceil}{2}} = 2^{\mathcal{O}(n)}$ labeled graphs of size $\lceil\sqrt{n}\rceil$, and at most $\binom{n}{\lceil\sqrt{n}\rceil} = 2^{\mathcal{O}(\sqrt{n}\log n)}$ different labelings of a graph of $\sqrt{n}$ nodes using n labels (so $\mathcal{G}^+$ is closed under isomorphisms).

Let $\mathcal{A}$ be an ϵ-error public-coin algorithm solving $\mathcal{G}^+$-WEAK-REC in $R(n)$ rounds and bandwidth $b(n)$, on input graphs of size n.

Consider now the following algorithm $\mathcal{B}$ that solves $\mathcal{U}$-WEAK-REC, where $\mathcal{U}$ is the set of all graphs: on input graph G of size n, each node $i \in [n]$ supposes that it is contained in a graph G^+ formed by G plus $n^2 - n$ isolated vertices with identifiers $(n+1), \ldots, n^2$. Note that G^+ belongs to $\mathcal{G}^+$. Then, node i simulates $\mathcal{A}$ as follows: at each round, node $i \in [n]$ produces the message of node i in G^+ according to $\mathcal{A}$. Note that the messages produced by nodes labeled $(n+1), \ldots, n^2$ do not depend on G, so they can be produced by any node of G without any extra communication. Since $\mathcal{A}$ solves $\mathcal{G}^+$-WEAK-REC, when the algorithm halts every node knows all the edges of G^+, so they reconstruct G ignoring vertices labeled $(n+1), \ldots, n^2$.

We deduce that algorithm $\mathcal{B}$ solves $\mathcal{U}$-WEAK-REC. Note that the cost of $\mathcal{B}$ is $R(n^2)b(n^2)$ on input graphs of size n. We deduce that $R(n^2)b(n^2) = \Omega(n)$, i.e., the cost of $\mathcal{A}$ is $\Omega(\sqrt{n})$. □

6 Discussion

Our result gives a straightforward, general strategy to solve arbitrary problems when the input graph belongs to some particular class of graphs. Hence, instead of designing ad-hoc algorithms to solve specific problems, we can reconstruct the input graph G and solve locally *any question* concerning G.

Even though in the congested clique model, by definition, the only complexity measure taken into account is communication, it is important to point out that the general algorithms we presented in this paper run in exponential local time.

However, note that, unless $P = NP$ (or even if stronger conjectures in computational complexity are false), this difficulty can not be overcome. In fact, for many graph classes $\mathcal{G}$, solving $\mathcal{G}$-STRONG-REC in polynomial local time is impossible.

Let us illustrate this with an example. Consider the *hereditary, sparse* class of 3-colorable planar graphs, that we denote **3-col-plan**. It is NP-complete to decide whether an arbitrary graph belongs to **3-col-plan** [7]. Any algorithm in the congested clique model that runs in polynomial local time can be simulated by a sequential algorithm that also runs in polynomial time: simply run the computation of each node one by one at each round. Therefore, unless $P = NP$, there is no algorithm running in polynomial local time solving **3-col-planar-**STRONG-REC.

References

1. Beame, P., Koutris, P., Suciu, D.: Communication steps for parallel query processing. J. ACM **64**(6), Article 40 (2017)
2. Becker, F., Matamala, M., Nisse, N., Rapaport, I., Suchan, K., Todinca, I: Adding a referee to an interconnection network: what can(not) be computed in one round. In: IPDPS 2011, pp. 508–514 (2011)
3. Brandstädt, A., Le, V.B., Spinrad, J.P.: Graph Classes: A Survey. SIAM, Philadelphia (1999)
4. Censor-Hillel, K., Kaski, P., Korhonen, J. H., Lenzen, C., Paz, A., Suomela, J.: Algebraic methods in the congested clique. In: PODC 2015, pp. 143–152 (2015)
5. Dean, J., Ghemawat, S.: MapReduce: simplified data processing on large clusters. Commun. ACM **51**(1), 107–113 (2008)
6. Drucker, A., Kuhn, F., Oshman, R.: On the power of the congested clique model. In: PODC 2014, pp. 367–376 (2014)
7. Garey, M.R., Johnson, D.S.: Computers and Intractability. W H Freeman, New York (2002)

8. Ghaffari, M.: An improved distributed algorithm for maximal independent set. In: SODA 2016, pp. 270–277 (2016)
9. Ghaffari, M., Parter, M.: MST in log-star rounds of congested clique. In: PODC 2016, pp. 19–28 (2016)
10. Ghaffari, M.: Distributed MIS via all-to-all communication. In: PODC 2017, pp. 141–149 (2017)
11. Hegeman, J.W., Pandurangan, G., Pemmaraju, S.V., Sardeshmukh, V., Scquizzato, M.: Toward optimal bounds in the congested clique: graph connectivity and MST. In: PODC 2015, pp. 91–100 (2015)
12. Hegeman, J.W., Pemmaraju, S.V.: Lessons from the congested clique applied to MapReduce. In: SIROCCO 2014, pp. 149–164 (2014)
13. Hegeman, J.W., Pemmaraju, S.V., Sardeshmukh, V.B.: Near-constant-time distributed algorithms on a congested clique. In: Kuhn, F. (ed.) DISC 2014. LNCS, vol. 8784, pp. 514–530. Springer, Heidelberg (2014). https://doi.org/10.1007/978-3-662-45174-8_35
14. Jurdzinski, T., Nowicki, K.: MST in O(1) rounds of the congested clique. In: SODA 2018, pp. 2620–2632 (2018)
15. Kari, J., Matamala, M., Rapaport, I., Salo, V.: Solving the INDUCED SUB-GRAPH problem in the randomized multiparty simultaneous messages model. In: Scheideler, C. (ed.) SIROCCO 2015. LNCS, vol. 9439, pp. 370–384. Springer, Cham (2015). https://doi.org/10.1007/978-3-319-25258-2_26
16. Karloff, H., Suri, S., Vassilvitskii, S.: A model of computation for MapReduce. In: SODA 2010, pp. 938–948 (2010)
17. Klauck, H., Nanongkai, D., Pandurangan, G., Robinson, P.: Distributed computation of large-scale graph problems. In: SODA 2015, pp. 391–410 (2015)
18. Lenzen, C.: Optimal deterministic routing and sorting on the congested clique. In: PODC 2013, pp. 42–50 (2013)
19. Lidl, R., Niederreiter, H.: Introduction to Finite Fields and Their Applications. Cambridge University Press, Cambridge (1994)
20. Lotker, Z., Patt-Shamir, B., Pavlov, E., Peleg, D.: Minimum-weight spanning tree construction in O (log log n) communication rounds. SIAM J. Comput. **35**(1), 120–131 (2005)
21. Malewicz, G., et al.: Pregel: a system for large-scale graph processing. In: SIGMOD 2010, pp. 135–146 (2010)
22. Montealegre, P., Todinca, I.: Brief announcement: deterministic graph connectivity in the broadcast congested clique. In: PODC 2016, pp. 245–247 (2016)
23. Reed, I.S., Solomon, G.: Polynomial codes over certain finite fields. J. Soc. Ind. Appl. Math. **8**(2), 300–304 (1960)
24. Scheinerman, E.R., Zito, J.: On the size of hereditary classes of graphs. J. Comb. Theory, Ser. B **61**(1), 16–39 (1994)
25. Schwartz, J.T.: Fast probabilistic algorithms for verification of polynomial identities. J. ACM **27**(4), 701–717 (1980)
26. White, T.: Hadoop: The definitive guide. O'Reilly Media Inc., Sebastopol (2012)

Space-Efficient Uniform Deployment of Mobile Agents in Asynchronous Unidirectional Rings

Masahiro Shibata[1(✉)], Hirotsugu Kakugawa[2], and Toshimitsu Masuzawa[2]

[1] Department of Computer Science and Electronics, Kyushu Institute of Technology,
680-4, Kawadu, Iizuka, Fukuoka 820-8502, Japan
shibata@cse.kyutech.ac.jp
[2] Graduate School of Information Science and Technology,
Osaka University, 1-5 Yamadaoka, Suita, Osaka 565-0871, Japan
{kakugawa,masuzawa}@ist.osaka-u.ac.jp

Abstract. In this paper, we consider the uniform deployment problem of mobile agents in asynchronous unidirectional ring networks. This problem requires agents to spread uniformly in the network. In this paper, we focus on the memory space per agent required to solve the problem. We consider two problem settings. The first setting assumes that agents have no multiplicity detection, that is, agents cannot detect whether another agent is staying at the same node or not. In this case, we show that each agent requires $\Omega(\log n)$ memory space to solve the problem, where n is the number of nodes. In addition, we propose an algorithm to solve the problem with $O(k + \log n)$ memory space per agent, where k is the number of agents. The second setting assumes that each agent is equipped with the weak multiplicity detection, that is, agents can detect another agent staying at the same node, but cannot learn the exact number. Then, we show that the memory space per agent can be reduced to $O(\log k + \log \log n)$. To the best of our knowledge, this is the first research considering the effect of the multiplicity detection on memory space required to solve problems.

Keywords: Distributed system · Mobile agent · Uniform deployment
Ring network · Space-efficient

1 Introduction

1.1 Background and Related Works

A *distributed system* consists of a set of computers (*nodes*) connected by communication links. As a promising design paradigm of distributed systems, (mobile)

A preliminary brief announcement of this work appeared in the proceedings of the 19th International Symposium on Stabilization, Safety, and Security of Distributed Systems (SSS 2017). This work was partially supported by JSPS KAKENHI Grant Number 16K00018, 17K19977, and 18K18031, and Japan Science and Technology Agency (JST) SICORP.

© Springer Nature Switzerland AG 2018
Z. Lotker and B. Patt-Shamir (Eds.): SIROCCO 2018, LNCS 11085, pp. 149–164, 2018.
https://doi.org/10.1007/978-3-030-01325-7_16

agents have attracted much attention [1]. Agents traverse the system carrying information collected at visited nodes and process tasks on each node using the information. In other words, agents encapsulate the process code and data, which simplifies design of distributed systems [2].

In this paper, we consider the *uniform deployment* (or *uniform scattering*) *problem* as a fundamental problem for coordination of agents. This problem requires agents to spread uniformly in the network. Uniform deployment is useful for network management. In a distributed system, it is necessary that regularly each node gets software updates and is checked whether some application installed on the node is running correctly or not [3]. Hence, considering agents with such services, uniform deployment guarantees that agents visit each node at short intervals and provide services. Uniform deployment might be useful also for a kind of load balancing. That is, considering agents with large-size database replicas, uniform deployment guarantees that not all nodes need to store the database but each node can quickly access the database [4]. Hence, we can see the uniform deployment problem as a kind of the resource allocation problem (e.g., the k-server problem).

As related works, the uniform deployment problem is considered in ring networks [5,6] and grid networks [7]. All of them assumed that agents are oblivious (or memoryless) but can observe multiple nodes within its visibility range. On the other hand, our previous work [8] considered uniform deployment in asynchronous unidirectional ring networks for agents that have memory but cannot observe nodes except for their currently visiting nodes.

1.2   Our Contribution

In this paper, we consider the uniform deployment problem in asynchronous unidirectional ring networks. Similarly to [8], we consider agents that have memory but cannot observe nodes except for their currently visiting nodes. While the previous work [8] considered uniform deployment with such agents for the first time and clarified the solvability, this work focuses on the memory space per agent required to solve the problem and aims to propose space-efficient algorithms in weaker models than that of [8]. That is, while agents in [8] assumed that they can send a message to the agents staying at the same node, agents in this paper do not have such ability. Instead, each agent initially has a token and can release it on a visited node, and agents can communicate only by the tokens. After a token is released, it cannot be removed. We also analyze the time complexity and the total number of moves. We assume that agents have knowledge of the number k of agents.

In Table 1, we compare our contributions with the results for agents with knowledge of k in [8]. We consider two problem settings. The first setting considers agents *without* multiplicity detection, that is, agents cannot detect whether another agent staying at the same node or not. In this model, we show that each agent requires $\Omega(\log n)$ memory space to solve the problem, where n is the number of nodes. In addition, we propose an algorithm to solve the problem with $O(k+\log n)$ memory space per agent, $O(n \log k)$ time, and $O(kn \log k)$ total number of moves. The second setting considers agents *with* the weak multiplicity detection, that is, agents can detect another agent staying at the same node, but cannot learn the exact number. In this setting, we also assume that agents know an upper bound $\log N$ of $\log n$ such that $\log N = O(\log n)$. Then, we propose an algorithm to reduce the memory space per agent to $O(\log k + \log \log n)$, but it uses $O(n^2 \log n)$ time and $O(kn^2 \log n)$ total number of moves. To the best of our knowledge, this is the first research considering the effect of the multiplicity detection on memory space required to solve problems.

Due to limitation of space, we omit several pseudocodes and proofs of theorems.

Table 1. Results for agents with knowledge of k (n, #nodes, k, #agents)

	Previous results [8]		Results of this paper	
	Result 1	Result 2	Model 1	Model 2
Communication	Messages	Messages	Unremovable tokens	Unremovable tokens
Multiplicity detection	Required	Required	Not required	Required
Agent memory	$O(k \log n)$	$O(\log n)$	$O(k + \log n)$	$O(\log k + \log \log n)$
Time complexity	$\Theta(n)$	$O(n \log k)$	$O(n \log k)$	$O(n^2 \log n)$
Total number of moves	$\Theta(kn)$	$\Theta(kn)$	$O(kn \log k)$	$O(kn^2 \log n)$

2 Preliminaries

2.1 System Model

We use almost the same model as that in [8]. A *unidirectional ring network* R is defined as 2-tuple $R = (V, E)$, where V is a set of anonymous nodes and E is a set of unidirectional links. We denote by $n (= |V|)$ the number of nodes, and let $V = \{v_0, v_1, \ldots, v_{n-1}\}$ and $E = \{e_0, e_1, \ldots, e_{n-1}\}$ ($e_i = (v_i, v_{(i+1) \bmod n})$). We define the direction from v_i to v_{i+1} as the *forward* direction. In addition, we define the i-th ($i \neq 0$) (forward) agent a' of agent a as the agent such that $i - 1$ agents exist between a and a' in a's forward direction. Moreover, the *distance* from node v_i to v_j is defined to be $(j - i) \bmod n$.

An agent is a state machine having an *initial state*. Let $A = \{a_0, a_1, \ldots, a_{k-1}\}$ be a set of $k (\leq n)$ anonymous agents. Since the ring is unidirectional, agents staying at v_i can move only to v_{i+1}. We assume that agents have knowledge of k. Each agent initially has a *token* and can release it on a visited node. After a token is released, it cannot be removed. The token on an agent can be realized by one bit memory and cannot carry any additional information. Hence, the tokens on a node represents just the number of the tokens and agents cannot recognize the owners of the tokens[1]. Moreover, we assume that agents move through a link in a FIFO manner, that is, when agent a_p leaves v_i after agent a_q, a_p reaches v_{i+1}

---

[1] In practice, each node can store more information, but it is sufficient to store information about tokens when considering anonymous agents.

after a_q. Note that such a FIFO assumption is natural because (1) agents are implemented as messages in practice, and (2) the FIFO assumption of messages is natural and can be easily realized using sequence numbers.

We consider two problem settings: agents *without multiplicity detection* and agents *with weak multiplicity detection*. While agents without multiplicity detection cannot detect whether another agent is staying at the same node or not, agents with weak multiplicity detection can detect another agent staying at the same node, but cannot learn the exact number[2]. Each agent a_i executes the following three operations in an atomic action: (1) Agent a_i reaches a node v (when a_i is in transit to v), or starts operations at v (when a_i stays at v), (2) agent a_i executes local computation, and (3) agent a_i leaves v if it decides to move. For the case with weak multiplicity detection, the local computation depends on whether another agent is staying at v or not. Note that these assumptions of atomic actions are also natural because they can be implemented by nodes with an incoming *buffer* that stores agents about to visit the node and makes them execute actions in a FIFO order. We consider an *asynchronous* system, that is, the time for each agent to transit to the next node or to wait until the next activation (when staying at a node) is finite but unbounded.

Table 2. Meaning of each element in configuration $C = (S, T, P, Q)$

Element	Meaning and example
$S = (s_0, s_1, \ldots, s_{k-1})$	Set of agent states (s_i: the state of agent a_i)
$T = (t_0, t_1, \ldots, t_{n-1})$	Set of node states (t_i: the number of tokens at node v_i)
$P = (p_0, p_1, \ldots, p_{n-1})$	Sets of agents staying at nodes (p_i: a set of agents staying at node v_i)
$Q = (q_0, q_1, \ldots, q_{n-1})$	Sets of agents residing on links (q_i: a sequence of agents in transit from v_{i-1} to v_i)

A (global) *configuration* C is defined as a 4-tuple $C = (S, T, P, Q)$ and the correspondence table is given in Table 2. Element S is a k-tuple $S = (s_0, s_1, \ldots, s_{k-1})$, where s_i is the state (including the state to denote whether it holds a token or not) of agent a_i. Element T is an n-tuple $T = (t_0, t_1, \ldots, t_{n-1})$, where t_i is the state (i.e., the number of tokens) of node v_i. The remaining elements P and Q represent the positions of agents. Element P is an n-tuple $P = (p_0, p_1, \ldots, p_{n-1})$, where p_i is a set of agents staying at node v_i. Element Q is an n-tuple $Q = (q_0, q_1, \ldots, q_{n-1})$, where q_i is a sequence of agents residing in the FIFO queue corresponding to link (v_{i-1}, v_i). Hence, agents in q_i are in transit from v_{i-1} to v_i.

We denote by $\mathcal{C}$ the set of all possible configurations. In *initial configuration* $C_0 \in \mathcal{C}$, all agents are in the initial state (where each has a token) and placed at

---

[2] This is why such multiplicity detection is called *weak*.

distinct nodes[3], and no node has any token. The node where agent a is located in C_0 is called the *home node* of a and is denoted by $v_{HOME}(a)$. For convenience, we assume that in C_0 agent a is stored at the incoming buffer of its home node $v_{HOME}(a)$. This assures that agent a starts the algorithm at $v_{HOME}(a)$ before any other agents make actions at $v_{HOME}(a)$.

A (sequential) *schedule* $X = \rho_0, \rho_1, \ldots$ is an infinite sequence of agents, intuitively which activates agents to execute their actions one by one. Schedule X is *fair* if every agent appears in X infinitely often. An infinite sequence of configurations $E = C_0, C_1, \ldots$ is called an *execution* from C_0 if there exists a fair schedule $X = \rho_0, \rho_1, \ldots$ that satisfies the following conditions for each h $(h > 0)$:

- If agent $\rho_{h-1} \in p_i$ (i.e., ρ_{h-1} is an agent staying at v_i) for some i in C_{h-1}, the states of ρ_{h-1} and v_i in C_{h-1} are changed in C_h by local computation of ρ_{h-1}. If ρ_{h-1} releases its token at v_i, the value of t_i increases by one. After this, if ρ_{h-1} decides to move to v_{i+1}, ρ_{h-1} is removed from p_i and is appended to the tail of q_{i+1}. If ρ_{h-1} decides to stay, ρ_{h-1} remains in p_i. The other elements in C_h are the same as those in C_{h-1}.
- If agent ρ_{h-1} is at the head of q_i (i.e., ρ_{h-1} is the next agent to reach v_i) for some i in C_{h-1}, ρ_{h-1} is removed from q_i and reaches v_i. Then, the states of ρ_{h-1} and v_i in C_{h-1} are changed in C_h by local computation of ρ_{h-1}. If ρ_{h-1} releases its token at v_i, the value of t_i increases by one. After this, if ρ_{h-1} decides to move to v_{i+1}, ρ_{h-1} is appended to the tail of q_{i+1}. If ρ_{h-1} decides to stay, ρ_{h-1} is inserted in p_i. The other elements in C_h are the same as those in C_{h-1}.

Note that if the activated agent ρ_{h-1} has no action, then C_{h-1} and C_h are identical. Actually after uniform deployment is achieved, the same configuration is repeated forever.

2.2 The Uniform Deployment Problem

The uniform deployment problem in a ring network requires k (≥ 2) agents to spread uniformly in the ring, that is, all the agents are located at distinct nodes and the distance between any two *adjacent agents* should be identical. Here, we say two agents are adjacent when there exists no agent between them. However, we should consider the case that n is not a multiple of k. In this case, we aim to distribute the agents so that the distance of any two adjacent agents should be $\lfloor n/k \rfloor$ or $\lceil n/k \rceil$.

We consider the uniform deployment problem *without termination detection*. In this case, *suspended states* are defined as follows. An agent stays at a node (not in a link) when it is at a suspended state. When agent a_i enters a suspended state, it neither changes its state nor leaves the current node v unless

---

[3] We assume this for simplicity, but even if two or more agents exist at the same node in C_0, agents can solve the problem similarly by using the number of tokens at each node and atomicity of execution.

the observable local configuration of v (i.e., existence of another agent or the number of tokens for agents with weak multiplicity detection, or the number of tokens for agents without multiplicity detection) changes. The uniform deployment problem without termination detection allows agents to stop in suspended states, which is also known as communication deadlock. We define the problem as follows.

Definition 1. *An algorithm solves the uniform deployment problem without termination detection if any execution satisfies the following conditions.*

- *All agents change their states to the suspended states in finite time.*
- *When all agents are in the suspended states, $q_i = \emptyset$ holds for any $q_i \in Q$ and the distance of each pair of adjacent agents is $\lfloor n/k \rfloor$ or $\lceil n/k \rceil$.* □

Next, we define the *time complexity* as the time required to solve the problem. Since there is no bound on time in asynchronous systems, it is impossible to measure the exact time. Instead we consider the *ideal time complexity*, which is defined as the execution time under the following assumptions: (1) The time for an agent to transit to the next node or to wait until the next activation is at most one, and (2) the time for local computation is ignored (i.e., zero)[4]. Note that these assumptions are introduced only to evaluate the complexity, that is, algorithms are required to work correctly without such assumptions. In the following, we use terms "time complexity" and "time" instead of "ideal time complexity".

3 Agents Without Multiplicity Detection

In this section, we consider uniform deployment for agents without multiplicity detection.

3.1 A Lower Bound of Memory Space per Agent

First, we show the following lower bound of memory space per agent.

Theorem 1. *For agents without multiplicity detection, the memory space per agent to solve the uniform deployment problem is $\Omega(\log n)$.* □

Proof. We show the theorem by contradiction. We assume that there exists an algorithm to solve the uniform deployment problem with at most $\log n - 2$ bit memory per agent. Then, each agent has at most $2^{(\log n-2)} = n/4$ states. Hence, when an agent enters a suspended state, it moved at most $n/4$ times after it last observed a token.

We consider the initial configuration such that two agents a_1 and a_2 are placed at neighboring nodes in a n-node ring. Then, the distance between the

---

[4] This definition is based on the ideal time complexity for asynchronous message-passing systems [9].

two agents in the final configuration should be $\lfloor n/2 \rfloor$ or $\lceil n/2 \rceil$. We assume that a_1 and a_2 move in a synchronous manner. Then, since they are placed at neighboring nodes and execute the same algorithm, they release tokens (if do) also at neighboring nodes. In addition, since a_1 and a_2 move at most $n/4$ times after they last observed a token and enter suspended states, the distance between them is at most $n/4 + 1 (\neq \lfloor n/2 \rfloor$ or $\lceil n/2 \rceil)$. However, this contradicts the condition of uniform deployment.   □

3.2 An Algorithm with $O(k + \log N)$ Memory Space per Agent

Next, we propose an algorithm to solve the uniform deployment problem with $O(k + \log n)$ memory space per agent, $O(n \log k)$ time, and $O(kn \log k)$ total number of moves. From Theorem 1, the algorithm is optimal in memory space per agent when $k = O(\log n)$. The algorithm consists of two phases as do the two algorithms in [8]: the selection phase and the deployment phase. In the selection phase, agents select some *base nodes*, which are the reference nodes for uniform deployment. In the deployment phase, based on the base nodes, each agent determines a *target node* where it should enters a suspended state and moves to the node. For simplicity, we assume $n = ck$ for some positive integer c since we can easily remove this assumption, but we omit the description.

3.2.1 Selection Phase
In this phase, some home nodes are selected as base nodes. The selected base nodes satisfy the following conditions called the **base node conditions**: (1) At least one base node exists, (2) the distance between every pair of *adjacent base nodes* is the same, and (3) the number of home nodes between every pair of adjacent base nodes is the same. We say that two base nodes are adjacent when there exists no base node between them. In Fig. 1, distances from $v_{HOME}(a_1)$ to $v_{HOME}(a_2)$, from $v_{HOME}(a_2)$ to $v_{HOME}(a_3)$, and from $v_{HOME}(a_3)$ to $v_{HOME}(a_1)$ are all 6, and the number of home nodes between $v_{HOME}(a_1)$ and $v_{HOME}(a_2)$, between $v_{HOME}(a_2)$ and $v_{HOME}(a_3)$, and between $v_{HOME}(a_3)$ and $v_{HOME}(a_1)$

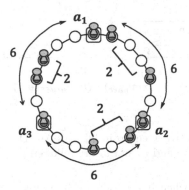

Fig. 1. An example of the base node conditions ($n = 18, k = 9$).

are all 2. Thus, $v_{HOME}(a_1)$, $v_{HOME}(a_2)$, and $v_{HOME}(a_3)$ satisfy the base node conditions. When the selection phase is completed, each agent stays at its home node and knows whether its home node is selected as a base node or not. We call an agent a *leader* (but probably not unique) when its home node is selected as a base node, and call it a *follower* otherwise. The state of an agent is active, leader or follower. Active agents are candidates for leaders, and initially all agents are active. We say that node v is active (resp., follower) when v is the home node of an active (resp., a follower) agent.

At first, we explain the basic idea of the selection phase in [8], which assumes weak multiplicity detection, and then we explain the way of applying the idea to the model in this section. In the selection phase of [8], agents use *IDs* (but probably not unique) and decrease the number of active agents. We explain the detail of the IDs later. At the beginning of the algorithm, each agent a_i releases its token at $v_{HOME}(a_i)$. The selection phase consists of several subphases. At the beginning of each subphase, each agent a_i stays at $v_{HOME}(a_i)$. During the subphase, if a_i is a follower, it keeps staying at $v_{HOME}(a_i)$. On the other hand, each active agent a_i travels once around the ring and gets its ID by the method described later[5]. Then, a_i compares its ID with IDs of other agents one by one (a_i gets them during the traversal of the ring) and determines the next behavior. Briefly, (a) if all active agents have the same ID, it means that home nodes of the active agents satisfy the base node conditions. Hence, the active agents become leaders and enter to the deployment phase. (b) If all agents do not have the same ID but a_i's ID is the maximum, it remains active and executes the next subphase. (c) If a_i does not satisfy (a) or (b), it becomes a follower. Agents execute such subphases until base nodes are selected.

Now, we explain the detail of the ID. The ID (not necessarily unique) of an active agent a_i is given in the form of $(fNum_i, d_i)$, where $fNum_i$ is the number of follower nodes between $v_{HOME}(a_i)$ and the next active node in the subphase, say v_{next}, and d_i is the distance from $v_{HOME}(a_i)$ to v_{next}. In Fig. 2(a), when agent a_i moves from its home node $v_j(= v_{HOME}(a_i))$ to the next active node $v'_j(= v_{next})$, it observes two follower nodes and visits four nodes. Hence, a_i gets its ID $ID_i = (2, 4)$. Note that since active agents traverse the ring and follower

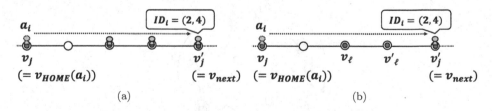

Fig. 2. (a): An ID of an active agent a_i in [8]. (b): An ID of an active agent a_i in this section (v_j and v'_j are active and v_ℓ and v'_ℓ are followers).

---

[5] Each agent can detect when it completes one circuit of the ring using knowledge of k.

agents stay at their home nodes, a_i can detect its arrival at the next active node when it visits a token node with no agent. This statement holds even in asynchronous systems by the FIFO property of links and the atomicity of execution (these facts are used in Sect. 4). Agents in [8] use $O(\log n)$ memory space to get such an ID and decide whether they remain active (or they have the lexicographically maximum ID) or not. Notice that an agent may get different IDs in different subphases.

In the following, we explain how to apply the previous idea to the model in this section (i.e., without multiplicity detection). Agents in this section cannot detect existence of other agents staying at the same node and cannot detect the arrival of the next active node using existence of an agent. To deal with this, each agent memorizes the state of all agents by using an array $Active_{now}$ of k bits. The value of $Active_{now}[i]$ is $true$ iff its i-th agent is active (otherwise it is a follower). Hence, agents can get an ID by going from node v to v' each of whose corresponding value of $Active_{now}$ is true. In Fig. 2(b), if v_j and v'_j are active and v_ℓ and v'_ℓ are followers, a_i can gets its ID $ID_i = (2, 4)$. In addition, each follower agent also moves in the ring instead of staying at its home node, and we explain this next.

Now, we explain implementation of the subphase. Each follower agent firstly moves to the nearest active node to simulate the behavior of the active agent. To do this, each agent has variable $nearActive_{now}$ that indicates the number of tokens to the nearest active node in the subphase (the values of $nearActive_{now}$ for active agents are 0). Then, each active or follower agent a_i travels once around the ring. While traveling, a_i executes the following actions:

(1) Get its ID $ID_i = (fNum_i, d_i)$: Agent a_i gets its ID ID_i by moving from the current node (i.e., $v_{HOME}(a_i)$ for active agent a_i or the nearest active node for follower agent a_i) to v_{next} with counting the numbers of followers and visited nodes (Fig. 2(b)).

(2) Compare ID_i with IDs of all active agents: During the traversal, a_i compares ID_i with IDs of all active agents one by one, and checks (1) whether ID_i is the lexicographically maximum and (2) whether the IDs of all active agents are the same. To check these, a_i keeps variables ID_{max} that is the largest ID among IDs a_i ever found, and $same$ ($same = true$ means that IDs a_i ever found are the same), and it updates the variables (if necessary) every time it finds an ID of another active agent. When ID_{max} is updated, a_i also updates the value of $nearActive_{next}$, indicating the number of tokens to the nearest active node in the next subphase.

When completing one circuit of the ring, a_i returns to $v_{HOME}(a_i)$ and determines its state for the next subphase. (a) If $same = true$, a_i (and all the other active agents) become leaders and completes the selection phase. (b) If $same = false$ and $ID_i = ID_{max}$, a_i remains its state (active or follower) and executes the next subphase. (c) If a_i does not satisfy (a) or (b), each active (resp., follower) agent becomes (resp., remains) a follower and executes the next subphase. By repeating such subphase at most $\lceil \log k \rceil$ times, all the remaining active agents become to have the same ID in some subphase and they are selected as leaders so that their home nodes should satisfy the base node conditions. Notice that $\lceil \log k \rceil$ subphases are sufficient, intuitively because (1) the largest ID increases every time a subphase completes, and thus (2) no pair of adjacent active agents remain active in every subphase.

Pseudocode is described in Algorithm 1. Each agent uses variable *preActive* for storing the position (i.e., the ordinary number) of the active node it visited for the last time before coming to the current node, and boolean array $Active_{next}$ of k bits for storing the states of all agents for the next subphase. In addition, agents use procedure *nextActive()* to move to the next active node. Note that, in each subphase each follower agent firstly moves to the nearest active node, travels once around the ring from the active node, and returns to its home node. Hence, each follower agent travels twice around the ring in each subphase and each active agent does so for simplicity. In addition, in Algorithm 1 each agent can get the number n of nodes when it finishes traveling once around the ring, but we omit the description.

3.2.2 Deployment Phase

In this phase, each agent determines its target node and moves to the node. At first, the nearest base node is first selected as the base node. Hence, if $v_{HOME}(a_i)$ is a base node (i.e., a_i is a leader), $v_{HOME}(a_i)$ is a_i's target node and a_i stays there. Otherwise (i.e., if a_i is a follower), a_i firstly moves until it observes $nearActive_{now}$ tokens to reach the nearest base node. After this, a_i moves $nearActive_{now} \times n/k$ times to reach its target node. When all agents move to their target nodes, the final configuration is a solution of the uniform deployment problem.

We have the following theorem for the proposed algorithm.

Theorem 2. *For agents without multiplicity detection, the proposed algorithm solves the uniform deployment problem with $O(k+\log n)$ memory space per agent, $O(n \log k)$ time, and $O(kn \log k)$ total number of moves.* □

4   Agents with Weak Multiplicity Detection

In this section, we consider agents with weak multiplicity detection, and propose an algorithm to solve the uniform deployment problem that reduces the memory space per agent to $O(\log k + \log \log n)$, but it uses $O(n^2 \log n)$ time and $O(kn^2 \log n)$ total number of moves. The algorithm consists of three phases: the selection phase, the collection phase, and the deployment phase. In the selection phase, agents select base nodes similarly to Sect. 3. In the collection phase, agents move in the ring so that they stay at consecutive nodes starting from the base nodes. In the deployment phase, agents move to their target nodes. In this section, we assume that agents know an upper bound $\log N$ of $\log n$ such that $\log N = O(\log n)$.

Algorithm 1. The behavior of active or follower agent a_i in the selection phase

Behavior of Agent a_i
1: /*selection phase*/
2: $phase = 1$, $nearActive_{now} = 0$, $nearActive_{next} = 0$, $preActive = 0$, $same = true$
3: **for** $j = 0; j < k - 1; j + +$ **do** $Active_{now}[j] = true$, $Active_{next}[j] = true$
4: release a token at its home node $v_{HOME}(a_i)$
5: **while** $phase \neq \lceil \log k \rceil$ **do**
6:     **if** a_i is a follower **then**
7:         move until it observes $nearActive_{now}$ tokens // reach the nearest active node
8:         $t = nearActive_{now}$
9:     **end if**
10:    execute $NextActive()$ and get the first ID $ID_i = (fNum_i, d_i)$, $ID_{max} = ID_i$
11:    **while** $t \neq nearActive_{now}$ **do**
12:        execute $NextActive()$ and get ID $ID_{oth} = (fNum_{oth}, d_{oth})$ of the next active agent
13:        **if** $ID_{oth} \neq ID_i$ **then** $same = false$
14:        **if** $ID_{max} > ID_{oth}$ **then** $Active_{next}[preActive] = false$
15:        **if** $ID_{max} < ID_{oth}$ **then**
16:            $ID_{max} = ID_{oth}$, $nearActive_{next} = preActive$
17:            **for** $j = 0; j < t - 1; j + +$ **do** $Active_{next}[j] = false$
18:        **end if**
19:    **end while**
20:    return to its home node $v_{HOME}(a_i)$
21:    **if** $same = true$ **then** // active nodes satisfy the base node conditions
22:        **if** a_i is active **then** enter a leader state
23:        terminate the selection phase and enter the deployment phase
24:    **end if**
25:    **if** $(a_i$ is active$) \wedge (ID_i \neq ID_{max})$ **then** enter a follower state
26:    $phase = phase+1$, $same = true$, $nearActive_{now} = nearActive_{next}$
27:    **for** $j = 0; j < k - 1; j + +$ **do** $Active_{now}[j] = Active_{next}[j]$
28: **end while**
29:
Procedure $NextActive()$
30: $preActive = t$
31: move to the next token node and set $t = (t + 1) \mod k$
32: **while** $Active_{now}[t] \neq true$ **do**
33:     move to the next token node and set $t = (t + 1) \mod k$
34: **end while**

4.1   Selection Phase

Similarly to Sect. 3, in this phase some home nodes are selected as base nodes. The basic idea is the same as that in Sect. 3, that is, agents use IDs and decrease the number of active agents. However, compared with the algorithm in Sect. 3, memory space is reduced to $O(\log k + \log \log n)$ from $O(k + \log n)$. We use two techniques for the reduction: (i) As in [8], a follower remains at its home node and informs an active agent of its state using the weak multiplicity detection: when an agent is detected at a node, it is recognized as a follower. This improves memory space from $O(k)$ to $O(\log k)$ since the algorithm in Sect. 3 requires $O(k)$ memory space to maintain the states of all agents. (ii) Distances are computed using Residue Number System (RNS) [10] that represents a large number as a set of small numbers. In particular, we use the technique called Chinese Remainder Theorem (CRT) [11]. The CRT says that for two positive integers n_1 and n_2 $(n_1, n_2 < n)$, if the remainders of the integers when divided by each of the

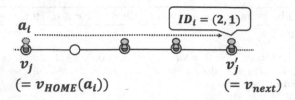

Fig. 3. An ID of an active agent a_i ($prime_l = 3$).

first $\log n$ prime numbers $2, 3, 5, \ldots, U$ are the same, then $n_1 = n_2$ holds [11]. The prime number theorem guarantees that the $(\log n)$-th prime U satisfies $U = O(\log^2 n)$. Thus, agents compare distances between adjacent active nodes using the CRT and reduce memory space from $O(\log n)$ to $O(\log \log n)$.

We explain the outline of the selection phase. As in Sect. 3, the state of an agent is active, leader, or follower, and initially all agents are active. At the beginning of the algorithm, each agent a_i releases its token at $v_{HOME}(a_i)$. The selection phase consists of at most $\lceil \log k \rceil$ subphases. As in Sect. 3, dropping out from active agents is realized by IDs each of which consists of the number of followers and the distance between active nodes. The only difference is that the distance part is compared using remainders by primes (Fig. 3). Each subphase consists of several iterations. At the beginning of each iteration, each agent a_i stays at $v_{HOME}(a_i)$. For the l-th iteration in each subphase, if a_i is a follower, different from Sect. 3, it keeps staying $v_{HOME}(a_i)$ to inform active agents visiting the node of its state. On the other hand, each active agent a_i travels once around the ring and gets the distance part d_l^{prime} of its ID as the remainder divided by the l-th prime $prime_l$. In Fig. 3, when $prime_l = 3$, a_i gets its ID $ID_i = (2, 1)$.

During the traversal, a_i lexicographically compares its ID ID_i with IDs of other active agents one by one, and it determines its next behavior when it returns to $v_{HOME}(a_i)$. As in Sect. 3, a_i uses variable $same$ ($same = true$ means that IDs a_i ever found are the same). Then, (a) if $same = true$ and $l = \log N$, it means that the distances between all the pairs of adjacent active nodes are the same, and these home nodes satisfy the base node conditions. Hence, the active agents become leaders and enter the collection phase without staying at its home node. (b) If $same = true$ but $l \neq \log N$, a_i executes the next $(l + 1)$-th iteration using the next prime $prime_{l+1}$. (c) If $same \doteq false$, they terminate the current subphase. If a_i has the maximum ID, a_i remains active and starts the next subphase. Otherwise, a_i becomes a follower. Each active agent executes such subphases at most $\lceil \log k \rceil$ times. Notice that the distances are compared using the CRT, which implies that the agents with the maximum distance among the agents with the maximum $fNum$ (the number of followers between adjacent active agents) do not necessarily remain active in the subphase. Hence, agents remaining active in the subphase may differ from those in the algorithm of Sect. 3. However, $\lceil \log k \rceil$ subphases are still sufficient as in Sect. 3, which is guaranteed by selecting active agents with the maximum $fNum$.

Algorithm 2. The behavior of active agent a_i in the selection phase

---

Behavior of Agent a_i
1: /*selection phase*/
2: $phase = 1$, $prime = 2$, $same = true$, $max = true$
3: release a token at its home node $v_{HOME}(a_i)$
4: **while** $(phase \neq \lceil \log k \rceil) \vee (prime \neq (\log N)\text{-th ptime})$ **do**
5:     move to the next active node and get its own ID $ID_i = (fNum_i, d_i^{prime})$
6:     **while** a_i is not at $v_{HOME}(a_i)$ **do**
7:         move to the next active node and get ID $ID_{oth} = (fNum_{oth}, d_{oth}^{prime})$ of the next active agent
8:         **if** $ID_{oth} \neq ID_i$ **then** $same = false$
9:         **if** $ID_{oth} > ID_i$ **then** $max = false$ // there exists an agent having a larger ID
10:     **end while**
11:     **if** $(same = true) \wedge (prime = (\log N)\text{-th prime})$ **then** terminate the selection phase, start the collection phase with a leader state, and leave the current node // all active agents have the same ID for all target primes
12:     **if** $(same = true) \wedge (prime \neq (\log N)\text{-th prime})$ **then** $prime = $ (next prime)
13:     **if** $max = false$ **then** terminate the selection phase and start the collection phase with a follower state
14:     **else** $phase = phase + 1$, $prime = 2$, $same = true$, $max = true$
15: **end while**

Pseudocode is described in Algorithm 2. Each agent a_i uses boolean variable max ($max = true$ means ID_i is the maximum among IDs a_i has ever found).

4.2   Collection Phase

In this phase, leader agents instruct follower agents so that they move to and stay at consecutive nodes starting from the base nodes. Concretely, each leader agent a_i firstly moves to the follower node v_j (i.e., the token node with another agent) so that a_i makes the follower agent to execute the collection phase. Then, a_i waits at v_j until the follower leaves v_j[6]. After this, a_i leaves v_j and moves to the next follower node. This process is repeated until a_i reaches the next leader node (i.e., the token node with no agent)[7]. On the other hand, each follower agent a_i waits at the current node until another agent (i.e., a leader) comes. Then, a_i firstly moves to the nearest leader node. After this, a_i moves until it reaches a node with no agent and stays there. When all agents finish their movements, the agents are divided into groups (possibly only one group) each of which consists of $fNum + 1$ agents, and the agents in a group are deployed at consecutive nodes starting from a base node.

For example, in Fig. 4 there exists one leader agent a_0 and two follower agents a_1 and a_2 between a_0 and its adjacent leader (i.e., $fNum = 2$). From (a) to (b), a_0 firstly moves to the nearest token node with an agent (i.e., follower node), and

---

[6] When an agent in the selection phase visits v_j, it leaves v_j without staying there by the atomicity of an action. Hence, the behavior of leader agent a_i can inform a follower agent of the beginning of the collection phase.

[7] By the atomicity of an action, when an agent moves to some leader node, the leader agent already starts its collection phase and leaves the leader node.

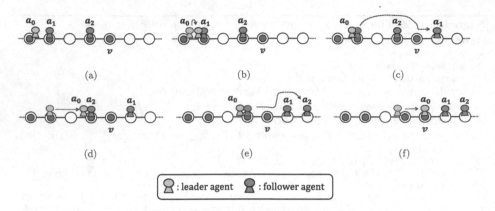

Fig. 4. An example of the collection phase (*fNum* = 2).

stays there until the follower agent leaves the node. From (b) to (c), a_1 detecting another agent a_0 firstly moves to the token node with no agent (i.e., leader node v), and then moves to the next node. From (c) to (d), a_0 similarly moves to the next follower node where agent a_2 exists. From (d) to (e), a_2 firstly moves to leader node v and moves until it visits a node with no agent. From (e) to (f), a_0 moves to leader node v and finishes the collection phase.

4.3 Deployment Phase

In this phase, leader agents control follower agents so that they should move to and stay at their target nodes to achieve uniform deployment. The basic idea is as follows. The deployment phase consists of several subphases, and the distance between every pair of adjacent agents in the same group is increased by one in each subphase. To realize it, each subphase consists of several iterations. For explanation of an iteration, consider a group where a_0 is a leader and followers $a_1, a_2, \ldots, a_{fNum}$ are following a_0 in this order. At the beginning of the first subphase, they stay at consecutive nodes. Each subphase consists of *fNum* iterations. In the l-th iteration, each of the l agents $a_{fNum-l+1}, a_{fNum-l+2}, \ldots, a_{fNum}$ moves to the next node. Consequently, in each subphase a_m moves m times and thus the distance between every pair of adjacent agents increases by one.

The l-the iteration is realized as follows. Leader agent a_0 firstly moves to the node where $a_{fNum-l+1}$ is staying and stays there until $a_{fNum-l+1}$ moves to the next node. Then, a_0 moves to the node where $a_{fNum-l+2}$ is staying to make $a_{fNum-l+2}$ to move to the next node. This process is repeated until a_{fNum} moves to the next node. After this, a_0 makes a remaining circuit of the ring, returns to the node where it started the deployment phase, say $v_{dep}(a_0)$, and terminates the l-th iteration. Then, a_0 checks if the locations of agents from $v_{dep}(a_0)$ to the next leader node are uniform or not using the CRT. If the locations are uniform, a_0 returns to $v_{dep}(a_0)$ and enters a suspended state. Otherwise, a_0 executes the next iteration. When a_0 executes the *fNum*-th iteration and the locations are not uniform, a_0 executes the next subphase.

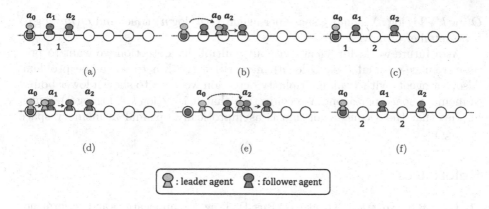

Fig. 5. An example of the deployment phase ($fNum = 2$).

For example, in Fig. 5 there exist one leader agent a_0 and two follower agents a_1 and a_2 (i.e., $fNum=2$). Let d_1 (resp., d_2) be the distance from a_0 to a_1 (resp., a_1 to a_2). In (a), $d_1 = d_2 = 1$ holds. From (a) to (b), as the first iteration in the first subphase a_0 moves to the node where the $fNum$-th follower agent (i.e., a_2) exists and stays there until a_2 moves to the next node. From (b) to (c), a_0 returns to the node $v_{dep}(a_0)$ where it started the deployment phase. Then, $d_1 = 1$ and $d_2 = 2$ hold. If a_0 recognizes that the locations of agents are not uniform, it executes the next iteration. From (c) to (d), as the second iteration in the first subphase a_0 moves to the node where the ($fNum - 1$)-th agent (i.e., a_1) exists and stays there until a_1 moves to the next node. From (d) to (e), a_0 moves to the next follower's (i.e., a_2's) node and stays there until a_2 moves to the next node. From (e) to (f), a_0 returns to node $v_{dep}(a_0)$. Then, $d_1 = d_2 = 2$ holds and a_i determines its next behavior depending on the location of agents. Each leader repeats such a behavior until it recognizes that the locations of agents are uniformly deployed.

We have the following theorem for the proposed algorithm.

Theorem 3. *For agents with weak multiplicity detection and knowledge of an upper bound $\log N$ of $\log n$ satisfying $\log N = O(\log n)$, the proposed algorithm solves the uniform deployment problem with $O(\log k + \log \log n)$ memory space per agent, $O(n^2 \log n)$ time, and $O(kn^2 \log n)$ total number of moves.*   □

5   Conclusion

In this paper, we proposed two space-efficient uniform deployment algorithms in asynchronous unidirectional ring networks. For agents without multiplicity detection, we showed that each agent requires $\Omega(\log n)$ memory space, and proposed an algorithm to solve the problem with $O(k + \log n)$ memory space per agent, $O(n \log k)$ time, and $O(kn \log k)$ total number of moves. This algorithm is optimal in memory space per agent when $k = O(\log n)$. For agents with weak multiplicity detection, we proposed an algorithm to solve the problem with

$O(\log k + \log \log n)$ memory space per agent, $O(n^2 \log n)$ time, and $O(kn^2 \log n)$ total number of moves.

As a future work, for agents without multiplicity detection we want to propose a space-optimal (i.e., $O(\log n)$ memory) algorithm to solve the problem. Also, for agents with weak multiplicity detection we want to show a lower bound of memory space per agent. We conjecture that it is $\Omega(\log k + \log \log n)$, which implies that the second algorithm is asymptotically optimal in memory space per agent.

References

1. Gray, R.S., Kotz, D., Cybenko, G., Rus, D.: D'agents: applications and performance of a mobile-agent system. Softw. Pract. Exper. **32**(6), 543–573 (2002)
2. Lange, D.B., Oshima, M.: Seven good reasons for mobile agents. CACM **42**(3), 88–89 (1999)
3. Kranakis, E., Krizanc, D.: An algorithmic theory of mobile agents. In: Montanari, U., Sannella, D., Bruni, R. (eds.) TGC 2006. LNCS, vol. 4661, pp. 86–97. Springer, Heidelberg (2007). https://doi.org/10.1007/978-3-540-75336-0_6
4. Cao, J., Sun, Y., Wang, X., Das, S.K.: Scalable load balancing on distributed web servers using mobile agents. JPDC **63**(10), 996–1005 (2003)
5. Flocchini, P., Prencipe, G., Santoro, N.: Self-deployment of mobile sensors on a ring. Theor. Comput. Sci. **402**(1), 67–80 (2008)
6. Yotam, E., Alfred, B.M.: Uniform multi-agent deployment on a ring. Theor. Comput. Sci. **412**(8), 783–795 (2011)
7. Barriere, L., Flocchini, P., Mesa-Barrameda, E., Santoro, N.: Uniform scattering of autonomous mobile robots in a grid. Int. J. Found. Comput. Sci. **22**(03), 679–697 (2011)
8. Shibata, M., Mega, T., Ooshita, F., Kakugawa, H., Masuzawa, T.: Uniform deployment of mobile agents in asynchronous rings. In: PODC, pp. 415–424 (2016)
9. Tel, G.: Introduction to Distributed Algorithms. Cambridge University Press, Cambridge (2000)
10. Amos, O.R., Benjamin, P.: Residue Number Systems: Theory and Implementation, vol. 2. World Scientific, Singapore (2007)
11. Pei, D., Salomaa, A., Ding, C.: Chinese Remainder Theorem: Applications in Computing, Coding, Cryptography. World Scientific, Singapore (1996)

Explorable Families of Graphs

Andrzej Pelc[✉]

Département d'informatique, Université du Québec en Outaouais,
Gatineau, Québec J8X 3X7, Canada
pelc@uqo.ca

Abstract. Graph exploration is one of the fundamental tasks performed by a mobile agent in a graph. An n-node graph has unlabeled nodes, and all ports at any node of degree d are arbitrarily numbered $0, \ldots, d-1$. A mobile agent, initially situated at some starting node v, has to visit all nodes of the graph and stop. In the absence of any initial knowledge of the graph the task of deterministic exploration is often impossible. On the other hand, for some families of graphs it is possible to design deterministic exploration algorithms working for any graph of the family. We call such families of graphs *explorable*. Examples of explorable families are all finite families of graphs, as well as the family of all trees.

In this paper we study the problem of which families of graphs are explorable. We characterize all such families, and then ask the question whether there exists a universal deterministic algorithm that, given an explorable family of graphs, explores any graph of this family, without knowing which graph of the family is being explored. The answer to this question turns out to depend on how the explorable family is given to the hypothetical universal algorithm. If the algorithm can get the answer to any yes/no question about the family, then such a universal algorithm can be constructed. If, on the other hand, the algorithm can be only given an algorithmic description of the input explorable family, then such a universal deterministic algorithm does not exist.

Keywords: Algorithm · Graph · Exploration · Mobile agent
Explorable family of graphs

1 Introduction

Network exploration is one of the fundamental tasks performed by mobile agents in networks. Depending on the application, the mobile agent may be a software agent that has to collect data placed at nodes of a communication network, or it may be a mobile robot collecting samples of ground in a contaminated building or mine whose corridors form links of a network, with corridor crossings represented by nodes.

Research supported in part by NSERC Discovery Grant 8136 − 2013 and by the Research Chair in Distributed Computing of the Université du Québec en Outaouais.

Z. Lotker and B. Patt-Shamir (Eds.): SIROCCO 2018, LNCS 11085, pp. 165–177, 2018.
https://doi.org/10.1007/978-3-030-01325-7_17

The network is modeled as a finite simple connected undirected graph $G = (V, E)$ with n nodes, called *graph* in the sequel. The number n of nodes is called the *size* of the graph. Nodes are unlabeled, and all ports at any node of degree d are arbitrarily numbered $0, \ldots, d-1$. The agent is initially situated at a starting node v of the graph. When the agent located at a current node u gets to a neighbor w of u by taking a port i, it learns the port j by which it enters node w and it learns the degree of w. The agent has to visit all nodes of the graph and stop. We assume that the agent is computationally unbounded (it is modeled as a Turing machine) and cannot mark the visited nodes.

It is well-known that, without any a priori information, the task of deterministic exploration is impossible to perform in arbitrary graphs. In fact, it is impossible even in quite simple and restricted families of graphs, such as the class of rings in which ports at all nodes are numbered 0,1 in clockwise order. Even if the agent knows that it is in some such ring (but does not know which), it cannot learn its size. If there existed a deterministic exploration algorithm for the class of such rings, not using any a priori knowledge, then the agent would have to stop after some t steps in every ring, and hence it would fail to explore a $(t + 2)$-node ring.

On the other hand, there exist classes of graphs for which deterministic exploration of any graph in the class is possible without knowing in which graph of the family the agent operates. Such are, for example, all finite classes of graphs. Knowing such a class, the algorithm can find an upper bound N on the size of all graphs in the family, and then apply, e.g., the algorithm from [19], based on universal exploration sequences, that visits all nodes of any graph of size at most N, regardless of the starting node. On the other hand, there are also infinite families of graphs that are possible to explore without any initial information. Such is, for example, the family of all trees. Indeed, any tree can be visited using the *basic walk* that consists in leaving the starting node by port 0 and then leaving every node w by port $(i + 1) \bmod d$, where i is the port by which the agent entered node w and d is the degree of w. Performing such a walk, the agent realizes when it made the full tour of the tree and got back to the starting node, where it stops.

The aim of this paper is to study families of graphs that have the property that any graph in the family can be deterministically explored without knowing in which graph the agent operates. We adopt the following definition.

A family $\mathcal{F}$ of graphs is *explorable*, if there exists a deterministic algorithm $\mathcal{A}(\mathcal{F})$ dedicated to this family, such that a mobile agent that executes algorithm $\mathcal{A}(\mathcal{F})$ starting from any node v of any graph $G \in \mathcal{F}$, visits all nodes of G and stops.

1.1 Our Results

We give an exact characterization of explorable families of graphs by formulating a condition C with the following properties. Given a family $\mathcal{F}$ of graphs that does not satisfy condition C, no deterministic algorithm can explore all graphs of $\mathcal{F}$. On the other hand, given any family $\mathcal{F}$ of graphs satisfying condition C we construct a deterministic algorithm $\mathcal{A}(\mathcal{F})$ that explores all graphs of $\mathcal{F}$.

The above algorithm $\mathcal{A}(\mathcal{F})$ used to explore graphs of a family $\mathcal{F}$ that has property C is *dedicated* to the family $\mathcal{F}$, i.e., it works only for graphs from $\mathcal{F}$, and for different families $\mathcal{F}$ different algorithms $\mathcal{A}(\mathcal{F})$ are used. Hence it is natural to ask if there exists a *universal* deterministic algorithm $\mathcal{U}$ that, given an explorable family $\mathcal{F}$ of graphs (i.e.. any family satisfying condition C), explores any graph of this family, without knowing which graph of the family it is exploring. The answer to this question turns out to depend on how the explorable family $\mathcal{F}$ is given to the hypothetical universal algorithm. (Since interesting explorable families are infinite, the input of $\mathcal{U}$ cannot be given as a finite object all at once, e.g., coded as a finite binary string). If the universal algorithm can get the answer to any yes/no question about the family, then such a universal deterministic algorithm $\mathcal{U}$ can be constructed. If, on the other hand, the universal algorithm can be only given an algorithmic description of the input explorable family, then such a universal deterministic algorithm does not exist.

1.2 Related Work

Exploration of unknown environments by mobile agents has been studied for many decades (cf. the survey [18]). The explored environment can be modeled in two distinct ways: either as a subset of the plane, e.g., an unknown terrain with convex obstacles [8], or a room with polygonal [10] or rectangular [4] obstacles, or as a graph, assuming that the agent may only move along its edges. The graph model can be further split into two different scenarios. One of them assumes that the graph is directed, in which case the agent can move only from tail to head of a directed edge [1,5,6]. The other scenario assumes that the graph is undirected and the agent can traverse edges in both directions [3,7,12,16]. Some authors impose further restrictions on the moves of the agent. In [3,7] it is assumed that the agent has a restricted tank, and thus has to periodically return to the base for refueling, while the authors of [12] assume that the agent is attached to the base by a cable of restricted length.

An important direction of research concerns exploration of anonymous graphs. In this case it is impossible to perform exploration with termination of arbitrary graphs in the absence of any a priori knowledge of the graph, if no marking of nodes is allowed. Hence some authors [5,6] allow *pebbles* which the agent can drop on nodes to recognize already visited ones, and then remove them and drop them in other nodes. A more restrictive scenario assumes that a stationary token is placed at the starting node of the agent [9,17]. Exploring anonymous graphs without the possibility of marking nodes (and thus possibly without stopping) is investigated, e.g., in [11,14]. In these papers the authors concentrate attention on the minimum amount of memory sufficient to carry out exploration. If marking of nodes is precluded, some knowledge about the graph is required in order to guarantee stopping after exploration, e.g., an upper bound on its size [2,9,19].

In [13,15], the authors study the problem of the minimum size of information that has to be given to the mobile agent in order to perform fast exploration. In [13], only exploration of trees is considered, and the algorithm performance is measured using the competitive approach. In [15], exploration of arbitrary graphs is studied, and the performance measure is the order of magnitude of exploration time.

2   Characterization of Explorable Families

In this section we provide a necessary and sufficient condition on the explorability of a family of graphs. From now on, we restrict attention to infinite families of graphs, since, as mentioned in the introduction, all finite families are trivially explorable. In order to formulate the condition we need the notion of a *truncated view* from a node v in a graph G. Let G be any graph, v a node in this graph and k a natural number. The truncated view from v in G of depth k, denoted $\mathcal{V}(v, G, k)$, is the tree of all simple paths of length at most k, starting from node v and coded as sequences of port numbers, where the rooted tree structure is defined by the prefix relation of sequences. This definition is equivalent to that from [20]. It follows from the above paper that the information that a mobile agent starting at node v in graph G can obtain after k steps is "included" in $\mathcal{V}(v, G, k + 1)$ in the following sense. For any deterministic algorithm $\mathcal{A}$ that works in graphs G and G' and any mobile agents A_1 and A_2 executing this algorithm, starting, respectively, from node v in G and from node v' in G', such that $\mathcal{V}(v, G, k + 1) = \mathcal{V}(v', G', k + 1)$, the behaviors of agents A_1 and A_2 during the first k steps of these executions, i.e., their trajectories and possible decisions to stop, are identical.

Since the set of (finite) graphs is countable, every (infinite) family $\mathcal{F}$ of graphs can be represented as a sequence $\{G_i : i \geq 1\}$, ordered so that no two graphs G_i and G_j are port-preserving isomorphic, and the sizes of the graphs are non-decreasing. In order to make the enumeration unambiguous, graphs of the same size are ordered lexicographically, using some fixed graph representation. The resulting ordering will be called *canonical* and used from now on.

The following condition C concerns a family $\mathcal{F} = \{G_i : i \geq 1\}$ of graphs:

> For every $i \geq 1$ and every node v in G_i, there exist positive integers k, m, such that $\mathcal{V}(v, G_i, k)$ is different from truncated views of depth k from all nodes in all graphs G_j, for $j > m$.

We now proceed to the proof that C is a necessary and sufficient condition on the explorability of $\mathcal{F}$. We first prove the necessity.

Lemma 1. *A family $\mathcal{F} = \{G_i : i \geq 1\}$ that does not satisfy the condition C is not explorable.*

Proof. Suppose that the condition C is not satisfied for a family $\mathcal{F} = \{G_i : i \geq 1\}$ of graphs. This implies that there exists a positive integer i and a node v in graph G_i, such that for all positive integers k, m, there exists an index

$j(k,m) > m$ and a node $v(k,m)$ in the graph $G_{j(k,m)}$, satisfying the equality $\mathcal{V}(v(k,m), G_{j(k,m)}, k) = \mathcal{V}(v, G_i, k)$.

Suppose, for a contradiction, that the family $\mathcal{F}$ is explorable. Hence there exists an algorithm $\mathcal{A}(\mathcal{F})$ that explores any graph of the family, starting from any node. Consider the execution of this algorithm by an agent A_1 on graph G_i, starting from node v. For some integer x, after x steps, the agent explores the graph G_i and stops. Take $k = x + 1$ and any m such that all graphs G_j, for $j > m$, have sizes larger than $x + 1$. Consider an index $j(k,m) > m$ and a node $v(k,m)$ in $G_{j(k,m)}$ for which $\mathcal{V}(v(k,m), G_{j(k,m)}, k) = \mathcal{V}(v, G_i, k)$. Now consider the execution of algorithm $\mathcal{A}(\mathcal{F})$ by an agent A_2 starting at node $v(k,m)$ in $G_{j(k,m)}$. Since $\mathcal{V}(v(k,m), G_{j(k,m)}, k) = \mathcal{V}(v, G_i, k)$, the behavior of this agent during the first x steps must be the same as the behavior of agent A_1. It follows that agent A_2 stops after x steps in the graph $G_{j(k,m)}$ as well. However, this graph has size larger than $x + 1$ and hence it cannot be explored in x steps. This contradiction proves the lemma.   □

We next proceed to the proof of the sufficiency of condition C. In order to do this we will construct an algorithm $Explo(\mathcal{F})$, dedicated to a family of graphs $\mathcal{F} = \{G_i : i \geq 1\}$ satisfying condition C, such that $Explo(\mathcal{F})$ explores all graphs in this family. For any $i \geq 1$ and every node v in G_i, let $(k(v,i), m(v,i))$ be the lexicographically first couple of integers (k, m), such that $\mathcal{V}(v, G_i, k)$ is different from truncated views of depth k from all nodes in all graphs G_j, for $j > m$. Call the integer $k(v,i)$ the *depth witness* of (v, i) and call the integer $m(v,i)$ the *range witness* of (v, i). For every graph, we define a *non-backtracking* path as a path in which the agent never exits a node by a port p immediately after entering it by the port p. The *reverse* of a path $P = (w_1, w_2, \ldots, w_s)$ is the path $\overline{P} = (w_s, w_{s-1}, \ldots, w_1)$.

We first describe Procedure Check (v, i), for a positive integer i and for a node v in G_i. The aim of this procedure is to check whether the truncated view of depth $k(i, v)$ from the unknown initial position of the agent in an unknown graph from $\mathcal{F}$ is equal to the truncated view $\mathcal{V}(v, G_i, (k(i, v))$. To this end, the agent traverses (in lexicographic order) all non-backtracking maximal paths from $\mathcal{V}(v, G_i, (k(i, v))$ starting at its initial node and returning to it after traversing each path, using the reverse of this path. In the case when a given non-backtracking path is impossible to traverse, either because the agent enters earlier a node of degree 1, or because the entry port at a given edge is different from that in $\mathcal{V}(v, G_i, (k(i, v))$, then the agent interrupts the traversal of this path and returns using the reverse of the path traversed till this point.

There are two possible terminations of Procedure Check (v, i). The first possibility is that all non-backtracking paths of length $k(v, i)$ are identical as in $\mathcal{V}(v, G_i, (k(v, i))$. This is called the *success* of Check (v, i). The second possibility, called the *failure* of Check (v, i), is that the above condition is not satisfied.

The next procedure is used to find a couple (v, i) for which Check (v, i) is terminated with success. Such a couple must exist: indeed, if the agent starts from node v in the graph G_i, then Check (v, i) is terminated with success.

```
Procedure Find Success


i := 1
result := failure
while result = failure do
      for all nodes v in G_i do
            Check (v, i)
            if Check (v, i) terminated with success then
                  result := success
      i := i + 1
return parameters v and i for which the variable result changed from
failure to success.
```

Let $R(N)$ be the procedure from [19] based on universal exploration sequences, that visits all nodes of any graph of size at most N, and stops, regardless of the starting node. Now the algorithm $Explo(\mathcal{F})$ can be succinctly formulated as follows.

```
Algorithm Explo(F)


Find Success
(v, i) := parameters returned by procedure Find Success
M := the maximum size of all graphs in the family {G_t : t ≤ m(v,i)}
R(M)
```

Lemma 2. *For every family $\mathcal{F}$ of graphs, satisfying condition C, Algorithm $Explo(\mathcal{F})$ correctly explores any graph of the family $\mathcal{F}$, starting at any node of this graph.*

Proof. The **while** loop in Procedure **Find Success** must be exited at some point, i.e., the variable *result* must be set to *success*. This happens at the latest in the execution of Procedure **Check** (v, i), where the graph from the family $\mathcal{F}$ in which the agent is operating is G_i and the starting node of the agent is v. Indeed, in this case, the truncated view from the initial position of the agent at any depth h is identical with the truncated view $\mathcal{V}(v, G_i, h)$. Hence the Procedure **Find Success** terminates and the agent learns parameters v and i for which the variable *result* changed from *failure* to *success*. By the definition of $m(v, i)$, the agent learns that it is in one of the graphs from the (finite) family $\{G_t : t \leq m(v, i)\}$. It follows that the execution of procedure $R(M)$, where M is the maximum size of all graphs in this family, must result in the exploration of the graph in which the agent operates, regardless of which graph of the family $\{G_t : t \leq m(v, i)\}$ it is and regardless of which node of this graph is the starting node of the agent. This proves the lemma. $\square$

Lemmas 1 and 2 imply the following characterization result.

Theorem 1. *A family of graphs is explorable if and only if it satisfies condition C.*

3   Universal Exploration Algorithm

In this section we use the characterization from Sect. 2 to investigate the following problem. Does there exist a *universal* algorithm $\mathcal{U}$, which when given as input an explorable family $\mathcal{F}$ of graphs, explores any graph of this family, starting from any initial node in it? Note that Algorithm $Explo(\mathcal{F})$ from the preceding section was *dedicated* to the exploration of graphs of one particular explorable family $\mathcal{F}$, i.e., it was supposed to work only for graphs of this family. Consequently, important information about the family, such as functions $k(v,i)$ and $m(v,i)$, could be included in the text of the algorithm. For a universal algorithm this is not the case: in contrast to dedicated algorithms, a universal algorithm must gain sufficient knowledge about the input explorable family $\mathcal{F}$, in order to be able to successfully explore any graph of this family.

Here comes the subtle issue of how the input explorable family is given. First recall that we may restrict attention to infinite families, as finite families are trivial to explore by applying the procedure $R(N)$ from [19], where N is an upper bound on the sizes of all graphs in the family. For a finite family, it can be given to the universal algorithm as a single finite input object, the algorithm can find N, apply $R(N)$ and we are done. By contrast, in the case of infinite explorable families $\mathcal{F}$, the family cannot be given to the hypothetical universal algorithm $\mathcal{U}$ as a single input object all at once. How then could it be given?

It seems reasonable to assume that the universal algorithm should be able to get knowledge about the input family $\mathcal{F}$ "piece by piece", i.e., it should be able to get items of information about this family as responses to queries. This idea can be implemented in at least two ways. We start with the more liberal way that intuitively allows the algorithm to get an answer to any yes/no query about the input family $\mathcal{F}$. This can be formalized as follows. Consider the set $\mathcal{X}$ of all infinite families of finite graphs. Observe that, while each family in $\mathcal{X}$ is countable, the set $\mathcal{X}$ which is the set of all these families is uncountable, but this fact has no impact on our formalization. Consider any definable subfamily $\varXi$ of $\mathcal{X}$, i.e., a family $\varXi = \{\mathcal{G} \in \mathcal{X} : \mathcal{G}$ satisfies $\varPhi\}$, where $\varPhi$ is some set-theoretic predicate. The questions that the universal algorithm is allowed to ask are of the form: "Is the input family $\mathcal{F}$ an element of $\varXi$?" This is of course equivalent to asking "Does the input family $\mathcal{F}$ satisfy the predicate $\varPhi$?" Such questions can be asked by the universal algorithm, using all possible predicates $\varPhi$, one at a time, and it is assumed that the algorithm will obtain a truthful answer to any such question. This formalization can be thought of as using an oracle that knows everything about the input family but answers only yes or no. Examples of questions that the algorithm can ask are: "Are there infinitely many planar graphs in $\mathcal{F}$?", "Is the 17-th graph in the canonical order of $\mathcal{F}$ a tree? or "Does there exist a tree in $\mathcal{F}$?".

It should be noted that although the allowed queries are only of yes/no type, they are very powerful, as the universal algorithm may get some information about the entire infinite input family all at once, as in the query "Does there exist a tree in $\mathcal{F}$?". A negative answer to such a query could not be obtained by looking at any finite part of $\mathcal{F}$. We will show that this powerful feature allows us to construct a correct universal exploration algorithm. Before doing it, we need some preparation.

Let $\{H_i : i \geq 1\}$ be the canonical enumeration of the family $\mathcal{X}$. Consider an explorable family $\mathcal{F}$ that is the input to the universal algorithm that we are going to describe. We first describe the procedure Find i-th graph that returns the graph that is the i-th element in the canonical enumeration of $\mathcal{F}$. The procedure asks the questions "Is the i-th element in the canonical enumeration of $\mathcal{F}$ equal to H_j?", for $j = 1, 2, ...$, until the answer yes is obtained, and returns H_j for which the positive answer is obtained.

The aim of the next two procedures is finding, respectively, the depth witness and the range witness of (v, i), where v is a node in the i-th element in the canonical enumeration of $\mathcal{F}$. Procedure Find the depth witness of (v, i) asks questions "Is the depth witness of (v, i) equal to j, for $j = 1, 2, ...$, until the answer yes is obtained, and returns the integer j for which the positive answer is obtained. Similarly, procedure Find the range witness of (v, i) asks questions "Is the range witness of (v, i) equal to j, for $j = 1, 2, ...$, until the answer yes is obtained, and returns the integer j for which the positive answer is obtained.

We will now modify the procedure Check (v, i) from Sect. 2 to make it work in the context of the universal algorithm. The modification, for any (v, i) consists in first applying procedure Find i-th graph and then applying procedure Find the depth witness of (v, i). Suppose that the first procedure returns the graph G_i and the second procedure returns the integer $k(v, i)$. The rest of procedure Check (v, i) is as in Sect. 2. Again, the procedure may terminate with success or failure, defined as previously. Procedure Find Success is as before, using the modified version of Check (v, i). We will call it Universal Find Success. Now our universal algorithm can be formulated as follows, assuming that the input explorable family given to the oracle that answers queries is $\mathcal{F}$.

Algorithm Universal Exploration

Universal Find Success
$(v, i) :=$ parameters returned by procedure Find Success
Find range witness of (i, v)
Let $m(i, v)$ be the range witness of (i, v)
for $t = 1$ **to** $m(i, v)$ **do**
    Find the t-th graph
    Let G_t be the graph returned by procedure Find the t-th graph
$M :=$ the maximum size of all graphs in the family $\{G_t : t \leq m(v, i)\}$
$R(M)$

The correctness of Algorithm Universal Exploration follows from the fact that it correctly finds the depth and range witnesses, as well as the graphs

$\{G_t : t \leq m(v,i)\}$. Other than that, the algorithm works like the dedicated algorithm $\mathcal{A}(\mathcal{F})$, and thus achieves exploration of any graph in the family. Hence we have the following theorem.

Theorem 2. *Algorithm* Universal Exploration *correctly explores any graph of an explorable family $\mathcal{F}$, starting at any node of this graph, if this family is given as input to an oracle that can answer all yes/no queries about it.*

The capability of getting an answer to any yes/no query about the input explorable family is very strong. It assumes the existence of an oracle that has a "magical" complete insight in this family. It can be argued that such an oracle could not exist in practice, and thus it would be desirable to design a universal exploration algorithm to which input explorable families would be given in a way possible to implement realistically. Here comes the second natural way in which a potential universal algorithm could get information about the input family. Suppose that the input explorable family $\mathcal{F}$ is recursively enumerable and let $\mathcal{E}(\mathcal{F})$ be the enumeration algorithm. More precisely, the algorithm $\mathcal{E}(\mathcal{F})$, given a positive integer i as input, returns the i-th graph of $\mathcal{F}$ in the canonical enumeration. The second natural way of providing information about the family $\mathcal{F}$ to the hypothetical universal exploration algorithm $\mathcal{U}$ would be to give to $\mathcal{U}$ the text of algorithm $\mathcal{E}(\mathcal{F})$ as input. Then the algorithm $\mathcal{U}$ would be able, for any positive integer i, to run $\mathcal{E}(\mathcal{F})$ on i and get the i-th graph of $\mathcal{F}$ in the canonical enumeration, returned by $\mathcal{E}(\mathcal{F})$. For simplicity, we may assume that finding this i-th graph is done in one step. This is reminiscent of the definition of Turing reducibility in which an algorithm A_1 reducible to A_2 may run A_2 on some input and receive the output in one step. In any case, in this paper we are not concerned with efficiency of exploration, only with the feasibility of this task.

We will say that a universal exploration algorithm *processes algorithmic input*, if it works as described above. Such an algorithm would be able to explore any graph of any (recursively enumerable) explorable family, given to it in this algorithmic way, without the help of any oracle. Unfortunately, we have the following negative result.

Theorem 3. *There does not exist a universal exploration algorithm that processes algorithmic input, which correctly explores any graph of any recursively enumerable explorable family.*

Proof. Suppose that there exists a universal exploration algorithm $\mathcal{U}$ that processes algorithmic input, which correctly explores any graph of any recursively enumerable explorable family. Denote by R_k, for $k \geq 3$, the ring of size k with ports at all nodes numbered 0,1 in the clockwise order. Let C_k denote the graph resulting from the ring R_k by attaching one node of degree 1 to one of the nodes of R_k. By definition, the edge joining the single node v of degree 3 with the single node of degree 1 corresponds to port number 2 at v. In C_k we will say that an agent goes clockwise if it leaves a node by port 1. For $i \geq 1$, let G_i be the graph C_{i+2}, and consider the family $\mathcal{F} = \{G_i : i \geq 1\}$ of graphs. Since the sizes of graphs G_i are strictly increasing, this is the canonical enumeration of $\mathcal{F}$.

We first observe that the family $\mathcal{F}$ is explorable. The (dedicated) exploration algorithm $\mathcal{A}(\mathcal{F})$ can be simply formulated as follows.

If the starting node is of degree 1 then take port 0, go clockwise until getting to a node of degree 3 and stop.

If the starting node is of degree 3 then take port 2, take port 0, go clockwise until getting to a node of degree 3 and stop.

If the starting node is of degree 2 then go clockwise until the second visit at a node of degree 3, take port 2 and stop.

Suppose that $\mathcal{E}(\mathcal{F})$ is an enumeration algorithm corresponding to $\mathcal{F}$, given to $\mathcal{U}$ as input. Consider the execution E_1 of $\mathcal{U}$ with this input, where the agent is initially placed at the single node v of degree 3 in graph G_1. Suppose that in execution E_1 the agent stops after k steps. Let r be the largest integer for which algorithm $\mathcal{E}(\mathcal{F})$ was called in the execution E_1 of $\mathcal{U}$.

For any positive integer j, define the following graph D_j (cf. Fig. 1).

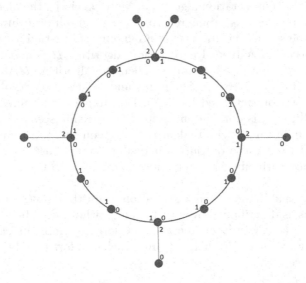

Fig. 1. The graph D_2

Consider the ring R_{6j}. Attach to every third node of R_{6j} a distinct node of degree 1. Finally, to one of the $2j$ resulting nodes of degree 3 attach another node of degree 1. Thus graph D_j has 1 node of degree 4 and $2j - 1$ nodes of degree 3, partitioning the ring R_{6j} from which the construction started into $2j$ segments of length 3. Moreover, D_j has $2j + 1$ nodes of degree 1. The port at each node of degree 3 corresponding to the edge joining it to a node of degree 1 has number 2. The ports at the single node of degree 4 corresponding to the edges joining it to nodes of degree 1 have numbers 2 and 3.

We now define the graphs H_i, for $i \geq 1$, as follows. For $1 \leq i \leq r$, let $H_i = G_i$; for $i > r$, let $H_i = D_i$. Finally, we consider the family $\mathcal{F}^* = \{H_i : i \geq 1\}$

of graphs. Since the sizes of graphs H_i are strictly increasing, this is the canonical enumeration of $\mathcal{F}^*$. We show that the family $\mathcal{F}^*$ is explorable. In order to formulate the (dedicated) algorithm $\mathcal{A}(\mathcal{F}^*)$, we first describe the following procedure that explores any graph D_j starting from a node of degree 3. (For this purpose, repeating until the second visit at a node of degree 4 would be enough, but we need the third visit to make it work within algorithm $\mathcal{A}(\mathcal{F}^*)$).

Procedure Go around

repeat
　　　　take port 1, take port 1, take port 2, take port 0
until the third visit at a node of degree 4
take port 3 and stop.

Now algorithm $\mathcal{A}(\mathcal{F}^*)$ can be described as follows. Its high level idea is to go clockwise around the ring in any graph H_j, sufficiently long to see if j is at most r or larger than r. In the first case, algorithm $\mathcal{A}(\mathcal{F})$ is applied because the graph is G_j, and in the second case procedure Go around is used to terminate exploration because the graph is D_j.

Algorithm $\mathcal{A}(\mathcal{F}^*)$

if the starting node is of degree 1 **then** take port 0
Go clockwise for $r + 1$ steps.
if no node of degree 4 is visited **then** apply the algorithm $\mathcal{A}(\mathcal{F})$
else
　　　　go clockwise to the closest node of degree 3
　　　　Go around

Let $m = \max(k, r) + 1$ and suppose that $\mathcal{E}(\mathcal{F}^*)$ is an enumeration algorithm corresponding to $\mathcal{F}^*$, given to $\mathcal{U}$ as input. Consider the execution E_2 of algorithm $\mathcal{U}$ with this input, where the agent is initially placed at the node w antipodal to the unique node of degree 4 in the graph $H_m = D_m$. Consider the first k steps of this execution. Observe that $\mathcal{V}(v, G_1, k) = \mathcal{V}(w, H_m, k)$, by construction of graphs G_i and H_i. By induction on the step number, in the first k steps of execution E_2, algorithm $\mathcal{U}$ calls the input enumeration algorithm for exactly the same integers as it does in execution E_1. For execution E_1 the input algorithm is $\mathcal{E}(\mathcal{F})$ and for execution E_2 it is $\mathcal{E}(\mathcal{F}^*)$, but these integers are at most r and the first r graphs in the canonical enumeration of $\mathcal{F}$ and of $\mathcal{F}^*$ are identical. Hence the returned graphs in these calls in both executions are identical. It follows that the k steps in both executions are identical. Since the agent stops after the k-th step of execution E_1, it must also stop after the k-th step of execution E_2. However, in execution E_2 it cannot explore the graph $H_m = D_m$ because this graph has more than $k + 1$ nodes. This concludes the proof.   □

One could hope for salvaging the idea of a universal exploration algorithm processing algorithmic input by restricting the class of input explorable families from recursively enumerable (as we did above) to recursive. In this case the hypothetical universal algorithm could be given as input the *decision* algorithm that answers, for any graph G, if this graph belongs to the family $\mathcal{F}$ of graphs that should be explored. Similarly as before, the hypothetical universal algorithm could run the decision algorithm on any graph of its choice and learn in one step if this graph belongs to the recursive family of graphs that should be explored. However, it is easy to see that even with this restriction our negative result still holds. Indeed, it is enough to modify the proof of Theorem 3 by defining r to be the *largest size of a graph* for which the decision algorithm was called in the execution E_1. The rest of the proof remains unchanged.

4   Conclusion

We gave a characterization of explorable families of graphs, and provided, for any such family, a *dedicated* exploration algorithm that explores any graph of the family starting from any node. Then we studied the issue of the existence of a deterministic universal exploration algorithm that, for any explorable family given as input, would explore any graph of the family starting from any node. Since such families may be infinite, it has to be made precise how would they be given to the hypothetical universal algorithm. We showed that a very liberal approach to this issue of providing the input, namely the assumption of an oracle that can answer any yes/no query asked by the universal algorithm about the input family, permits us to construct such a universal algorithm. This approach, however, is arguably unrealistic. Hence we defined the way of presenting the input family to the hypothetical universal algorithm in a more restrictive but more realistic way: by giving the universal algorithm either the enumeration algorithm of the input family of graphs, in the case when this family is recursively enumerable, or giving it the decision algorithm for the input family, if the latter is recursive. We showed that this more realistic way of presenting the input explorable family precludes the existence of a deterministic universal exploration algorithm.

A similar idea to that in the proof of Theorem 3 can be used to prove that the problem of explorability of a family of graphs is undecidable in the following sense. There does not exist a decision algorithm that, given as input an enumeration algorithm of a recursively enumerable family $\mathcal{G}$ of graphs, can decide whether the family $\mathcal{G}$ is explorable.

In this paper we concentrated on the issue of feasibility of exploration for families of graphs, rather than on the efficiency of exploration. An open, probably very challenging problem yielded by our study is to find, for any explorable family of graphs, a dedicated algorithm which would explore any graph of this family in optimal time.

References

1. Albers, S., Henzinger, M.R.: Exploring unknown environments. SIAM J. Comput. **29**, 1164–1188 (2000)
2. Aleliunas, R., Karp, R., Lipton, R., Lovasz, L., Rackoff, C.: Random walks, universal traversal sequences, and the complexity of maze problems. In: Proceedings of 20th Annual IEEE Symposium on Foundations of Computer Science (FOCS 1979), pp. 218–223 (1979)
3. Awerbuch, B., Betke, M., Rivest, R.L., Singh, M.: Piecemeal graph exploration by a mobile robot. Inf. Comput. **152**, 155–172 (1999)
4. Bar-Eli, E., Berman, P., Fiat, A., Yan, R.: On-line navigation in a room. J. Algorithms **17**, 319–341 (1994)
5. Bender, M.A., Fernandez, A., Ron, D., Sahai, A., Vadhan, S.P.: The power of a pebble: exploring and mapping directed graphs. Inf. Comput. **176**, 1–21 (2002)
6. Bender, M.A., Slonim, D.: The power of team exploration: two robots can learn unlabeled directed graphs. In: Proceedings of 35th Annual Symposium on Foundations of Computer Science (FOCS 1994), pp. 75–85 (1994)
7. Betke, M., Rivest, R., Singh, M.: Piecemeal learning of an unknown environment. Mach. Learn. **18**, 231–254 (1995)
8. Blum, A., Raghavan, P., Schieber, B.: Navigating in unfamiliar geometric terrain. SIAM J. Comput. **26**, 110–137 (1997)
9. Chalopin, J., Das, S., Kosowski, A.: Constructing a map of an anonymous graph: applications of universal sequences. In: Lu, C., Masuzawa, T., Mosbah, M. (eds.) OPODIS 2010. LNCS, vol. 6490, pp. 119–134. Springer, Heidelberg (2010). https://doi.org/10.1007/978-3-642-17653-1_10
10. Deng, X., Kameda, T., Papadimitriou, C.H.: How to learn an unknown environment I: the rectilinear case. J. ACM **45**, 215–245 (1998)
11. Diks, K., Fraigniaud, P., Kranakis, E., Pelc, A.: Tree exploration with little memory. J. Algorithms **51**, 38–63 (2004)
12. Duncan, C.A., Kobourov, S.G., Anil Kumar, V.S.: Optimal constrained graph exploration. ACM Trans. Algorithms **2**, 380–402 (2006)
13. Fraigniaud, P., Ilcinkas, D., Pelc, A.: Tree exploration with advice. Inf. Comput. **206**, 1276–1287 (2008)
14. Fraigniaud, P., Ilcinkas, D.: Directed graphs exploration with little memory. In: Proceedings of 21st Symposium on Theoretical Aspects of Computer Science (STACS 2004), pp. 246–257 (2004)
15. Gorain, B., Pelc, A.: Deterministic graph exploration with advice. In: Proceedings of 44th International Colloquium on Automata, Languages and Programming (ICALP 2017), pp. 132:1–132:14 (2017)
16. Panaite, P., Pelc, A.: Exploring unknown undirected graphs. J. Algorithms **33**, 281–295 (1999)
17. Pelc, A., Tiane, A.: Efficient grid exploration with a stationary token. Int. J. Found. Comput. Sci. **25**, 247–262 (2014)
18. Rao, N.S.V., Kareti, S., Shi, W., Iyengar, S.S.: Robot navigation in unknown terrains: introductory survey of non-heuristic algorithms, Technical report ORNL/TM-12410, Oak Ridge National Laboratory, July 1993
19. Reingold, O.: Undirected connectivity in log-space. J. ACM **55**, 17:1–17:24 (2008)
20. Yamashita, M., Kameda, T.: Computing on anonymous networks: Part I - characterizing the solvable cases. IEEE Trans. Parallel Distrib. Syst. **7**, 69–89 (1996)

A Characterization of t-Resilient Colorless Task Anonymous Solvability

Carole Delporte-Gallet[1], Hugues Fauconnier[1], Sergio Rajsbaum[2]([✉]),
and Nayuta Yanagisawa[3]

[1] IRIF-GANG-Université Paris-Diderot, Paris, France
`{cd,hf}@irif.fr`
[2] Instituto de Matemáticas, UNAM, Mexico City, Mexico
`rajsbaum@math.unam.mx`
[3] Department of Mathematics, Graduate School of Science,
Kyoto University, Kyoto, Japan
`nayuta87@math.kyoto-u.ac.jp`

Abstract. One of the central questions in distributed computability is characterizing the tasks that are solvable in a given system model. In the *anonymous* case, where processes have no identifiers and communicate through multi-writer/multi-reader registers, there is a recent topological characterization (Yanagisawa 2017) of the *colorless* tasks that are solvable when any number of asynchronous processes may crash. In this paper, we consider the case where at most t asynchronous processes may crash, where $1 \leq t < n$. We prove that a colorless task is t-resilient solvable anonymously if and only if it is t-resilient solvable non-anonymously. We obtain our results through various reductions and simulations that explore how to extend techniques for non-anonymous computation to anonymous one.

Keywords: MWMR registers · Anonymity · Distributed task Topology

1 Introduction

One of the central questions in distributed computability is characterizing the tasks which are solvable in a given system model. A *task* is the distributed equivalent of a function in sequential computing: each process starts with a private input value, communicates with other processes, and eventually decides an output value, such that the vector of output values is valid for the vector of input values according to the task specification.

The *asynchronous computability theorem* (ACT) [26] is one of the central results in distributed computability. It characterizes the tasks that are solvable

C. Delporte-Gallet and H. Fauconnier—Supported by LiDiCo.
S. Rajsbaum—Supported by UNAM-PAPIIT IN109917. Part of this work was done while visiting Université Paris-Diderot.

© Springer Nature Switzerland AG 2018
Z. Lotker and B. Patt-Shamir (Eds.): SIROCCO 2018, LNCS 11085, pp. 178–192, 2018.
https://doi.org/10.1007/978-3-030-01325-7_18

in shared-memory systems where n processes that may fail by crashing communicate by reading and writing shared registers. It is sometimes called the *wait-free* characterization, because any number of processes may crash and the processes are asynchronous (run at arbitrary speeds, independent from each other). The characterization is of an algebraic topological nature. In terms of algebraic topology, a *task* is represented as a relation Δ between an input complex $\mathcal{I}$ and an output complex $\mathcal{O}$. Each simplex σ in $\mathcal{I}$ is a set that specifies the initial inputs to the processes in some execution. The processes communicate with each other, and eventually decide output values that form a simplex τ in $\mathcal{O}$. The computation is correct if τ is in $\Delta(\sigma)$. The complex $\mathcal{I}$ (resp. $\mathcal{O}$) is *chromatic* because each simplex specifies not only input values, but also which process gets which input (resp. output) value. Roughly, the ACT characterization states that the task is solvable if and only if there is a chromatic simplicial map δ from a chromatic subdivision of $\mathcal{I}$ to $\mathcal{O}$ respecting Δ.

The ACT is the basis to obtain a characterization of distributed task computability in the case where at most t asynchronous processes may crash, where $1 \leq t < n$. It is also the basis to study other failure, timing, and communication models, and even mobile robot models [30]. There are basically two ways of extending the results from the wait-free model to other models. One is by directly generalizing the algorithmic and topological techniques, and the other is by reduction to other models using simulations (either algorithmic [6] or topological [24]). An overview of results in this area can be found in the book [20].

The theory of distributed computing presented in [20] assumes that the processes, $p_0, \ldots, p_{n-1}$, communicate using single-writer/multi-reader (SWMR) registers, $R_0, \ldots, R_{N-1}$. Thus, p_i knows that it is the i-th process and it can write exclusively to R_i while the size of the namespace, N, is assumed to be much bigger than the number of the process, n. In this situation, preallocating a register for each identifier would lead to a distributed algorithm with a very large space complexity, namely N registers. Instead, it is shown in [13] that n multi-writer/multi-reader (MWMR) registers are sufficient to solve any read-write wait-free solvable task.

However, in some distributed systems, processes are *anonymous*; they have no ids at all or they cannot make use of their identifiers (e.g., due to privacy issues). In such a system, processes run identical programs, and the means by which processes access the shared memory are identical to all processes. A process cannot have a private register to which only this process may write, and hence the shared memory consists only of MWMR registers. This anonymous shared memory model of asynchronous distributed computing has been studied since early on [3, 29], in the case where processes do not fail.

In an anonymous system, *colorless* tasks are natural, because they are defined only in terms of input and output values without stating which process receives which input value or which process produces which output value. Furthermore, the class of colorless tasks includes various important tasks, such as consensus and set agreement, and is rich enough to be undecidable even for three processes [17, 21].

Colorless tasks have been well studied in shared-memory and message passing models, where each process has a distinct identifier [20], but less so in anonymous systems. Only recently, the ACT has been extended to the anonymous case [31], providing a characterization of wait-free anonymous computability of colorless tasks. The characterization implies that the anonymity does not reduce the computational power of the asynchronous shared-memory model as far as colorless tasks are concerned. In consequence, the topological characterization is in terms of input and output complexes which are not chromatic.

Results. Our main result is an extension of the wait-free characterization of [31] to the case where at most t processes may crash, where $1 \leq t < n$. We prove that a colorless task is t-resilient solvable anonymously if and only if it is t-resilient solvable non-anonymously. This implies a complete characterization of t-resilient, asynchronous, and anonymous computability of colorless tasks.

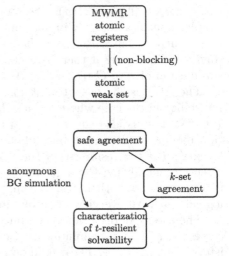

The result is obtained through a series of reductions depicted in the figure below. First, we design an anonymous non-blocking implementation of an atomic *weak set object* with n registers. The construction is based on the non-blocking atomic snapshot of [15,19]. Then, we build a wait-free implementation of a safe agreement object for an arbitrary value set V. Our implementation is a generalization of the anonymous consensus algorithm proposed in [3]. We describe two ways of deriving the t-resilient anonymous solvability characterization. One way is through a novel anonymous implementation of the BG-simulation [6], which we use to simulate a non-anonymous system by an anonymous system, both t-resilient. The other way is to use the safe-agreement object to solve k-set agreement and then do the topological style of analysis [24,31].

Related work. Colorless tasks were first identified in [6]. They include fundamental tasks such as consensus [16], set agreement [11], and loop agreement [22], and have been widely studied in the non-anonymous setting. The first part of the book [20] is devoted to colorless tasks. Not all tasks of interest are colorless though, and general tasks can be much harder to study, notably renaming [9,10].

A characterization of the colorless tasks that are solvable in the presence of processes that can crash in a dependent way is provided in [23], and a characterization when several processes can run solo is provided in [25]. Both encompass the wait-free colorless task solvability characterization, and the former encompasses the t-resilient characterization that we use in this paper.

A certain kind of anonymity has been considered in [26] to establish the anonymous computability theorem. However, they allow the use of SWMR registers while we assume a fully anonymous model with only MWMR registers.

Anonymous distributed computing remains an active research area since the shared-memory seminal papers [3,29] and the message-passing paper [1]. For some recent papers and references therein see, e.g. [8,18].

Closer to our paper is [19] where the anonymous asynchronous MWMR fault-tolerant shared-memory model is considered. Our weak set object uses n MWMR registers and is non-blocking; it provides an enhanced atomic implementation of the weak set object supporting non-atomic operations presented in [12]. A wait-free implementation of a weak set object using $2n$ registers is in [14]. A set object that also supports a remove operation, but satisfies a weaker consistency condition, called per-element sequential consistency is presented in [4,5].

Organization. In Sect. 2 we briefly recall some of the notions used in this paper, about the model of computation and the topology tools, both of which are standard. In Sect. 3 we present the anonymous implementation of an atomic weak set object from MWMR registers. In Sect. 4 we present the safe agreement implementation. In Sect. 5 we derive our anonymous characterization of the t-resilient solvability of colorless tasks. Some proofs are omitted from this extended abstract for lack of space.

2   Preliminaries

We recall here briefly some standard notions of concurrent programming, for more precise definitions see [27]. We assume a standard *anonymous asynchronous shared-memory model* e.g., [19] consisting of n sequential processes that have no identifiers and execute an identical code. We assume that at most t of the processes may fail by crashing, where $1 \leq t < n$. Processes are asynchronous, i.e., they run at arbitrary speeds, independent from each other. We consider *linearizable* implementations where each operation appears to take effect instantaneously at some point between its invocation and response [28]. A *non-blocking* algorithm guarantees system-wide progress, while a *wait-free* also guarantees per-process progress. The processes communicate via multi-writer/multi-reader (MWMR) registers. Let $R[0 \ldots m - 1]$ denote an array of m registers. The read operation, denoted by READ(i), returns the state of $R[i]$. The write operation, denoted by WRITE(i, v), changes the state of $R[i]$ to v and returns *ack*. The registers are assumed to be atomic (linearizable). We sometimes refer to the processes by unique names p_0, ..., p_{n-1} for the convenience of exposition, but processes themselves have no means to access these names. Let $\Pi = \{p_0, \ldots, p_{n-1}\}$.

3   Atomic Weak Set

Here, we present an anonymous implementation of an atomic weak set object on an arbitrary value set V.

3.1   Specification and Algorithm

An *atomic weak set* object, denoted by SET, is an atomic object used for storing a set of values. The object supports only two operations, ADD() and GET(), and has no remove operation, which is why it is called "weak." The ADD(v) operation takes an argument $v \in V$ and returns ACK. The GET() operation, takes no argument and returns the set of values that have appeared as arguments in all the ADD() operations preceding the GET() operation. We assume that SET initially holds no values, i.e., it holds $\emptyset$.

We assume that a non-blocking n-component atomic *snapshot* object is available. An implementation in an anonymous setting with n registers is described in [15,19]. The snapshot object exports two operations, UPDATE() and SCAN(). Informally, an UPDATE(i, v) updates the i-th component of the object with the value v and a SCAN() returns an array of n values, which are contained in the n components at some point in time between the invocation and the response of the SCAN() operation.

We present an anonymous non-blocking implementation of the atomic weak set object on an n-component atomic snapshot object. The pseudocode of the implementation appears in Fig. 1. If $Snap$ is an array of n cells, we define $vals(Snap) = \cup_{i \in \{0,...,n-1\}} Snap[i]$. The idea of the algorithm is as follows. To execute an ADD(v) operation, the algorithm repeatedly tries to store the value v in each one of the n components of the snapshot object, using an *update* operation (line 5) until it detects that v appears in all the components. In each iteration, the algorithm deposits in the snapshot object not only v but $View$ containing all the values known to be in the set so far. Once v is detected to be in all components of the snapshot object, the ADD(v) terminates. The GET() operation is similar, except that now the $View$ of the process has to appear in all the components of the snapshot for the operation to terminate. Intuitively, once a value v (or a set of values) appears in all n components of the snapshot object, it cannot be overwritten and cannot go unnoticed by other processes. This is because the other processes can be covering (about to overwrite) at most $n - 1$ components.

Theorem 1. *The algorithm of Fig. 1 is an anonymous non-blocking implementation of an atomic weak set object using n MWMR registers.*

Here is a sketch of the proof.

Given an operation op, $invoc(op)$ denotes its invocation and $resp(op)$ its response. Let H be a history of the algorithm as defined in [28]. Let H' be the history H in which some of the operations that are invoked by a process that crashes during the operation and doesn't get a response are removed. H_{seq} denotes the sequential history in which each operation of H' appears as if it has been executed at a single point (the linearization) of the time line.

Safety. For the safety part, we have to define linearization points and prove that

- the linearization point of each operation GET() and ADD() appear between the beginning and the end of this operation;

– the sequential history that we get with these points respects the sequential specification of the weak set.

Consider a history H, let v be a value or a set of values. Define time τ_v as the first time, if any, that v belongs to all components of R. When there is no such time, τ_v is $\bot$.

Shared variable :
    n-component atomic snapshot object: R

CODE FOR A PROCESS
Local variable:
    array of value sets $Snap[0 \ldots n - 1]$
    set of Values $View$ init $\emptyset$
    integer $next$

Macro:
    $vals(Snap) = \cup_{i=0}^{n-1} Snap[i]$

ADD(v):
1    $next = 0$
2    $Snap = R.\text{SCAN}()$
3    $View = View \cup vals(Snap) \cup \{v\}$
4   **while** $(\#\{r | v \text{ in } Snap[r]\} < n)$
5        $R.\text{UPDATE}(next, View)$
6        $next = (next + 1) \bmod n$
7        $Snap = R.\text{SCAN}()$
8        $View = vals(Snap) \cup View$
9   **return** ACK

GET:
10    $next = 0$
11    $Snap = R.\text{SCAN}()$
12    $View = vals(Snap) \cup View$
13   **while** $(\#\{r | View = Snap[r]\} < n)$
14        $R.\text{UPDATE}(next, View)$
15        $next = (next + 1) \bmod n$
16        $Snap = R.\text{SCAN}()$
17        $View = vals(Snap) \cup View$
18   **return** $View$

Fig. 1. Non-blocking implementation of atomic weak set for n processes.

Lemma 1. *If the operation* ADD(v) *terminates, then v belongs to all components of R at some time instance before the end of this operation. If the operation* GET() *terminates and returns V, then V belongs to all components of R before the end of this operation.*

By Lemma 1, τ_v is not $\bot$ for each operation ADD(v) that terminates and τ_V is also not $\bot$ for each operation GET() that terminates and returns V.

It can be shown that the linearization points for operations ADD() and GET() are as follows

- $op = $ ADD(v): If $\tau_v \neq \bot$, the linearization point τ_{op} of an operation $op = $ ADD(v) is $max\{\tau_v, invoc(op)\}$. If $\tau_v = \bot$, the operation op does not terminate and is not linearized.
- $op = $ GET(): The linearization point τ_{op} of an operation $op = $ GET() that returns V is $max\{\tau_V, invoc(op)\}$. A GET() operation that does not terminate is not linearized.

The main safety claim is the following.

Lemma 2. H_{seq} *satisfies the sequential specification of the weak set.*

Liveness. We prove that the algorithm is non-blocking; namely, if processes perform operations forever, an infinite number of operations terminate. By contradiction, assume that there is only a finite number of operations GET() and ADD() that terminate and some operations made by correct processes do not terminate.

Operations ADD() or GET() may not terminate because the termination conditions of the while loop are not satisfied (Lines 4 or 13): for an ADD(v) operation, in each SCAN() made by the process, v is not in at least one of the components of R, and for a GET() operation, in each *snap*, all the components are not equal to the view of the process.

There is a time τ_0 after which there is no new process crash and all processes that terminate GET() or ADD() operations in the run have already terminated. Consider the set N of processes alive after time τ_0 that do not terminate operations in the run. Notice that after time τ_0 only processes in N take steps. Also, as no process in N may crash, each process in N takes an infinite number of steps.

Notice that all values in variables $View$ have been proposed by some ADD(). If there is a finite number of operations, then all variables $View$ are subsets of a finite set of values. The main idea of the liveness proof is to analyze *stable views*. That is, the sequence of views of each process is non-decreasing, each two consecutive views satisfy $view \subseteq view'$. Thus, there is a time $\tau_1 > \tau_0$ after which the view of each process p in N converges to a *stable view* $SView_p$: forever after time τ_1 the view of p is $SView_p$. Let $SV = \{SView_p | p \in N\}$, be the set of all stable views for processes in N. It can be shown that there is no minimal stable view, proving that $SV = \emptyset$ and also $N = \emptyset$, a contradiction.

4   Safe Agreement Object

A *safe agreement object* [6] on a set V provides two operations, *propose* and *resolve*. A propose operation, denoted by PROPOSE(v), takes an argument $v \in V$ and returns ACK. A resolve operation, denoted by RESOLVE(), takes no argument and returns $u \in V$ or $\bot$, where $\bot \notin V$. An execution is *well-formed* if each process invokes at most one propose operation and no process invokes a propose or resolve operation before its previous operation has terminated. In any well-formed execution, the object satisfies the following four conditions [2,7]:

Validity. Any non-$\perp$ value returned by a resolve operation is an argument of some propose operation;

Agreement. If two resolve operations return non-$\perp$ values v and v', then $v = v'$;

Termination. Every operation invoked by a non-faulty process eventually terminates;

Nontriviality. If no process fails while performing its propose operations, every resolve operation started after some time instance returns a non-$\perp$ value.

An anonymous wait-free implementation of a safe agreement object for an arbitrary value set V is presented in Fig. 2. The implementation makes use of an array of n weak set objects, denoted by $SET[0 \ldots n-1]$. To perform a propose operation, each process first assigns its input value to a local variable *view*. Then, the process repeats the following procedure for $i = 0, \ldots, n-1$: it adds *view* to $SET[i]$; if $SET[i]$ holds a set of cardinality more than one and *view* is the minimum value of the set, it returns ACK and immediately breaks the loop; otherwise, it assigns the minimum value of the set to *view*. To perform a resolve operation, each process checks the set held by $SET[n-1]$. If the set is not empty, the process returns the minimum value in the set. Otherwise, the process returns $\perp$.

Our implementation is a generalization of the anonymous consensus algorithm proposed by Attiya et al. [3]. Bouzid and Corentin [7] have proposed an anonymous implementation of a safe agreement object for the case of $V = \{0, 1\}$, also based on [3]. However, their implementation does not (directly) extend to the case of an infinite value set.

We now sketch the correctness proof of the algorithm of Fig. 2. Recall that, although we refer to the processes by unique names $p_0, \ldots, p_{n-1}$, processes do not know these names and that $\Pi = \{p_0, \ldots, p_{n-1}\}$.

Lemma 3. *Fix a well-formed execution of the algorithm of Fig. 2. Let V_i be the set of all the values that are added to $SET[i]$ in the execution. Then, $V_i \supseteq V_{i+1}$ holds for all $i = 0, \ldots, n-2$.*

Lemma 4 (Validity). *The algorithm of Fig. 2 satisfies the validity condition.*

Lemma 5. *Fix a well-formed execution of the algorithm of Fig. 2. Let V_i be the set of the all values that are added to $SET[i]$ in the execution. Let us define*

$$\Pi_i = \{p \in \Pi \mid p \text{ performs } \text{PROPOSE}() \text{ and adds some } v \in V_i \setminus \{\min V_i\} \text{ to } SET[i]\}.$$

Then, $\Pi_i \supseteq \Pi_{i+1}$ holds for all $i = 0, \ldots, n-2$.

Lemma 6 (Agreement). *The algorithm of Fig. 2 satisfies the agreement condition.*

Shared variable :
   array of atomic weak set objects : $SET[0 \ldots n-1]$

CODE FOR A PROCESS

Local variable:
   Value $view$ init $\perp$
   Integer i init 0
   set of Values $Snap$ init $\emptyset$

operation PROPOSE(v):
```
1   view = v
2   for i = 0, ..., n − 1 do
3       SET[i].ADD(view)
4       Snap = SET[i].GET()
5       if #Snap ≥ 2 && view = min(Snap) then
6           return ACK
7       else
8           view = min(Snap)
9   return ACK
```

operation RESOLVE():
```
10  Snap = SET[n − 1].GET()
11  if Snap ≠ ∅ then
12      return min(Snap)
13  else
14      return ⊥
```

Fig. 2. Anonymous implementation of safe agreement object

Lemma 7 (Termination). *The algorithm of Fig. 2 satisfies the termination condition.*

Lemma 8 (Nontriviality). *The algorithm of Fig. 2 satisfies the nontriviality condition.*

By Lemmas 4, 6, 7, and 8, we obtain the following theorem. Furthermore, notice that the algorithm uses n atomic registers, because an arbitrary finite number of atomic weak set objects can be simulated on top of a single atomic weak set object.

Theorem 2. *The algorithm of Fig. 2 is an anonymous wait-free implementation of safe agreement object, using n atomic registers.*

5 t-Resilient Solvable Colorless Tasks

We give a characterization of t-resilient solvable colorless tasks (see formal definitions below) in the anonymous shared-memory model.

Theorem 3. *A colorless task is t-resilient solvable in the anonymous shared-memory model if and only if it is t-resilient solvable in the non-anonymous one. Moreover, if a colorless task is t-resilient solvable by n anonymous processes, it can be solved by n shared atomic registers.*

The only if part of the theorem is immediate because every anonymous protocol can be executed by non-anonymous processes. We next describe the if part by two different approaches, a topological one and an operational one.

5.1 Topological Approach

We briefly recall some notions of combinatorial topology for distributed computing, additional details can be found in [20].

Let I and O be complexes. A *carrier map* from $\mathcal{I}$ to $\mathcal{O}$ is a mapping $\Delta : \mathcal{I} \to 2^{\mathcal{O}}$ such that, for each $s \in \mathcal{I}$, $\Delta(s)$ is a subcomplex of $\mathcal{O}$ and $s' \subseteq s$ implies $\Delta(s') \subseteq \Delta(s)$. If a continuous map $f : |\mathcal{I}| \to |\mathcal{O}|$ satisfies $f(|\sigma|) \subseteq |\Delta(\sigma)|$ for all $\sigma \in \mathcal{I}$, we say that f is carried by Δ. If a simplicial map $\delta : \mathrm{bary}^b \mathcal{I} \to \mathcal{O}$ satisfies $\delta(\mathrm{bary}^b \sigma) \subseteq \Delta(\sigma)$ for all $\sigma \in \mathcal{I}$, we say that δ is *carried by* Δ. As an immediate consequence of Lemma 3.7.8. of [20], the following lemma holds.

Lemma 9. *If $\Delta : \mathcal{I} \to 2^{\mathcal{O}}$ is a carrier map and $f : |\mathcal{I}| \to |\mathcal{O}|$ is a continuous map carried by Δ, then there is a non-negative integer b and a simplicial map $\delta : \mathrm{bary}^b \mathcal{I} \to \mathcal{O}$ carried by Δ.*

A *colorless task* is a triple $T = (\mathcal{I}, \mathcal{O}, \Delta)$, where $\mathcal{I}$ and $\mathcal{O}$ are simplicial complexes and Δ is a carrier map. A colorless task T is solvable, if for each input simplex $s \in \mathcal{I}$, whenever each process p_i starts with input value $v_i \in s$ (different processes may start with the same value), eventually it decides an output value v_i', such that the set of output values form a simplex $s' \in \Delta(s)$. The colorless tasks that are fundamental to the present paper are *b-iterated barycentric agreement* and *k-set agreement*. The b-iterated barycentric agreement task is a colorless task $T = (\mathcal{I}, \mathrm{bary}^b \mathcal{I}, \mathrm{bary}^b)$, where we write by bary^b the carrier map that maps $s \in \mathcal{I}$ to $\mathrm{bary}^b s$ for an abuse of notation. The k-set agreement task is a colorless task $T_k = (\mathcal{I}, \mathrm{skel}^k \mathcal{I}, \mathrm{skel}^k)$, where skel^k denotes the carrier map that maps a simplex $s \in \mathcal{I}$ to the subcomplex $\mathrm{skel}^k \mathcal{I}$.

To prove the if part of Theorem 3, we first show that the $(t+1)$-set agreement task is t-resilient solvable by n anonymous processes. An algorithm of Fig. 3 presents an anonymous t-resilient protocol for the $(t + 1)$-set agreement task. In the protocol, input value to $SA[i]$ for $i = 0, \ldots, t$. Then, the process repeatedly performs a RESOLVE() operation to all $SA[i]$ in the round-robin manner until it gets non-$\bot$ value. Once the process gets non-$\bot$ value, the process returns the value.

Theorem 4. *The algorithm of Fig. 3 is a t-resilient anonymous protocol for the $(t + 1)$-set agreement task.*

Proof. Termination: It suffices to show that the while loop of Line 4–6 eventually terminates. In the protocol, each process performs PROPOSE() operations to

---

Shared variable :
   array of safe agreement objects : $SA[0 \ldots t]$

CODE FOR A PROCESS

Local variable:
   Integer i init 0
   Value $result$ init $\perp$

SETAGREE(v):
1   **for** $i = 0, \ldots, t$ **do**
2        $SA[i]$.PROPOSE(v)
3    $i = 0$
4   **while** $result = \perp$ **do**
5        $result = SA[i]$.RESOLVE()
6        $i = i + 1 \mod t + 1$
7   **return** $result$

Fig. 3. Anonymous t-resilient $(t+1)$-set agreement protocol

$SA[0], \ldots, SA[t]$ sequentially. Thus, even if t processes fail, there is at least one safe agreement object such that no process fails while performing a PROPOSE() operation on the object. By the nontriviality property of safe agreement objects, after some time instance, RESOLVE() operations on some safe agreement object return non-$\perp$ value and thus the while loop eventually terminates.

Validity: Every argument of a PROPOSE() operation is a proposed value. Because of the validity property of safe agreement objects, a non-$\perp$ value returned by some RESOLVE() operation is one of the arguments of PROPOSE() operations. Thus, the validity condition holds.

k-Agreement: There are $t+1$ distinct safe agreement objects. Thus, by the agreement property of safe agreement objects, at most $t+1$ distinct values are decided.

As the b-iterated barycentric agreement task is wait-free solvable by anonymous processes [31], the following lemma holds.

Lemma 10. *Let $T = (\mathcal{I}, \mathcal{O}, \Delta)$ be a colorless task. If there exists a continuous map $f : |skel^t \mathcal{I}| \to |\mathcal{O}|$ carried by Δ, T is t-resilient solvable by n anonymous processes.*

Proof. By Lemma 9, there is an integer b and a simplicial map $\delta : \text{bary}^b skel^t \mathcal{I} \to \mathcal{O}$ that satisfies $\delta(\text{bary}^b \sigma) \subseteq \Delta(\sigma)$ for every $\sigma \in skel^t \mathcal{I}$.

The following anonymous protocol solves the colorless task. Suppose that the set of all inputs to the processes is $s \in \mathcal{I}$. Execute first the anonymous $(t+1)$-set agreement protocol so that the processes all choose vertices that form a simplex sigma in $skel^t \mathcal{I}$, and then the b-iterated barycentric agreement protocol (for a sufficiently large value of b). Each process chooses a vertex of a common simplex of $\text{bary}^b skel^t \sigma$. Finally, each process determines its output by applying δ to the vertex it chose.

Shared variable :
    atomic weak set : SET
    array of safe agreement objects : $SA[0\ldots][0\ldots n-1]$

CODE FOR A PROCESS

Local variable:
    Value $view_i$ init $\perp$ for $i = 0,\ldots,n-1$
    Integer $round_i$ init 0 for $i = 0,\ldots,n-1$
    Integer i init 0
    Value $snap$ init $\perp$

Simulation(v):
```
1    for i = 0,...,n − 1 do
2        SA[0][i].PROPOSE(v)
3    while true do
4        for i = 0,...,n − 1 do
5            view_i = SA[round_i][i].RESOLVE()
6            if view_i is a termination state of P_i then
7                return f(view_i)
8            elseif view_i ≠⊥ then
9                SET.ADD((P_i, view_i, round_i))
10               snap = SET.GET()
11               view_i = latest_views(snap)
12               round_i = latest_round_i(snap) + 1
13               SA[round_i][i].PROPOSE(view_i)
```

Fig. 4. n anonymous processes simulates n non-anonymous processes

The if part of Theorem 3 follows from Lemma 10 and the following theorem by Herlihy and Rajsbaum:

Theorem 5 ([23, Theorem 4.3]). *A colorless task $T = (I, O, \Delta)$ is t-resilient solvable by n non-anonymous processes if and only if there exists a continuous map $f : |skel^t I| \rightarrow |O|$ carried by Δ.*

Note that the protocol in the proof of Lemma 10 only makes use of a finite number of atomic weak set objects, which are constructed on top of a single atomic weak set object. Thus, every colorless task that is t-resilient solvable by n anonymous processes is solved with n atomic registers. The space complexity upper bound of Theorem 3 follows.

5.2   Simulation-Based Approach

We now prove the if part of Theorem 3 by a simulation, which is an anonymous variant of the BG-simulation [6]. More precisely, we show that n anonymous t-resilient processes with atomic weak set objects can simulate n non-anonymous t-resilient processes with *atomic snapshot objects*. We denote the anonymous

simulators by $p_0, \ldots, p_{n-1}$ and the non-anonymous simulated processes by $P_0,$
$\ldots, P_{n-1}$. Without loss of generality, we may assume that non-anonymous processes communicate via a single n-ary atomic snapshot object and execute a
full-information protocol. In the protocol, the process P_i repeatedly writes its
local state to the i-th component of the array, takes a snapshot of the whole
array and updates its state by the result of the snapshot until it reaches a termination state. When the process reaches the termination state, it decides on
the value obtained by applying some predefined function f to the state.

Our simulation algorithm for each simulator is presented in Fig. 4. The
algorithm makes use of a two dimensional array of safe agreement objects
$SA[0 \ldots][0 \ldots n-1]$, where the column $SA[0 \ldots][i]$ is for storing simulated states
of the process P_i. The local variables $view_i$ and $round_i$ stand for the current
simulated state and the current simulated round of P_i respectively. The function *latest_views* maps a set of tuples consisting of a process name, its simulated
state, and its simulated round to the array whose i-th component is the simulated view of P_i associated with the largest simulated round number of P_i. The
function *latest_round_i* maps a set of the same kind to the latest round number
of P_i.

In the algorithm, each simulator first proposes its input value to $SA[0][i]$
for all $P_0, \ldots, P_{n-1}$. Then, a simulator repeats the following procedure for
$P_0, \ldots, P_{n-1}$ in the round-robin manner until one of $P_0, \ldots, P_{n-1}$ reach a termination state: it performs RESOLVE() operation on $SA[round_i][i]$; if the return
value of the RESOLVE() operation is not $\perp$, the simulator adds the return value,
with the name P_i and its current simulated round, to SET, updates simulated
state and round, and proposes the new simulated state of P_i to $SA[round_i][i]$.

By the use of safe agreement objects, simulators can agree on the return value
of each simulated snapshot. Note that there is no need to use a safe agreement
object on each simulated update because each value to be updated is deterministically determined by the return value of the preceding simulated snapshot. In
the algorithm of Fig. 4, each simulator performs PROPOSE() operations sequentially. Thus, even though t simulators crash, they block at most t simulated
processes by the nontriviality property of the safe agreement object. By these
observations, we establish the following lemma:

Lemma 11. *If a colorless task is t-resilient solvable by n non-anonymous processes with atomic snapshot objects, it is also t-resilient solvable by n anonymous
processes with atomic weak set objects.*

The proof of the lemma is similar to the proof of Theorem 5 in [6], while we
omit the proof.

The space complexity of the simulation of Fig. 4 is exactly n atomic registers
because a single atomic weak set object can simulate, in the non-blocking manner, an arbitrary finite number of atomic weak set objects and safe agreement
objects. This establishes the space complexity upper bound of Theorem 3.

6 Conclusion

In this paper, we have extended the wait-free colorless task solvability of [31] to the case where at most t processes may crash, where $1 \leq t < n$. Furthermore, we have shown that any t-resilient solvable colorless task can be t-resilient solvable anonymously using only n MWMR registers. We have derived our result through a series of reductions that seem interesting in themselves, to study anonymous computability. We hope they are useful to study further long-lived objects (as opposed to tasks), perhaps using a wait-free implementation of the weak set object [14], and uniform solvability (instead of a fixed number of processes n). Also, it would be interesting to look for lower bounds on the number of MWMR registers needed to solve specific colorless tasks.

References

1. Angluin, D.: Local and global properties in networks of processors. In: 12th Annual ACM Symposium on Theory of Computing (STOC), pp. 82–93 (1980)
2. Attiya, H.: Adapting to point contention with long-lived safe agreement. In: Flocchini, P., Gąsieniec, L. (eds.) SIROCCO 2006. LNCS, vol. 4056, pp. 10–23. Springer, Heidelberg (2006). https://doi.org/10.1007/11780823_2
3. Attiya, H., Gorbach, A., Moran, S.: Computing in totally anonymous asynchronous shared memory systems. Inf. Comput. **173**(2), 162–183 (2002)
4. Baldoni, R., Bonomi, S., Raynal, M.: Value-based sequential consistency for set objects in dynamic distributed systems. In: D'Ambra, P., Guarracino, M., Talia, D. (eds.) Euro-Par 2010. LNCS, vol. 6271, pp. 523–534. Springer, Heidelberg (2010). https://doi.org/10.1007/978-3-642-15277-1_50
5. Baldoni, R., Bonomi, S., Raynal, M.: Implementing set objects in dynamic distributed systems. J. Comput. Syst. Sci. **82**(5), 654–689 (2016)
6. Borowsky, E., Gafni, E., Lynch, N., Rajsbaum, S.: The BG distributed simulation algorithm. Distrib. Comput. **14**(3), 127–146 (2001)
7. Bouzid, Z., Travers, C.: Anonymity-preserving failure detectors. In: Gavoille, C., Ilcinkas, D. (eds.) DISC 2016. LNCS, vol. 9888, pp. 173–186. Springer, Heidelberg (2016). https://doi.org/10.1007/978-3-662-53426-7_13
8. Capdevielle, C., Johnen, C., Kuznetsov, P., Milani, A.: On the uncontended complexity of anonymous agreement. Distrib. Comput. **30**(6), 459–468 (2017)
9. Castañeda, A., Rajsbaum, S., Raynal, M.: The renaming problem in shared memory systems: an introduction. Comput. Sci. Rev. **5**(3), 229–251 (2011)
10. Castañeda, A., Imbs, D., Rajsbaum, S., Raynal, M.: Generalized symmetry breaking tasks and nondeterminism in concurrent objects. SIAM J. Comput. **45**(2), 379–414 (2016)
11. Chaudhuri, S.: More choices allow more faults: set consensus problems in totally asynchronous systems. Inf. Comput. **105**(1), 132–158 (1993)
12. Delporte-Gallet, C., Fauconnier, H.: Two consensus algorithms with atomic registers and failure detector Ω. In: Garg, V., Wattenhofer, R., Kothapalli, K. (eds.) ICDCN 2009. LNCS, vol. 5408, pp. 251–262. Springer, Heidelberg (2008). https://doi.org/10.1007/978-3-540-92295-7_31
13. Delporte-Gallet, C., Fauconnier, H., Gafni, E., Rajsbaum, S.: Linear space bootstrap communication schemes. Theor. Comput. Sci. **561**(Pt. B), 122–133 (2015). Special Issue on Distributed Computing and Networking

14. Delporte-Gallet, C., Fauconnier, H., Rajsbaum, S., Yanagisawa, N.: An anonymous wait-free weak-set object implementation. In: 6th International Conference on Networked Systems (NETYS). LNCS (2018, to appear)

15. Ellen, F., Fatourou, P., Ruppert, E.: The space complexity of unbounded timestamps. Distrib. Comput. **21**(2), 103–115 (2008)

16. Fischer, M.J., Lynch, N.A., Paterson, M.S.: Impossibility of distributed consensus with one faulty process. J. ACM **32**(2), 374–382 (1985)

17. Gafni, E., Koutsoupias, E.: Three-processor tasks are undecidable. SIAM J. Comput. **28**(3), 970–983 (1999)

18. Gelashvili, R.: On the optimal space complexity of consensus for anonymous processes. In: Moses, Y. (ed.) DISC 2015. LNCS, vol. 9363, pp. 452–466. Springer, Heidelberg (2015). https://doi.org/10.1007/978-3-662-48653-5_30

19. Guerraoui, R., Ruppert, E.: Anonymous and fault-tolerant shared-memory computing. Distrib. Comput. **20**(3), 165–177 (2007)

20. Herlihy, M., Kozlov, D., Rajsbaum, S.: Distributed Computing Through Combinatorial Topology. Morgan Kaufmann, San Francisco (2013)

21. Herlihy, M., Rajsbaum, S.: The decidability of distributed decision tasks (extended abstract). In: 29th Annual ACM Symposium on Theory of Computing (STOC), pp. 589–598 (1997)

22. Herlihy, M., Rajsbaum, S.: A classification of wait-free loop agreement tasks. Theor. Comput. Sci. **291**(1), 55–77 (2003)

23. Herlihy, M., Rajsbaum, S.: The topology of shared-memory adversaries. In: 29th ACM Symposium on Principles of Distributed Computing (PODC), pp. 105–113 (2010)

24. Herlihy, M., Rajsbaum, S.: Simulations and reductions for colorless tasks. In: 31st ACM Symposium on Principles of Distributed Computing, PODC 2012, pp. 253–260. ACM, New York (2012)

25. Herlihy, M., Rajsbaum, S., Raynal, M., Stainer, J.: From wait-free to arbitrary concurrent solo executions in colorless distributed computing. Theor. Comput. Sci. **683**, 1–21 (2017)

26. Herlihy, M., Shavit, N.: The topological structure of asynchronous computability. J. ACM **46**(6), 858–923 (1999)

27. Herlihy, M., Shavit, N.: The Art of Multiprocessor Programming. Morgan Kaufmann, San Francisco (2008)

28. Herlihy, M.P., Wing, J.M.: Linearizability: a correctness condition for concurrent objects. ACM Trans. Program. Lang. Syst. **12**(3), 463–492 (1990)

29. Jayanti, P., Toueg, S.: Wakeup under read/write atomicity. In: 4th International Workshop on Distributed Algorithms, pp. 277–288 (1991)

30. Rajsbaum, S., Castañeda, A., Flores-Peñaloza, D., Alcantara, M.: Fault-tolerant robot gathering problems on graphs with arbitrary appearing times. In: 31st IEEE International Parallel and Distributed Processing Symposium (IPDPS), pp. 493–502, May 2017

31. Yanagisawa, N.: Wait-free solvability of colorless tasks in anonymous shared-memory model. In: Theory of Computing Systems, pp. 1–18 (2017)

Deterministic Distributed Ruling Sets of Line Graphs

Fabian Kuhn$^{(\boxtimes)}$, Yannic Maus$^{(\boxtimes)}$, and Simon Weidner$^{(\boxtimes)}$

Department of Computer Science, University of Freiburg, 79110 Freiburg, Germany
{kuhn,yannic.maus,simon.weidner}@cs.uni-freiburg.de

Abstract. An (α, β)-ruling set of a graph $G = (V, E)$ is a set $R \subseteq V$ such that for any node $v \in V$ there is a node $u \in R$ in distance at most β from v and such that any two nodes in R are at distance at least α from each other. The concept of ruling sets can naturally be extended to edges, i.e., a subset $F \subseteq E$ is an (α, β)-*ruling edge set* of a graph $G = (V, E)$ if the corresponding nodes form an (α, β)-ruling set in the line graph of G. This paper presents a simple deterministic, distributed algorithm, in the CONGEST model, for computing $(2, 2)$-ruling edge sets in $O(\log^* n)$ rounds. Furthermore, we extend the algorithm to compute ruling sets of graphs with bounded *diversity*. Roughly speaking, the diversity of a graph is the maximum number of maximal cliques a vertex belongs to. We devise $(2, O(\mathcal{D}))$-ruling sets on graphs with diversity $\mathcal{D}$ in $O(\mathcal{D} + \log^* n)$ rounds. This also implies a fast, deterministic $(2, O(\ell))$-ruling edge set algorithm for hypergraphs with rank at most ℓ.

Furthermore, we provide a ruling set algorithm for general graphs that for any $B \geq 2$ computes an $(\alpha, \alpha \lceil \log_B n \rceil)$-ruling set in $O(\alpha \cdot B \cdot \log_B n)$ rounds in the CONGEST model. The algorithm can be modified to compute a $(2, \beta)$-ruling set in $O(\beta \Delta^{2/\beta} + \log^* n)$ rounds in the CONGEST model, which matches the currently best known such algorithm in the more general LOCAL model.

Keywords: Ruling set · Ruling edge set · Congest · Bounded diversity

1 Introduction, Motivation and Related Work

This paper presents fast and simple deterministic distributed algorithms, in the CONGEST model, for computing ruling sets of graphs, line graphs, line graphs of hypergraphs, and graphs of bounded diversity as introduced in [3].

The CONGEST *Model of Distributed Computing* [30]. The graph is abstracted as an n-node network $G = (V, E)$ with maximum degree at most Δ. Each node is assumed to have a unique $O(\log n)$-bit ID. Communication happens in synchronous rounds. Per round, each node can send one message of at most $O(\log n)$ bits to each of its neighbors and perform (unbounded) local computations[1].

Supported by ERC Grant No. 336495 (ACDC).

[1] All our algorithms only use local computations that require at most polynomial time.

© Springer Nature Switzerland AG 2018
Z. Lotker and B. Patt-Shamir (Eds.): SIROCCO 2018, LNCS 11085, pp. 193–208, 2018.
https://doi.org/10.1007/978-3-030-01325-7_19

At the end, each node should know its own part of the output, e.g., whether it belongs to the ruling set or not. The time complexity of an algorithm is the number of rounds it requires to terminate.

Ruling Sets. A (α, β)-*ruling set* of a graph $G = (V, E)$ is a subset $R \subseteq V$ of the nodes such that any two nodes in R are at distance at least α in G and for every node $v \in V \setminus R$, there is a node in R within distance β [2]. That is, R is an independent set in $G^{\alpha-1}$, where G^r denotes the graph with node set V and where two nodes u, v are connected by an edge if $d_G(u, v) \leq r$. Typically, α is called the *independence parameter* and r the *domination parameter* of the ruling set R. If $\alpha = 2$, one often also simply calls R a β-*ruling set*. The concept of ruling sets can naturally be extended to edges, i.e., a subset $F \subseteq E$ is an (α, β)-*ruling edge set* of a graph $G = (V, E)$ (or a hypergraph $H = (V, E)$) if the corresponding nodes form an (α, β)-ruling set in the line graph of G (or H). In the present paper, we concentrate on deterministic algorithms for computing ruling sets in the CONGEST model. We will specifically see that edge ruling sets of graphs and low-rank hypergraphs can be computed particularly efficiently.

The Relevance of Ruling Sets. The distributed computation of ruling sets is a simple and clean symmetry breaking problem. In particular, ruling sets are a generalization of maximal independent sets (MIS), arguably one of the most central and best studied distributed symmetry breaking problems. A $(2, 1)$-ruling set is an MIS of G and more generally, a $(r + 1, r)$-ruling set is an MIS of G^r. For $\beta \geq 1$, a $(2, \beta)$-ruling set of G is therefore a strict relaxation of an MIS of G, where the problem becomes weaker with larger values of β. The parameter β thus allows to study a trade-off between the strength of the symmetry breaking requirement and the complexity of computing a ruling set.

Ruling sets have been introduced by Awerbuch, Goldberg, Luby, and Plotkin as a building block to efficiently construct a so-called network decomposition (a partition of a graph into clusters of small diameter together with a coloring of the cluster graph with a small number of colors) [2]. Since then, ruling sets have been used as a powerful tool in various distributed graph algorithms. Computing ruling sets can often replace computing the more stringent and harder to obtain maximal independent sets. Ruling sets have for example been used in order to compute network decompositions [2,29], graph colorings [28], maximal independent sets [21], or shortest paths [16]. Ruling sets are also used as a subroutine to obtain the state-of-the-art randomized distributed algorithms for many of the classic distributed graph problems, such as distributed coloring [7,15], maximal independent set [11], or maximal matching [4]. These algorithms are based on the so-called graph shattering technique, which was originally introduced by Beck in [5]. Using an efficient randomized algorithm, the problem is solved on most of the graph such that the only unsolved remaining parts are components of size at most $\text{poly}(\Delta \cdot \log n)$, where Δ is the maximum degree of G. Using existing ruling set algorithms, one can then further reduce the problems on the remaining components to problems on graphs of size $\text{poly}\log n$.

Previous Work on Distributed Ruling Set Algorithms. As mentioned before the first appearance of ruling sets was in the work of Awerbuch, Goldberg, Luby, and Plotkin [2], who provided a deterministic distributed algorithm to compute an $(\alpha, O(\alpha \log n))$-ruling set in $O(\alpha \log n)$ rounds. Their algorithm uses the bit representation of the node IDs to recursively compute ruling sets. For each of the $O(\log n)$ bits of the IDs, the nodes are divided into two parts according to the value of the current bit and ruling sets are computed recursively for the two parts. The two recursively computed ruling sets S_0 and S_1 are merged to a single ruling set by keeping all nodes in S_0 and all nodes in S_1 that do not have an S_0-node in their $(\alpha - 1)$-neighborhood. Note that the algorithm loses an additive α in the domination for each of the $O(\log n)$ recursion levels.

Schneider and Wattenhofer refine the ideas of [2] to deterministically compute $(2, \beta)$-ruling sets in time $O(\beta \Delta^{2/\beta} + \log^* n)$ [31]. At the cost of an increased running time, they use a larger branching factor than Awerbuch et al. to decrease the recursive depth and thus the domination parameter. Further, for small values of β, the best known deterministic algorithm requires time $2^{O(\sqrt{\log n})}$, even for $\beta = 1$. It is based on first computing a network decomposition using the algorithm of [29] and to then use this decomposition to compute the ruling set. All these algorithms work in the LOCAL model, where the size of messages is unbounded. In [16], Henzinger, Krinninger, Nanongkai sketch how the algorithm of [2] can be adapted to compute a $(\alpha, O(\alpha \log n))$-ruling set in $O(\alpha \log n)$ rounds in the CONGEST model. Any $(2, \beta)$-ruling set algorithm applied to $G^{\alpha-1}$ implies a $(\alpha, (\alpha - 1)\beta)$-ruling set algorithm on G, e.g., [31] can be used to compute $(\alpha, O(\alpha^2 \cdot \beta))$-ruling sets in time $O(\alpha^2 \beta \Delta^{2/\beta} + \alpha \cdot \log^* n)$. However, the black box simulation of an algorithm on $G^{\alpha-1}$ heavily relies on the LOCAL model.

In contrast to the few deterministic ruling set algorithms there are many randomized algorithms for the problem. In particular, there is a long history of efficient randomized algorithms for computing an MIS. The famous algorithms by Luby and Alon, Babi, and Itai allow to compute an MIS (and thus a $(2, 1)$-ruling set) in time $O(\log n)$ [1,25]. In [11], Ghaffari improved the randomized running time of computing an MIS to $O(\log \Delta) + 2^{O(\sqrt{\log \log n})}$. There are also more efficient randomized algorithms that directly target the computation of $(2, \beta)$-ruling sets for $\beta > 1$. Gfeller and Vicari found an algorithm that finds a $(1, O(\log \log \Delta))$-ruling set in time $O(\log \log \Delta)$ such that the degree in the graph induced by the ruling set nodes is $O(\log^5 n)$ [10]. Together with the ruling set algorithms of [2,31], the algorithm allows to compute a $(2, O(\log \log n))$-ruling set in time $O(\log \log n)$. In [4], Barenboim et al. used the graph shattering technique to compute $(2, \beta)$-ruling set in time $O(\beta \log^{1/(\beta-1/2)}) + 2^{O(\sqrt{\log \log n})}$. This was later improved by Ghaffari to compute $(2, \beta)$-ruling sets in $O(\beta \log^{1/\beta}) + O(2^{\sqrt{\log \log n}})$ rounds [11]. Kothapalli and Pemmaraju showed how to compute $(2, 2)$-ruling sets in $O(\log^{3/4} n)$ rounds [19]. The core idea is a randomized sparsification process that reduces the degree while maintaining some domination property. Afterwards the algorithm of [4] is applied to the sparsified graph. The same authors presented a randomized algorithm that computes $(2, \beta)$-ruling sets in time $O(\beta \log^{1/(\beta-1)} n)$ if

$\beta \le \sqrt{\log \log n}$ and in time $O(\sqrt{\log \log n})$ for arbitrary β [6]. Pai et al. showed how to compute 3-ruling sets in $O(\log n / \log \log n)$ rounds and 2-ruling sets in $O(\log(\Delta) \cdot (\log n)^{1/2+\epsilon} + \log n \ \epsilon \log \log n)$ rounds in the CONGEST model [26]. Further, the work deals with the *message complexity* of ruling set algorithms. They provide a $\Omega(n^2)$ lower bound on the message complexity for MIS if nodes have no knowledge about their neighbors and in contrast present a 2-ruling set algorithm that uses $O(n \log^2 n)$ messages and runs in $O(\Delta \log n)$ rounds.

We are not aware of any work that explicitly studies ruling edge sets. However, there is substantial work on computing maximal matchings and maximal hypergraph matchings, i.e., on computing $(2,1)$-ruling edge sets. While for the MIS problem no polylogarithmic-time deterministic algorithm is known, in [13], Hańćkowiak, Karoński and Panconesi showed that a maximal matching can be computed in $O(\log^7 n)$ rounds deterministically. They improved the algorithm to $O(\log^4 n)$ rounds in [14]. The current best algorithm is by Fischer and it computes a maximal matching in time $O(\log^2 \Delta \cdot \log n)$ [8]. Fischer, Ghaffari and Kuhn have recently shown that maximal matchings can even be computed efficiently in low-rank hypergraphs. For hypergraphs of rank at most r (i.e., every hyperedge consists of at most r nodes), they presented a deterministic algorithm to compute hypergraph maximal matching in $O(\log^{O(\log r)} \Delta \cdot \log n)$ [9]. Later in [12], the dependency on the rank was improved; the paper obtains a runtime of $\Delta^{O(r)} + O(r \log^* n)$ to compute a hypergraph maximal matching. Furthermore, [4] contains a randomized algorithm that (combined with [8]) computes a maximal matching in $O(\log \Delta + \log^3 \log n)$ rounds.

Finally, we note that the $\Omega(\log^* n)$ lower bound of [24] that was designed for coloring and MIS on a ring network also holds for computing ruling sets. On a ring, given a β-ruling (edge) set, an MIS can be computed in time $O(\beta)$. Maximal matchings have a lower bound of $O(\sqrt{\log n / \log \log n})$ rounds [20].

Contributions. The ruling set algorithms for general graphs by Awerbuch et al. [2] and Schneider et al. [31] only works in the LOCAL model. In [16] Henzinger, Krinninger, and Nanongkai sketch a variant of the algorithm that achieves a ruling set of the same quality as Awerbuch et al. but that also works in the CONGEST model. In the full version of this paper [22], we provide a formal analysis and a generalization of the algorithm of [16]. Further, slightly beyond [16], our simple deterministic distributed algorithm also levels Schneider et al.'s work in the CONGEST model.

Theorem 1. *Let α be a positive integer. For any $B \ge 2$ there exists a deterministic distributed algorithm that computes a $(\alpha, (\alpha - 1)\lceil \log_B n \rceil)$-ruling set of G in $O(\alpha \cdot B \cdot \log_B n)$ rounds in the CONGEST model.*

The n in the runtime of Theorem 1 stems from the size of the ID-space. To compute a $(2, \beta)$-ruling set it is sufficient to use the colors of a $c\Delta^2$-coloring computed with Linial's algorithm [23] as IDs; setting $B = c \cdot \Delta^{2/\beta}$ implies the same trade-off as in [31].

Corollary 1. *Let $\beta > 2$ be an integer. There exists a deterministic distributed algorithm that computes a $(2, \beta)$-ruling set of G in $O(\beta \Delta^{2/\beta} + \log^* n)$ rounds in the* CONGEST *model.*

Algorithm Sketch (consult [22] for details): The simple algorithm of Theorem 1 begins with a tentative ruling set $S = V$ and iteratively sparsifies S until it is an independent set. In iteration i of $O(\log_B n)$ iterations it removes nodes from S such that S is independent with regard to the i'th digit of the B-ary representation of the ids—the set S is called *independent with regard to digit i* if both endpoints of each edge of $G[S]$ have the same value at the i'th digit. Then, if all bits are independent S is an independent set; the node removal in each iteration is such that the domination increases by at most two in each iteration.

All further contributions center around the fast computation of ruling sets in line graphs and their generalizations. The main result is the computation of 2-ruling edge sets in $O(\log^* n)$ rounds.

Theorem 2. *There exists a deterministic distributed algorithm that computes a 2-ruling edge set of G in $\Theta(\log^* n)$ rounds in the* CONGEST *model.*

The main idea of the algorithm can actually be explained in a few lines: In the first step each node sends a proposal along one of its incident edges; in the second step each node that received a proposal adds exactly one of the edges through which it received a proposal to a set F; in the third step nodes compute a matching on the graph induced by the edges in F and add matching edges to the ruling edge set. One can show that the graph that is induced by the edges in F has maximum degree at most two and thus the computation of the matching only takes $O(\log^* n)$ rounds. The resulting ruling edge set is a 3-ruling edge set and we use our following result to transform it into a 2-ruling edge set.

The *proposal technique* is similar the one of Israeli et al. in [17] for the randomized computation of maximal matchings. For multiple phases they first reduce the maximal degree of the graph using a randomized version of the proposal algorithm. Then they randomly add certain edges from the reduced graph to the matching and remove their adjacent edges. They show that the algorithm removes a constant fraction of the edges in each phase and thus they obtain a maximal matching in $O(\log n)$ rounds.

Theorem 3. *Let $\beta \geq 2$. Any β-ruling edge set can be transformed into a 2-ruling edge set in $O(\beta)$ rounds of communication in the* CONGEST *model.*

We emphasize that a further reduction of the domination parameter, i.e., to 1-ruling edge sets or equivalently to maximal matchings, cannot be done in less than $O(\sqrt{\log n / \log \log n})$ rounds due to the lower bound of [20].

The algorithm from Theorem 2 can be seen as a 2-ruling set algorithm on line graphs and it is significantly faster than any known algorithm to compute 2-ruling sets on general graphs. Our third contribution extends the ideas of the algorithm for line graphs to a much larger class of graphs, i.e., graphs with bounded diversity. For a graph G, a *clique edge cover* Q is a collection of cliques of G such that every edge (and node) of G is contained in at least one of the cliques.

The *diversity* of a pair (G, Q) where Q is a clique edge cover of G is $\mathcal{D}$ if any node is contained in at most $\mathcal{D}$ distinct cliques of the clique edge cover. The diversity of a graph G is the minimum diversity of all (G, Q) where Q is an arbitrary clique edge cover of G. The concept of diversity was introduced in [3]. In the following we always assume that the clique edge cover Q is known by all nodes, i.e., each node knows all the cliques in which it is contained. Note that for many graphs, e.g., for line graphs of graphs or hypergraphs of small rank a clique edge cover with a small diversity is obtained in a single round of communication by taking a clique for each node v consisting of the set of all the edges containing v. Note that in [3], the definition of diversity is defined by using maximal cliques. However, we do not require maximality in our algorithms.

Theorem 4. *There exists a algorithm that, given a graph with a clique edge cover of diversity $\mathcal{D}$, computes a $(\mathcal{D}+4)$-ruling set in $O(\mathcal{D} + \log^* n)$ rounds in the* CONGEST *model.*

Line graphs have diversity two and (non-uniform) hypergraphs with rank ℓ have diversity ℓ. The corresponding clique edge covers can be computed in a single round which implies the following corollary.

Corollary 2. *There exists a algorithm that computes $(\ell+4)$-ruling edge sets in $O(\ell + \log^* n)$ rounds in (non uniform) hypergraphs with rank at most ℓ.*

Outline. Section 2 focuses on ruling edge sets. We believe that it is helpful to read this section to understand the more involved algorithm in Sect. 3 which extends results to graphs of bounded diversity and line graphs of hypergraphs.

2   Ruling Edge Sets of Simple Graphs

We provide an algorithm to compute ruling edge sets with an asymptotically optimal runtime. Even though the same asymptotic runtime can be obtained with the more general algorithm in Sect. 3 (the algorithm works for graphs of bounded diversity — line graphs have diversity two) we believe that this section is simpler and more straightforward. This section also helps to understand the more involved algorithm in Sect. 3. In Sect. 2.1 we show how to compute 3-ruling edge sets in $O(\log^* n)$ rounds and in Sect. 2.2 we show how any β-ruling edge set can be transformed into a 2-ruling edge set in $O(\beta)$ rounds.

The distance $\mathrm{dist}(e, f)$ between two edges e and f is defined as the distance of the corresponding nodes in the line graph. The graph $G = (V, E)$ induced by a set of vertices $U \subseteq V$ is defined as $G[U] = (U, \{\{u, v\} \mid \{u, v\} \in E, u, v \in U\}$ and the graph induced by a set of edges as $F \subseteq E$ is $G[F] = (V, F)$. We extend the definition of ruling (vertex) sets to *ruling edge sets*.

Definition 1 *(Ruling Edge Set).* An (α, β)-ruling edge set $R \subseteq E$ of a graph $G = (V, E)$ is a subset of edges such that the distance between any two edges in R is at least α and for every edge $e \in E$ there is an edge $f \in R$ with $\mathrm{dist}(e, f) \leq \beta$.

In the LOCAL model the computation of ruling edge sets is equivalent to computing ruling sets on line graphs. Note that line graphs have many additional properties, e.g., bounded diversity (cf. Sect. 3), and not every graph can appear as a line graph of another graph. We focus on ruling edge sets with distance two between any two 'selected' edges. The next remark indicates why: any independence greater than two immediately leads to results for ruling vertex sets.

Remark 1. Any (α, β)-ruling edge set with $\alpha \geq 2$, $\beta \geq 1$ directly leads to a $(\alpha - 1, \beta + 1)$-ruling set.

Proof. Let S be a (α, β)-ruling edge set. For each edge $\{v_1, v_2\} \in S$ add one of the nodes, e.g., v_1 to the node set R. Isolated nodes are also added to R.

Independence. Let v_1 and v_1' be two nodes in R. By construction there are two distinct edges $e = \{v_1, v_2\}$ and $e' = \{v_1', v_2'\}$ in S. As the distance between e and e' is at least α the shortest path p that contains both edges e and e' has at least $\alpha + 1$ edges. Hence the distance between v_1 and v_2 is at least $\alpha - 1$.

Domination. Let $v \in V$ be a node with incident edge $e = \{v, w\}$. Then there is an edge $f = \{v_f, w_f\} \in S$ with $\text{dist}(e, f) \leq \beta$. Either v_f or w_f is contained in R and thus there is a node in distance $\beta + 1$ to v. □

Maximal matchings are $(2, 1)$-ruling edges sets. These can be computed in poly $\log n$ time with [8] or with a large dependency on the maximum degree and only a $O(\log^* n)$ dependence on n.

Corollary 3 ([27]). *Maximal matchings in graphs with maximum degree at most Δ can be computed in $O(\Delta + \log^* n)$ deterministic distributed time.*

2.1 Proposal Technique for Simple Graphs

In the first step of our ruling edge set algorithm we compute, in a constant number of rounds, a subset $F \subseteq E$ of the edges such that (1) for every edge $e \in E$ there is an edge $f \in F$ such that the distance between e and f is small and (2) the graph $G[F]$ has small maximum degree. In the second step we apply any (known) ruling edge set algorithm on the edges of $G[F]$, e.g., the algorithm from Corollary 3. We call a set F with these properties an *edge-kernel*.

Definition 2 (Edge-kernel). *Let $G = (V, E)$ be a graph. A (d, r)-edge-kernel $F \subseteq E$ is a subset of edges, so that the degree of the induced graph $G[F]$ is at most d and for every edge $e \in E$ there exists an edge $f \in F$ with $\text{dist}_G(e, f) \leq r$.*

The core idea of our algorithm is the proposal technique of the next lemma.

Lemma 1. *There is a deterministic two round CONGEST algorithm to compute a $(2, 2)$-edge-kernel.*

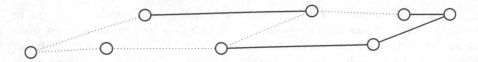

Fig. 1. Non-dotted lines form a $(2,2)$-edge-kernel.

Algorithm: Each node proposes one of its incident edges and in the next step each node accepts a single of its incident edges that were proposed by other nodes. Return the set F of accepted edges.

Algorithm 1 Proposal Technique

1: **for each** node n **in parallel do**
2:     Propose one incident edge to all neighbor nodes
3:     Arbitrarily add one of the edges that are proposed by neighbors to the set F
4: **return** F

Proof. The algorithm requires two rounds in the CONGEST model. We claim is that F is a (d,r)-edge-kernel with $d, r \leq 2$.

$d \leq 2$. Any node $v \in V$ has at most two incident edges in F: the edge that v proposed itself if it was accepted by the corresponding neighbor and the edge v accepted. This concludes that $\Delta(G[F]) \leq 2$.

$r \leq 2$. Consider any edge $e = \{v, u\} \in E$, the vertex v proposed some edge $f = \{v, w\}$ in the first step of the algorithm. Thus w has at least one incident edge that was proposed by a neighbor. Let $g = \{w, w'\} \in F$ be the edge that is accepted by w. Then the distance between f and g is at most 2 through the path e, f, g. Note that the distance is even smaller if v proposed edge e or if w accepted edge f. $\qquad\square$

Computing a ruling edge set on an edge-kernel provides a ruling edge set of the original graph whose domination parameter is the sum of the domination parameters of the edge-kernel and the ruling edge set.

Lemma 2. *Let d, r_1 and r_2 be positive integers. Given an (d, r_1)-edge-kernel $F \subseteq E$ of a graph $G = (V, E)$ and an r_2-ruling edge set algorithm with runtime $T(n, max\ degree)$, one can compute an $(r_1 + r_2)$-ruling edge set of G in time $T(n, d)$.*

Algorithm: Apply the r_2 ruling edge set algorithm on the graph $G[F]$; let $R \subseteq F$ be its output.

Proof. R is independent on G: Let e and f be two edges in $R \subseteq F$. They are not adjacent in $G[F]$ by the guarantees of the algorithm. Further, if they were adjacent in G then they would, by the definition of the induced graph, also be adjacent in $G[F]$, a contradiction.

R is $r_1 + r_2$ dominating on G: Let $e \in E$ be an arbitrary edge of G. Due to the edge-kernel properties there is an edge $f \in F$ with $\text{dist}_G(e, f) \leq r_1$. As R is an r_2-ruling edge set in $G[F]$ there is an edge $g \in R$ with $\text{dist}_{G[F]}(f, g) \leq r_2$. This implies that $\text{dist}_G(e, g) \leq \text{dist}_G(e, f) + \text{dist}_G(f, g) \leq r_1 + r_2$.   □

The bottleneck when computing a maximal matching is the maximum degree (cf. Corollary 3). An (d, r)-edge-kernel reduces the degree to d. By first computing an $(2, 2)$-edge-kernel and thereafter running our matching algorithm on it we obtain a $(2, 3)$-ruling edge set.

Proof. (of Theorem 2). First compute a $(2, 2)$-edge-kernel F with Lemma 1 in $O(1)$ rounds. Thereafter run the matching algorithm Corollary 3 on $G[F]$ in $O(\log^* n + 2) = O(\log^* n)$ rounds and return the matching. If we formulate these steps in the language of line graphs to apply Lemma 2 we obtain that the returned set is a 3-ruling edge set of G.

Use Theorem 3 to reduce the domination from 3 to 2 in $O(1)$ rounds. The lower bound of $\Omega(\log^* n)$ follows from Linial's lower bound [24] as the line graph of the ring forms an isomorphic ring.   □

The concept of edge-kernels as introduced in Definition 2 is not helpful to compute ruling edge sets with independence parameter $\alpha > 2$. Given an edge-kernel F, we use that we can handle the connected components of $G[F]$ separately as the distance between connected components is at least two. If one was to compute ruling edge sets with independence $\alpha > 2$ one had to ensure that the distance between connected components is at least α. Note that Remark 1 implies that such an algorithm would immediately imply an (unknown) algorithm for the computation of a (non trivial) ruling set of G.

2.2   From β-Ruling Edge Sets to 2-Ruling Edges Sets

In this subsection we show how to decrease the domination parameter of ruling edge sets from β to 2 within $O(\beta)$ rounds. In particular, we show how to transform 3-ruling edge sets into 2-ruling edge sets in $O(1)$ rounds and essentially repeating the algorithm β times leads to the result for general β.

The core idea is adding additional edges to a β-ruling edge set R to decrease its domination parameter: Let E_2 be the set of edges whose shortest distance to an edge in R is two. We carefully select an independent set $I \subseteq E_2$ such that every edge in distance three to R has an edge in distance at most two in I. Then $R \cup I$ forms the desired 2-ruling edge set. Note that adding all edges with distance two (or any other fixed distance) to R cannot be done without losing independence. Furthermore the induced graph $G[E_2]$ might have degree up to Δ. Thus, to obtain constant runtime, we cannot apply any of the known algorithms with non-constant runtime in a black box fashion to $G[E_2]$. We first need one very simple but also very useful observation.

Observation 1. *The distance of any pair of incident edges to the closest edge in an ruling edge set differs at most by one (cf. Fig. 2).*

Lemma 3. *A 3-ruling edge set can be transferred into a 2-ruling edge set in $O(1)$ CONGEST rounds.*

Algorithm: Given a 3-ruling edge set $R \subseteq E$, we compute a 2-ruling edge set $R \subseteq S \subseteq E$. First, split the edges E into four sets $E = E_0 \cup E_1 \cup \cdots \cup E_3$ according to their distance to an edge in R. This can be done in three rounds. Then every node that is adjacent to at least one edge from E_2 and at least one edge from E_3 selects a single of its incident edges $e_2 \in E_2$ as a *candidate edge*. Now, each node with at least one incident candidate edge that also has an incident edge in E_1 chooses one of its incident candidate edges and adds it to the set I. Finally return the set $S = R \cup I$.

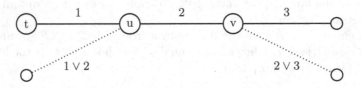

Fig. 2. Neighborhood of a candidate edge $e = \{u, v\} \in E_2$ proposed by node v. Solid edges exist in any graph, dotted edges may exist in any cardinality.

Proof. *S is independent.* It is helpful to keep the essence of Observation 1 in mind which implies that the set of nodes that propose an edge cannot be connected by an edge in E_2. For contradiction, assume such an edge $e = \{u, v\} \in E_2$ exists. As both nodes propose an edge they both have incident edges in E_3. However, then both nodes do not have an incident edge in E_1 which contradicts that $e \in E_2$ (cf. Fig. 2).

As the set I of added edges is a subset of E_2 none of them is adjacent to edge edge in E_0. Thus we only need to prove that no two edges in I are adjacent. Assume there are two edges $e = \{u, v\}, f = \{u, w\}$ in $I \subseteq E_2$ that are adjacent. Then v proposed e and w proposed f because if any of the edges would have been proposed by u the other edge could not be proposed at all by our previous observation that proposed nodes cannot be connected by an edge in E_2. As the proposals of e and f compete at u and u can only accept one of them not both edges can be contained in I.

S has domination two. Edges in $E \setminus E_3$ are still dominated in distance two as $R \subseteq S$. Let $e = \{u, v\} \in E_3$. At least one of its endpoints has an incident edge f in E_2 that it proposed. W.l.o.g. assume that u proposed edge $f = \{u, w\} \in E_2$. Either f is accepted by w which implies that e is dominated or w accepted some other edge g which dominates e in distance two. $\square$

We use the same idea to improve the domination parameter.

Lemma 4. *Any β-ruling edge set with $\beta \geq 3$ of a simple graph can be transformed into an $\beta - 1$-ruling edge set in $O(1)$ rounds of communication in the* CONGEST *model.*

Proof. Let R be the given β-ruling edge set. First, for each edge $e \in E$ we check whether its 3-neighborhood contains an edge of R and split the edges E into four sets according to their distance to R, i.e., $E = E_0 \cup E_1 \ldots E_3 \cup E_{\geq 4}$. Let $H = G[E_{\leq 3}]$ be the subgraph induced by the edges $E_{\leq 3} = E_0 \cup E_1 \cup E_2 \cup E_3$ and apply Lemma 3 to transform R into a 2-ruling edge set S of H in $O(1)$ rounds.

We observe that this 2-ruling edge set of H is a $(\beta - 1)$-ruling edge set of G: In the graph G the shortest path $p = e_\beta, \ldots, e_3, e_2, e_1, e_0$ from any edge e_β (with distance β to R) to $E_0 = R$ contains an edge of E_3; the indices of the path edges correspond to their distances to e_0. As e_3 has an edge at distance at most 2 in S, the edge e_β has an edge in S at distance at most $\beta - 1$. $\square$

Lemma 4 reduces the domination of a β-ruling edge set in a constant number of rounds, independent from β. Particularly, the reduction even works in constant time if no node knows how far it is from the closest ruling edge before the algorithm starts. Iteratively applying Lemma 4 implies the following theorem.

Proof (of Theorem 3). Apply Lemma 4 $\beta - 2$ times iteratively reducing β to 2 in $O(\beta)$ rounds. $\square$

3   Ruling Sets of Bounded Diversity Graphs

In Sect. 2, we have seen that the computation of ruling sets on line graphs seems to be much easier than on general graphs. In this section, we identify graph properties that allow us to essentially apply the same algorithm as in Sect. 2 to a much more general class of graphs, in particular to bounded diversity graphs. Bounded diversity was introduced in [3]. Given a graph $G = (V, E)$ and a *clique edge cover* Q, i.e., a set of cliques (where each clique is a subgraph of G) such that any node of G is contained in at least one clique and for any two nodes $u, v \in V$ that are adjacent in G there exists a clique $C \in Q$ in which u and v are adjacent. The *diversity* with respect to the cover Q is the maximal number of cliques a vertex is contained in. The diversity of a graph is the minimum diversity over all clique edge covers. We show that, given such a cover with diversity $\mathcal{D}$, we can compute an $O(\mathcal{D})$-ruling set in time $O(\mathcal{D} + \log^* n)$. In many cases, e.g., in line graphs and line graphs of hypergraphs clique edge covers with very low diversity can be computed in constant time even in the CONGEST model.

Definition 3 (Diversity). *Given a graph $G = (V, E)$ and a clique edge cover Q, the* diversity *of (G, Q) is defined as $\max_{v \in V} |\{C \in Q \mid v \in C\}|$. The diversity of G is the minimum of the diversities of (G, Q) over all clique edge covers Q.*

Definition 3 is slightly different from the definition in [3] where the cliques are required to be maximal. However, none of our algorithms use this property and going without it might lead to covers with smaller diversity and hence faster runtimes. One downside of both definitions is that (so far) algorithms rely on a globally known cover that, in the best case, levels the diversity of the graph. In both models of computation it is not clear that computing such a cover can always be done efficiently. However, in the LOCAL model it is straightforward to compute a cover with diversity Δ: For each node, add all maximal cliques that it is contained in. In the full version of this paper, we provide an example that shows that this is not necessarily optimal [22]. In the CONGEST model the seemingly hard problem of triangle detection (see e.g., [18]) can be reduced to the problem of identifying maximal cliques. Often it is not difficult to compute a clique edge cover with a small diversity, e.g., a cover with diversity two in line graphs can be computed in constant time in the CONGEST model. For a further discussion of the computability of such covers consult [3].

A *hypergraph* H is a tuple (V, E) of vertices and edges and each edge is a set of vertices. The *rank* of a hypergraph is the maximum number of vertices that are contained in an edge. One way to define a distributed algorithm on a hypergraph is that, in one round, each vertex v of the hypergraph broadcasts one message on each of its incidents hyperedges e (the messages can be different for different hyperedges but all nodes in the same hyperedge receive the same message from v on that edge) and receives the messages sent by its neighbors. The diversity of line graphs of hypergraphs is bounded above by the rank of the hypergraph. Simple graphs are (uniform) hypergraphs of rank two and hence have diversity of at most two.

Remark 2. The diversity of the line graph of a hypergraph of rank at most ℓ is at most ℓ and a corresponding cover can be computed in constant time.

Proof. Let $H = (V, E)$ be a hypergraph and $L = \mathcal{L}(H)$ its line graph. For each vertex v we define the (constant time computable) clique $C_v = \{e \mid e \in E, v \in e\}$ and $Q = \{C_v \mid v \in V\}$. Then Q is a clique edge cover of L with diversity ℓ as an edge $e = \{v_1, \ldots, v_\ell\}$ is only contained in the cliques $C_{v_1}, \ldots, C_{v_\ell}$. □

In this section we show how to adapt the proposal technique of Lemma 1 in Sect. 2. Recall that we used the proposal technique to compute (d, r)-edge-kernels of a graph. In this section we use vertex-kernels.

Definition 4 (Vertex-kernel). *Let* $G = (V, E)$ *be a graph. A* (d, r)-*vertex-kernel* $A \subseteq V$ *is a subset of nodes, so that the degree of the induced graph* $G[A]$ *is at most* d *and for every node* $v \in V$ *there exists a node* $u \in A$ *with* $dist_G(v, u) \leq r$.

First, we rephrase the proposal technique of Lemma 1 directly on the line graph. Each node of the original graph can be identified with a clique in the line graph. The 'proposing an edge' in Algorithm 1 corresponds to proposing a single node from each such clique. Then, 'accepting a proposed edge' corresponds to accepting one of the proposed nodes in the clique. In Lemma 1 we showed that at most two nodes per clique *survive* this process on the line graph. In general many more nodes can survive a single step of this proposal and accepting technique. We repeat the process to sparsify the selected nodes further and further.

Lemma 5. *There exists an $O(\mathcal{D})$ time algorithm in the* CONGEST *model that, given a graph $G = (V, E)$ and a set of cliques Q with diversity $\mathcal{D}$, computes a subset of nodes $\mathcal{A}$ with the following properties:*

(small degree) For all cliques $C \in Q : |\mathcal{A} \cap C| \leq \mathcal{D}$,

(domination) For all nodes $v \in V$ there is a node $p \in \mathcal{A}$ with $dist_G(v, p) \leq \mathcal{D}$.

Moreover, $\mathcal{A}$ is a $(\mathcal{D}^2, \mathcal{D})$-vertex-kernel of the graph.

Algorithm: At the start each node is set *active*. Then, in each of $\mathcal{D}$ phases, each clique proposes one of its active nodes that it has not proposed in any phase before (if such a node exists). Any node that was active before, has not been proposed in the current phase but has a neighbor that is proposed in the current phase is set *inactive*. In the end we return the set of active nodes. Confer the full version of this paper for detailed pseudocode [22].

For the correctness of the algorithm we show that any node in $\mathcal{A} \cap C$ has been proposed by clique C and as each clique proposes at most one node in each of the $\mathcal{D}$ iterations the claim *(small degree)* follows. The second property follows as a node is only set inactive if it has a neighbor that is active in the next phase.

Proof. For $i = 1, \ldots, \mathcal{D}$ let $\mathcal{A}_i$ denote the set of nodes that are active at the end of phase i, S_i the set of nodes that are proposed in phase i, W_i the set of nodes that are active at the end of phase i and do not have a neighbor that is proposed in phase i and R_i^C be the set of nodes that have been proposed by clique C until phase i. To prove the lemma we first prove the following property: (1) $S_1 \supseteq S_2 \supseteq \ldots \supseteq S_{\mathcal{D}}$. Assume for contradiction, that $v \notin S_j$ and $v \in S_{j+1}$ for some $j < d$. Let C be a clique that proposes v in phase $j + 1$. In phase $j + 1$ only nodes in $\mathcal{A}_j$ can be proposed. Thus v is contained in $\mathcal{A}_j = W_j \cup S_j$. As v is not contained in S_j we deduce that $v \in W_j$, i.e., v does not have a neighbor that is proposed in phase j. In particular, C does not propose a neighbor of v in round j, i.e., either C proposed v in phase j or v does not propose any node in phase j at all. In both cases C cannot propose v in phase $j + 1$, a contradiction.

Fix a clique $C \in Q$. We show that each node in $P \cap C$ has been proposed by C. As C proposes at most $\mathcal{D}$ nodes the claim *(small degree)* follows. If there is an $i < d$ with $(\mathcal{A}_i \cap C) \setminus R_i^C = \emptyset$ the claim holds because $\mathcal{A} \cap C \subseteq \mathcal{A}_i \cap C$ and clique C already proposed all nodes in $\mathcal{A}_i \cap C$ in the first i rounds. So assume that $(\mathcal{A}_{\mathcal{D}-1} \cap C) \setminus R_{\mathcal{D}-1}^C \neq \emptyset$ and let v be the node that C proposes in the last phase. All nodes in C that are not proposed in phase $\mathcal{D}$ are set inactive as their neighbor v is proposed. Thus any node in $\mathcal{A} \cap C$ is a node in $S_{\mathcal{D}}$, i.e., any such node is proposed in phase $\mathcal{D}$ by some clique and due to Property (1) also in each phase before. Thus any node in $\mathcal{A} \cap C$ has been proposed by $\mathcal{D}$ many cliques. As no clique can propose a node twice and each node is in at most $\mathcal{D}$ cliques (including clique C) each such node has been proposed by C.

Domination. At the start every node has an active neighbor (i.e., a neighbor in $\mathcal{A}$). A node is only set inactive (i.e., removed from $\mathcal{A}$) in some phase if it has neighbor that is proposed in the phase. Thus the domination distance increases at most by one per phase which proves the claim.

The Algorithm Runs in $O(\mathcal{D})$ Rounds in the CONGEST *Model.* In a single phase, removing the non proposed nodes with a proposed neighbor from the set of active nodes can be done in a single round. Thus the runtime is in the order of the number of phases, i.e., it is $O(\mathcal{D})$.

$\mathcal{A}$ *is a* $(\mathcal{D}^2, \mathcal{D})$-*vertex-kernel.* Due to diversity $\mathcal{D}$ any v can only be part of $\mathcal{D}$ distinct cliques. Due to the *(small degree)* property it has at most $\mathcal{D}-1$ neighbors in each clique. As the cliques cover every edge of G this implies that the maximum degree of $G[\mathcal{A}]$ is upper bounded by $\mathcal{D}(\mathcal{D}-1) \leq \mathcal{D}^2$. The domination follows immediately from the second property.   □

Analogously to Lemma 2 one can prove the following lemma.

Lemma 6. *Let d, r_1 and r_2 be positive integers. Given an (d, r_1)-vertex-kernel $S \subseteq V$ of a graph $G = (V, E)$ and an r_2-ruling set algorithm with runtime $T(n, max\ degree)$ one can compute an $(r_1 + r_2)$-ruling set of G in time $T(n, d)$.*

Lemmas 5 and 6 and the ruling set algorithm from Corollary 1 imply the main result of the section.

Proof. (of Theorem 4). First use Lemma 5 to compute a $(\mathcal{D}^2, \mathcal{D})$-vertex-kernel $\mathcal{A}$ of G in $O(\mathcal{D})$ rounds. Then run the ruling set algorithm from Corollary 1 with $\beta = 4$ on $G[\mathcal{A}]$ in $O(\mathcal{D} + \log^* n)$. With Lemma 6 this yields a $(\mathcal{D}+4)$-ruling set.   □

As hypergraphs of rank ℓ have diversity ℓ and we can efficiently compute a corresponding clique decomposition and we obtain Corollary 2.

References

1. Alon, N., Babai, L., Itai, A.: A fast and simple randomized parallel algorithm for the maximal independent set problem. J. Algorithms **7**(4), 567–583 (1986)
2. Awerbuch, B., Goldberg, A.V., Luby, M., Plotkin, S.A.: Network decomposition and locality in distributed computation. In: Proceedings of the IEEE Symposium on Foundations of Computer Science (FOCS), pp. 364–369 (1989)
3. Barenboim, L., Elkin, M., Maimon, T.: Deterministic distributed $(\Delta + o(\Delta))$-edge-coloring, and vertex-coloring of graphs with bounded diversity. In: Proceedings of ACM Symposium on Principles of Distributed Computing (PODC), pp. 175–184 (2017)
4. Barenboim, L., Elkin, M., Pettie, S., Schneider, J.: The locality of distributed symmetry breaking. J. ACM **63**(3), 20 (2016)
5. Beck, J.: An algorithmic approach to the lovász local lemma. I. Random Struct. Algorithms **2**(4), 343–365 (1991)
6. Bisht, T., Kothapalli, K., Pemmaraju, S.: Brief announcement: Super-fast t-ruling sets. In: Proceedings of ACM Symposium on Principles of Distributed Computing (PODC), pp. 379–381 (2014)

7. Chang, Y.J., Li, W., Pettie, S.: An optimal distributed $(\Delta+1)$-coloring algorithm? In: Proceedings of 50th ACM Symposium on Theory of Computing (STOC) (2018)
8. Fischer, M.: Improved deterministic distributed matching via rounding. In: Proceedings of Symposium on Distributed Computing (DISC). LIPIcs, vol. 91, pp. 17:1–17:15 (2017)
9. Fischer, M., Ghaffari, M., Kuhn, F.: Deterministic distributed edge-coloring via hypergraph maximal matching. In: Proceedings of IEEE Symposium on Foundations of Computer Science (FOCS), pp. 180–191 (2017)
10. Gfeller, B., Vicari, E.: A randomized distributed algorithm for the maximal independent set problem in growth-bounded graphs. In: Proceedings of ACM Symposium on Principles of Distributed Computing (PODC), pp. 53–60 (2007)
11. Ghaffari, M.: An improved distributed algorithm for maximal independent set. In: Proceedings of ACM-SIAM Symposium on Discrete Algorithms (SODA), pp. 270–277 (2016)
12. Ghaffari, M., Harris, D.G., Kuhn, F.: On derandomizing local distributed algorithms. CoRR abs/1711.02194 (2017)
13. Hańćkowiak, M., Karoński, M., Panconesi, A.: On the distributed complexity of computing maximal matchings. In: Proceedings of ACM-SIAM Symposium on Discrete Algorithms (SODA), pp. 219–225 (1998)
14. Hańćkowiak, M., Karoński, M., Panconesi, A.: A faster distributed algorithm for computing maximal matchings deterministically. In: Proceedings of ACM Symposium on Principles of Distributed Computing (PODC), pp. 219–228 (1999)
15. Harris, D.G., Schneider, J., Su, H.H.: Distributed $(\Delta+1)$-coloring in sublogarithmic rounds. In: Proceedings of the 48th ACM Symposium on Theory of Computing (STOC), pp. 465–478 (2016)
16. Henzinger, M., Krinninger, S., Nanongkai, D.: A deterministic almost-tight distributed algorithm for approximating single-source shortest paths. In: Proceedings of ACM Symposium on Theory of Computing (STOC), pp. 489–498 (2016)
17. Israeli, A., Itai, A.: A fast and simple randomized parallel algorithm for maximal matching. Inf. Process. Lett. **22**(2), 77–80 (1986)
18. Izumi, T., Le Gall, F.: Triangle finding and listing in CONGEST networks. In: PODC, pp. 381–389 (2017)
19. Kothapalli, K., Pemmaraju, S.V.: Super-fast 3-ruling sets. In: FSTTCS. LIPIcs, vol. 18, pp. 136–147 (2012)
20. Kuhn, F., Moscibroda, T., Wattenhofer, R.: Local computation: lower and upper bounds. J. ACM **63**(2), 17:1–17:44 (2016)
21. Kuhn, F., Moscriboda, T., Nieberg, T., Wattenhofer, R.: Fast deterministic distributed maximal independent set computation on growth-bounded graphs. In: Proceedings of the 19th International Conference on Distributed Computing (DISC), pp. 273–287 (2005)
22. Kuhn, F., Maus, Y., Weidner, S.: Deterministic distributed ruling sets of line graphs (2018)
23. Linial, N.: Distributive graph algorithms global solutions from local data. In: Proceedings of IEEE Symposium on Foundations of Computer Science (FOCS), pp. 331–335 (1987)
24. Linial, N.: Locality in distributed graph algorithms. SIAM J. Comput. **21**(1), 193–201 (1992)
25. Luby, M.: A simple parallel algorithm for the maximal independent set problem. SIAM J. Comput. **15**(4), 1036–1053 (1986)

26. Pai, S., Pandurangan, G., Pemmaraju, S.V., Riaz, T., Robinson, P.: Symmetry breaking in the congest model: time- and message-efficient algorithms for ruling sets. In: Proceedings of Symposium on Distributed Computing (DISC). LIPIcs, vol. 91, pp. 38:1–38:16 (2017)
27. Panconesi, A., Rizzi, R.: Some simple distributed algorithms for sparse networks. Distrib. Comput. **14**(2), 97–100 (2001)
28. Panconesi, A., Srinivasan, A.: The local nature of Δ-coloring and its algorithmic applications. Combinatorica **15**(2), 255–280 (1995)
29. Panconesi, A., Srinivasan, A.: On the complexity of distributed network decomposition. J. Algorithms **20**(2), 581–592 (1995)
30. Peleg, D.: Distributed computing: a locality sensitive approach. SIAM (2000)
31. Schneider, J., Elkin, M., Wattenhofer, R.: Symmetry breaking depending on the chromatic number or the neighborhood growth. Theor. Comput. Sci. **509**, 40–50 (2013)

Broadcast with Energy-Exchanging Mobile Agents Distributed on a Tree

Jurek Czyzowicz[1]([✉]), Krzysztof Diks[2], Jean Moussi[1], and Wojciech Rytter[2]

[1] Département d'informatique, Université du Québec en Outaouais, Gatineau, Québec, Canada
{jurek,Jean.Moussi}@uqo.ca
[2] Faculty of Mathematics, Informatics and Mechanics, University of Warsaw, Warsaw, Poland
{diks,rytter}@mimuw.edu.pl

Abstract. Mobile agents are deployed at selected nodes of an edge-weighted tree network. Each agent originally possesses an amount of energy, possibly different for all agents. Initially, in a given *source node* of the network is placed a piece of information (data packet) that must be broadcast to all other nodes. Such transfer of the packet needs to be achieved with aid of collaborating mobile agents, which may transport copies of the packet to all nodes.

Agents travel in the network spending energy proportionally to the distance traversed. They can stop only at network nodes. If two agents are present at a node at the same time, they can exchange any amount of currently possessed energy. Our goal is to verify if there exists a sequence of agents' movements (a schedule) and energy exchanges between meeting agents, which results in the packet reaching all nodes of the tree network.

Our algorithm produces an optimal centralized scheduler as we assume that the central authority knows everything about the network and controls the movement of the agents, which do not need to possess any knowledge. In this sense it is a semi-distributed algorithm.

The important part of our algorithm uses dynamic programming in order to compute an optimal *agents migration flow* of every edge of the network, i.e. the number of agents traversing every edge in each direction. The approach is far from being straightforward, as its correctness is based on multiple complex interactions between the subtrees obtained by removal of any given edge.

It is known that, if energy exchange is not allowed, the broadcasting problem for trees (even for lines) is NP-complete.

Keywords: Mobile agents · Data delivery · Broadcast
Energy exchange · Tree network

© Springer Nature Switzerland AG 2018
Z. Lotker and B. Patt-Shamir (Eds.): SIROCCO 2018, LNCS 11085, pp. 209–225, 2018.
https://doi.org/10.1007/978-3-030-01325-7_20

1   Introduction

We are given a weighted tree T of n nodes and its distinguished source node r. An edge weight represents its length, i.e., the distance between its endpoints along the edge. There is a collection of k mobile agents that are deployed at selected nodes of the tree. Initially, each agent possesses an amount of energy (possibly distinct for different agents). If the agent has enough energy it can move to a neighbouring node - its energy is then reduced by the length of the traversed edge. Agents can stop only at nodes. When two agents meet at a same node, one of them can transfer a portion of currently possessed energy to another one. We assume that the source node r of the tree possesses a piece of data (packet). This data must be broadcast to all other tree nodes using mobile agents for its transportation. Initially no agent possesses the knowledge of the packet and the only node knowing the packet is the source. A node of the tree learns the packet when visited by an agent possessing its knowledge. An agent learns the packet by visiting the source node or any other node, which was previously visited by an agent possessing its knowledge.

Our goal is to solve the following decision problem:

> **Data broadcasting decision problem:** We are given a source node r of an n-node tree T, and a configuration of k agents, placed at selected nodes, each having some initial amount of energy (possibly different for all agents). Is it possible to schedule the moves of the agents and energy transfers so that the initial packet of information placed at node r reaches all nodes of T?

We will look for schedules of agents' movements that will not only result in completing the broadcast, but also attempt to maximize the energy, which is eventually brought to the root. We call such schedules *optimal*. Consequently, our approach permits to solve a more general optimization problem:

> **Data broadcasting optimization problem:** What is the largest amount of energy, which may be deposited at source r while some sequence of moves of the agents and energy transfers result in the initial packet reaching all nodes of T?

1.1   Related Work

The research questions in mobile agents computing concern problems for which a collection of individual mobile processors collaborate in order to solve efficiently an assigned task. Most studied problems for mobile agents involve some sort of environment search or exploration (cf. [2,9,14–16]). In the case of a team of collaborating mobile agents, the challenge is to balance the workload among the agents in order to minimize the exploration time. However this task is often hard (cf. [17]), even in the case of two agents in a tree [5]. The tree exploration by energy-constrained mobile agents has been considered in [14].

The task of broadcast consists in communicating the information available to some designed processor to all other processors of the network. This problem, in the case of stationary processors, has been previously studied in the case of the wireless model (cf. [10]) and the message passing model (e.g. [3]). The problem of data communication performed by collaborating data-transporting mobile agents has been recently studied in [1,6–8,11]. The mobile agents of [1] perform efficient convergecast and broadcast for line networks. However, in the case of trees, both problems turn up to be strongly *NP*-hard (cf. [3]). The related problem of *data delivery*, when the data packet has to be transmitted by energy constrained agents between two given network nodes, has been studied in [11]. It was proved in [11] that data delivery is *NP*-hard already in the case of line networks. A number of positive and negative results concerning the data delivery problem has been provided in the PhD Thesis of Bärtschi [8] (see also [6,7]). The question is closely related to the *vehicle routing problem*, e.g. see [18].

The result of [11] implies that, in the case when agents cannot exchange energy, the data broadcasting problem is *NP*-hard for line networks. The present paper shows that the situation is quite different if the agents are allowed to transfer energy. The communication problems for energy-exchanging mobile agents in tree networks were studied in [13] (the case of data delivery and convergecast) and in [12] (the case of broadcast by agents starting from the same node). The setting where the agents are originally distributed at arbitrary positions turns up to be much more involved.

1.2 Preliminairies

In the remainder of the paper we assume that the tree T is rooted at its source r.

In our algorithm we propose a specific treatment for each tree node, related to its number of children, presence or absence of an agent and weight of the incident edges. To ease its understanding, we convert the given tree to another one, in which every node will have only one property that needs to be taken into account by our algorithm. The structure of the tree and the lengths of the corresponding weighted paths remain the same for the converted tree, hence the movements of the agents performing broadcasting produced by our algorithm may be reconverted back to the original tree.

We observe first that using the standard *folklore* technique, by adding extra nodes and edges of zero weight, the original weighted tree may be converted to a *binary* tree, in which all weighted path lengths between the corresponding tree nodes are the same. Although the depth of such converted binary tree is increased, its complexity remains $O(n)$ and it may be obtained in $O(n)$ time. Hence w.l.o.g. we can assume that the given initial tree is binary.

Lemma 1. *By adding extra nodes and edges of zero weight we can convert a binary tree T with agents placed at its nodes into a tree with the following four types of nodes: (see Fig. 1):*

(a) if v is a terminal node, then it initially contains no agents,

(b) if v is a parent of two children, then v contains no agents and both children are accessible by edges of zero weight,

(c) if v originally contains one or more agents, it has exactly one child accessible by an edge of zero weight, or

(d) node v, which has one child and no agents (this is the only type of node that may have an edge of non-zero weight incoming from a child).

Moreover, the converted tree is of $O(n)$ size and the paths between the corresponding nodes in the converted tree are of the same weight as in the original one.

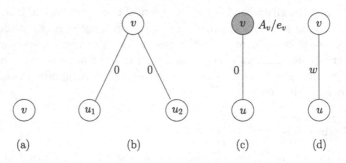

Fig. 1. Four cases of node v in tree T: (a) terminal node, (b) node with two children, (c) node containing A_v agents having total energy e_v, (d) node with an incoming edge of non-zero weight.

The proof of the lemma is easy and we omit it. It is possible to make such conversion in linear time with respect to the size of the original tree. For convenience we add to such converted tree one extra node r of type (d), which we make parent of the root of T, using zero-weight edge (cf. Fig. 2). In the remainder of the paper we assume that T denote the converted, rooted version of the tree. Despite the fact that tree T is undirected, as agents move along the edges in particular direction, we consider also directed version of the edges. In particular, for two adjacent nodes v, w, we denote by (v, w) an undirected edge between v, w and by $v \rightarrow w$ and $w \rightarrow v$ the two directed edges. For any node $v \neq r$, we denote by T_v a subtree of T, rooted at v, and containing all descendants of v in T. If $w = parent(v)$ we call $v \rightarrow w$ the *exit edge* of tree T_v. Hence all exit edges point in the direction of the root.

2   Testing Feasibility of Broadcast

In this section we describe an algorithm, which computes the largest amount of energy that may be deposited at the root r by the agents performing a successful packet broadcasting from r to all remaining nodes. The algorithm needs to manipulate its three resources in the form of information (packet), agents and energy. Consequently, energy might be transported by agents from some parts of the tree T to other parts, where it is more needed. Similarly, agents might be more useful when moving from some parts of T so that they finish their walks in other parts. The main idea of our algorithm is to find a combination of transfers of energy and agents across the tree, resulting in the largest unused energy, which is eventually deposited at the root. This is realized by a dynamic programming approach.

Consider an isolated subproblem of the broadcast of the packet present at node v to all nodes of the sub-tree T_v. Such a broadcast might be successfully performed using only agents originally present in T_v (and their energy). However, we may also use agents originally from outside T_v and/or extra energy, incoming via its reversed exit edge. On the other hand, unused energy and/or agents which do not need to terminate their walks inside T_v may be transferred through its exit edge to other parts of tree T, where they turn up to be more useful. We will show that with respect to any given exit edge $v \to w$, an optimal schedule might consist of three steps (in this order):

1. If the total energy available inside T_v is sufficiently large, an agent will first traverse edge $v \to w$ bringing an excessive energy to node w. Such energy may be subsequently transferred and deposited at the root r. Depending on the distribution of energy inside T_v this step may or may not exist.
2. An agent A will traverse edge $w \to v$ in order to transport the packet into T_v.
3. Then
   (a) Either a number of other agents traverse together edge $w \to v$. Then all these agents, together with agent A and the agents initially present inside T_v will transport the packet to all nodes of T_v.
   (b) Or, a number of other agents traverse together edge $v \to w$. Before exiting from T_v these agents together with agent A and the agents initially present inside T_v will transport the packet to all nodes of T_v.
   (c) Or, no other agent traverses either of the edges $w \to v$ and $v \to w$. In this case agent A together with agents initially present inside will transport the packet to all nodes of T_v, eventually terminating their walks inside T_v.

Suppose for a moment that we know some optimal schedule. Suppose also that, for this optimal schedule and for every subtree T_v, the integer i_v denotes the difference between the number of agents' walks traversing the exit edge $v \to w$ and the number of agents' walks traversing the reverse edge $w \to v$. For a given schedule, we call such i_v the *agents migration flow* of the edge $v \to w$ or shortly its M-flow. For any possible value of M-flow i_v, we denote by $\mathcal{B}_v[i_v]$ the *energy potential* of the exit edge $v \rightarrow w$. The value $\mathcal{B}_v[i_v]$ equals the largest energy that could be deposited at node v, which would permit a successful completion of the broadcast from node v to all nodes of T_v, assuming i_v being the M-flow of the edge $v \rightarrow w$. More exactly, if $\mathcal{B}_v[i_v] \geq 0$, then it represents the maximum energy that may be left at node v after a successful broadcast from v to all nodes of T_v. Similarly, if $\mathcal{B}_v[i_v] < 0$, then $-\mathcal{B}_v[i_v]$ represents the minimum energy that must be added at node v, after which it would be possible to perform a successful broadcast inside T_v.

It is assumed, that if $i_v \geq 0$, there will be a total of i_v agents available at node v after the broadcast and ready to be used outside T_v. Similarly, it is assumed that if $i_v < 0$, the total number of $-i_v$ extra agents that were not initially placed inside T_v will be used to terminate their broadcasting walks inside T_v.

Figure 2 illustrates the values of all tables $\mathcal{B}$ computed for an example tree. At this stage, it may not be completely clear to the reader how the values of tables $\mathcal{B}$ are computed. This is the goal of Algorithm TEST-BROADCAST-FROM-ROOT given below. However, it is interesting to note, that each table contains a non-increasing sequence of values. This is due to the fact that energy and agents are, to some extent, exchangeable resources. Consider a tree with "heavy" terminal edges. If we send one agent to broadcast a packet inside a given subtree T_v, this agent needs to traverse all but one heavy terminal edges twice (once in each direction). If we have more agents available, they may terminate their walks inside T_v, so that many heavy edges will be traversed once only and energy will be saved.

As in fact we do not know in advance any optimal schedule, our algorithm will compute energy potential for any possible M-flow of every edge. It is worth noting that more than one value of M-flow of a given edge may lead to an optimal schedule. The most important part of our algorithm is the computation of M-flows for all edges of T, resulting from some optimal broadcast schedule. This process is done in bottom-up order (i.e. postorder) of tree T. For each exit edge of some tree T_v we compute the array $\mathcal{B}_v[]$ by calculating the energy potentials for all possible M-flows through this edge.

The following algorithm computes the upper bound on the amount of energy that may be brought to the root of T, so that the broadcast may be still performed successfully. We use the procedure COMPUTE $\mathcal{B}(v)$ computing the array $\mathcal{B}_v[]$ for each node v, assuming that the arrays for the children (or child) of v has been previously computed. For nodes of type (a) and (c) the computation straightforward. For a node v of type (b) the procedure computes for each component $\mathcal{B}_v[i]$ what is the agent migration flow between v and its children, which results in the optimal energy potential (i.e. larger gain or the smallest deficit). Finally, for a node of type (d), the procedure considers the energy expense along the weighted edge (u, v), taking into account that there may be an agent transferring an excessive energy along $u \rightarrow v$ towards the root and there must be one agent travelling along $v \rightarrow u$ that brings the packet into tree T_u.

ALGORITHM TEST-BROADCAST-FROM-ROOT(T);
{ input: tree T rooted at r; if node v contains agents, then A_v is
the number of agents present at v and e_v is their total energy }

1. **for** all $v \in T$ and all $-k \leq i \leq k$ **do**

2.   Initialise $\mathcal{B}_v[i] := -\infty$;

3. **for** all $v \in T$ taken in an ascending order with respect to *root* **do**

4.   COMPUTE $\mathcal{B}(v)$;

5. **if** $\mathcal{B}_r[0] \geq 0$

6.   **then** Report r as a possible broadcast node;

7.   Report $\mathcal{B}_r[0]$ as the maximum energy which may be left
                at r during a succesful broadcast from r;

8.   **else** Report broadcast from r is infeasible;

PROCEDURE COMPUTE $\mathcal{B}(v)$;

1.   **for** $i := k$ **to** $-k$ **do**

2.   **case** type of node v **of**

3.   (a) **if** $i \leq 0$ **then** $\mathcal{B}_v[i] := 0$;

4.   (b) $\mathcal{B}_v[i] := \max\{\mathcal{B}_{u_1}[j] + \mathcal{B}_{u_2}[h] : j + h = i\}$;

5.   (c) **if** $(i - A_v \geq -k)$

6.   **then** $\mathcal{B}_v[i] := \mathcal{B}_u[i - A_v] + e_v$

7.   **else** $\mathcal{B}_v[i] := \mathcal{B}_u[-k] + e_v$;

8.   (d) **if** $i \geq 0 \lor \mathcal{B}_u[i] > 2 \cdot weight(u, v)$

9.   **then** $\mathcal{B}_v[i] := \mathcal{B}_u[i] - (|i| + 2) \cdot weight(u, v)$

10.   **else** $\mathcal{B}_v[i] := i \cdot weight(u, v)$;

11. **if** $(i < k) \land (\mathcal{B}_v[i] < \mathcal{B}_v[i + 1])$ **then** $\mathcal{B}_v[i] := \mathcal{B}_v[i + 1]$;

Figure 2(b) contains the tables $\mathcal{B}$ computed for all nodes of converted tree
from Fig. 2(a). The following lemma proves that for each node v the sequence of
elements of the table $\mathcal{B}_v[\,]$ is nondecreasing.

Lemma 2. *For any node v and any index i^*, s.t. $-k < i^* \leq k$, we have*
$\mathcal{B}_v[i^* - 1] \geq \mathcal{B}_v[i^*]$.

Proof. The proof goes by induction on the height of node v. The claim is clearly
true for each node of depth 0 (node of type (a)), for which $\mathcal{B}_v[i] = 0$ if $i \leq 0$ and
$\mathcal{B}_v[i] = -\infty$ if $i > 0$ (cf. line 2 of algorithm TEST-BROADCAST-FROM-ROOT(T)
and line 3 of procedure COMPUTE $\mathcal{B}(v)$).

Suppose that the claim of the lemma is true for each node of height at most
h and consider any node v of height $h + 1$. Three cases are possible:

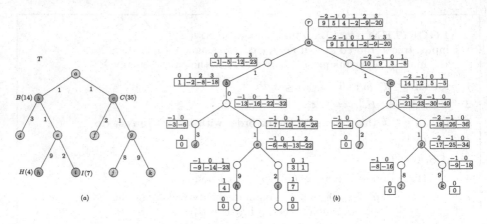

Fig. 2. Part (a) contains an example tree T. At nodes b, c, h, i there are present agents B, C, H, I having the initial amounts of energy $14, 35, 4, 7$, respectively. Part (b) illustrates the converted tree (for clarity we assume that all its unweighted edges have weight 0). Each tree node of the converted tree has a $\mathcal{B}$ table associated to it. Only the significant entries of each table $\mathcal{B}$ are illustrated: for larger indices of each table all entries are equal to $-\infty$; for smaller indices of each table the entries are the same as the first one illustrated. As an example note that $\mathcal{B}_b[1] = -2$ corresponds to agent B trajectory $bdbeh$ of length 16 and agent I trajectory $ieieb$ of length 7. Agent H does not move and its energy is lost in this case. $\mathcal{B}_b[1] = -2$ means that 2 extra units of energy are needed initially at b so that the broadcast resulting in 1 agent eventually arriving at b.

Case 1 (node v is of type (b), cf. Fig. 1). Take any index i^* and suppose that j^* and k^* are such that $i^* = j^* + k^*$ and

$$\mathcal{B}_v[i^*] = \max\{\mathcal{B}_{u_1}[j] + \mathcal{B}_{u_2}[k] : j + k = i^*\} = \mathcal{B}_{u_1}[j^*] + \mathcal{B}_{u_2}[k^*]$$

according to line 4 of procedure COMPUTE $\mathcal{B}(v)$. As nodes u_1 and u_2, which are the children of v, are both of height at most h, by the inductive hypothesis we have

$$\mathcal{B}_v[i^*] = \max\{\mathcal{B}_{u_1}[j] + \mathcal{B}_{u_2}[k] : j + k = i^*\} = \mathcal{B}_{u_1}[j^*] + \mathcal{B}_{u_2}[k^*]$$
$$\leq \mathcal{B}_{u_1}[j^* - 1] + \mathcal{B}_{u_2}[k^*] \leq \max\{\mathcal{B}_{u_1}[j] + \mathcal{B}_{u_2}[k] : j + k = i^* - 1\} = \mathcal{B}_v[i^* - 1]$$

Case 2 (node v is of type (c)). Suppose first that $i^* - A_v - 1 \geq -k$. Then, using the inductive hypothesis, according to line 6 of procedure COMPUTE $\mathcal{B}(v)$ we have

$$\mathcal{B}_v[i^*] = \mathcal{B}_u[i^* - A_v] + e_v \leq \mathcal{B}_u[i^* - A_v - 1] + e_v = \mathcal{B}_v[i^* - 1]$$

Consider the remaining case, when $i^* - A_v - 1 < -k$. Then we have $\mathcal{B}_v[A_v - k] = \mathcal{B}_u[-k] + e_v$ and for which $i < A_v - k$, the value of $\mathcal{B}_v[i]$ is computed in line 7 of procedure COMPUTE $\mathcal{B}(v)$. In this case we have

$$\mathcal{B}_v[A_v - k] = \mathcal{B}_v[A_v - k - 1] = \cdots = [-k]$$

hence the claim of the lemma in this case is also true.

Case 3 (node v is of type (d)). The claim of the lemma follows directly from line 11 of procedure COMPUTE $\mathcal{B}(v)$. Indeed, in this case either we have $\mathcal{B}_v[i^* - 1] \geq \mathcal{B}_v[i^*]$ or the condition of the if-clause from line 11 is true and after its execution we obtain $\mathcal{B}_v[i^* - 1] = \mathcal{B}_v[i^*]$.

This completes the proof. $\square$

Before proving that no broadcasting algorithm can bring to the root more energy than Algorithm TEST-BROADCAST-FROM-ROOT(T) in line 7, we give some intuition.

For any node v of type (b), the computation of each component $\mathcal{B}_v[i]$ results in a choice of respective indices j, h for nodes u_1, u_2, which maximize the energy that may be transferred towards the root. This implies the optimal choice of M-flows through the exit edges of T_{u_1} and T_{u_2}. Considering exit edges in the top-down order, we can compute then the index i_v of each table $\mathcal{B}_v[]$ obtaining the M-flows of all edges of tree T, which result in the deposit of $\mathcal{B}_r[0]$ energy at the root. Each such value i_v is the difference between the number of agents traversing the exit edge $v \to w$ of T_v and the reverse edge $w \to v$. We call i_v an *optimal flow index* of node v.

The broadcasting schedule induces some directed multigraph T_M, built over the set of nodes of T: each directed edge $x \to y \in T_M$ corresponds to a traversal by some agent of the edge $(x, y) \in T$. We can partition the edges of T_M into three classes: *packet transfer* edges T_p, *energy transfer* edges T_e and *agent migration* edges T_a. The class T_p is composed of one copy of each edge $x \to y$, such that x is parent of y in T. As the packet kept in r needs to be broadcast to all nodes of T, each such multi-edge is clearly needed in T_M.

The class T_e contains multi-edges going up the tree T. Their goal is to conjointly move portions of energy in the direction of the root. Clearly only some edges of T induce multi-edges belonging to T_e, as energy is moved towards the root only when some excess of it is available. Although the subgraph T_e may turn out to be disconnected we create in T_e a multi-edge $v \to y$ whenever $\mathcal{B}[i_v] > 2 \cdot weight(v, y)$ or when $i_v \geq 0$ and $\mathcal{B}[i_v] > weight(v, y)$.

All remaining edges of T_M belong to class T_a and their number and direction observe the M-flow of the corresponding edge of T. In particular, consider any exit edge $v \to y$ of T_v. Let $k = 1$ if there exists a multi-edge $v \to y \in T_e$, otherwise $k = 0$. Then if the optimal flow index $i_v \geq 0$, there are $i_v - k$ copies of $v \to y$ multiedge in T_e. Otherwise, if $i_v < 0$, we have $k - 1 - i_v$ copies of the reverse, $y \to v$ multiedge in T_e.

In the sequel, we denote by $f(x \to y)$ the element of the M-flow being the number of agents that traverse the edge $x \to y$. Consequently, if x is a child of y in T, we have

$$i_x = f(x \to y) - f(y \to x)$$

The following lemma shows that the value of $\mathcal{B}_r[0]$, as computed by the algorithm TEST-BROADCAST-FROM-ROOT(T), is the upper bound on the amount of energy, which may be left at the root of T, when executing a successful broadcasting algorithm.

Lemma 3. *There exists no schedule of agents' movements which performs successful broadcasting and results in depositing at the root r of tree T an amount of energy larger than $\mathcal{B}_r[0]$.*

Proof. Suppose that the claim of the lemma is not true and consider a broadcasting schedule S^* depositing at r an amount of energy $E^* > \mathcal{B}_r[0]$. Let M^* denote the agent migration flow of such schedule and M_v denote (for any node v) the amount of energy E_v, deposited by S^* at node v. We prove by induction on the height of node v that $E_v \leq \mathcal{B}_v[i_v]$, where i_v is the agents' migration flow through the exit edge of T_v, as computed by the algorithm TEST-BROADCAST-FROM-ROOT(T). The claim is clearly true for each node v of height 0 (leaf). Suppose that the claim is true for any node of height h and consider a node v of height $h + 1$. Let i_v be the flow of the exit edge of T_v in M^*. Three cases are possible:

Case 1 (node v is of type (b)), see Fig. 1. Let i_{u_1} and i_{u_2} denote the flows of M^* through the edges (u_1, v) and (u_2, v), respectively. As $weight(u_1, v) = weight(u_2, v) = 0$, we can assume that in the optimal schedule S^* no agent finishes its walk at node v. Moreover, as no agent was initially present at v we have $i_v = i_{u_1} + i_{u_2}$. By the inductive hypothesis $E_{u_1} \leq \mathcal{B}_{u_1}[i_{u_1}]$ and $E_{u_2} \leq \mathcal{B}_{u_2}[i_{u_2}]$. Hence,

$$E_v \leq E_{u_1} + E_{u_2} \leq \mathcal{B}_{u_1}[i_{u_1}] + \mathcal{B}_{u_2}[i_{u_2}] \leq \max\{\mathcal{B}_{u_1}[j] + \mathcal{B}_{u_2}[k] : j + k = i_v\} = \mathcal{B}_v[i_v]$$

For the remaining Cases 2 and 3 we denote by i_u the flow of M^* along the exit edge of T_u and we assume, by the inductive hypothesis, that $E_u \leq \mathcal{B}_u[i_u]$.

Case 2 (node v is of type (c)). As at node v there are A_v new agents with total energy e_v, we have $i_v \leq i_u + A_v$, otherwise flow M^* is not feasible. Therefore, by line 6 of the procedure we have

$$E_v \leq E_u + e_v \leq \mathcal{B}_u[i_u - A_v] + e_v = \mathcal{B}_v[i_v]$$

Case 3 (node v is of type (d)). Denote $w = weight(u, v)$. In order to show the bound on E_v, we need to evaluate the cost of energy $\tilde{E}$ spend by the agents traversing the edge $e = (u, v)$ in both directions. Clearly

$$\tilde{E} \geq w \cdot (f(u \to v) + f(v \to u))$$

There is at least one agent which has to traverse edge e in the direction from v to u, namely the agent which brings the broadcast packet to u. Therefore, $f(v \to u) \geq 1$. We will consider 3 sub-cases of Case 3.

Sub-case 3a ($i_u \geq 0$). In this situation, $i_u + 1$ agents have to traverse the edge $u \to v$, i.e. $f(u \to v) = i_u + 1$ and

$$\tilde{E} \geq w \cdot (f(u \to v) + f(v \to u)) = (|i_u| + 2) \cdot w$$

However, this case is treated at line 9 of procedure COMPUTE $\mathcal{B}(v)$, according to which

$$E_v = E_u - \tilde{E} \leq \mathcal{B}_u[i_v] - (i_u + 2) \cdot w = \mathcal{B}_v[i_v] \tag{1}$$

Sub-case 3b $((i_u < 0) \wedge (\mathcal{B}_u[i_u] \geq 2w))$, i.e. there exists an amount of energy that is not needed to perform a local broadcasting inside T_u. If $f(u \to v) = 0$, then no agent can transport this energy so it may be subsequently potentially transferred to the root r of T. In such a case this energy is lost. We can then assume that one agent transports the amount of $\mathcal{B}_u[i_u] - 2w$ units of energy from u to its parent v and later there is an agent that returns to u with the packet, so that the broadcasting inside T_u may be successfully completed. Note that such energy may later reach the root or lost (if the cost of its subsequent transfer is to large), but its transfer from u to v would never imply that the energy deposited at the root is diminished.

Consequently, we can assume that $f(u \to v) = 1$, hence $f(v \to u) = -i_u - 1$ and the Eq. (1) holds in this sub-case as well.

Sub-case 3c $((i_u < 0) \wedge (\mathcal{B}_u[i_u] < 2w))$. In this case, even if $\mathcal{B}_u[i_u] > 0$ the extra energy available at u cannot be transferred towards the root without a loss of energy deposited at the root. Indeed, as $i_u < 0$, an attempt to transfer energy from u to v results in $f(u \to v) \geq 1$, which in turn implies $\tilde{E} \geq |i_u + 2|w$. In such a case we would obtain

$$\mathcal{B}_v[i_v] \leq \mathcal{B}_u[i_u] - \tilde{E} \leq \mathcal{B}_u[i_u] - |i_u + 2|w < -|i_u| \cdot w = i_u \cdot w$$

where the last amount equals the energy potential computed for this case in line 10 of procedure COMPUTE $\mathcal{B}(v)$. We conclude by induction that $E_r \leq \mathcal{B}_r[0]$, which completes the proof. $\square$

3   Constructing Broadcast Schedule

Lemma 3 shows the upper bound on the amount of energy which may be deposited at the root of the tree from which a broadcast may be performed. We present now an algorithm generating a broadcasting schedule which succeeds in depositing such maximal amount of energy. Obviously, if this amount of energy is negative, there exists no broadcasting schedule for the given tree. The rough idea of the algorithm is the following. Firstly, we compute an optimal M-flow of T. Then, using this M-flow, in each sub-tree, the excessive energy is transferred up the tree towards the root, deposited there and never used. Finally, a recursive procedure transferring the broadcast packet from the root r is called. This procedure performs the agents' transfer according to the optimal M-flow computed before. Interestingly, the algorithm performs four traversals of tree T, which are alternately bottom-up (1st and 3rd) and top-down (2nd and 4th). Each of them may be given as a recursive procedure, but, for clarity, we chose only the last one to be recursive. Before giving our algorithm we describe below its idea in more details. Because of the lack of space the proof of correctness of the algorithm is deferred to the Appendix.

ALGORITHM BROADCAST-IN-TREE(T);

STEP 1: TEST-BROADCAST-FROM-ROOT(T);

STEP 2: $r.Mindex := 0$;

        for each edge $u \to v$ of T taken in top-down order **do**

           $u.Mindex :=$ the value used to obtain $v.Mindex$ when

                computing $\mathcal{B}_v$ in TEST-BROADCAST-FROM-ROOT(T)

STEP 3: **for** each edge $u \to v$ of T taken in an bottom-up order **do**

        if $\mathcal{B}_u[u.Mindex] > 0$ **then** MOVE-EXTRA-ENERGY-UP(u)

STEP 4: REMOVE-ENERGY($r, \mathcal{B}_r[0]$);

        INFORM(r);

In the first step of the algorithm we call procedure TEST-BROADCAST-FROM-ROOT(T) computing for each node v its table $\mathcal{B}_v$.

The second step performs a top-down traversal of tree T and, using the tables $\mathcal{B}$ compute an optimal distribution of agent flows between any given node and its children. Note that, when a node v is of type (b) for any already computed optimal flow i_v through the exit edge of tree T_v, we obtain feasible optimal flows i_{u_1} and i_{u_2}. This step computes the entire M-flow of tree T, which results in achieving the amount of $\mathcal{B}_r[0]$ energy deposited at the root. The rest of the algorithm refers to this M-flow.

The agents' moves forming an optimal broadcasting schedule are generated in the third and the fourth step of the algorithm. The third step generates the moves corresponding to the energy transfer edges T_e of the multigraph T_M. This step makes a bottom-up traversal of T and identifies the subtrees containing excessive energy, i.e. the energy which is not needed to perform the local broadcast inside the subtree. Such amounts of energy are moved up the tree by mobile agents and the amount of $\mathcal{B}_r[0]$ energy is eventually accumulated at the root. Observe that a bottom-up traversal ensures that, whenever two or more different agents arrive to the same node, they wait until the last such agent appear at the node. Then a single agent collects the energy of all other agents and alone continues its upward walk. Note that, it may happen that some energy of the system may be lost (cf. energy of agent H from Fig. 2) because transferring it up the tree results in a bigger cost than the amount of energy to transport. Moreover, the set of edges T_e may be disconnected, hence some energy that is moved up might not reach the root. The edges used to transfer energy are marked, so that in the rest of the algorithm the number of agents traversing them according to the computed M-flow is diminished by 1.

The final, fourth step starts when one of the agents walking up the tree reaches the root and deposits there the amount of $\mathcal{B}_r[0]$ energy. This agent starts the procedure of distributing it down the tree. The diffusion is performed by the recursive procedure INFORM. When a recursive call of INFORM is made to a node with two children, a child with a larger flow is visited first. Indeed, it might be necessary to bring superfluous agents exiting from this child in order to use them in its sibling sub-tree.

Procedure INFORM generates the sequence of agents' moves resulting in moving the packet from the root down to all other nodes. It also controls the migration of the agents according to the optimal M-flow computed earlier.

PROCEDURE INFORM(v);

1.   **case** type of node v **of**
2.   (a) **Exit** ;
3.   (b) **if** $u_1.Mindex \geq u_2.Mindex$ **then** $f_1 := u_1, f_2 := u_2$
        else $f_1 := u_2, f_2 := u_1$;
4.   MOVE-DOWN(u_1); INFORM(u_1); MOVE-UP(u_1);
5.   MOVE-DOWN(u_2); INFORM(u_2); MOVE-UP(u_2);
6.   (c), (d) MOVE-DOWN(u); INFORM(u); MOVE-UP(u);

Procedure INFORM calls two procedures MOVE-DOWN(u) and MOVE-UP(u) to execute the travel according to the flow of the exit edge $u \to v$, for $v = parent(u)$. These procedures call the function MOVE, whose purpose is to generate the schedule of the agents' moves.

Procedure MOVE-EXTRA-ENERGY-UP(u) tests whether the excessive energy available in T_u is sufficiently large so that the transferring it along edge $u \to v$ is not too costly.

PROCEDURE MOVE-EXTRA-ENERGY-UP(u);

1. $v := parent(u)$; $w := weight(u, v)$; $i_u := u.Mindex$;
2.   **if** $(i_u < 0 \wedge \mathcal{B}_u[i_u] > 2w) \vee \mathcal{B}_u[i_u] > w$ **then**
3.   MOVE-ENERGY(u); MARK-EDGE($u \to v$);

Procedures MOVE-DOWN(u) and MOVE-UP(u) generate the moves of the groups of agents, respectively, up and down a given edge $u \to parent(u)$. We assume that, when the sequence of such moves is made from the function MOVE-UP(u), or from the function MOVE-DOWN(u) for which u is an only child, they carry the entire energy available. Otherwise, if u has a sibling (i.e. it's parent is a node of type (b)), the energy is split according to the calls to MOVE-DOWN(u_1) and MOVE-DOWN(u_2). Procedure MOVE-ENERGY(u) creates a move of a single agent up the tree from node u, carrying entire energy (brought to u by all agents present there).

PROCEDURE MOVE-DOWN(u);

1. $v := parent(u)$; $i_u := u.Mindex$;
2.   MOVE($v \to u$); {move of the agent transfering the packet}
3.   **if** edge $u \to v$ marked **then** $e := -i_u$ **else** $e := -(i_u + 1)$;
4.   **for** $i := 1$ **to** e **do** MOVE($v \to u$); {moves of other agents according to M-flow}

PROCEDURE MOVE-UP(u);

1. $v := parent(u)$; $i_u := u.Mindex$;

2.   **if** edge $u \to v$ marked **then** $e := i_u - 1$ **else** $e := i_u$; {e agents to exit T_u;
       if edge marked, an energy-transferring agent moved earlier}

3.   **for** $i := 1$ **to** e **do** MOVE($u \to v$); {agents exiting according to M-flow}

Lemma 4. *The excessive energy of $\mathcal{B}_r[0]$ is accumulated at the root r at the completion of STEP 2 of algorithm* BROADCAST-IN-TREE.

Proof. We prove that, after each iteration of the for-loop from STEP 2, the maximal energy $\mathcal{B}_x[x.Mindex]$, available inside T_x but not used for the broadcast from x into T_x, is present at node x. Moreover, if $\mathcal{B}_x[x.Mindex] > 0$ at least one agent is present at x. The prove goes by induction on the height of T_x. The statement is clearly true for x being a leaf of T. Suppose that the statement of the lemma is true for trees of height h. Take edge $x \to y$ leading to node y of height $h + 1$. If $\mathcal{B}_x[x.Mindex] \leq w = weight(x, y)$ then no energy may be transferred to node y, as the cost of this transfer equals w. Otherwise, if $i = x.Mindex < 0$, two cases are possible: either i agents enter T_x and no agent exits it, so no energy transfer along edge $x \to y$ is made, or, one agent exits T_x transporting excessive energy along edge $x \to y$ and $i + 1$ agents enter T_x. The total cost of all traversals of edge $x \to y$ is then iw in the former case and $(i+2)w$ in the latter one. The transfer of such energy is then profitable only in the case when it exceeds $2w$, otherwise it is lost and never used in the schedule. This is exactly what is done in lines 2 and 3 of procedure MOVE-EXTRA-ENERGY-UP. Observe that, as $\mathcal{B}_x[x.Mindex] > 0$, by inductive assumption, the energy required for the transfer is already available at x as well as an agent necessary to perform the transfer is present at x. $\square$

We omit the proof of the following fact.

Lemma 5. *All agent actions generated by the calls of procedure* INFORM(y) *are feasible, i.e. when an agent move is generated along an edge $x \to y$ (or along an edge $y \to x$) then, there is an agent available at node x (or y) and there is at least $weight(x, y)$ energy available at x (or y), so that such move may be successfully completed.*

Theorem 1. *If $\mathcal{B}_r[0] \geq 0$, the algorithm* BROADCAST-IN-TREE *produces a correct broadcasting schedule, which deposits $\mathcal{B}_r[0]$ energy at the root of T.*

Proof. By Lemma 4 the excessive energy $\mathcal{B}_r[0]$ is indeed accumulated at the root r at the end of the for-loop from STEP 2. This energy is removed by the function REMOVE-ENERGY.

The remaining part of the algorithm is done by the function INFORM. By Lemma 5, all agent moves generated from the main call of INFORM(r) are feasible. Observe, that the sequence of produced agent moves contains a subsequence which results in transferring the information from the root to all other nodes. Indeed, procedure INFORM is called for the root r and then recursively for all other nodes of the tree in the top-down order. Each call of INFORM to any node v contains calls to the children of v, at the same time generating agent moves along edges leading to these children. This results in the transfer of the broadcast information to all nodes. This completes the proof. $\square$

Theorem 2. *The optimal broadcasting on a tree of n nodes having k agents can be computed in $O(nk^2)$ time. The size of the optimal schedule is $O(nk)$.*

Proof. Observe that for any given tree of n nodes its converted version (cf. Fig. 2) is of size $\Theta(n)$ and it may be obtained in $O(n)$ time. The time complexity of algorithm TEST-BROADCAST-FROM-ROOT is dominated by the call of BROADCAST-IN-TREE(T), which calls function COMPUTE $\mathcal{B}$ $O(n)$ times. The function COMPUTE $\mathcal{B}$ consists of a for-loop executed $O(k)$ times. The most expensive case is when an iteration of the loop executes the max operation at line 4 (the case of node of type (b)) which takes $O(k)$ time, resulting in the overall time of $O(nk^2)$. The complexity of for-loops from lines 2 and 4 of the algorithm TEST-BROADCAST-FROM-ROOT equal $O(n)$ as each iteration is carried in constant time. Finally, the recursive call to function INFORM in STEP 4 takes $O(nk)$ time. Indeed, there are $O(n)$ total number of calls of function INFORM. Every call to INFORM executes functions MOVE-UP and MOVE-DOWN, each one containing a loop executed the number of times equal to the flow of the corresponding parameter. As each flow is bound by k, we have $O(nk)$ complexity of the function INFORM and the same complexity for the number of agent moves generated by our algorithm. $\square$

4   Final Remarks

Our approach may be extended as follows:

Theorem 3. *In $O(nk^2)$ time we can find all nodes from which we can perform a successful broadcasting in a given tree T of n nodes with k agents.*

Proof. (sketch) Consider a converted version of T. The algorithm BROADCAST-IN-TREE(T) computes the optimal energy, which can be deposited by the successful broadcast at its root r. The algorithm considers the set of directed edges $u \rightarrow v$, for all nodes u, v such that the path from u to r in T contains v. The algorithm computes tables $\mathcal{B}_u$ performing a bottom-up traversal of this set of edges. Observe that we can consider the set of all directed edges of the tree (in both directions), which is only twice larger. It is easy to see, that this set may be ordered so that for any edge $u \rightarrow v$, each edge $y \rightarrow u$, for $u \neq y$ is earlier in this order. We can run slightly modified our algorithm BROADCAST-IN-TREE for such set of edges. Consequently, for any node x of the tree, we will have available the tables $\mathcal{B}_u$ for u being children of x (x has at most two children for the converted version of T). Having them, in constant time we can compute the largest energy which may be delivered to x, from which the broadcast needs to be performed. $\square$

We pose as an open question an improvement of the complexity of our algorithm. Other open questions concern the design of polynomial-time algorithms for the problem of broadcasting from a set of many source nodes and for the gossiping problem for trees (in energy exchange setting).

Acknowledgment. The work of K. Diks and W. Rytter was supported by National Science Centre of Poland grant NCN2014/13/B/ST6/00770. The work of J. Czyzowicz was supported by NSERC.

References

1. Anaya, J., Chalopin, J., Czyzowicz, J., Labourel, A., Pelc, A., Vaxès, Y.: Collecting information by power-aware mobile agents. In: Aguilera, M.K. (ed.) DISC 2012. LNCS, vol. 7611, pp. 46–60. Springer, Heidelberg (2012). https://doi.org/10.1007/978-3-642-33651-5_4
2. Albers, S., Henzinger, M.R.: Exploring unknown environments. SIAM J. Comput. **29**(4), 1164–1188 (2000)
3. Awerbuch, B., Goldreich, O., Peleg, D., Vainish, R.: A trade-off between information and communication in broadcast protocols. J. ACM **37**(2), 238–256 (1990)
4. Annamalai, V., Gupta, S.K.S., Schwiebert, L.: On tree-based convergecasting in wireless sensor networks. IEEE Wirel. Commun. Netw. **3**, 1942–1947 (2003)
5. Averbakh, I., Berman, O.: A heuristic with worst-case analysis for minimax routing of two traveling salesmen on a tree. Discr. Appl. Math. **68**, 17–32 (1996)
6. Bärtschi, A., Chalopin, J., Das, S., Disser, Y., Geissmann, B., Graf, D., Labourel, A., Mihalák, M.: Collaborative delivery with energy-constrained mobile robots. In: Suomela, J. (ed.) SIROCCO 2016. LNCS, vol. 9988, pp. 258–274. Springer, Cham (2016). https://doi.org/10.1007/978-3-319-48314-6_17
7. Bärtschi, A., Chalopin, J., Das, S., Disser, Y., Graf, D., Hackfeld, J., Penna, P.: Energy-efficient delivery by heterogeneous mobile agents. In: STACS , pp. 10:1–10:14 (2017)
8. Bärtschi, A.: Efficient Delivery with Mobile Agents (Ph.D. thesis), ETH Zurich (2017)
9. Baeza-Yates, R.A., Schott, R.: Parallel searching in the plane. Comput. Geom. **5**, 143–154 (1995)
10. Bar-Yehuda, R., Goldreich, O., Itai, A.: On the time-complexity of broadcast in multi-hop radio networks: an exponential gap between determinism and randomization. J. Comput. Syst. Sci. **45**(1), 104–126 (1992)
11. Chalopin, J., Jacob, R., Mihalák, M., Widmayer, P.: Data delivery by energy-constrained mobile agents on a line. In: Esparza, J., Fraigniaud, P., Husfeldt, T., Koutsoupias, E. (eds.) ICALP 2014. LNCS, vol. 8573, pp. 423–434. Springer, Heidelberg (2014). https://doi.org/10.1007/978-3-662-43951-7_36
12. Czyzowicz, J., Diks, K., Moussi, J., Rytter, W.: Energy-optimal broadcast in a tree with mobile agents. In: Fernández Anta, A., Jurdzinski, T., Mosteiro, M.A., Zhang, Y. (eds.) ALGOSENSORS 2017. LNCS, vol. 10718, pp. 98–113. Springer, Cham (2017). https://doi.org/10.1007/978-3-319-72751-6_8
13. Czyzowicz, J., Diks, K., Moussi, J., Rytter, W.: Communication problems for mobile agents exchanging energy. In: Suomela, J. (ed.) SIROCCO 2016. LNCS, vol. 9988, pp. 275–288. Springer, Cham (2016). https://doi.org/10.1007/978-3-319-48314-6_18

14. Das, S., Dereniowski, D., Karousatou, C.: Collaborative exploration by energy-constrained mobile robots. In: Scheideler, C. (ed.) Structural Information and Communication Complexity. LNCS, vol. 9439, pp. 357–369. Springer, Cham (2015). https://doi.org/10.1007/978-3-319-25258-2_25

15. Dynia, M., Korzeniowski, M., Schindelhauer, C.: Power-aware collective tree exploration. In: Grass, W., Sick, B., Waldschmidt, K. (eds.) ARCS 2006. LNCS, vol. 3894, pp. 341–351. Springer, Heidelberg (2006). https://doi.org/10.1007/11682127_24

16. Fraigniaud, P., Gasieniec, L., Kowalski, D.R., Pelc, A.: Collective tree exploration. In: Farach-Colton, M. (ed.) LATIN 2004. LNCS, vol. 2976, pp. 141–151. Springer, Heidelberg (2004). https://doi.org/10.1007/978-3-540-24698-5_18

17. Frederickson, G., Hecht, M., Kim, C.: Approximation algorithms for some routing problems. SIAM J. Comput. **7**, 178–193 (1978)

18. Toth, P., Vigo, D.: Vehicle Routing. Problems, Methods and Applications. MOS-SIAM series on Optimization, 2nd edn. (2014)

A Deterministic Distributed 2-Approximation for Weighted Vertex Cover in $O(\log N \log \Delta / \log^2 \log \Delta)$ Rounds

Ran Ben-Basat[1]($\boxtimes$), Guy Even[2], Ken-ichi Kawarabayashi[3], and Gregory Schwartzman[3]

[1] Department of Computer Science, Technion, Haifa, Israel
sran@cs.technion.ac.il
[2] School of Electrical Engineering, Tel Aviv University, Tel Aviv, Israel
guy@eng.tau.ac.il
[3] NII, Tokyo, Japan
{k_keniti,greg}@nii.ac.jp

Abstract. We present a deterministic distributed 2-approximation algorithm for the Minimum Weight Vertex Cover problem in the CONGEST model whose round complexity is $O(\log n \log \Delta / \log^2 \log \Delta)$. This improves over the currently best known deterministic 2-approximation implied by [KVY94]. Our solution generalizes the $(2 + \epsilon)$-approximation algorithm of [BCS17], improving the dependency on ϵ^{-1} from linear to logarithmic. In addition, for every $\epsilon = (\log \Delta)^{-c}$, where $c \geq 1$ is a constant, our algorithm computes a $(2 + \epsilon)$-approximation in $O(\log \Delta / \log \log \Delta)$ rounds (which is asymptotically optimal).

1 Introduction

The Minimum Weight Vertex Cover Problem (MWVC) is defined as follows. The input is a graph $G = (V, E)$ with nonnegative vertex weights $w(v)$. A subset $U \subseteq V$ is a *vertex cover* if, for every edge $e = \{u, v\}$, the intersection $U \cap \{u, v\}$ is not empty. The weight of a subset of vertices U is $\sum_{v \in U} w(v)$. The goal is to find a minimum weight vertex cover. This problem is one of the classical NP-hard problems [Kar72].

In this paper we deal with distributed deterministic approximation algorithms for MWVC. We focus on the CONGEST model of distributed computation in which the communication network is the graph G itself.[1] Computation

---

R. Ben-Basat—This work was partially sponsored by the Technion-HPI research school.

K. Kawarabayashi and G. Schwartzman—This work was supported by JST ERATO Grant Number JPMJER1201, Japan.

[1] In the CONGEST model vertices have distinct IDs (that are polynomial in $|V|$), however, as in [BCS17], our algorithm works also in the case of anonymous vertices.

© Springer Nature Switzerland AG 2018
Z. Lotker and B. Patt-Shamir (Eds.): SIROCCO 2018, LNCS 11085, pp. 226–236, 2018.
https://doi.org/10.1007/978-3-030-01325-7_21

proceeds in synchronous rounds. Each round consists of three parts: each vertex receives messages from its neighbors, performs a local computation, and sends messages to its neighbors. The sent messages arrive at their destination in the beginning of the next round. In the CONGEST model, message lengths are bounded by $O(\log |V|)$. In order to send vertex weights, we assume that all the vertex weights are positive integers bounded by polynomial in $n \triangleq |V|$. See [BCS17, ÅS10] for detailed overviews of distributed algorithms for MWVC.

Let Δ denote the maximum vertex degree in G. Two of the most relevant results in this setting to our paper are the lower bound of [KMW16] and the upper bound of [BCS17]. The lower bound of Kuhn et al. [KMW16] states that every constant approximation algorithm for MWVC requires at least $\Omega(\log \Delta / \log \log \Delta)$ rounds of communication. The upper bound of Bar-Yehuda et al. [BCS17] presents a deterministic distributed $(2 + \epsilon)$-approximation algorithm (BCS Algorithm) that requires $O(\log \Delta / (\epsilon \cdot \log \log \Delta))$ rounds for $\epsilon \in (0, 1)$. For $\epsilon = \Omega(\log \log \Delta / \log \Delta)$, the running time is $O(\log \Delta / \log \log \Delta)$, with no dependence on ϵ, and is optimal according to [KMW16].

In this paper, we present a generalization of the BCS Algorithm with improved guarantees on the running time for certain ranges of ϵ. We focus on decreasing the dependency of the number of rounds on ϵ. Since the round complexity of the BCS Algorithm is optimal for constant values of ϵ (and even $\epsilon = \Omega(\log \log \Delta / \log \Delta)$), we consider values of ϵ that depend on Δ.

Our main result[2] is a deterministic distributed $(2 + \epsilon)$-approximation algorithm in which the number of rounds is bounded by

$$O\left(\frac{\log \Delta}{\log \log \Delta} + \frac{\log \epsilon^{-1} \log \Delta}{\log^2 \log \Delta} \right).$$

This result assumes that all the vertices know Δ or an estimate that is a polynomial in Δ. This result leads to the following consequences:

1. If $\epsilon^{-1} = (\log \Delta)^c$, for a constant $c > 0$, then the number of rounds asymptotically matches the lower bound, and is thus optimal. In [BCS17] the same asymptotic running time is guaranteed only for $\epsilon^{-1} = O(\log \Delta / \log \log \Delta)$.
2. If $\epsilon^{-1} = (\log \Delta)^{\omega(1)}$, then the dependency of the round complexity on $1/\epsilon$ is reduced from linear to logarithmic. In addition, the round complexity is decreased by an additional factor of $\log \log \Delta$.
3. Every $(2 + \epsilon)$-approximation is a 2-approximation if $\epsilon < 1/(nW)$, where $W = \max_v w(v)$. Since we assume that $W = n^{O(1)}$, where $n = |V|$, we obtain a 2-approximate deterministic distributed algorithm for MWVC with round complexity $O(\log n \cdot \log \Delta / \log^2 \log \Delta)$. This improves over the 2-approximation in $O(\log^2 n)$ rounds implied by [KVY94][3] (which has the lowest round complexity for deterministic 2-approximation to the best of our knowledge).

---

[2] All logarithms are base 2 unless the basis is written explicitly.
[3] The actual result is stated as a $(2 + \epsilon)$-approximation in $O(\log \epsilon^{-1} \log n)$ rounds, from which we infer a 2-approximation by setting $\epsilon = 1/nW$.

Our round complexity increases for the case that the maximum degree Δ is unknown to the vertices of the graphs. We propose two alternatives for the case that Δ is unknown. The first alternative holds for every $\epsilon \in (0,1)$, and achieves a $(2 + \epsilon)$-approximate solution with a round complexity of $O\left(\frac{\log \epsilon^{-1} \log \Delta}{\log \log \Delta}\right)$. The second alternative holds for $\epsilon > (\log \Delta)^q$, where $q > 0$ is a constant. In the second alternative, a $(2+\epsilon)$-approximation is achieved with an optimal asymptotic round complexity of $O(\log \Delta / \log \log \Delta)$.

Our algorithm builds on the BCS Algorithm [BCS17]. This algorithm adapts the *local ratio* framework [BE85] to the distributed setting, with several improvements that provide the desired speedup. The BCS Algorithm can be also interpreted as the following "primal-dual" algorithm. Essentially the algorithm aims to increase the edge variables (i.e., *dual*) such that the following holds:

1. The sum of edge variables incident to every node does not exceed its weight (*feasibility of edge variables*).
2. The set of vertices whose edge variable sum is at least $(1 - \epsilon)$-fraction of the vertex weight constitute a *vertex cover*.

The above conditions yield a $(2 + \epsilon)$-approximation for MWVC.

The challenge in the above framework is to maintain feasibility of the edge variables while converging as fast as possible to a vertex cover. To increase the edge variables, vertices send *offers* to their neighbors. The neighbors respond to these offers in a way that guarantees feasibility of the edge variables. This requires a coordination mechanism in the distributed setting, as a vertex both sends and receives offers simultaneously. To this end, the weight of every vertex is divided into two parts: *vault* and *bank*. Offers are allocated from the vault, while responses are allocated from the bank, respectively. Hence the agreed upon increases to the edge variables do not violate the feasibility of the edge variables. The BCS algorithm sets the vault to be an ϵ-fraction of the vertex weight (and the bank to be the remainder). This leads to a running time of $O(\epsilon^{-1} \log \Delta / \log \log \Delta)$ and $O(\log \Delta / \log \log \Delta)$ if $\epsilon = \Omega(\log \log \Delta / \log \Delta)$.

Our algorithm introduces three modifications to the BCS Algorithm, which allows us to improve the round complexity. First, we attach *levels* to the vertices that measure by how much the remaining weight of a vertex has decreased. Second, the size of the vault decreases as the level of the vertex increases. Third, offers are not sent to all the neighbors. Instead, offers are sent only to the neighbors whose level is the smallest level among the remaining neighbors.

Related Work. An excellent overview of the related work is presented in [BCS17, ÅS10] which we summarize hereinafter. Minimum vertex cover is one of Karp's 21 NP-hard problems [Kar72]. A simple 2-approximation for the unweighted version can be achieved by a reduction from maximal matching (see, e.g., [CLRS09, GJ79]). For the weighted case, [BE81] achieves the first linear-time 2-approximation algorithm using the primal-dual schema, while [BE85] achieves the same result using the local-ratio technique. Prior to that, the first polynomial-time 2-approximation algorithm was due to [NJ75] and observed by [Hoc82]. For any constant $\epsilon > 0$, if the Unique Games conjecture holds, no polynomial-time algorithm can compute a $(2-\epsilon)$ approximation of the minimum vertex cover [KR08].

Let us now turn our attention to the distributed setting. Let us start from the unweighted case. A 2-approximation can be found in $O(\log^4 n)$ rounds [HKP01] and in $O(\Delta+\log^* n)$ rounds [PR01]. Completely local algorithms with no dependence on n are presented in [ÅFP+09] which gives an $O(\Delta^2)$-round 2-approximation algorithm, and in [PS09] which gives an $O(\Delta)$-round 3-approximation algorithm. Using the maximal matching algorithm of [BEPS12] gives a 2-approximation algorithm for vertex cover in $O(\log \Delta + (\log \log n)^4)$ rounds. This can be made into a $(2 + 1/\text{poly}\Delta)$-approximation algorithm within $O(\log \Delta)$ rounds [Pet16].

For the weighted case, [GKP08] presents a randomized 2-approximation algorithm in $O(\log n + \log W)$ rounds (where W is a bound on the vertex weights). In [KY11] the first (randomized) 2-approximation algorithm running in $O(\log n)$ rounds is presented (note that the running time is logarithmic in n and independent of the weights). A deterministic 2-approximation algorithm in $O(\Delta+\log^* n)$ rounds is given within [PR01]. In [KVY94], a deterministic $(2+\epsilon)$-approximation algorithm is given within $O(\log \epsilon^{-1} \log n)$ rounds. As for deterministic algorithms independent of n, [KMW06] presents a $(2 + \epsilon)$-approximation algorithm in $O(\epsilon^{-4} \log \Delta)$ rounds and [ÅFP+09] presents a 2-approximation algorithm in $O(1)$ rounds for $\Delta \leq 3$, while [ÅS10] presents a 2-approximation algorithm in $O(\Delta + \log^* W)$ rounds (where $W \triangleq \max_v w(v)$). Finally in [BCS17] a deterministic $(2 + \epsilon)$-approximation which runs in $O(\epsilon \log \Delta / \log \log \Delta)$ rounds is given. In [Sol18] a $(2 + \epsilon)$-approximation in $O(\epsilon^{-1} \log(\alpha/\epsilon) / \log \log(\alpha/\epsilon))$ rounds for graphs of arboricity bounded by α.

As the result of [Sol18] uses the algorithm of [BCS17] as a black box, plugging $\Delta = \alpha/\epsilon$, our results can also be used. This means all of results stated in this paper also hold for bounded arboricity graphs setting $\Delta = \alpha/\epsilon$. We list the previous results and the results of this paper in Table 1 (Adapted from [ÅS10]).

2   The MWVC Local Ratio Template

In this section we overview [BCS17]'s local ratio paradigm for approximating MWVC. We note that the template does not assume anything about the model of computation and that our algorithms will fit into this framework. This template can also be viewed via the primal-dual schema.

Let $G = (V, E)$ denote a graph with a vertex-weight function $w : V \to \mathbb{R}^+$. An edge-weight function $\delta : E \to \mathbb{R}^+$ is G-valid if for every vertex v the incident edges weight sum does not exceed $w(v)$; that is, δ is G-valid if $\forall v \in V : \sum_{v \ni e} \delta(e) \leq w(v)$. (In fact, a G-valid function δ is a feasible solution to the dual edge packing LP.)

Next, for a G-valid function δ, define the vertex-weight function $\tilde{w}_\delta : V \to \mathbb{R}^+$ by $\tilde{w}_\delta(v) = \sum_{e:v \in e} \delta(e)$. Let $S_\delta = \{v \in V \mid w(v) - \tilde{w}_\delta(v) \leq \epsilon' w(v)\}$ be the set of vertices for which w and $\tilde{w}$ differ by at most $\epsilon' w(v)$, for $\epsilon' = \epsilon/(2 + \epsilon)$. We refer

Table 1. In the table (adapted from [ÅS10]), $n = |V|$ and $\epsilon \in (0,1)$. The running times are stated for the case of unit weight vertices. For randomized algorithms the running times hold in expectation or with high probability.

Deterministic	Weighted	Approximation	Time ($W = 1$)	Algorithm
no	yes	2	$O(\log n)$	[GKP08]
no	yes	2	$O(\log n)$	[KY09]
yes	no	3	$O(\Delta)$	[PS09]
yes	no	2	$O(\log^4 n)$	[HKP01]
yes	no	2	$O(\Delta^2)$	[ÅFP+09]
yes	yes	$2 + \epsilon$	$O(\log \epsilon^{-1} \log n)$	[KVY94]
yes	yes	2	$O(\log^2 n)$	[KVY94]
yes	yes	$2 + \epsilon$	$O(\epsilon^{-4} \log \Delta)$	[Hoc82, KMW06]
yes	yes	2	$O(\Delta + \log^* n)$	[PR01]
yes	yes	2	$O(\Delta)$	[ÅS10]
yes	yes	2	$O(1)$ for $\Delta \le 3$	[ÅFP+09]
yes	yes	$2 + \epsilon$	$O(\epsilon^{-1} \log \Delta / \log \log \Delta)$	[BCS17]
yes	yes	$2 + \frac{\log \log \Delta}{\log \Delta}$	$O(\log \Delta / \log \log \Delta)$	[BCS17]
yes	yes	$2 + \epsilon$	$O\left(\frac{\log \Delta}{\log \log \Delta} + \frac{\log \epsilon^{-1} \log \Delta}{\log^2 \log \Delta}\right)$	This work
yes	yes	$2 + (\log \Delta)^{-c}$	$O(\log \Delta / \log \log \Delta)$	This work, $\forall c = O(1)$
yes	yes	2	$O(\log n \log \Delta / \log^2 \log \Delta)$	This work

to vertices in S_δ as ϵ'-*tight* vertices. The essence of the template consists of two parts: (1) The sum of the weights of the vertices in S_δ is at most $(2+\epsilon)$ times the weight of an optimal solution to MWVC. (2) When the algorithm terminates, S_δ is a vertex cover.

Theorem 1. ([BCS17]) *Fix $\epsilon > 0$ and let δ be a G-valid function. Let OPT be the sum of weights of vertices in a minimum weight vertex cover S_{OPT} of G. Then $\sum_{v \in S_\delta} w(v) \le (2 + \epsilon)OPT$. In particular, if S_δ is a vertex cover, then it is a $(2 + \epsilon)$-approximation for MWVC for G.*

3   A Fast Distributed Implementation

In this section, we present a modification of the distributed algorithm for MWVC of Bar-Yehuda *et al.* [BCS17]. The pseudo-code for our algorithm is given in Algorithm 1. In this section we assume that the maximal degree Δ is known to all vertices.[4] In Sect. 4 we provide an algorithm with a slightly higher running time in which this assumption is lifted.

For clarity of presentation, we first describe an implementation for the LOCAL model. This algorithm can be easily adapted to the CONGEST model using the techniques of [BCS17].

---

[4] A polynomial upper bound of $\Delta^{O(1)}$ would yield the same asymptotic bound on the number of rounds.

Overview of Algorithm 1. The algorithm uses the following parameters: (i) $\epsilon' \triangleq \epsilon/(2+\epsilon)$, (ii) $\gamma \in (0,1)$, (iii) $z \triangleq \lceil \log_\gamma \epsilon' \rceil$. The parameter ϵ' is used for defining tightness of the dual packing constraint. The parameter γ is used for defining levels. Loosely speaking, in every iteration the weight of a vertex is reduced, and the level of a vertex is proportional to $\log_\gamma(w_i(v)/w_0(v))$. The parameter z is used to bound the number of levels till a vertex becomes ϵ'-tight, meaning that $w_i(v)/w_0(v) \leq \epsilon'$.

Our algorithm, listed as Algorithm 1, is a variation of the Algorithm of Bar-Yehuda *et al.* [BCS17] with a few modifications. We begin with a description of the common features. In the course of the algorithm, the weight of each vertex is reduced. Once a vertex v becomes ϵ'-tight (i.e., the reduced weight is an ϵ'-fraction of its original weight) it decides to join the vertex cover and terminates after sending the message $(v, cover)$ to its remaining neighbors. The message $(v, cover)$ causes the neighbors of v to erase v from their list of remaining neighbors. If a vertex v loses all its neighbors (i.e., becomes isolated), it decides that v is not in the vertex cover, and terminates. Upon termination, the ϵ'-tight vertices constitute a vertex cover.

The handling of offers is as in [BCS17]. Vertex v sends an irrevocable offer $request_i(v, u)$ to every $u \in N_i(v)$. The offers are allocated from the vault. The responses to the offers are allocated greedily from the bank, namely v's responses satisfy: $budget_i(v, u) \leq request_i(u, v)$ and $\sum_u budget_i(v, u) \leq bank_i(v)$. The updating of the weights can be interpreted as follows. For every edge $e = \{u, v\}$ the dual edge packing variable $\delta(e)$ is increased by $budget_i(u, v) + budget_i(v, u)$. The remaining weight satisfies $w_{i+1}(v) = w_0(v) - \sum_{e \ni v} \delta(e)$. Note that each iteration of the while-loop requires a constant number of communication rounds.

The first modification is that we attach a *level* to each vertex as follows. Let $w_i(v)$ denote the weight of v in the beginning of iteration i of the while-loop. The level of v in iteration i satisfies $\ell_i(v) = 1 + \left\lfloor \log_\gamma \frac{w_i(v)}{w_0(v)} \right\rfloor$. Note that the initial level is one, and that if the level of v is greater than z, then v is ϵ'-tight (see Claim 3.1).

The second modification is how we partition $w_i(v)$ between the vault and the bank. Instead of using a fixed fraction of the initial weight for the vault, our vault decreases as the level of the vertex increases. Formally, $vault_i(v) \triangleq w_0(v) \cdot \gamma^{\ell_i(v)}$. The bank is the rest of the weight, namely, $bank_i(v) \triangleq w_i(v) - vault_i(v)$.

The third modification is that in each iteration, every vertex v only sends offers to its remaining neighbors with the smallest level. Let $N_i(v)$ denote the set of remaining neighbors of v in the beginning of the ith iteration. The smallest level of the neighbors of v is defined by $\ell'_i(v) \triangleq \min\{\ell_i(u) \mid u \in N_i(v)\}$. The set of neighbors of lowest level is defined by $N'_i(v) \triangleq \{u \in N_i(v) \mid \ell_i(u) = \ell'_i(v)\}$. Let $d'_i(v) = |N'_i(v)|$. The size of each offer sent is $vault_i(v)/d'_i(v)$.

Note that if $\gamma = \epsilon'$, then Algorithm 1 reduces to the BCS Algorithm because there is just one level, and the vault size is fixed and equals to $\epsilon' \cdot w_0(v)$. On the other hand, if $\gamma = 1/2$, then there are $O(\log 1/\epsilon)$ levels. Per level $\ell_i(v)$, the algorithm can be viewed as a version of the BCS algorithm with $\epsilon' = 1/2^{\ell_i(v)}$. This also explains why our algorithm may be adapted to the CONGEST model of distributed computation using the techniques of [BCS17]. In essence they give an adaptation for a single level of our algorithm, which can easily be extended to multiple levels.

We now state the main theorem of this work.

Theorem 2. *Algorithm 1 (with $\gamma = \frac{1}{\sqrt{\log \Delta}}$ if $\Delta > 16$ and $\gamma = 0.5$ otherwise) is a deterministic distributed $(2 + \epsilon)$-approximation algorithm for MWVC. The number of rounds required for the algorithm to terminate is $O\left(\frac{\log \Delta}{\log \log \Delta} + \frac{\log \epsilon^{-1} \log \Delta}{\log^2 \log \Delta}\right)$ if $\Delta > 16$ and $O(\log \epsilon^{-1})$ otherwise.*

3.1   Proof of Theorem 2

Notation. In the analysis we use $w_i(v), \ell_i(v), d_i'(v)$ to denote the value of these variables at the beginning of the ith iteration.

The following claim states an invariant that Algorithm 1 satisfies.

Claim. The following invariant holds in every iteration of the while-loop:

$$\gamma^{\ell_i(v)} < \frac{w_i(v)}{w_0(v)} \le \gamma^{\ell_i(v)-1} \tag{1}$$

Hence, (i) $vault_i(v) < w_i(v)$ and (ii) if $\ell_{i+1}(v) \ge z + 1$, then $\frac{w_{i+1}(v)}{w_0(v)} \le \epsilon'$.

This invariant of Claim 3.1 implies, among other things, that every vertex that decides to join the vertex cover is ϵ'-tight. This property, together with the fact that the set of vertices that join the vertex cover constitute a vertex cover leads to the proof that Algorithm 1 is a $(2 + \epsilon)$-approximation algorithm. An analogous lemma and its proof also appears in [BCS17]. We remark that termination of the algorithm is implied by the upper bound on the number of iterations of the while-loop proved in the sequel.

Lemma 1. ([BCS17, Lemma 3.2]) *For every $\epsilon, \gamma \in (0,1)$, upon termination Algorithm 1 computes a $(2 + \epsilon)$-approximate solution to MWVC.*

In the following lemma we show that, for every vertex v and every iteration of the while-loop, either many of v's neighbors of the smallest level have increased their level or v's weight has decreased significantly.

Lemma 2. *Let $K > 1$. Let i be an iteration of the while-loop in the execution of Algorithm 1 by vertex v in which v does not join the cover. At least one of following conditions must hold:*

1. *At least $d_i'(v)(1 - 1/K)$ of the neighbors of v of the lowest level have increased their level. Formally, If $\ell_{i+1}'(v) = \ell_i'(v)$, then $d_{i+1}'(v) < d_i'(v)/K$.*
2. *$w_{i+1}(v) \le w_i(v) - w_0(v)\gamma^{\ell_i(v)}/K$.*

Algorithm 1. A distributed $(2 + \epsilon)$-approximation algorithm for MWVC, code for vertex v. (Listing taken from [BCS17] and edited to include our modifications.)

---

1 γ = parameter in the interval $(0, 1)$.
2 $\epsilon' = \epsilon/(2 + \epsilon)$
3 $z = \lceil \log_\gamma \epsilon' \rceil$
4 $w_0(v) = w(v)$
5 $\ell_0(v) = 1$
6 $N_0(v) = N(v), d_i(v) \triangleq |N_i(v)|$
7 //Let $N_i'(v)$ be the set of neighbors of lowest level in iteration i and
   $d_i'(v) \triangleq |N_i'(v)|$
8 $i = 0$
9 **while** *true* **do**
10     $vault_i(v) = w_0(v) \cdot \gamma^{\ell_i(v)}$
11     $bank_i(v) = w_i(v) - vault_i(v)$
12     $w_{i+1}(v) = w_i(v)$ and $\ell_{i+1}(v) \leftarrow \ell_i(v)$
13   **foreach** $u \in N_i'(v)$ **do**
14        $request_i(v, u) = vault_i(v)/d_i'(v)$
15   Send $request_i(v, u)$ to u
16   Let $budget_i(u, v)$ be the response from u
17        $w_{i+1}(v) = w_{i+1}(v) - budget_i(u, v)$

18   Let $u_1 \ldots u_{m_i}$ be an arbitrary order of neighbors that sent requests in this iteration
19   **foreach** $k = 1, \ldots, m_i$ **do**
20   Let $request_i(u_k, v)$ be received from $u_k \in N_i'(v)$
21        $budget_i(v, u_k) = \min\{request_i(u_k, v), bank_i(v) - \sum_{t=1}^{k-1} budget_i(v, u_t)\}$
22   Send $budget_i(v, u_k)$ to u_k

23     $w_{i+1}(v) = w_{i+1}(v) - \sum_{k=1}^{d_i(v)} budget_i(v, u_k)$
24   **if** $w_{i+1}(v) \neq 0$ *and* $w_{i+1}(v) \leq vault_i(v)$ **then**
25        $\ell_{i+1}(v) = 1 + \left\lfloor \log_\gamma \frac{w_{i+1}(v)}{w_0(v)} \right\rfloor$

26   **if** $w_{i+1}(v) = 0$ *or* $\ell_{i+1}(v) \geq z + 1$ **then**
27   Send $(v, cover)$ to all neighbors
28   Return InCover

29   **foreach** $(u, cover)$ *received from* $u \in N_i(v)$ **do**
30        $N_i(v) = N_i(v) \setminus \{u\}$

31   **if** $d_i(v) = 0$ **then**
32   Return NotInCover

33     $N_{i+1}(v) = N_i(v)$
34     $i = i + 1$

Proof. Assume that $\ell'_{i+1}(v) = \ell'_i(v)$ and $d'_{i+1}(v) \geq d'_i(v)/K$. Note that if the level of a vertex remains unchanged (i.e., $\ell'_{i+1}(v) = \ell'_i(v)$), then either $w_{i+1}(v) = 0$ or $w_{i+1}(v) > vault_i(v)$. If $w_{i+1}(v) = 0$, then v joins the cover, a contradiction. If $w_{i+1}(v) > vault_i(v)$, then the bank was not exhausted and $budget_i(u, v) = request_i(v, u)$. To conclude, at least $d'_i(v)/K$ vertices $u \in N'_i(v)$ responded with $budget_i(u, v) = request_i(v, u)$. This implies that

$$w_i(v) - w_{i+1}(v) \geq \frac{d'_i(v)}{K} \cdot \frac{vault_i(v)}{d'_i(v)}$$

$$= w_0(v)\gamma^{\ell_i(v)}/K.$$

Lemma 3. *For every $\gamma \in (0, 1)$ and $K > 1$, the number of iterations of the while-loop for every vertex v is bounded by:*

$$z \cdot \left(\frac{K}{\gamma} + \frac{\log d(v)}{\log K} \right).$$

Proof. The number of levels is bounded by z. Hence it suffices to prove that the number of rounds per level is at most $K/\gamma + \log_K d(v)$. Indeed, the number of rounds that satisfy Condition 1 per level is bounded by $\log_K d(v)$ because $d'_i(v)$ is divided by at least K in each such iteration.

We now bound the number of iterations that satisfies Condition 2 per level. By Claim 3.1, $w_0(v) \cdot \gamma^{\ell_i(v)-1} \geq w_i(v)$. Hence, the number of iterations that satisfies Condition 2 is bounded by K/γ, as required, and the lemma follows.

We now prove Theorem 2.

Proof. First, consider the case where $\Delta \leq 16$. We set $\gamma = 0.5$ (hence, $z = O(\log \epsilon^{-1})$) and $K = 2$. Lemma 3 immediately shows that the termination time is $O(\log \epsilon^{-1})$. Next, assume that $\Delta > 16$ (thus hereafter: $\log \log \Delta > 2$, $\log \Delta / \log \log \Delta > 2$, $1/\sqrt{\log \Delta} < 1/2$, and $\sqrt{\log \Delta}/\log \log \Delta > 1$). We set $\gamma = 1/\sqrt{\log \Delta}$ and $K = \sqrt{\log \Delta}/\log \log \Delta$. Now we can express the running time as:

$$z \cdot \left(\frac{K}{\gamma} + \frac{\log d(v)}{\log K} \right) \leq z \left(K\gamma^{-1} + \frac{\log \Delta}{\log K} \right) = z \left(\frac{\log \Delta}{\log \log \Delta} + \frac{\log \Delta}{0.5 \log \log \Delta - \log \log \log \Delta} \right)$$

$$= O \left(\frac{z \log \Delta}{\log \log \Delta} \right).$$

Let us analyze the running time according to the values of ϵ. First, consider the case where $\epsilon^{-1} = \log^{O(1)} \Delta$. Since $\epsilon \in (0, 1)$, it follows that $\epsilon' = \Theta(\epsilon)$. We get that $z < 1 + \log_\gamma \epsilon' = O(1 + \log \epsilon^{-1}/\log \gamma^{-1}) = O(1 + \log \log \Delta / \log \log \Delta) = O(1)$. Thus, the total running time for this case is $O(\log \Delta / \log \log \Delta)$. Next we consider the complementary case, where $\epsilon^{-1} = \log^{\omega(1)} \Delta$. This means that $\log \epsilon^{-1} = \omega(\log \log \Delta)$. Therefore, $z = O(\log \epsilon^{-1}/\log \gamma^{-1}) = O(\log \epsilon^{-1}/\log \log \Delta)$, and the running time for the second case is given by $O(\log \epsilon^{-1} \log \Delta / \log^2 \log \Delta)$. Thus, we may express the running time of our algorithm asymptotically as:

$$O(\log \Delta / \log \log \Delta + \log \epsilon^{-1} \log \Delta / \log^2 \log \Delta).$$

The number of rounds is bounded as required, and the theorem follows.

4   An Algorithm Without Knowing Δ

The bound on the round complexity in Theorem 2 assumes that every vertex knows the maximum degree Δ (or a polynomial upper bound on Δ). This is required in order to determine the value of γ. In this section we consider the setting in which Δ is unknown to the vertices.

Note that the analysis in Lemma 3 is per a vertex. Hence, in the analysis of the round complexity, we may use a different K per a vertex. Let K_v denote the value of K that is used in the analysis for bounding the round complexity of v.

We propose two alternatives for the setting of unknown maximum degree, as follows:

1. In the first setting, we simply set $\gamma = 1/2$ in the algorithm. For the analysis, we consider $K_v = \frac{\log d(v)}{\log \log d(v)}$, where $d(v)$ denotes the degree of v. Plugging in these parameters in Lemma 3 gives a round complexity of $O\left(\frac{\log \epsilon^{-1} \log d(v)}{\log \log d(v)}\right)$.

2. For any $q = O(1)$, we can set $\gamma = \epsilon^{1/2q}$ (hence, $z = O(1)$). An analysis with $K_v = \frac{\gamma \log d(v)}{\log \log d(v)}$ shows that v terminates in the optimal $O\left(\log d(v) / \log \log d(v)\right)$ rounds for any $\epsilon > (\log \Delta)^{-q}$. This is because

$$K_v = \frac{\epsilon^{1/2q} \log d(v)}{\log \log d(v)} > \frac{\log^{-0.5} d(v) \log d(v)}{\log \log d(v)} = \frac{\log^{0.5} d(v)}{\log \log d(v)}.$$

That allows us to express the running time as

$$z\left(K_v \gamma^{-1} + \log d(v) / \log K_v\right) = O\left(\log d(v) / \log \log d(v)\right).$$

References

[ÅFP+09]   Åstrand, M., Floréen, P., Polishchuk, V., Rybicki, J., Suomela, J., Uitto, J.: A local 2-approximation algorithm for the vertex cover problem. In: Keidar, I. (ed.) DISC 2009. LNCS, vol. 5805, pp. 191–205. Springer, Heidelberg (2009). https://doi.org/10.1007/978-3-642-04355-0_21

[ÅS10]   Åstrand, M., Suomela, J.: Fast distributed approximation algorithms for vertex cover and set cover in anonymous networks. In: SPAA 2010 Proceedings of the 22nd Annual ACM Symposium on Parallelism in Algorithms and Architectures, Thira, Santorini, Greece, pp. 294–302, 13–15 June 2010

[BCS17]   Bar-Yehuda, R., Censor-Hillel, K., Schwartzman, G.: A distributed $(2+\epsilon)$-approximation for vertex cover in o(log Δ / ϵ log log Δ) rounds. J. ACM **64**(3), 23:1–23:11 (2017)

[BE81]   Bar-Yehuda, R., Even, S.: A linear-time approximation algorithm for the weighted vertex cover problem. J. Algorithms **2**(2), 198–203 (1981)

[BE85]   Bar-Yehuda, R., Even, S.: A local-ratio theorem for approximating the weighted vertex cover problem. N.-Holland Math. Stud. **109**, 27–45 (1985)

[BEPS12]   Barenboim, L., Elkin, M., Pettie, S., Schneider, J.: The locality of distributed symmetry breaking. In: 53rd Annual IEEE Symposium on Foundations of Computer Science, FOCS 2012, New Brunswick, NJ, USA, pp. 321–330, 20–23 October 2012

[CLRS09] Cormen, T.H., Leiserson, C.E., Rivest, R.L., Stein, C.: Introduction to Algorithms, 3rd Edition. The MIT Press, Cambridge (2009)

[GJ79] Garey, M.R., Johnson, D.S., Freeman, W.H.: Computers and Intractability: A Guide to the Theory of NP-Completeness (1979)

[GKP08] Grandoni, F., Könemann, J., Panconesi, A.: Distributed weighted vertex cover via maximal matchings. ACM Trans. Algorithms $5(1)$ (2008)

[HKP01] Hanckowiak, M., Karonski, M., Panconesi, A.: On the distributed complexity of computing maximal matchings. SIAM J. Discrete Math. $15(1)$, 41–57 (2001)

[Hoc82] Hochbaum, D.S.: Approximation algorithms for the set covering and vertex cover problems. SIAM J. Comput. $11(3)$, 555–556 (1982)

[Kar72] Karp, T.M.: Reducibility among combinatorial problems. In: Proceedings of a Symposium on the Complexity of Computer Computations. The IBM Thomas J. Watson Research Center, Yorktown Heights. Springer, New York, pp. 85–103, 20–22 March 1972. https://doi.org/10.1007/978-1-4684-2001-2_9

[KMW06] Kuhn, F., Moscibroda, T., Wattenhofer, R.: The price of being near-sighted. In: Proceedings of the Seventeenth Annual ACM-SIAM Symposium on Discrete Algorithms, SODA 2006, Miami, Florida, USA, pp. 980–989, 22–26 January 2006

[KMW16] Kuhn, F., Moscibroda, T., Wattenhofer, R.: Local computation: lower and upper bounds. J. ACM $63(2)$, 17:1–17:44 (2016)

[KR08] Khot, S., Regev, O.: Vertex cover might be hard to approximate to within 2-epsilon. J. Comput. Syst. Sci. $74(3)$, 335–349 (2008)

[KVY94] Khuller, S., Vishkin, U., Young, N.E.: A primal-dual parallel approximation technique applied to weighted set and vertex covers. J. Algorithms $17(2)$, 280–289 (1994)

[KY09] Koufogiannakis, C., Young, N.E.: Distributed and parallel algorithms for weighted vertex cover and other covering problems. In: Proceedings of the 28th ACM Symposium on Principles of Distributed Computing, PODC 2009, pp. 171–179. ACM, New York (2009)

[KY11] Koufogiannakis, C., Young, N.E.: Distributed algorithms for covering, packing and maximum weighted matching. Distrib. Comput. $24(1)$, 45–63 (2011)

[NJ75] Nemhauser, G.L., Trotter Jr., L.E.: Vertex packings: structural properties and algorithms. Math. Program. $8(1)$, 232–248 (1975)

[Pet16] Pettie, S.: Personal communication (2016)

[PR01] Panconesi, A., Rizzi, R.: Some simple distributed algorithms for sparse networks. Distrib. Comput. $14(2)$, 97–100 (2001)

[PS09] Polishchuk, V., Suomela, J.: A simple local 3-approximation algorithm for vertex cover. Inf. Process. Lett. $109(12)$, 642–645 (2009)

[Sol18] Solomon, S.: Local algorithms for bounded degree sparsifiers in sparse graphs. In: ITCS, volume 94 of LIPIcs. Schloss Dagstuhl - Leibniz-Zentrum fuer Informatik, pp. 52:1–52:19 (2018)

Online Service with Delay on a Line

Marcin Bienkowski$^{(\boxtimes)}$ ⓘ, Artur Kraska, and Paweł Schmidt

Institute of Computer Science, University of Wrocław, Wrocław, Poland
{marcin.bienkowski,artur.kraska,pawel.schmidt}@cs.uni.wroc.pl

Abstract. In the *Online Service with Delay* (OSD) problem, introduced recently by Azar et al. (STOC 2017), there are an n-point metric space and a server occupying some point. Points request service over time and these requests need to be eventually served by moving the server to these points. To exploit spatial locality of requests, a service may be delayed and requests may be served in batches. However, there are certain penalties associated with the delays, e.g., such penalty may be proportional to the waiting time of a given request. The goal is to minimize the sum of the total distance traveled by the server and all delay penalties. The OSD problem is closely related to widely studied optimization problems, such as the reordering buffer management and the multi-level aggregation. Azar et al. (STOC 2017) gave a randomized online $O(\log^4 n)$-competitive algorithm for general metric spaces. In this paper, we present a *deterministic* $O(\log n)$-competitive algorithm for the case when the metric space is a line consisting of n equidistant points.

Keywords: Delayed service · Server problems · Online algorithms
Competitive analysis

> *"They also serve who only stand and wait"*
> — John Milton, On His Blindness

1 Introduction

In the *Online Service with Delay* (OSD) problem, there are n points in the metric space and a server occupying some point. During runtime, points request service and these requests have to be eventually served. To this end, an algorithm has to move its server to the requesting point. To minimize the traveled distance, an algorithm may serve the requests not immediately, e.g., it may travel to a remote location once the number of requests there is sufficiently large. This however incurs a *waiting cost*, which is a non-decreasing function of the delay between the time a request is issued and the time it is served by an algorithm. The goal is to minimize the total cost, defined as the sum of total distance traveled by the server and the waiting costs of all issued requests.

Supported by Polish National Science Centre grant 2016/22/E/ST6/00499.

Z. Lotker and B. Patt-Shamir (Eds.): SIROCCO 2018, LNCS 11085, pp. 237–248, 2018.
https://doi.org/10.1007/978-3-030-01325-7_22

The OSD problem is inherently online: an algorithm learns about a request only once it is issued and has to make decisions about server movement without knowing the future requests. The cost of an online algorithm ALG is then compared to the cost of an optimal *offline* solution OPT for the same instance; the ratio between their costs is called *the competitive ratio* [12]. We say that ALG is α-competitive if its competitive ratio is at most α. The OSD problem has been recently introduced by Azar et al. [6], who presented an $O(\log^4 n)$-competitive randomized solution.

For a real-life example modeled by this problem, consider a technician (the server) who needs to respond to repair requests from clients (points in the metric space). The speed of the technician is not restricted and once she arrives at the scene, she fixes the problem immediately. The waiting cost function for a request represents its urgency, e.g., it may depend on the importance of a given client.

1.1   Related Problems

The solution given by Azar et al. for the OSD problem [6] is an $O(h^3)$-competitive deterministic algorithm for any hierarchically separated tree of depth h. As any metric space on n points can be randomly approximated by an HST of depth $O(\log n)$ with the expected distance distortion of $O(\log n)$ [7,19], this result yields a randomized $O(\log^4 n)$-competitive algorithm for any metric space.

Essentially, the OSD problem studies a trade-off between serving requests in batches (and saving because they are located close to each other) and minimizing the delays of particular requests. Similar trade-offs occur naturally in many areas of logistics and planning, scheduling or supply chain management.

While the OSD problem has been defined only recently, it is closely related to the reordering buffer management (RBM) problem [1–5,9,17,18,20–22] and the multi-level aggregation (MLA) problem [8,10,11,13–15]. The MLA problem can be seen as a special variant of the OSD problem on a tree. In this variant, the server is initially stored at the root. At any time, the server may choose a set of requests, serve them by navigating along a minimum sub-tree spanning these requests and the root, and then return to the root.

In contrast, in the RBM problem, requests do not have waiting costs, but at any time there may be at most b unserved requests. The currently best (randomized) algorithm for general metrics, due to Englert and Räcke [16], is $O(\log n \cdot \log b)$-competitive. An important observation that separates the OSD and the RBM problems is the performance of the following "rent-or-buy" strategy. Assume that an online algorithm waits till there is a subset of requests whose total waiting cost becomes equal to the total cost of serving them. An algorithm for OSD that simply serves only these requests would be $\Omega(n)$-competitive even on simple metrics such as weighted stars [6], while an analogous algorithm for RBM would be $O(h \cdot \log b)$-competitive for trees of depth h [17].

1.2   Line Metric: Our Contribution

In this paper, we study the OSD problem on a line consisting of n equidistant points. Apart from the theoretical importance, the original motivation for studying such metric comes from minimizing the movement of a hard disk head: in this scenario, a request is a write demand to a particular cylinder of the disk (where a cylinder is represented by a point on a line) [20].

Refined results are known for the line metrics both for the MLA and RBM problems. For the RBM problem, Gamzu and Segev constructed a deterministic $O(\log n)$-competitive algorithm MOVING PARTITION [20]. Beating this competitive ratio is a long standing challenge: surprisingly, $O(\log n)$ is also the best known approximation ratio for the offline variant and the best known lower bound on the competitive ratio is only 2.154 [20]. The MLA problem is better understood in such spaces: Bienkowski et al. presented a 5-competitive algorithm for a line [11] and no algorithm can beat the ratio of 4 [8].

Our contribution. In this paper, we present a deterministic $O(\log n)$-competitive algorithm BCKT (short for BUCKET) for the OSD problem on a line. Our algorithm combines ideas from the MOVING PARTITION [20] and the preemptive service approach used for the OSD and MLA problems [6,8]. Namely, once our algorithm identifies an interval I of a line, such that the total waiting cost of requests pending in I is comparable to the cost of travelling to I, it puts an additional work into serving not only requests from I, but also from its surrounding. As we prove, this extra work significantly reduces the algorithm cost in the future.

A bit surprisingly, our algorithm is *non-clairvoyant*: when a request is presented, the algorithm does not need to know its waiting cost function upfront. (We require that the waiting cost functions are continuous, though.) This stands in contrast to the bound presented by Azar et al. for weighted stars [6]: they show that for such metrics, the competitive ratio of any non-clairvoyant deterministic algorithm is at least $\Omega(\Delta)$, where Δ is the aspect ratio of the metric. (This lower bounds holds even if waiting cost functions are continuous.)

1.3   Preliminaries

In the OSD problem on a line, the metric space is a line consisting of n equidistant points (each consecutive pair is connected with an edge of length 1). The server of an algorithm starts at a position chosen by the adversary. An input consists of requests and each request is a triple (τ, p, f), where τ is its arrival time, p is the point an algorithm has to visit to serve the request and $f : \mathbb{R}_{\geq 0} \to \mathbb{R}_{\geq 0}$ is an arbitrary continuous non-decreasing function, such that $f(0) = 0$. If, at time $\tau' \geq \tau$, an algorithm serves the request (τ, p, f), i.e., moves its server to point p, the request incurs the *waiting cost* of $f(\tau' - \tau)$. The *service cost* of an algorithm is defined as the sum of distances traveled by the server. The goal is to serve all requests and minimize the sum of the service cost and all waiting costs.

2   The Algorithm

We present a deterministic algorithm BCKT solving the OSD problem for a line metric comprising n equidistant points. BCKT balances service and waiting costs. It works in phases, each consisting of a *waiting subphase* and a *serving subphase*. In the waiting subphase, BCKT does not move the server and waits until there is a group of requests whose overall waiting cost becomes roughly the distance to this group from the current position of the algorithm's server. In the subsequent serving subphase, BCKT serves this group of requests along with its surrounding (to be defined later). The serving subphase is immediate, i.e., the waiting cost is accrued only in the waiting subphases.

2.1   Algorithm Definition

More concretely, at the beginning of the waiting subphase, BCKT (re)numbers the points on the line from left to right with consecutive integers, so that the position of its server is at point 0. Next, BCKT splits the line points into $O(\log n)$ *buckets*. For $i \geq 1$, the i-th right (left) bucket consists of 2^{i-1} points that lie to the right (left) from the server's position and whose distances from server's position are in the range $[2^{i-1}, 2^i - 1]$. By B_{+i} and B_{-i} we denote the i-th right and left bucket, respectively; we say that their *indexes* are equal to i. The *size* $|B|$ of a bucket B is the number of points it contains, i.e., $|B_{+i}| = |B_{-i}| = 2^{i-1}$.

Note that the position of the server does not belong to any bucket and all requests arriving at this point are immediately served for free. For each bucket B and time τ, we define its *weight* $w_\tau(B)$ as the total waiting cost incurred by requests in B still pending for BCKT at time τ. We will omit τ and write $w(B)$ whenever it does not lead to ambiguity. We say that a bucket B is *full* if $w(B) \geq |B|$.

Fix any phase S. The waiting subphase lasts until some bucket B becomes full. (As the waiting cost functions are continuous, we may then assume that for such bucket $w(B)$ becomes exactly equal to $|B|$.) At that moment, BCKT considers *quarter-full* buckets, i.e., buckets B' satisfying $w(B') \geq |B'|/4$. Let $B_{\pm r}$ (for $r \geq 1$) be the quarter-full bucket farthest from the server, i.e., the one with the largest index r. We call this bucket *critical* for phase S. (If there are two such buckets, left and right, we pick an arbitrary one of them to be critical.)

On the basis of the critical bucket index, r, we define two notions: the *phase label* becomes equal to r and the *cleaning area* $C(S)$ is defined as the region $\biguplus_{j=1}^{r+1}(B_{-j} \uplus B_{+j}) = [-(2^{r+1}-1), 2^{r+1}-1]$. An example is presented in Figure 1.

In the serving subphase, BCKT serves all requests pending in the cleaning area $C(S)$. To this end, BCKT chooses its new position to be the closest point of the critical bucket (the point $\pm 2^{r-1}$). BCKT's server follows then the shortest route that visits both endpoints of the cleaning area and ends at point $\pm 2^{r-1}$. For simplicity, we make BCKT visit each point of $C(S)$ even if there is no request waiting at that point. Note that the corresponding service cost is at most twice the size of the cleaning area. A pseudocode of BCKT in a single phase is given in Algorithm 1.

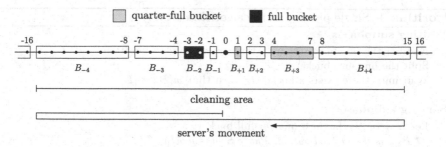

Fig. 1. An example phase (of label 3) of BCKT. After a waiting subphase, B_{+3} is the critical bucket (the quarter-full bucket with the largest index). Within a serving subphase, BCKT moves its server, so that it visits all the points from the cleaning area and finishes at the closest point of B_{+3}.

2.2 Correctness

We start with proving that BCKT is defined properly, i.e., right after a serving phase ends and the algorithm splits the line into new buckets, no bucket is full. Intuitively, buckets that are close to the new server's position are contained in the cleaning area of the serving phase and are now empty. On the other hand, buckets that are far from the server's position are properly contained in two consecutive buckets (of the previous phase) whose weight was small.

Lemma 1. *No bucket is full at the beginning of a phase of* BCKT.

Proof. Fix a phase S and let $r \geq 1$ be its label. The cleaning area of the phase is then $C(S) = [-(2^{r+1} - 1), 2^{r+1} - 1]$. Assume, without loss of generality, that the critical bucket of S is the right bucket, $B_{+r} = [2^{r-1}, 2^r - 1]$, i.e., BCKT ends the phase with its server at the point $s = 2^{r-1}$.

We denote the (old) buckets of the serving phase by $B_{\pm i}$ and the buckets of the new waiting phase, i.e., constructed relative to the point s, by $B'_{\pm i}$. We will show that all buckets $B'_{\pm i}$ satisfy $w(B'_{\pm i}) < |B'_{\pm i}|$.

First, we observe that buckets $B'_{-(r+1)}, B'_{-r}, \ldots, B'_{-1}$ and $B'_{+1}, \ldots, B'_{+r}$ are fully contained in the cleaning area $C(S)$. As all the requests in these buckets were served, the weights of these buckets are now equal to 0. Second, all buckets $B'_{\pm i}$ are shifted by 2^{r-1} to the right relative to $B_{\pm i}$. Thus, for any $i \geq r + 1$, bucket B'_{+i} is contained in the union $B_{+i} \uplus B_{+(i+1)}$. Since B_{+r} was the farthest quarter-full bucket, the weights of buckets B_{+i} and $B_{+(i+1)}$ are less than a quarter of their sizes. Hence, $w(B'_{+i}) \leq w(B_{+i}) + w(B_{+(i+1)}) < 2^{i-1}/4 + 2^i/4 < 2^{i-1} = |B'_{+i}|$. Similarly, for $i \geq r + 2$, bucket B'_{-i} is contained in $B_{-i} \uplus B_{-(i-1)}$, i.e., the union of two buckets that are not quarter-full, and therefore $w(B'_{-i}) \leq w(B_{-i}) + w(B_{-(i-1)}) < 2^{i-1}/4 + 2^{i-2}/4 < 2^{i-1} = |B'_{-i}|$. This concludes the proof. □

Algorithm 1. Single phase S of the algorithm BCKT.

Waiting subphase:

    Number the points relatively to the current server's position.

    Split the line into buckets.

    Wait until there exists a bucket B, such that $w(B) = |B|$.

Serving subphase:

    Let $B_{\pm r}$ be the farthest quarter-full bucket.

    /* $B_{\pm r}$ is the critical bucket. Phase S gets label r. */

    Let $C(S) = [-(2^{r+1} - 1), 2^{r+1} - 1]$

    if $B_{\pm r}$ is a left bucket **then**

        /* Go right and then left. */

        Move to $2^{r+1} - 1$ serving all requests on the way.

        Move to $-(2^{r+1} - 1)$ serving all requests on the way.

        Move to -2^{r-1}.

    else

        /* Go left and then right. */

        Move to $-(2^{r+1} - 1)$ serving all requests on the way.

        Move to $2^{r+1} - 1$ serving all requests on the way.

        Move to 2^{r-1}.

---

3 Competitiveness

In this section, we prove that BCKT is $O(\log n)$-competitive, where n is the number of points on the line. In our reasoning, we do not aim at minimizing the constants, but rather at the simplicity of the argument.

3.1 Waiting and Service Costs

We start by showing that our algorithm balances its waiting and service costs, which allows us to focus only on bounding the latter. By $BCKT_{\text{wait}}$ and $BCKT_{\text{serv}}$ we denote, respectively, the waiting and the service costs of BCKT.

Lemma 2. *It holds that* $BCKT_{\text{wait}} \leq BCKT_{\text{serv}}$.

Proof. Fix any phase S and let $r \geq 1$ be its label, i.e., the cleaning area of the phase is equal to $C(S) = [-(2^{r+1} - 1), 2^{r+1} - 1]$. As the server starts in the middle of the cleaning area and has to visit its both endpoints, the service cost incurred in the serving subphase is at least $3 \cdot (2^{r+1} - 1) > 2 \cdot 2^{r+1}$.

On the other hand, by the definition of BCKT, the weight of each bucket contained in $C(S)$ is at most the size of this bucket. Hence, the total waiting cost incurred by requests served in phase S is $\sum_{i=1}^{r+1}(w(B_{-i}) + w(B_{+i})) \leq \sum_{i=1}^{r+1}(|B_{-i}| + |B_{+i}|) < 2 \cdot 2^{r+1}$.

Each request is eventually served by BCKT, and therefore summing the waiting and service costs of all phases yields the lemma. □

By Lemma 2, it suffices to bound $\text{BCKT}_{\text{serv}}$, the total distance traveled by the BCKT's server. In our proof later we would like to use the following local argument: "if our algorithm traverses an edge, then it moves towards requests (whose total waiting cost was sufficiently large)". Unfortunately, this is not true for all edges from the route traversed by BCKT, especially for the edges near the borders of the cleaning area.

Therefore (for the analysis purposes only), we define the following "virtual" algorithm BCKT_{eff} that operates almost in the same way as BCKT does, and we analyze BCKT_{eff} instead of BCKT. BCKT_{eff} has identical notion of waiting subphases and buckets as BCKT. The only difference is that in the serving subphase, BCKT_{eff} moves the server directly to the closest point of the critical bucket (i.e., to the final position of BCKT's server in the serving subphase).

We will pretend that such server movement of BCKT_{eff} serves all the pending requests in the cleaning area $C(S)$, i.e., serves the same set of requests as BCKT does. Furthermore, we will assume that BCKT_{eff} is charged only for server movement, i.e., it does not pay for the waiting costs. It can be easily observed that replacing BCKT by BCKT_{eff} changes the cost at most by a constant factor.

Lemma 3. *It holds that* $\text{BCKT} \leq 32 \cdot \text{BCKT}_{\text{eff}}$.

Proof. Consider a single phase S and let $r \geq 1$ be its label. The cost of BCKT_{eff} on this phase is 2^{r-1} while BCKT pays at most $2 \cdot 2^{r+2}$ for the server movement (the server visits each point of the cleaning area at most twice). This implies that $\text{BCKT}_{\text{serv}} \leq 16 \cdot \text{BCKT}_{\text{eff}}$ for a single phase, and hence also for the entire input sequence. Combining this relation with Lemma 2 concludes the proof. □

3.2   Critical Requests and Freshness Property

Fix any phase S and let r be its label. The set of requests for phase S which is served in the critical bucket $B_{\pm r}$ is called a *critical set* and denoted $R(S)$. The following properties of critical sets will become useful once we define our charging scheme in Sect. 3.4.

Lemma 4. *Fix any edge e and consider two phases S and S' of the same label r during which the server of BCKT_{eff} moves along e in the same direction. The following two properties hold.*

Weight property. *The total waiting cost of requests from $R(S')$ is at least 2^{r-3}.*
Freshness property. *All requests from $R(S')$ appeared after phase S ended.*

Proof. The first condition of the lemma follows immediately by the definition of the algorithm: $R(S')$ is the set of requests from the bucket $B'_{\pm r}$, which is critical and hence quarter-full. That is, $w(B_{\pm r}) \geq |B_{\pm r}|/4 = 2^{r-3}$.

For the second condition, we assume, without loss of generality, that within both phases BCKT_{eff} moves to the right. Let B_{+r} and B'_{+r} be the critical buckets in phases S and S', respectively. Observe that $C(S)$, the cleaning area of S, covers all edges that BCKT_{eff} traverses and also $2^{r+1} - 2^{r-1}$ points to the right of this path. Hence, $C(S)$ covers at least $2^{r+1} - 2^{r-1}$ points to the right of edge e.

In phase S', the bucket B'_{+r} is also to the right of edge e and the distance between e and the farthest point of B'_{+r} is at most 2^r. Hence, $B'_{+r} \subseteq C(S)$, which means that all the requests at points from B'_{+r} are served in phase S. Therefore, all the requests from B'_{+r} that are served by BCKT_{eff} during phase S' must have arrived after the end of phase S.   □

3.3 Moving Towards and Away from OPT

From now on, we fix any optimal solution OPT and analyze both BCKT_{eff} and OPT running on the same input simultaneously. Roughly speaking, we would like to argue that if BCKT_{eff} traverses an edge e many times, then OPT has to either move along e or pay large waiting costs. However, if BCKT_{eff} traverses edge e towards OPT, we cannot say anything about the relative position of OPT and the requests BCKT_{eff} serves (it might even happen that OPT serves them for free). Therefore, we split the moves of BCKT_{eff} into $\text{BCKT}_{\text{away}}$— the moves in the direction away from the current position of OPT's server and $\text{BCKT}_{\text{toward}}$— the moves towards it. Note that $\text{BCKT}_{\text{away}}$ and $\text{BCKT}_{\text{toward}}$ are defined only in the analysis. It turns out that by focusing only on $\text{BCKT}_{\text{away}}$ we lose only a constant factor in the competitive ratio.

Lemma 5. *It holds that* $\text{BCKT}_{\text{toward}} \leq \text{BCKT}_{\text{away}} + \text{OPT}$.

Proof. Fix any edge e. Let $\text{BCKT}_{\text{toward}}(e)$ and $\text{BCKT}_{\text{away}}(e)$ be the total cost of moves of BCKT_{eff}'s server along edge e in the directions toward and away the OPT's server, respectively. Let $\text{OPT}(e)$ be the total costs of OPT traversals of e. It is sufficient to show that

$$\text{BCKT}_{\text{toward}}(e) \leq \text{BCKT}_{\text{away}}(e) + \text{OPT}(e). \tag{1}$$

To prove this inequality, we analyze how its both sides evolve in time. Clearly, at the beginning, both sides are equal to zero. Since BCKT_{eff} and OPT start at the same side of e, before BCKT_{eff} moves toward OPT for the first time (i.e., the left hand side of (1) increases for the first time), either BCKT_{eff} needs to move away from OPT along e or OPT needs to traverse e (i.e., the right hand side of (1) increases for the first time).

Furthermore, between any two consecutive increases of $\text{BCKT}_{\text{toward}}(e)$ (two consecutive increments of the left hand side of (1)) either BCKT_{eff} traverses e in the direction away from OPT or OPT traverses e (the right hand side of (1) increments).   □

3.4 Charging Scheme

The plan for the remaining part of the analysis is as follows. We look at any increment of the value $\text{BCKT}_{\text{away}}$ over time. It corresponds to a traversal of some edge e performed by the server of BCKT_{eff} in the direction away from OPT. We call any such movement an *away-traversal of edge e*.

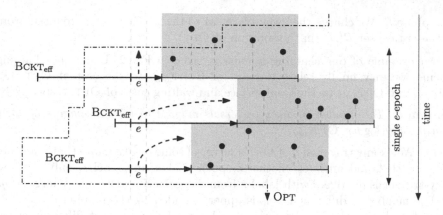

Fig. 2. Illustration of the charging scheme for away-traversals of edge e performed by the algorithm BCKT_{eff}. The figure contains a single e-epoch defined by the times at which OPT traversed edge e. The route of OPT's server is drawn as a dash-dotted line. All depicted away-traversals of edge e performed by BCKT_{eff} have the same label r: each corresponding movement of BCKT_{eff} (thick right arrows) consists of 2^{r-1} away-traversals. The shaded regions contain critical sets of requests (represented by small discs) that are served by BCKT_{eff} at the ends of the phases in critical buckets (of sizes 2^{r-1}). Dashed arrows denote our charging of away-traversals.

We will map (charge) all away-traversals to some action(s) of OPT. Some away-traversals will be charged directly to the moves of OPT's server. Others will be charged to the waiting of requests; we will show that these requests incurred sufficiently large cost in the solution of OPT. Then, we will estimate, for any action of OPT, how many away-traversals are mapped to them.

To formally define the charging, we now focus entirely on a single edge e. We split the execution of both BCKT_{eff} and OPT into e-epochs using the moves of OPT along e: when OPT traverses edge e in any direction, an e-epoch ends and the next one begins. Fix any such e-epoch. Assume, without loss of generality, that in this e-epoch the server of OPT remains on the left side of edge e. In this case, all away-traversals of e within the e-epoch are movements of BCKT_{eff}'s server to the right. Any such away-traversal of e is a part of a path traveled by BCKT_{eff} in the serving subphase of S. The away-traversal receives the same label that is assigned to phase S.

In a given e-epoch, for each label value r, we will charge away-traversals of e in the following way.

The first away-traversal of e with label r. We charge this away-traversal either to the movement of OPT's server along e that initiates the current e-epoch or to the movement that finishes it. Such move always exists as OPT eventually serves all requests.

Subsequent away-traversals of e with label r. Recall that this away-traversal is a part of a movement performed by the BCKT_{eff}'s server within a phase S. We charge the away-traversal to the waiting of the requests from the critical set $R(S)$ (in the solution of OPT).

An example of our charging scheme is given in Fig. 2. Using the charging scheme, we may finally relate the cost of $\text{BCKT}_{\text{away}}$ to the cost of OPT. Let OPT_{serv} and OPT_{wait} be the total service and waiting costs of OPT, respectively.

Lemma 6. *The number of away-traversals mapped to the movements of* OPT *is at most* $O(\log n) \cdot \text{OPT}_{\text{serv}}$.

Proof. We fix any traversal q of OPT's server of some edge e. Note that q finishes one e-epoch $\mathcal{E}$ and initiates another e-epoch $\mathcal{E}'$. For any label r, only the first away-traversals of edge e with label r from e-epochs $\mathcal{E}$ and $\mathcal{E}'$ may charge to q.

The number of different labels is upper bounded by the number of possible labels for phases, that is, by $O(\log n)$. Hence, the number of different away-traversals that may charge to q is at most $2 \cdot O(\log n)$. The lemma follows by summing over all edge traversals performed by OPT.   □

Lemma 7. *The number of away-traversals mapped to the waiting of requests in the solution of* OPT *is at most* $4 \cdot \text{OPT}_{\text{wait}}$.

Proof. We now fix any phase S of the algorithm BCKT_{eff}, let r be its label and let $R(S)$ be the corresponding critical set of requests (served in the serving subphase of S). For succinctness, we define the waiting cost of $R(S)$ as the total waiting cost of all requests from $R(S)$. We will now claim that the number of away-traversals that charge to $R(S)$ is at most 4 times larger than the waiting cost of $R(S)$ incurred by OPT. The lemma will then follow by summing over all critical set of requests (they are clearly disjoint).

If no away-traversals charge to $R(S)$, then the claim follows trivially.

We may therefore assume that there exists an away-traversal of some edge e that charges to $R(S)$. Without loss of generality, we assume that BCKT_{eff} traverses e to the right, and within the corresponding phase OPT remains to the left of edge e. Note that the label of the away-traversal has to be the same as the label of phase S, i.e., equal to r. By the definition of our charging, the preceding away-traversal of edge e with label r belongs to the same e-epoch; let S_p be the phase during which this preceding away-traversal is performed. By the freshness property of Lemma 4, all requests from the critical set $R(S)$ arrived after S_p. By the weight property of the same lemma, the waiting cost of $R(S)$ incurred on BCKT_{eff} is at least 2^{r-3}.

As OPT remains on the left side of edge e for the whole e-epoch, it cannot serve the requests of $R(S)$ earlier than BCKT_{eff}, and therefore the waiting cost of $R(S)$ incurred on OPT is also at least 2^{r-3}. The claim follows by observing that only edges belonging to the BCKT_{eff}'s server movement in phase S (2^{r-1} edges) may charge to $R(S)$.   □

Recall that $\text{BCKT}_{\text{away}}$ is the set of all away-traversals. As all away-traversals are mapped by our charging scheme, using Lemmas 6 and 7, we immediately obtain the following corollary.

Corollary 1. *It holds that* $\text{BCKT}_{\text{away}} \leq 4 \cdot \text{OPT}_{\text{wait}} + O(\log n) \cdot \text{OPT}_{\text{serv}}$.

3.5   The Competitive Ratio

Theorem 1. BCKT *is* $O(\log n)$-*competitive.*

Proof. The theorem follows by a straightforward combination of the definitions and lemmas proven in this section.

$$
\begin{aligned}
\text{BCKT} &\leq 32 \cdot \text{BCKT}_{\text{eff}} && \text{(by Lemma 3)}\\
&= 32 \cdot (\text{BCKT}_{\text{toward}} + \text{BCKT}_{\text{away}})\\
&\leq 32 \cdot (2 \cdot \text{BCKT}_{\text{away}} + \text{OPT}) && \text{(by Lemma 5)}\\
&= O(1) \cdot \text{OPT}_{\text{wait}} + O(\log n) \cdot \text{OPT}_{\text{serv}} && \text{(by Corollary 1)}\\
&= O(\log n) \cdot \text{OPT}
\end{aligned}
$$

$\square$

4   Final Remarks

Our algorithm BCKT can easily be adapted, with virtually no changes, to the setting where points are not necessarily equidistant. The competitive ratio becomes then $O(\log \Delta)$, where Δ is the aspect ratio of the metric (the ratio between the largest and the smallest distances). The details will appear in the full version of the paper.

References

1. Aboud, A.: Correlation clustering with penalties and approximating the reordering buffer management problem. Master's thesis, Computer Science Department, The Technion Israel Institute of Technology (2008)
2. Adamaszek, A., Czumaj, A., Englert, M., Räcke, H.: Almost tight bounds for reordering buffer management. In: Proceedings 43rd ACM Symposium on Theory of Computing (STOC), pp. 607–616 (2011)
3. Avigdor-Elgrabli, N., Im, S., Moseley, B., Rabani, Y.: On the randomized competitive ratio of reordering buffer management with non-uniform costs. In: Proceedings 42nd International Colloquium on Automata, Languages and Programming (ICALP), pp. 78–90 (2015)
4. Avigdor-Elgrabli, N., Rabani, Y.: An improved competitive algorithm for reordering buffer management. In: Proceedings 21st ACM-SIAM Symposium on Discrete Algorithms (SODA), pp. 13–21 (2010)
5. Avigdor-Elgrabli, N., Rabani, Y.: An optimal randomized online algorithm for reordering buffer management. In: Proceedings 54th IEEE Symposium on Foundations of Computer Science (FOCS), pp. 1–10 (2013)
6. Azar, Y., Ganesh, A., Ge, R., Panigrahi, D.: Online service with delay. In: Proceedings 49th ACM Symposium on Theory of Computing (STOC), pp. 551–563 (2017)
7. Bansal, N., Buchbinder, N., Mądry, A., Naor, J.: A polylogarithmic-competitive algorithm for the k-server problem. J. ACM **62**(5), 40:1–40:49 (2015)

8. Bienkowski, M., et al.: Online algorithms for multi-level aggregation. In: Proceedings 24th European Symposium on Algorithms (ESA), pp. 12:1–12:17 (2016)
9. Bienkowski, M., et al.: Logarithmic price of buffer downscaling on line metrics. Theor. Comput. Sci. **707**, 89–93 (2018)
10. Bienkowski, M., Byrka, J., Chrobak, M., Jeż, L., Nogneng, D., Sgall, J.: Better approximation bounds for the joint replenishment problem. In: Proceedings 25th ACM-SIAM Symposium on Discrete Algorithms (SODA), pp. 42–54 (2014)
11. Bienkowski, M., Byrka, J., Chrobak, M., Jeż, Ł., Sgall, J., Stachowiak, G.: Online control message aggregation in chain networks. In: Proceedings 13th International Workshop on Algorithms and Data Structures (WADS), pp. 133–145 (2013)
12. Borodin, A., El-Yaniv, R.: Online Computation and Competitive Analysis. Cambridge University Press, New York (1998)
13. Brito, C., Koutsoupias, E., Vaya, S.: Competitive analysis of organization networks or multicast acknowledgement: how much to wait? Algorithmica **64**(4), 584–605 (2012)
14. Buchbinder, N., Feldman, M., Naor, J.S., Talmon, O.: O(depth)-competitive algorithm for online multi-level aggregation. In: Proceedings 28th ACM-SIAM Symposium on Discrete Algorithms (SODA), pp. 1235–1244 (2017)
15. Buchbinder, N., Kimbrel, T., Levi, R., Makarychev, K., Sviridenko, M.: Online make-to-order joint replenishment model: primal dual competitive algorithms. In: Proceedings 19th ACM-SIAM Symposium on Discrete Algorithms (SODA), pp. 952–961 (2008)
16. Englert, M., Räcke, H.: Reordering buffers with logarithmic diameter dependency for trees. In: Proceedings 28th ACM-SIAM Symposium on Discrete Algorithms (SODA), pp. 1224–1234 (2017)
17. Englert, M., Räcke, H., Westermann, M.: Reordering buffers for general metric spaces. Theory Comput. Syst. **6**(1), 27–46 (2010)
18. Englert, M., Westermann, M.: Reordering buffer management for non-uniform cost models. In: Caires, L., Italiano, G.F., Monteiro, L., Palamidessi, C., Yung, M. (eds.) ICALP 2005. LNCS, vol. 3580, pp. 627–638. Springer, Heidelberg (2005). https://doi.org/10.1007/11523468_51
19. Fakcharoenphol, J., Rao, S., Talwar, K.: A tight bound on approximating arbitrary metrics by tree metrics. J. Comput. Syst. Sci. **69**(3), 485–497 (2004)
20. Gamzu, I., Segev, D.: Improved online algorithms for the sorting buffer problem on line metrics. ACM Trans. Algorithms **6**(1), 15:1–15:14 (2009)
21. Khandekar, R., Pandit, V.: Online and offline algorithms for the sorting buffers problem on the line metric. J. Discret. Algorithms **8**(1), 24–35 (2010)
22. Räcke, H., Sohler, C., Westermann, M.: Online scheduling for sorting buffers. In: Möhring, R., Raman, R. (eds.) ESA 2002. LNCS, vol. 2461, pp. 820–832. Springer, Heidelberg (2002). https://doi.org/10.1007/3-540-45749-6_71

Mixed Fault Tolerance in Server Assignment: Combining Reinforcement and Backup

Tal Navon and David Peleg[✉]

The Weizmann Institute, Rehovot, Israel
david.peleg@weizmann.ac.il

Abstract. We study the mixed approach to fault tolerance in the general context of server assignment in networks. The approach is based on mixing two different existing strategies, namely, *reinforcement* and *backup*. The former strategy protects clients by reinforcing the servers assigned to them and making them fault-resistant (at a possibly high cost), while the latter protects clients by assigning to them alternate low price backup servers that can replace their primary servers in case those fail. Applying the mixed approach to fault tolerance gives rise to new fault-tolerant variations of known server assignment problems. We introduce several NP-hard problems of this type, including the *mixed fault-tolerant dominating set* problem, the *mixed fault-tolerant centers* problem, and the *mixed fault-tolerant facility location* problem, and present polynomial time approximation algorithms for them, demonstrating the viability of the mixed strategy for server assignment problems.

Keywords: Approximation · Fault tolerance · Backup
Reinforcement · Server assignment problems · Dominating set
Centers · Facility location

1 Introduction

Background and Motivation. An important aspect of network design concerns coping with potential failures, such as link disconnections or vertex crashes. Many different approaches were proposed in the literature for avoiding or overcoming failures. A natural approach that's been applied in numerous settings relies on introducing *redundancy* to the system, by employing more resources than needed, as *backup*. An alternative approach is to *reinforce* some of the network components (possibly at a high cost) so as to prevent their failure, or at least lower their failure probability.

The current paper deals with a typical setting of a computer network with clients and failure-prone servers. The goal is to cope with situations where (up to f) servers might crash simultaneously, thus depriving certain clients of nearby service. Previous studies of such problems (cf. [6,10]) employed redundancy-based fault tolerance, relying on selecting sufficiently many backup servers. Note,

Z. Lotker and B. Patt-Shamir (Eds.): SIROCCO 2018, LNCS 11085, pp. 249–263, 2018.
https://doi.org/10.1007/978-3-030-01325-7_23

however, that the backup-reinforcement tradeoff occurs naturally in this setting, as one can cope with potential server failures either by buying some extra servers as backup, or alternatively, by using resilient servers that are less likely to fail but are more expensive. Depending on the relative prices and the network topology, intermediate (mixed) solutions may turn out to be desirable.

To illustrate this approach, we consider a number of NP-hard network design problems involving clients and servers, and introduce new fault-tolerant versions of these problems that employ a mixed strategy based on using a combination of both redundancy and reinforcement. (These new versions are proper extensions of the original problems, hence are also NP-hard.)

As a "warm-up", we first look at an easy but illustrative example, namely, the *minimum dominating set (DS)* problem, defined as follows. Given a graph G, it is required to find a smallest possible set of vertices D such that each vertex in G is either in D or has a neighbor in D. Intuitively, thinking of the vertices of D as locations where we want to place certain services, the goal is to ensure that every vertex in G has nearby service (at most one hop away from it). Hence this is one of the simplest examples of a server assignment problem.

A redundancy-based fault-tolerant version of this problem, named the *minimum f-fault-tolerant dominating set* (or f-FT-DS) problem, aims to cope with the failure of (up to f) vertices in D, thus leaving certain vertices without a dominating neighbor in D, by taking an appropriately larger set D, such that each vertex is either in D or has $f+1$ neighbors in D. Intuitively, the redundancy introduced into the solution ensures that even if f vertices of D have failed, each surviving vertex will still have a functioning neighbor in D.

In order to study mixed fault-tolerant solutions, based on both backup and reinforcement, we define the *minimum cost f-mixed fault-tolerant dominating set-pair* (or f-MFT-DSP) problem as follows. Intuitively, each client vertex can be protected in one of two ways: either it has a reinforced neighbor (including possibly itself) that cannot fail, or it has $f+1$ normal neighbors that are chosen as servers. Hence given a graph G and prices p for a normal server and $\tilde{p}$ for a reinforced server, it is required to find a cheapest possible pair of sets of vertices (normal servers D and reinforced $\tilde{D}$ servers), such that each vertex in G either belongs to one of the sets, or has $f+1$ neighbors in D or one neighbor in $\tilde{D}$. (Precise definitions to all the problems studied are given in the respective sections.) As it turns out, this problem can be easily approximated with ratio similar to the (ordinary or redundancy-based fault-tolerant) dominating set problem, using the same methods.

We then turn to our main problems. We first study fault-tolerant versions of the *k-centers* problem. Two redundancy-based fault-tolerant versions of the problem were studied in the past. The first is the *f-neighbor k-centers* problem. Again we make the assumption that no more than f vertices might fail simultaneously. Given a weighted graph G and a set D of k vertices, each client v will always use the nearest available server in D. If the f closest servers to v have crashed, then v is forced to turn to the next $((f+1)\text{st})$ closest server. Let $\delta^f(v, D)$ denote the distance from v to that server. The client $\hat{v}$ most affected by the failures is the one with the largest distance $\delta^f(\hat{v}, D)$. Hence the problem is to find the set D of k vertices that minimizes this value. The second version is the *f-all-neighbor k-centers* problem, defined similarly except that the vertices in D itself are also considered as clients, and need to be serviced even when the server located in them crashes. In other words, we want the vertices in D whose server has failed to also have a close vertex in D that did not fail.

To study mixed fault-tolerant solutions to the k-centers problem, based on mixing backup and reinforcement, we define the *f-MFT neighbor centers* problem as follows. Given a weighted graph G, prices p and $\tilde{p}$ for normal and reinforced servers respectively, and a budget B, it is required to find a pair of vertex sets D and $\tilde{D}$ such that the vertex that is not in $D \cup \tilde{D}$ and is the farthest away from the closest $f + 1$ vertices in D and the closest vertex in $\tilde{D}$, is as close as possible. We define the *f-MFT all-neighbor centers* problem similarly, except that the vertices selected to D are also clients that require service.

The last (and most elaborate) problem we study is the *uncapacitated facility location* (UFL) problem. The redundancy-based fault-tolerant version of the fault-tolerant facility location (FT-FL) problem, studied in the past, again aims to cope with crashes by providing clients with nearby backup servers. The problem is defined as follows. We are given a weighted graph G, an opening price for each vertex, and an individual "backup demand" r_i for each vertex i (such that i must be protected against the possible failure of up to $r_i - 1$ of its servers). It is required thus to find a set of vertices D and assign them to the vertices in G such that, each client vertex i in G is assigned to at least r_i servers in D, and the sum of distances from all the clients to their servers in D plus the opening prices of all the vertices in D is as small as possible. (Note that in this version of the problem, a client needs to pay for all r_i servers it is assigned to.)

To study mixed fault-tolerant solutions for this problem, we define the *mixed fault-tolerant facility location* (MFT-FL) problem as follows. Given a weighted graph G, an opening price for each vertex, and a backup demand r_i for each vertex i, it is required to find a pair of vertex sets (a reinforcement set $\tilde{D}$ and a backup set D) and assign them to the vertices in the graph such that each vertex is assigned either to at least r_i vertices in D or to at least one vertex in $\tilde{D}$, and the sum of distances from each client to the servers it is assigned to (in D or $\tilde{D}$) plus the opening prices of all the vertices in D and $\tilde{D}$ is as small as possible.

Related Work. The minimum dominating set problem is well-studied and one of the first problems shown to be NP-hard [1]. Observe that this problem is a special case of minimum set cover. By [5], the minimum set cover problem has a polynomial time $\log n$ approximation greedy algorithm. Hence, the minimum dominating set problem also has a polynomial time $\log n$ approximation algorithm. More generally, every integer linear program (ILP) with nonnegative integer values (ILP^+ as defined in Sect. 2) enjoys submodularity and hence has a logarithmic factor approximation by the greedy algorithm by [11]. The minimum f-FT DS problem can be defined as an ILP^+, hence it too has a logarithmic factor approximation.

Gonzalez presented a polynomial time 2-approximation greedy algorithm for the k-centers problem [2]. Hochbaum and Shmoys also gave a polynomial time 2-approximation algorithm for that problem, based on finding maximal independent sets in (the powers of) suitable subgraphs of the given graph, and prove that this is the best approximation possible in polynomial time assuming $NP \neq P$ [4]. Based on the latter, Khuller, Pless, and Sussmann gave a polynomial time 2-approximation algorithm for the f-neighbor k-centers problem (which is the best possible assuming $NP \neq P$ since the problem is an extension of the basic k-centers problem) and a polynomial time 3-approximation (2-approximation for $f \leq 3$) algorithm for the f-all-neighbor k-centers problem [6].

The uncapacitated facility location (UFL) problem is widely studied as well. An approximation algorithm for the problem is presented in [9]. In [10], Swamy and Shmoys give a 2.076-approximation polynomial time algorithm for the fault-tolerant facility location (FT-FL) problem. They also give a simpler 4-approximation algorithm for the problem.

The mixed approach to fault tolerance, based on combining redundancy and reinforcement, was utilized before in the context of fault-tolerant network structures. Specifically, the problem of constructing fault-tolerant BFS structures in graphs was studied in [7].

Results. Our main contribution is in demonstrating that redundancy-based fault tolerance can be augmented into mixed fault tolerance in the context of server assignment problems in networks. In particular, our concrete results involve extending known approximation algorithms for the redundancy-based fault-tolerant versions of the problems discussed above into approximation algorithms for the corresponding mixed fault-tolerant versions. While each of these algorithms is based on a different approach (a greedy algorithm for the f-MFT-DSP problem, combinatorial graph algorithms for the FT-centers problems, and a linear programming based algorithm for the MFT-FL problem), all of them exhibit a dependence on the relative prices, and in particular, on whether or not $\tilde{p} \leq (f+1)p$.

In more detail, in Sect. 2 we extend the known polynomial time $\log(n(f+1)) + 1$ approximation algorithm for the minimum f-FT dominating set problem into a $\log(n(f+1)) + 1$ approximation algorithm for the minimum cost f-MFT dominating set-pair problem. In Sect. 3 we extend the algorithms of [6] into a polynomial time 3-approximation algorithm for the f-MFT all-neighbor centers problem and a polynomial time 8-approximation algorithm for the f-MFT neighbor centers problem. In Sect. 4 we extend the 4-approximation algorithm of [10] into a (rather involved) polynomial time $(3 + \max\{3, 2r\})$-approximation algorithm (where $r = \max\{r_i\}$ is the maximum demand for all the vertices) for the mixed fault tolernt facility location (MFT-FL) problem.

Discussion. The dependence of the approximation ratio of our algorithm for the MFT-FL problem on r is dissatisfactory. It seems plausible that the problem enjoys an approximation algorithm of constant ratio independent of r. This question is left for future study.

A broader intriguing question left for future study is whether there may be a general relationship between redundancy-based and mixed fault-tolerant versions of server assignment (as well as other) NP-hard problems. In particular, it would be useful to have a general transformation technique that will allow taking a solution for a redundancy-based fault tolerance problem and using it to generate a solution for a mixed fault tolerance version of the problem, with roughly similar approximation qualities. While our results seem to hint that such a relationship may exist, if only for some suitably restricted scope of problems, it is unclear to us at the moment how to formally define and establish it in a general setting.

Another possible extension is to consider a *hierarchy* of component types $T_1, \ldots, T_m$ (rather than just two types), where components of type T_{i+1} are more expensive than components of type T_i, but also more resilient (e.g., have lower probability of failing). This extension may lead to a number of intriguing modeling and optimization problems, where the goal may be, for instance, to maximize the resilience achievable with a given budget, or to minimize the cost of achieving a specified resilience. These problems are also left for future study.

2   Warm-Up: Mixed Fault-Tolerant Dominating Sets

In this section we illustrate the mixed approach on a simple example problem, namely, the minimum dominating set problem and its mixed fault-tolerant variant. Consider a graph $G = (V, E)$. A vertex set $D \subseteq V$ is a *dominating set* in G if every vertex $v \notin D$ has a neighbor in D. The basic *minimum dominating set* problem is to find a minimum size dominating set for a given graph G. This problem is NP-hard and has a $\log |V| + 1$ approximation algorithm [5,11]. The redundancy-based fault-tolerant version of the problem, called the minimum fault-tolerant dominating set (FT-DS) problem, can be approximated in a similar manner.

We now introduce our new mixed variant of the fault-tolerant dominating set problem, which allows both backup and reinforcement. Assume that at most f of the vertices can fail (and be removed from the graph). A solution consists of a set-pair $(\tilde{D}, D)$, where the *backup vertices* in D host normal (failure-prone) servers of low price p and the *reinforcement vertices* in $\tilde{D}$ host stronger servers of high price $\tilde{p}$ that cannot fail. The pair $(\tilde{D}, D)$ is an *f-MFT dominating set-pair* in $G = (V, E)$ if for every $F \subseteq V$ s.t. $|F| \leq f$, $(D \setminus F) \cup \tilde{D}$ is a dominating set in $G \setminus F$. The cost of the set-pair is $Cost(\tilde{D}, D) = p|D| + \tilde{p}|\tilde{D}|$.

Observation 1. *The pair $(\tilde{D}, D)$ is an f-MFT dominating set-pair in G iff every vertex $v \notin D \cup \tilde{D}$ has either a neighbor in $\tilde{D}$ or $f + 1$ neighbors in D.*

We next formulate the problem of finding a minimum cost f-MFT dominating set-pair as an integer linear program in the class ILP^+, as follows. Given a graph $G = (V, E)$, where $|V| = n$, define a matrix $A_{2n \times n}$ of nonnegative integers s.t.

for $1 \leq i \leq n$:

$$A_{ij} = \begin{cases} f + 1, & \text{if } i = j, \\ 1, & \text{if } (v_i, v_j) \in E, \\ 0, & \text{otherwise,} \end{cases} \qquad A_{i(n+j)} = \begin{cases} f + 1, & \text{if } i = j, \\ f + 1, & \text{if } (v_i, v_j) \in E, \\ 0, & \text{otherwise.} \end{cases}$$

Also define a vector b of length n s.t. $b_i = f + 1$ for $1 \leq i \leq n$. Finally, define a vector c of length $2n$ s.t. $c_i = p$ if $1 \leq i \leq n$ and $c_i = \tilde{p}$ otherwise. The problem is to minimize cx subject to $Ax \geq b$ for $x \in \{0, 1\}^{2n}$.

Given $\tilde{D}, D \subset V$ we represent these sets by defining a vector x of length $2n$ s.t. for $1 \leq i \leq n$, $x_i = 1$ if $v_i \in D$, and 0 otherwise, and $x_{n+i} = 1$ if $v_i \in \tilde{D}$, and 0, otherwise. It is easy to verify that the set-pair $(\tilde{D}, D)$ is f-FT dominating in G iff $Ax \geq b$, and has minimum cost iff cx is minimum.

According to [11], optimization problems that can be formulated as an ILP^+ (hence enjoy submodularity) have a $\log(\sum_i b_i) + 1$ approximation algorithm. This implies:

Lemma 1. *The problem of finding a minimum cost f-MFT dominating set-pair has a polynomial time greedy $\log(n(f + 1)) + 1$ approximation algorithm.*

3 Mixed Fault-Tolerant Centers

In this section we consider the k-centers problem and its mixed fault-tolerant variants. Throughout this section, we consider a weighted graph $G = (V, E, \omega)$, with the weight function ω. Define a distance function d on G as follows. The length of a path P in G is the sum of the weights of all the edges in P. The distance $d_{v,u}$ is the length of the shortest path between v and u.

The basic k-centers problem is defined as follows. Given G, find a subset $D \subseteq V$ such that $|D| = k$ and the maximum distance from a vertex in V to the closest vertex in D is a as small as possible. Formally, define the distance from a vertex v to a set of vertices D to be $\delta(u, D) = \min_{v \in D}\{d_{u,v}\}$. The *radius* of G w.r.t D is $rad(D) = \max_{u \in V}\{\delta(u, D)\}$. We look for a subset D^* of size at most k attaining a minimum radius, $D^* = \arg\min_{D \subseteq V, |D| \leq k}\{rad(D)\}$. This problem is NP-hard. It has a 2-approximation algorithm by [3] and this is known to be the best possible in polynomial time assuming $NP \neq P$ [4].

Introducing redundancy, the f-neighbor k-centers problem requires, given $G = (V, E, \omega)$, to find a subset $D \subseteq V$, $|D| = k$, s.t. the maximum distance from and $v \in V \setminus D$ to the set of $f + 1$ closest vertices in D is as small as possible. (Note that only vertices of $V \setminus D$ are considered as clients, i.e., the vertices of D are used only as servers.) Formally, let $\delta^f(u, D) = \min_{A \subseteq D, |A| = f+1} \max_{a \in A}\{d_{u,a}\}$. The f-radius f_rad^- of a set D in G is defined as $f\_rad^-(D) = \max_{u \in V \setminus D}\{\delta^f(u, D)\}$. We look for a subset D^* of size at most k attaining a minimum f_rad^- value, i.e., $D^* = \arg\min_{D \subseteq V, ||D| \leq k}\{f\_rad^-(D)\}$. This problem has a 2-approximation algorithm [6] and being an extension of the basic k-centers problem, it is the best possible in polynomial time assuming $NP \neq P$.

The f-all-neighbor k-centers problem is similar, except the vertices of D are also clients. Formally, the f-radius is now $f\_rad^+(D) = \max_{u \in V}\{\delta^f(u, D)\}$. We look for $D^* = \arg\min_{D \subseteq V, ||D| \leq k}\{f\_rad^+(D)\}$ attaining a minimum f_rad^+. This problem has a 3-approximation, and a 2-approximation for $f < 4$ [6].

Turning to mixed fault tolerance, let us define the f-MFT neighbor centers problem as follows. Given $G = (V, E, \omega)$ and constants p (the price of setting up a normal center at a vertex), $\tilde{p}$ (the price of setting up a reinforced center at a vertex) and B (the total budget), find subsets $\tilde{D}, D \subseteq V$ such that $Cost(\tilde{D}, D) = |\tilde{D}|\tilde{p} + |D|p \leq B$ (i.e., the centers set-up cost is within the budget) and the maximum distance from a vertex in $V \setminus (\tilde{D} \cup D)$ to a vertex in $\tilde{D}$ or a set of $f + 1$ vertices in D is a as small as possible. Formally, let $\delta^f(u, \tilde{D}, D) = \min\{\delta^f(u, D), \delta(u, \tilde{D})\}$. The mixed radius $\texttt{Mrad}^-$ of a set-pair $(\tilde{D}, D)$ in G is $\texttt{Mrad}^-(\tilde{D}, D) = \max_{u \in V \setminus (\tilde{D} \cup D)}\{\delta^f(u, \tilde{D}, D)\}$. We look for a set-pair $(\tilde{D}, D)$ of $Cost(\tilde{D}, D)$ at most B attaining a minimum $\texttt{Mrad}^-$ value,

$$(\tilde{D}^*, D^*) = \arg\min\{\texttt{Mrad}^-(\tilde{D}, D) \mid \tilde{D}, D \subseteq V, \ Cost(\tilde{D}, D) \leq B\} \ .$$

Let $opt^-(G)$ be the mixed radius of the optimal solution $(\tilde{D}^*, D^*)$ in G.

Define the f-MFT all-neighbor centers problem similarly, except that the vertices of D are also clients. Formally, the mixed radius $\texttt{Mrad}^+$ of a set-pair $(\tilde{D}, D)$ in G is $\texttt{Mrad}^+(\tilde{D}, D) = \max_{u \in V}\{\delta^f(u, \tilde{D}, D)\}$. We look for a set-pair $(\tilde{D}, D)$ of $Cost(\tilde{D}, D)$ at most B attaining a minimum $\texttt{Mrad}^+$ value,

$$(\tilde{D}^*, D^*) = \arg\min\{\texttt{Mrad}^+(\tilde{D}, D) \mid \tilde{D}, D \subseteq V, \ Cost(\tilde{D}, D) \leq B\} \ .$$

Let $opt^+(G)$ be the mixed radius of the optimal solution $(\tilde{D}^*, D^*)$ in G.

3.1   Approximating the f-MFT All-Neighbor Centers Problem

In this section we give a 3-approximation algorithm for f-MFT all-neighbor centers that relies on ideas of the algorithms for the ordinary and f-all-neighbor k-centers problems by [4,6] and extends them to handle also reinforcement.

A solution to the problem must specify, for each client v, either a node in $\tilde{D}$ or $f + 1$ nodes in D that serve it. We use the following notation. Denote by $S_{(\tilde{D},D)}(v)$ the node or nodes that "serves" v (i.e., the closest $f + 1$ nodes in D or the closest node in $\tilde{D}$, if it is closer than the $(f + 1)$st closest node in D. If there is a tie between the closest node in $\tilde{D}$ and the $(f + 1)$st closest node, then $S_{(\tilde{D},D)}(v)$ will contain the closest node in $\tilde{D}$. If there is a tie between nodes in D or between nodes in $\tilde{D}$ we break it arbitrarily, say, by picking the nodes with the smaller indices.) For $v \in \tilde{D} \cup D$, let us denote the client nodes that v "serves" by $C_{(\tilde{D},D)}(v) = \{u \mid v \in S_{(\tilde{D},D)}(u)\}$. For any graph $G = (V, E)$, define the power graph $G^2 = (V, \{(v, u) \mid \exists w \in V \text{ s.t. } (v, w), (w, u) \in E\})$.

We can assume the graph is complete (if it is not, then we can add a new edge (u, v) for every $(u, v) \notin E$, and define $\omega(u, v)$ as the weight of the shortest path between u and v). Sort the edges in a nondecreasing order $e_1, e_2, ..., e_{|E|}$ and define $E_i = \{e_1, e_2, ..., e_i\}$ and $G_i = (V, E_i)$ (the unweighted graph that contains all the edges in G whose weight is less than or equal to e_i).

Algorithm MFT_all_centers uses a procedure named MIS_centers. It runs the procedure on G_i for $i = 1, ..., m$, and halts once the procedure returns a set pair $(\tilde{D}, D)$ of cost at most B. (If this never happens, then the procedure returns "No feasible solution".)

Invoked on the subgraph G_i, Procedure MIS_centers computes a maximal independent set I in G_i^2 (using a simple greedy algorithm). If $(f+1)p \geq \tilde{p}$, then the procedure sets $D \leftarrow \emptyset$ and $\tilde{D} \leftarrow I$. On the other hand, if $(f+1)p < \tilde{p}$, then the procedure looks at the subset $I' \subseteq I$ of vertices with at most $f - 1$ neighbors in G_i, sets $\tilde{D} \leftarrow I'$ and for each $v \in (I \setminus I')$ it adds to D the node v and f of its neighbors in G_i (chosen arbitrarily). The procedure then returns $(\tilde{D}, D)$. Formal code for our algorithms is given next.

Main algorithm MFT_all_centers
for $i \leftarrow 1$ **to** m **do**
    Run $(\tilde{D}, D) \leftarrow$ MIS_centers(G_i)
    if $Cost(\tilde{D}, D) \leq B$ **then**   stop and return the solution $(\tilde{D}, D)$.
end
return *"no feasible solution"*

Procedure MIS_centers
1. Find a maximal independent vertex set I in G_i^2.
2. Let $I' \subseteq I$ be the set of vertices with at most $f - 1$ neighbors in G_i.
3. **if** $(f+1)p < \tilde{p}$ **then do:**
    $\tilde{D} \leftarrow I'$; $D \leftarrow \emptyset$
    for each $v \in (I \setminus I')$ **do**
        $D \leftarrow D \cup \{v\} \cup \{f$ arbitrarily chosen neighbors of v in $G_i\}$
4. **else** set $D \leftarrow \emptyset$; $\tilde{D} \leftarrow I$
5. Return $(\tilde{D}, D)$

The analysis is based on the following claims (some proofs are deferred to the full paper).

Observation 2. *There exist a feasible solution iff either $B \geq \tilde{p}$ or $|V| \geq f + 1$ and $B \geq (f+1)p$.*

Consider iteration i of the algorithm. Let $(\tilde{D}_i, D_i)$ be the solution returned by MIS_centers on the graph G_i, and let $(\tilde{D}'_i, D'_i)$ the solution with minimum cost possible with radius $\omega(e_i)$. The following two observations are used to establish the approximation ratio of the algorithm.

Lemma 2. $Cost(\tilde{D}_i, D_i) \leq Cost(\tilde{D}'_i, D'_i)$.

Proof. Let I be the maximal independent set selected by procedure MIS_centers in iteration i. Let us first observe that for every $v, u \in I$ such that $v \neq u$, their servers are disjoint, namely, $S_{(\tilde{D}'_i, D'_i)}(v) \cap S_{(\tilde{D}'_i, D'_i)}(u) = \emptyset$. Otherwise there is a node $w \in S_{(\tilde{D}'_i, D'_i)}(v) \cap S_{(\tilde{D}'_i, D'_i)}(u)$ which means $\omega(v, w) \leq \omega(e_i)$ and $\omega(u, w) \leq \omega(e_i)$ (since $\omega(e_i)$ is the radius of $(\tilde{D}'_i, D'_i)$), hence u and w are neighbors in G_i and v and w are neighbors in G_i, which means u and v are neighbors in $G_i{}^2$, in contradiction to the fact that I is an independent vertex set in $G_i{}^2$. Hence for each node v in I there must be at least one distinct node in $\tilde{D}'_i$ or $f+1$ distinct nodes in D'_i (serving v and no other vertices in I). This implies that there must be at least x nodes in $\tilde{D}'_i$ and $(f+1)(|I| - x)$ nodes in D'_i for some $0 \leq x \leq |I|$.

If $(f+1)p \geq \tilde{p}$, then procedure MIS_centers returns $(\tilde{D}_i, D_i) = (I, \emptyset)$, hence,

$$Cost(\tilde{D}_i, D_i) = |\tilde{D}_i|\tilde{p} + |D_i|p = |I|\tilde{p} = x\tilde{p} + (|I| - x)\tilde{p}$$
$$\leq x\tilde{p} + (f+1)(|I| - x)p = |\tilde{D}'_i|\tilde{p} + |D'_i|p = Cost(\tilde{D}'_i, D'_i) .$$

Now consider the case $(f+1)p < \tilde{p}$. In this case, it is cheaper to serve a client using $f+1$ backup servers than using a reinforced server, if at all possible. Note, however, that for each $v \in I'$, using backup servers is infeasible, hence $S_{(\tilde{D}'_i, D'_i)}(v) \in \tilde{D}'_i$. (By definition the nodes in I' have strictly fewer than f neighbors in G at distance smaller or equal to $\omega(e_i)$). Hence, there must be at least $|I'|$ nodes in $\tilde{D}'_i$, i.e., $|I'| \leq x \leq |I|$. Therefore

$$Cost(\tilde{D}_i, D_i) = |\tilde{D}_i|\tilde{p} + |D_i|p = |I'|\tilde{p} + (f+1)|I \setminus I'|p$$
$$= |I'|\tilde{p} + (f+1)(|I| - x)p + (f+1)(x - |I'|)p$$
$$\leq |I'|\tilde{p} + (f+1)(|I| - x)p + (x - |I'|)\tilde{p} = x\tilde{p} + (f+1)(|I| - x)p$$
$$= |\tilde{D}'_i|\tilde{p} + |D'_i|p = Cost(\tilde{D}'_i, D'_i).$$
□

Lemma 3. *Procedure* MIS_centers *on G_i returns a solution $(\tilde{D}_i, D_i)$ with radius* $\mathtt{Mrad}^+(\tilde{D}_i, D_i) \leq 3\omega(e_i)$.

Proof. For every vertex v in I, either v is in $\tilde{D}_i$ or v is in D_i and is a neighbor in G_i of at least f vertices in D_i. Since I is a maximal independent set in G_i^2, all the vertices in the graph G_i are at distance at most 2 from some vertex in I. Hence all the vertices in the graph G_i are at distance at most 2 from a vertex in $\tilde{D}_i$ or at distance at most 3 from at least $f+1$ vertices in D_i. And since all the edges in G_i have weights at most $\omega(e_i)$ in G, the distance (in G) between any vertex in G to the closest vertex to it in D_i or the closest $f+1$ vertices to it in $\tilde{D}_i$ is no more then $3\omega(e_i)$. Hence the mixed radius of $(\tilde{D}_i, D_i)$ is no more than $3\omega(e_i)$.
□

Lemma 4. *Algorithm* MFT_all_centers *returns a solution of cost at most B and radius at most $3opt^+(G)$ (assuming a feasible solution exists).*

Proof. Let $(\tilde{D}', D')$ be a solution with cost at most B, let $r = \mathtt{Mrad}^+(\tilde{D}', D')$ be the mixed radius of that solution, and let e_i be an edge in the graph such that $\omega(e_i) \geq r$. If the algorithm does not reach the i'th iteration it must mean that the algorithm found a solution with cost smaller than B and returned it in an earlier iteration. Else in the i'th iteration algorithm MFT_all_centers will run Procedure MIS_centers on G_i and by Lemma 2 Procedure MIS_centers will return a solution $(\tilde{D}, D)$ satisfying $Cost(\tilde{D}, D) = cost(\tilde{D}_i, D_i) \leq cost(\tilde{D}'_i, D'_i) \leq Cost(\tilde{D}', D') \leq B$, and algorithm MFT_all_centers will stop and return that solution.

Let $(\tilde{D}^*, D^*)$ be the optimal solution in G (a solution with cost at most B and the smallest possible radius under the restriction), and let $r = \omega(e_i)$ be the radius of that solution (there must be an edge e_i of that weight). If the algorithm stops at iteration j such that $j \leq i$, then it will return the solution given by procedure MIS_centers on G_j. By Lemma 3 that solution has a radius of no more than $3\omega(e_j) \leq 3\omega(e_i) = 3opt^+(G)$. Else the algorithm will reach the i'th iteration and will run procedure MIS_centers on G_i. By Lemma 3 that solution has a radius of no more than $3\omega(e_i) = 3opt^+(G)$. Since $(\tilde{D}^*, D^*)$ has cost at most B and radius $\omega(e_i)$, by Lemma 2 the solution returned by procedure MIS_centers has cost at most B, hence algorithm MFT_all_centers will stop and return it. □

3.2 Approximating the f-MFT Neighbor Centers Problem

We next present an 8-approximation algorithm named MFT_centers for f-MFT neighbor centers, making use of two existing 2-approximation algorithms. The first is Procedure Actr for the basic k-centers problem [3]. For a graph G and integer k, $\texttt{Actr}(G, k)$ returns a vertex set D of size k s.t. $rad(D) \leq 2rad(D')$, where D' is the optimal solution for the problem. The second is Procedure Afneig_ctr for the f-neighbor k-centers problem [6]. For a graph G and integers f and k, $\texttt{Afneig\_ctr}(G, f, k)$ returns a vertex set D of size k s.t. $f\_rad^-(D) \leq 2f\_rad^-(D'')$, where D'' is the optimal solution for the problem.

Main algorithm MFT_centers

if $(f+1)p < \tilde{p}$ **then**   /* no need to use reinforced centers */
$\quad k \leftarrow \lfloor B/p \rfloor$
$\quad$ run $D \leftarrow \texttt{Afneig\_ctr}(G, f, k)$
$\quad$ **return** $(\emptyset, D)$
else   /* $(f+1)p \geq \tilde{p}$ */
$\quad$ **for** $i \leftarrow 1$ **to** m **do**
$\qquad D_i^{low} \leftarrow \emptyset; \quad S \leftarrow V;$
$\qquad$ **while** *there exists some* $v \in S$ *such that* $|\Gamma_{(S, E_i)^2}(v)| < \tilde{p}/p$ **do**
$\qquad\qquad D_i^{low} \leftarrow D_i^{low} \cup \{v\}; \quad S \leftarrow S \setminus \{v\}$
$\qquad$ **end**
$\qquad \hat{G}_i \leftarrow G \setminus D_i^{low}; \quad k \leftarrow \lfloor \frac{B - |D_i^{low}|p}{\tilde{p}} \rfloor$
$\qquad$ run $\tilde{D}_i \leftarrow \texttt{Actr}(G_i, k)$
$\quad$ **end**
$\quad i_{min} = \arg \min_{0 \leq i \leq m} \texttt{Mrad}^-(\tilde{D}_i, D_{i_{min}}^{low})$
$\quad$ **return** $(\tilde{D}_{i_{min}}, D_{i_{min}}^{low})$
end

As in Sect. 3.1, assume G is complete and define E_i and G_i for $1 \le i \le m$.

If $(f+1)p < \tilde{p}$, then there is no need to use reinforced centers, so the algorithm sets $k \leftarrow \lfloor B/p \rfloor$, runs $D \leftarrow \texttt{Afneig\_ctr}(G, f, k)$, and returns the set pair $(\emptyset, D)$.

In case $(f+1)p \ge \tilde{p}$, the algorithm generates a set pair $(\tilde{D}_i, D_i^{low})$ for every $i \in \{1, ..., m\}$, based on G_i. The algorithm first selects the set of ordinary servers, D_i^{low}, as follows. It sets $D_i^{low} \leftarrow \emptyset$ and $S \leftarrow V$. Then, while there exists some $v \in S$ such that $|\Gamma_{(S,E_i)^2}(v)| < \tilde{p}/p$, the algorithm moves v from S to D_i^{low}. Once D_i^{low} is fixed, the algorithm removes its vertices from G, remaining with a subgraph $\hat{G}_i$. It then sets $k \leftarrow \lfloor (B - |D_i^{low}|p)/\tilde{p} \rfloor$ and runs Procedure $\texttt{Actr}$ to get $\tilde{D}_i \leftarrow \texttt{Actr}(G_i, k)$. Finally, the algorithm returns the set pair $(\tilde{D}_i, D_i^{low})$ with the lowest $\texttt{Mrad}^-$ value.

Observation 3. *The problem has a feasible solution iff $B \ge \min\{(f+1)p, |V|p, \tilde{p}\}$.*

Lemma 5. *Assume the problem admits a feasible solution. Then Alg. MFT_centers returns a solution with cost at most B. Moreover, the solution has radius at most $4opt^-(G)$ if $(f+1)p < \tilde{p}$, and at most $8opt^-(G)$ if $(f+1)p \ge \tilde{p}$.*

4   Mixed Fault-Tolerant Facility Location

This section presents an approximation algorithm for the mixed fault-tolerant facility location (MFT-FL) problem, which generalizes the fault-tolerant facility location (FT-FL) problem. The FT-FL problem concerns a graph $G = (V, E)$ where each node v_i represents both a client (with backup demand r_i) and a potential facility (with opening price p_i). We slightly simplify notation by denoting the vertices by $1, \ldots, n$ and the distance between nodes i and j by d_{ij}. The goal is to select a set of vertices D, open a facility in each of these vertices, and assign each client i to r_i of them, such that the total cost, composed of the sum of all the opening prices of the open facilities plus the sum of all the distances d_{ij} from each client i to a facility j it's assigned to, is as small as possible.

The MFT-FL problem is defined in a similar setting, except that in each site i it is possible to open either a normal facility (with opening price p_i) or a reinforced facility (with opening price $\tilde{p}_i$). The goal is to open a set D of normal facilities and a set $\tilde{D}$ of reinforced facilities, and to assign each client i to either r_i of the normal facilities or one of the reinforced facilities, such that the total cost, namely, the sum of all the opening costs of the open normal and reinforced facilities plus the sum of all the distances d_{ij} from each client i to a facility j it's assigned to is as small as possible.

Our algorithm (which is more involved than the previous ones) is based on the ideas of the algorithm presented in Sect. 2 of [10] for the usual (redundancy-based) fault-tolerant facility location problem, modified to fit our problem. We represent the MFT-FL problem as the following integer linear program, where $y_i = 1$ (respectively, $\tilde{y}_i = 1$) means that a normal (resp., reinforced) facility is opened at node i, namely, $i \in D$ (resp., $i \in \tilde{D}$), and $x_{ij} = 1$ (resp., $\tilde{x}_{ij} = 1$) means the client j is assigned to the normal (resp., reinforced) facility of node i.

ILP: minimize $C(x, y, \tilde{x}, \tilde{y}) = \sum\limits_{i=1}^{n} p_i y_i + \sum\limits_{i=1}^{n} \tilde{p}_i \tilde{y}_i + \sum\limits_{j=1}^{n} \sum\limits_{i=1}^{n} d_{ij} x_{ij} + \sum\limits_{j=1}^{n} \sum\limits_{i=1}^{n} d_{ij} \tilde{x}_{ij}$

subject to the constraints

(ILP1)	$\sum_{i=1}^{n}(x_{ij} + \tilde{x}_{ij} r_j) \geq r_j$	for every $1 \leq j \leq n$
(ILP2)	$x_{ij} \leq y_i$	for every $1 \leq i, j \leq n$
(ILP3)	$\tilde{x}_{ij} \leq \tilde{y}_i$	for every $1 \leq i, j \leq n$
(ILP4)	$y_i, \tilde{y}_i, x_{ij}, \tilde{x}_{ij} \in \{0, 1\}$	for every $1 \leq i, j \leq n$

Towards an approximation algorithm for our problems, we define the linear program **LP**, which is a relaxed version of problem ILP, and its dual **DP**.

LP: minimize $C(x, y, \tilde{x}, \tilde{y})$ subject to the constraints

(LP1)	$\sum_{i=1}^{n}(x_{ij} + \tilde{x}_{ij} r_j) \geq r_j$	for every $1 \leq j \leq n$
(LP2)	$x_{ij} \leq y_i$	for every $1 \leq i, j \leq n$
(LP3)	$\tilde{x}_{ij} \leq \tilde{y}_i$	for every $1 \leq i, j \leq n$
(LP4)	$y_i \leq 1$	for every $1 \leq i \leq n$
(LP5)	$\tilde{y}_i \leq 1$	for every $1 \leq i \leq n$
(LP6)	$y_i, \tilde{y}_i, x_{ij}, \tilde{x}_{ij} \geq 0$	for every $1 \leq i, j \leq n$

DP: maximize $\sum\limits_{j=1}^{n} r_j \alpha_i - \sum\limits_{i=1}^{n} (z_i + \tilde{z}_i)$ subject to the constraints

(DP1)	$\alpha_j \leq \beta_{ij} + d_{ij}$	for every $1 \leq i, j \leq n$
(DP2)	$\alpha_j \leq \tilde{\beta}_{ij} + d_{ij}$	for every $1 \leq i, j \leq n$
(DP3)	$\sum_{j=1}^{n} \beta_{ij} \leq p_i + z_i$	for every $1 \leq i \leq n$
(DP4)	$\sum_{j=1}^{n} \tilde{\beta}_{ij} \leq \tilde{p}_i + \tilde{z}_i$	for every $1 \leq i \leq n$
(DP5)	$\alpha_i, \beta_{ij}, \tilde{\beta}_{ij}, z_i, \tilde{z}_i \geq 0$	for every $1 \leq i, j \leq n$

The following lemma summarizes some basic connections between the variables of the LP program and its dual DP, derived by the complementary slackness theorem (see Chap. 7.9 of [8]).

Lemma 6.

(1) $x_{ij} > 0 \Rightarrow \alpha_j = \beta_{ij} + d_{ij}$ *for every* $1 \leq i, j \leq n$.
(2) $\tilde{x}_{ij} > 0 \Rightarrow \alpha_j = \tilde{\beta}_{ij} + d_{ij}$ *for every* $1 \leq i, j \leq n$.

(3) $y_i > 0 \Rightarrow \sum_{j=1}^{n} \beta_{ij} = p_i + z_i$ for every $1 \leq i \leq n$.

(4) $\tilde{y}_i > 0 \Rightarrow \sum_{j=1}^{n} \tilde{\beta}_{ij} = \tilde{p}_i + \tilde{z}_i$ for every $1 \leq i \leq n$.

(5) $\alpha_j > 0 \Rightarrow \sum_{i=1}^{n} (x_{ij} + \tilde{x}_{ij} r_j) = r_j$ for every $1 \leq j \leq n$.

(6) $\beta_{ij} > 0 \Rightarrow x_{ij} = y_i$ for every $1 \leq i, j \leq n$.

(7) $\tilde{\beta}_{ij} > 0 \Rightarrow \tilde{x}_{ij} = \tilde{y}_i$ for every $1 \leq i, j \leq n$.

(8) $z_i > 0 \Rightarrow y_i = 1$ for every $1 \leq i \leq n$.

(9) $\tilde{z}_i > 0 \Rightarrow \tilde{y}_i = 1$ for every $1 \leq i \leq n$.

Lemma 7. *Consider a solution* $(x, y, \tilde{x}, \tilde{y})$ *for LP. Without loss of generality, we may assume that for every* $1 \leq j \leq n$,

(1) $\sum_i (x_{ij} + \tilde{x}_{ij} r_j) = r_j$,

(2) *there cannot be more than one* i *such that* $0 < x_{ij} < y_i$ *or* $0 < \tilde{x}_{ij} < \tilde{y}_i$,

(3) *there cannot be any* i *such that both* $0 < x_{ij} < y_i$ *and* $0 < \tilde{x}_{ij} < \tilde{y}_i$.

The algorithm operates in three stages. In the preliminary Stage 0, it solves the linear program LP and obtains its optimal fractional solution $(x, \tilde{x}, y, \tilde{y})$.

In Stage 1, the algorithm selects initial sets $L = \{i \mid y_i = 1\}$ of normal centers and $\tilde{L} = \{i \mid \tilde{y}_i = 1\}$ of reinforced centers, and opens normal facilities at the nodes of L and reinforced facilities at the nodes of $\tilde{L}$. In addition, for every client $1 \leq j \leq n$, the algorithm identifies the sets $L_j = \{i \in L \mid x_{ij} > 0\}$ of normal servers that serve j in the fractional solution, and $\tilde{L}_j = \{i \in \tilde{L} \mid \tilde{x}_{ij} > 0\}$, and assigns j to every center in $L_j \cup \tilde{L}_j$. Denote the number of normal (respectively, reinforced) facilities serving j by $n_j = |L_j|$ (resp., $\tilde{n}_j = |\tilde{L}_j|$). Note that $\tilde{n}_j$ must be 0 or 1. If $\tilde{n}_j = 1$, then the residual (unsatisfied) backup demand of j is $\hat{r}_j = 0$. Otherwise, $\hat{r}_j = r_j - n_j$. As the final step of this stage, the algorithm prepares the sets of potential servers for j that were not used yet, $F_j = \{i \mid y_i < 1, x_{ij} > 0\}$ and $\tilde{F}_j = \{i \mid \tilde{y}_i < 1, \tilde{x}_{ij} > 0\}$.

Stage 2 aims at handling the set $S = \{j \mid \hat{r}_j \geq 1\}$ of unsatisfied clients, starting from $D = L$ and $\tilde{D} = \tilde{L}$ and building up. This is done in successive iterations, as long as $S \neq \emptyset$. Iteration l focuses on the client $j \in S$ such that α_j is minimal (later denoted $J(l)$ in the analysis) and tries to satisfy its rquirements. This is done in one of two ways.

Case (a): The client j under consideration satisfies $\sum_{i \in \tilde{F}_j} \tilde{y}_i \geq 1/(2r_j)$. In this case, the algorithm picks the node with the cheapest price $\tilde{p}_i$ in $\tilde{F}_j$, $i_{min} = argmin_{i \in \tilde{F}_j} \{\tilde{p}_i\}$, and opens a reinforced facility at i_{min}. Then, for each client k such that $\tilde{F}_k \cap \tilde{F}_j \neq \emptyset$, it assigns the reinforced i_{min} to serve k, and de-assigns k from any other (normal) facility assigned to it in previous steps. As the client k is now satisfied, it sets $\hat{r}_k = 0$. The algorithm now removes the facilities in $\tilde{F}_j$ from the input (setting the value $\tilde{y}_{i'}$ of each such node $i' \in \tilde{F}_j$ to 0, and for all $1 \leq j' \leq n$ setting $\tilde{x}_{i'j'} = 0$ and removing i' from $\tilde{F}_{j'}$).

Case (b): The client j has $\sum_{i \in \tilde{F}_j} \tilde{y}_i < 1/(2r_j)$. In this case, the algorithm will satisfy j's requirements using a set M_j of normal facilities. It first sorts the facilities by nondecreasing opening prices. If $\sum_{i \in F_j} y_i \leq \hat{r}_j$, then taking $M_j \leftarrow F_j$

satisfies j. If, on the other hand, $\sum_{i \in F_j} y_i > \hat{r}_j$, then a more elaborate selection is necessary. We first pick the elements of F_j into M_j in nondecreasing order of opening prices until $\sum_{i \in M_j} y_i \geq \hat{r}_j$. If $\sum_{i \in M_j} y_i > \hat{r}_j$, then we replace the last facility i in M_j with two copies, i_1 and i_2, and set $y_{i_1} = \hat{r}_j - \sum_{i' \in M_j \setminus \{i\}} y'_i$, $y_{i_2} = y_i - y_{i_1}$, $\tilde{y}_{i_1} = 0$ and $\tilde{y}_{i_2} = \tilde{y}_i$. Next, for each client k (including j) such that $x_{ik} > 0$, we set $x_{i_1 k}$ and $x_{i_2 k}$ such that $x_{i_1 k} \leq y_{i1}$, $x_{i_2 k} \leq y_{i2}$, $x_{i_1 k} + x_{i_2 k} = x_{ik}$. For each k (including j) such that $\tilde{x}_{ik} > 0$, we set $\tilde{x}_{i_1 k} \leftarrow \tilde{x}_{ik}$ and $\tilde{x}_{i_2 k} \leftarrow 0$. Finally, the algorithm includes i_1 in M_j.

Note that once M_j is selected, it satisfies $\sum_{i' \in M_j} y_{i'} = \hat{r}_j$. The algorithm now opens $\hat{r}_j$ cheapest (normal) facilities in M_j. These facilities are used to serve every client k (including j) such that $F_k \cap M_j \neq \emptyset$. Specifically, for each such k, the algorithm assigns $\min\{\hat{r}_k, \hat{r}_j\}$ of the open facilities to serve k, and then sets $\hat{r}_k = \hat{r}_k - \min\{\hat{r}_k, \hat{r}_j\}$ and $F_k \leftarrow F_k \setminus M_j$. Once this is done, tha algorithm removes the facilities in M_j from the input. This completes the iteration.

At the end of each iteration, the algorithm updates the set $S \leftarrow \{j \mid \hat{r}_j \geq 1\}$. The algorithm terminates when $S = \emptyset$, returning the obtained solution.

We now sketch the analysis (leaving a detailed description of the algorithm and some of the proofs to the full paper).

Denote the cost of stage 1 facility selection and client assignments by $Cost_1$. This cost is bounded as follows.

Lemma 8. $Cost_1 \leq \sum_{j=1}^{n} (n_j + \tilde{n}_j)\alpha_j - \sum_{i=1}^{n} (z_i + \tilde{z}_i)$.

To analyze Stage 2, we look at the iterations of the main loop. For every client k, let $\hat{r}_k(l)$ be the value of $\hat{r}_k$ at the end of iteration l, and let $\hat{r}_k(0)$ be the value of $\hat{r}_k$ at the beginning of stage 2. Let $F_k(l)$ and $\tilde{F}_k(l)$ be the values of F_k and $\tilde{F}_k$ at the end of iteration l, and let $F_k(0)$ and $\tilde{F}_k(0)$ be the values of F_k and $\tilde{F}_k$ at the beginning of stage 2.

Claim. At the beginning of stage 2, for every client k,

$$\sum_{i \in F_k(0)} y_i + r_k \sum_{i \in \tilde{F}_k(0)} \tilde{y}_i \geq \hat{r}_k(0) .$$

Lemma 9. After iteration l, for every client k, $\sum_{i \in F_k(l)} y_i + r_k \sum_{i \in \tilde{F}_k(l)} \tilde{y}_i \geq \hat{r}_k(l)$.

To establish the correctness of the algorithm, we show the following.

Lemma 10. *The final solution obtained by our algorithm satisfies constraints (ILP1) - (ILP4) of the ILP problem.*

It remains to analyze the approximation ratio. Denote the cost of the service between the clients and the facilities that are opened in stage 2 by $Cost_{2ser}$, the cost of opening the facilities in iteration l of stage 2 by $Cost_{2op}(l)$, and the cost of opening the facilities in stage 2 by $Cost_{2op} = \sum_{l=1}^{I} Cost_{2op}(l)$, where

I is the number of iterations in stage 2. Finally, denote the cost of the optimal solution to the LP version by OPT_{LP}, and set $r = \max_j\{r_j\}$.

Our analysis relies on the key observation that for each iteration l, either $Cost_{2op}(l) \leq 2r\sum_{i\in\tilde{F}_{J(l)}} \tilde{y}_i\tilde{p}_i$ and the servers of $\tilde{F}_{J(l)}$ are removed from the input, or $Cost_{2op}(l) \leq 3\sum_{i\in M_{J(l)}} y_ip_i$ and the servers of $M_{J(l)}$ are removed from the input. This allows us to bound the cost as follows.

Lemma 11. $Cost_{2op} \leq \max\{3, 2r\} \cdot (\sum_{i=0}^{n} p_iy_i + \sum_{i=0}^{n} \tilde{p}_i\tilde{y}_i)$.

Lemma 12. Let k be a client assigned to a facility i in stage 2. Then $d_{ik} \leq 3\alpha_k$.

Lemma 13. The cost of the service between the clients and the facilities that are opened in stage 2 satisfies $Cost_{2ser} \leq 3\sum_{j=1}^{n} \hat{r}_j\alpha_j$.

Combining, we get, letting $C_{1+2}^{-} = Cost_{2ser} + Cost_1$,

Lemma 14. The cost of service between the clients and the facilities that are opened in stage 2, plus the cost of stage 1, satisfies $C_{1+2}^{-} \leq 3 \cdot OPT_{LP}$.

Lemma 15. The cost of the overall solution is $Cost \leq (3+\max\{3, 2r\})\cdot OPT_{LP}$.

References

1. Garey, M.R., Johnson, D.S.: Computers and Intractability: A Guide to the Theory of NP-Completeness. W.H. Freeman and Co., New York (1979)
2. Gonzalez, T.F.: Clustering to minimize the maximum intercluster distance. Theor. Comput. Sci. **38**, 293–306 (1985)
3. Hochbaum, D.S., Shmoys, D.B.: A best possible heuristic for the k-center problem. Math. Oper. Res. **10**(2), 180–184 (1985)
4. Hochbaum, D.S., Shmoys, D.B.: A unified approach to approximation algorithms for bottleneck problems. J. ACM **33**(3), 533–550 (1986)
5. Johnson, D.S.: Approximation algorithms for combinatorial problems. J. Comput. Syst. Sci. **9**(3), 256–278 (1974)
6. Khuller, S., Pless, R., Sussmann, Y.J.: Fault tolerant k-center problems. Theor. Comput. Sci. **242**(1), 237–245 (2000)
7. Parter, M., Peleg, D.: Fault tolerant BFS structures: a reinforcement-backup trade-off. In: 27th ACM Symposium on Parallel Algorithms and Architectures (2015)
8. Schrijver, A.: Theory of Linear and Integer Programming. Wiley, Wiley-Interscience Series in Discrete Mathematics and Optimization (1999)
9. Sviridenko, M.: An improved approximation algorithm for the metric uncapacitated facility location problem. In: Cook, W.J., Schulz, A.S. (eds.) IPCO 2002. LNCS, vol. 2337, pp. 240–257. Springer, Heidelberg (2002). https://doi.org/10.1007/3-540-47867-1_18
10. Swamy, C., Shmoys, D.B.: Fault-tolerant facility location. ACM Trans. Algorithms (TALG) **4**(4), 51 (2008)
11. Wolsey, L.A.: An analysis of the greedy algorithm for the submodular set covering problem. Combinatorica **2**(4), 385–393 (1982)

Communication Complexity in Vertex Partition Whiteboard Model

Tomasz Jurdzinski$^{(\boxtimes)}$, Krzysztof Lorys, and Krzysztof Nowicki

Institute of Computer Science, University of Wroclaw, Wrocław, Poland
{tju,lorys,knowicki}@cs.uni.wroc.pl

Abstract. We study the multi-party communication model, where players correspond to the nodes of a graph and each player knows its neighbors in the input graph. The players can send messages on a *whiteboard* which are immediately available to each player. Eventually, the referee which knows only messages on the whiteboard is supposed to give a solution to the considered (graph) problem. We distinguish between *oblivious* and *adaptive* variant of the model. The former model is related to simultaneous multi-party communication complexity, while the latter is closely related to so-called broadcast congested clique.

Communication complexity is the maximum over all nodes of the sizes of messages put on the whiteboard by a node. Our goal is to study the impact of adaptivity on communication complexity of graph problems. We show that there exists an infinite hierarchy of problems with respect to the number of rounds for constant size messages. Moreover, motivated by unsuccessful attempts to establish non-adaptive communication complexity of graph connectivity in recent years, we study the connectivity problem in the severely restricted class of two-regular graphs We determine an asymptotically tight bound on communication complexity in the oblivious model and provide $\omega(1)$ lower bound on the number of rounds in the adaptive model for some message size $b(n) = \omega(1)$.

1 Introduction

Simultaneous two-party communication with referee model was proposed in [16]. In this paper we focus on the simultaneous multi-party communication model with referee, a generalization to the larger number of players. Additionally we consider two variants of the model – *oblivious* (*nonadaptive*) and *adaptive*.

In the oblivious two-party communication model, each player sends one message to a referee, who computes output of the protocol. Instead of considering referee we may think that all players simultaneously write messages on the whiteboard and, based on the state of the whiteboard, the result of the protocol is computed. The adaptive two-party communication model corresponds to the recently studied broadcast congested clique model model with referee. In this

This work was supported by the National Science Centre, Poland grant 2017/25/B/ST6/02010 model, each protocol may consist of multiple rounds. In each round, all players write messages simultaneously on the whiteboard, but each message might depend on the private input of the player and the state of the whiteboard after the previous round. At the end, the result of the protocol is computed based on the final content of the whiteboard.

Additionally, for graph problems we may consider two variants of input partition: edge partition and vertex partition. In the former variant the set of edges of the input graph is arbitrary distributed among players, therefore each player knows some subset of the set of edges of the graph. The latter variant (*vertex partition*) requires that each player is associated with a fixed subset of the set of vertices of the graph, which means that this player knows all the edges adjacent to the nodes associated to it. Significantly, in the vertex partition model, each edge (u, v) of the input graph is known to two players: the player associated with u and the player associated with v. The vertex partition model is potentially more powerful than the edge partition model. This strength makes it difficult to prove non-trivial lower bounds in the vertex partition model for some specific problems, especially for connectivity. We focus on the vertex partition simultaneous multi-party communication model, both adaptive and oblivious, in the context of graph problems.

The whiteboard n-party model with vertex-partition of an input graph closely corresponds with recently studied broadcast congested clique. The broadcast congested clique in turn is a restricted variant of the unicast congested clique, which attracted interest since its close relationships with more realistic models of computing on big data as k-machine big data model [12] or MapReduce [7,11].

1.1   Related Work

There is quite a broad research related to the connectivity of the input graph in the whiteboard model as well as in the congested clique. This includes algorithm research on the *connectivity* problem (CONN), where the goal is only to determine whether the input graph is connected and the *connected components* (CC) problem, where the partition into connected components has to be determined. The difference between CC and CONN, which we want to emphasize, is that CC problem trivially requires $\Omega(n \log n)$ bits to be sent on the whiteboard in the worst case (even in the vertex partition of input) in order to distinguish all partitions of the set of n nodes in disjoint connected components.

For edge partition 2-party communication model, it is known for quite long time that the connectivity problem (CONN) requires $\Omega(n \log n)$ bits of communication as well. More recently, it was shown that any one-round protocol with public randomness requires $\Omega(nk/\log^2 k)$ bits to be written on the whiteboard in the k-party simultaneous communication model [15]. This result implies that, for $k = n$ parties, $\Omega(n/\log^2 n)$-bit messages are necessary to solve the CONN problem.

Vertex partition version of simultaneous multi-party communication model is more powerful. In [1] authors give a non-adaptive protocol finding CC with high probability using public randomness with communication complexity $O(\log^3 n)$, which is way below $\Omega(n/\log^2 n)$ bits required in the edge partition model. There are also randomized algorithms which solve MST (and CC) in $O(1)$ rounds and $O(\log n)$-size messages in the unicast congested clique [5,6,8]. Moreover, deterministic algorithms solving CC in $O(1)$ rounds with $O(n^\varepsilon)$-bit messages for $\varepsilon < 1$ and in $O(\log/\log\log n)$ rounds with $O(\log n)$-bit messages have been obtained recently in the broadcast congested clique [9,14].

Proving lower bounds in the vertex partition whiteboard model is much harder than in the edge partition version, as it is impossible to manipulate with input of one player without changing the input of other players (as information about each edge is shared by two players). However, there are quite recent lower bound results, including subgraph detection lower bounds [2,4,10].

1.2 Model Definition and Complexity Classes

We consider a model of distributed computation and communication, where $k \leq n$ players work in synchronous rounds. At the beginning of each round, all players perform some internal computation. Then, all players simultaneously write messages of limited size b on the *whiteboard*. At the end of the round, each player can see (and store locally) all messages written by all other players so far. The joint input is an undirected n-node (for $n \geq k$) graph $G(V,E)$. The set of nodes V of G is split into k subsets and the ith player for $i \in [k]$ is associated with the ith subset of V. Moreover, each player has as input information about all edges adjacent to nodes of G associated with that player.

We define $\mathrm{WB}_r^k(b)$ as the set of problems, which can be decided by r-round, b-bit, k-party protocols working in the above described model. We assume that a protocol is parametrized by the size of the input graph n, while k, r, b might be either constants or functions of n.

The most popular variant of the above defined general model called *broadcast congested clique* assumes that $n = k$, i.e., each party is associated with one vertex of the input graph. Moreover, special emphasis has been made on one-round protocols; this model is called the *whiteboard communication* in some papers. Therefore, we use the following simplified notations for specific variants of $\mathrm{WB}_r^k(b)$:

- $\mathrm{WB}_r(b) = \mathrm{WB}_r^n(b)$. That is, $\mathrm{WB}_r(b)$ is the set of problems, which can be decided by r-round, b-bit, n-party protocols,
- $\mathrm{WB}_r = \bigcup_{b \in \mathbb{N}} \mathrm{WB}_r(b)$. That is WB_r is the class of problems which can be decided by r-round, $O(1)$-bits, n-party protocols.
- $\mathrm{WB}(b) = \mathrm{WB}_1(b)$ (i.e., $\mathrm{WB}(b) = \mathrm{WB}_1(b) = \mathrm{WB}_1^n(b)$. That is $\mathrm{WB}(b)$ is the class of problems which can be decided by 1-round, b-bits, n-party protocols

The model with only one round of communication and n parties, i.e., corresponding to $\bigcup_b \mathrm{WB}_1(b)$, will be called *non-adaptive (or oblivious) whiteboard model*. The *adaptive whiteboard model* corresponds to the situation that the allowed number of rounds of communication is larger than 1.

1.3 Our Results

The results of the paper are twofold. On one hand, we establish complexity theoretic results separating classes of problems solvable for various parameters b (message size) and r (the number of rounds). Moreover, we focus on (limited variants of) the connectivity problem, one of the most studied problems in context of the considered model. Below, we discuss the specific results split in two parts: round hierarchy results and lower bounds for (restricted variants of) connectivity. (Due to limited space, some proofs are deferred to the full version of this paper.)

Round hierarchy theorem. There exists an infinite sequence of natural numbers $r_1 < r_2 < r_3 < \cdots$ such that, the family of problems solvable in r_{i+1} rounds strictly includes the family of problems solvable in r_i rounds for each $i > 0$ and constant b: $\mathrm{WB}_{r_i}(b) \subsetneq \mathrm{WB}_{r_{i+1}}(b)$. Moreover, smaller number of rounds cannot be overcome by larger size of messages, i.e., for each $b = O(1)$, $\mathrm{WB}_{r_{i+1}} = \mathrm{WB}_{r_{i+1}}(1) \not\subset \mathrm{WB}_{r_i}(b)$.

Lower bounds for connectivity and its restricted variants. We show that communication complexity of CONN is $\Theta(n \log n)$, even in 2-party computation. As a simple corollary, we obtain logarithmic lower bound on communication complexity of CONN in one-round n-party model, i.e., CONN $\notin \mathrm{WB}(c \log n)$.

As the above lower bound does not apply for 2-degree graphs, we then focus on CONN problem in n-party whiteboard for 2-regular graphs. We call it the HAM problem, since connectivity of a 2-regular graphs corresponds to the fact that a graph is just a Hamiltonial cycle. We show that $\Theta(\log n)$ is the optimal size of messages for deterministic one-round protocols solving CONN. By adjusting our proof technique to multi-round protocols, we also show that HAM cannot be decided in $O(1)$ rounds with $O(1)$-size messages. The result is even stronger, i.e., HAM $\notin \mathrm{WB}_{f(n)}(g(n))$ for some $f(n), g(n) = \omega(1)$.

1.4 Notations

For natural numbers $a \leq b$, $[a, b]$ denotes the set of integers $\{a, a+1, \ldots, b\}$ and $[a] = [1, a]$. The set of neighbors of a vertex $v \in V$ in a graph $G(V, E)$ is denoted by $N_G(v)$ or shortly $N(v)$. For a one-round n party whiteboard protocol P, a node v and a set $X \subset V$, $P_v(X)$ denotes the message of v in P when $N(v) = X$. We say that a set of edges E is a *perfect bipartite matching* between disjoint sets V_1 and V_2 if edge edge from E has an endpoint in V_1 and an endpoint in V_2 and each node from $V_1 \cup V_2$ is incident to exactly one edge from E.

2 Two-party Communication

In this section we consider connectivity (CONN) in the two-party non-adaptive whiteboard model. We show that $\Theta(n \log n)$ bits are necessary and sufficient to solve CONN, i.e., CONN $\in \mathrm{WB}_1^n(O(n \log n)$ and CONN $\notin \mathrm{WB}_1^n(b(n))$ for each $b(n) = o(n \log n)$. Then, as corollaries, we establish lower bounds for message sizes necessary to solve CONN in the non-adaptive whiteboard model with arbitrary number $k \leq n$ parties.

Theorem 1. *In the k-party communication model for $k \geq 2$, $O(n \log n)$ bits per player are sufficient to decide whether an input graph is connected, $\Omega(\frac{n}{k} \log n)$ bits are necessary.*

In the k-party communication model, $O(n \log n)$ bits per player are sufficient to decide whether an input graph is connected, $\Omega(\frac{n}{k} \log n)$ bits are necessary.

3   Non-adaptive n-Party Communication

In this section we focus on the non-adaptive n-party whiteboard model. More precisely, we analyze communication complexity of the connectivity problem restricted to graphs with maximal degree two.

3.1   Non-adaptive Complexity of HAM

In this section we consider the HAM problem, i.e., the connectivity problem for two-regular graphs. Our goal is to establish communication complexity of this problem for the non-adaptive n-party whiteboard model. Formally, the HAM problem is to decide whether a given input graph is a Hamiltonian cycle. Note that the lower bound $\Omega(\log n)$ on the message size for the connectivity problem and $k = n$ players from Theorem 1 does not apply to the HAM problem, since the graphs analyzed in the proof of Theorem 1 might have nodes of degree three.

First, let us make a simple observation.

Fact 1. *The graph is a Hamiltonian cycle if and only if every node has two neighbors and graph is connected.*

Given the above property, one can easily solve HAM by a $O(\log n)$-bit n-party protocol. It is sufficient that each party sends a bit encoding information whether its degree is two or not and then each node of degree two sends $2 \log n$ bits with IDs of its neighbors. (By a simple trick generalized in [3] one can even limit the actual size of messages to $2 + \log n$.)

In the following theorem, we give an $\Omega(\log n)$ lower bound on asymptotic complexity of HAM for the n-party non-adaptive whiteboard model. Theorem 2 combined with the above described algorithm show that communication complexity of HAM in n-party non-adaptive whiteboard model is $\Theta(\log n)$.

Theorem 2. *Each one-round protocol solving HAM in the n-party whiteboard model requires messages of size at least $\frac{(\log n) - 3}{3}$.*

We give the proof of the above theorem in the remaining part of this section. The idea of the proof is to show that the protocol P, using messages of size smaller than $\frac{(\log n) - 1}{3}$, does not distinguish a graph G_1 which is a Hamiltonian cycle and a graph G_2 which is a sum of disjoint cycles. For a node v of the input graph we can define an auxiliary *conflict graph* i.e., a graph in which we put an edge between u and w, if message sent by v *has to* differentiate between the case that $\{v, u\}$ and the case that $\{v, w\}$ is an edge of the input graph. Then we can argue that the graph H_v has to be 2^b colorable, if the protocol solves HAM problem. To prove that there exists a node v for which H_v requires large number of colors (for correct node coloring), we show that each pair of nodes $\{u, w\}$ is connected by an edge in H_v for many nodes v of the input graph. This allows, by counting argument, to show that there is v such that H_v has large number of edges. Then, we use a specific lower bound on the chromatic number of a graph as a function of the number of the edges of the graph. This gives the lower bound on the number of colors 2^b of H_v with many edges and in turn we obtain a lower bound on the size of messages b of the protocol P.

The proof is given by an analysis of P on graphs from a specific family $\mathcal{G}$. For a fixed $n \in \mathbb{N}$, consider graphs on the set of vertices $V = \{v_1, \ldots, v_{4n}\}$ split into L_1, L_2, L_3, L_4 such that L_i (or *layer i*) is equal to $\{v_{(i-1)n+1}, v_{(i-1)n+2}, \ldots, v_{in}\}$ for $i \in [4]$. Let $\mathcal{G}$ be a family of such graphs, such that each graph $G(V, E)$ from $\mathcal{G}$ satisfies the following constraints:

- v_i is connected by an edge with v_{i+2n} for each $i \in [2n]$; that is each node v_i from $L_1 \cup L_2$ is connected by an edge with its *mirror* v_{i+2n};
- E contains a perfect matching between L_1 and L_2; that is each $v \in L_1$ ($v \in L_2$, resp.) has exactly one neighbor in L_2 (L_1, resp.);
- E contains a perfect matching between L_3 and L_4; that is each $v \in L_3$ ($v \in L_4$, resp.) has exactly one neighbor in L_4 (L_3, resp.);
- there are no more edges in G except of those described above.

For brevity of notation, we will denote nodes by their indices, i.e., the node v_i will be denoted just as i. For a node $u \in L_1$ ($u \in L_2$, resp) and $w \in L_2$ ($w \in L_1$, resp.), let $P_u(w)$ denote the message of u in the protocol P when $N(u) = \{w, u + 2n\}$. That is, $P_u(w) = P_u(\{w, u + 2n\})$.

For given $u, v \in L_2$, we split L_1 into 2^{2b} buckets according to messages, which u and v send on the whiteboard, for various graphs from $\mathcal{G}$. More precisely, for binary strings M_u, M_v of length b, a node $w \in L_1$ belongs to the bucket B_{M_u, M_v} iff $P_u(w) = M_u$ and $P_v(w) = M_v$. That is, all elements of a bucket are indistinguishable for u and v which means that u and v send the same message on the whiteboard for each element of the bucket being their neighbor.

Now, for each $w \in L_1$, we define the auxiliary graph $H_w(L_2, E_w)$ such that $\{u, v\} \in E_w$ iff $P_w(u) \neq P_w(v)$. That is, the edge $\{u, v\}$ says that the message w sends on the whiteboard when u is its neighbor and the message w sends on the whiteboard when v is its neighbor are different.

Below, we show that, if the messages written on the whiteboard are small, each edge $\{u, v\}$ for $u, v \in L_2$ appears in almost all graphs $H_w(L_2, E_w)$.

Proposition 1. *Let u, v be arbitrary elements of L_2 and M_u, M_v be b-bit strings. Then, for all but (at most) one of elements $w \in B_{M_u, M_v}$, (u, v) is an edge of $H_w(L_2, E_w)$.*

Proof. Contrary, assume that $\{u,v\} \notin E_{w_1}$ and $\{u,v\} \notin E_{w_2}$ for some $u, v \in L_2$, $w_1, w_2 \in B_{M_u, M_v}$.

Then, for any G_1, G_2 from the family $\mathcal{G}$ such that

- $\{u, w_1\}$ and $\{v, w_2\}$ are edges of G_1 and are not among edges of G_2,
- $\{u, w_2\}$ and $\{v, w_1\}$ are edges of G_2 and are not among edges of G_1,
- the remaining edges of G_1 and G_2 are identical (i.e., $E_{G_1} \setminus \{\{u, w_1\}, \{v, w_2\}\} = E_{G_2} \setminus \{\{u, w_2\}, \{v, w_1\}\}$),

G_1 and G_2 are indistinguishable by the algorithm, i.e., they give the same output on the whiteboard. We call such G_1 and G_2 *twins*.

Now, let us choose arbitrary $1 < i < n-1$. Consider a permutation of L_1 and L_3 in which w_1 and w_2 are on the ith and $(i+1)$st position respectively. Similarly, consider a permutation of L_2 and L_4 in which u and v are on the ith and $(i+1)$st position respectively. Let $v'_j, v'_{n+j}, v'_{2n+j}, v'_{3n+j}$ denote the jth node of L_1, L_2, L_3 and L_4 respectively after the application of these permutations. Now, we will fix a set of edges E' such that graphs G_1 with edges $E' \cup \{\{u, w_1\}, \{v, w_2\}\}$ and G_2 with edges $E' \cup \{\{u, w_2\}, \{v, w_1\}\}$ are twins. Moreover, we ensure that G_1 forms two separate cycles while the set of edges of G_2 is a Hamilton cycle.

Thus, we get a contradiction with the assumption that the algorithm solves the HAM problem: we get the same result for G_1 and G_2, while G_2 is in HAM and G_1 is not in HAM. □

Given Proposition 1, we are ready to prove Theorem 2. As each pair u, v from L_2 splits L_1 in at most 2^{2b} buckets, for each $u, v \in L_2$, the edge $\{u, v\}$ appears in at least $n - 2^{2b}$ of graphs $H_{v_1}(L_2, E_{v_1})$, $H_{v_2}(L_2, E_{v_2}), \ldots, H_{v_n}(L_2, E_{v_n})$, by Proposition 1. This in turn implies (by the pigeonhole principle) that H_{v_i} contains at least $\binom{n}{2} \frac{n-2^{2b}}{n}$ edges for some $i \in [n]$. On the other hand, by assigning the color $c(u) = (P_{v_i}(u))$ for each $u \in L_2$, we obtain a correct coloring of H_{v_i} with at most 2^b colors. Indeed, $c(u) = (P_{v_i}(u)) \neq (P_{v_i}(v)) = c(v)$ for each u, v connected by an edge in H_{v_i}, according to the definition of H_{v_i}. Now, we can apply the following lemma establishing a relationship between the number of edges of an n-node graph and its chromatic number.

Lemma 1. *The following property holds for any natural numbers x and b: if an n-node graph with at least $\binom{n}{2} \frac{n-2^{xb}}{n}$ edges is 2^b colorable, then $b \geq \frac{(\log n)-1}{x+1}$*

By Lemma 1 applied to H_{v_i}, $b \geq \frac{(\log n)-1}{3}$. This gives us that any protocol solving HAM requires messages of size $\frac{(\log n)-1}{3}$ for $4n$-node graphs. Therefore for an n-node graph, it would require messages of size $\frac{(\log(n/4))-1}{3} = \frac{(\log n)-3}{3}$.

4   Lower Bounds and Hierarchy Result for Adaptive Whiteboard Model

In this section we consider adaptive n-party whiteboard model, i.e., protocols with many (more than one) rounds. In Sect. 4.3 we show that HAM cannot be solved in $O(1)$ rounds with messages of size $O(1)$. Then, in Sect. 4.4, we show that there exist an infinite hierarchy of problems with respect to the number of rounds with $O(1)$-message size. The hierarchy result is obtained by providing an algorithm and a lower bound for the PATH$_d$ problem for $d \in \mathbb{N}$, where we ask whether the input graph contains a connected component equal to a path of length d.

Firstly, in Sects. 4.1 and 4.2, we introduce auxiliary notions of *matching sensitivity* and grid graphs with gadgets. They will serve as the main tools in lower bounds presented in Sects. 4.3 and 4.4.

Since we are considering protocols with many rounds, we need to introduce some new notations. Let $P^{(i)}(G)$ denote the state of the whiteboard of the protocol P after the ith round for an input graph G. Moreover, let us fix a decision problem $\mathcal{A}$ on graphs. We will say that a family of graphs $\mathcal{G}$ is (P, i)-fooling (with respect to $\mathcal{A}$), if $P^{(i)}(G_1) = P^{(i)}(G_2)$ for each $G_1, G_2 \in \mathcal{G}$, while $\mathcal{G}$ contains at least one graph for which the output for $\mathcal{A}$ is equal to true and $\mathcal{G}$ contains at least one graph for which the output for $\mathcal{A}$ is equal to false.

4.1   MATCHING Sensitivity

Let $\mathcal{G}$ be a family of graphs containing (among others) the vertices $v_1, \ldots, v_{2n}$ such that:

- the set of edges with one endpoint in $\{v_1, \ldots, v_{2n}\}$ is fixed (the same in all graphs from $\mathcal{G}$),
- the edges connecting nodes from $\{v_1, \ldots, v_{2n}\}$ form a perfect bipartite matching between $V_1 = \{v_1, v_2, \ldots, v_n\}$ and $V_2 = \{v_{n+1}, v_{n+2}, \ldots, v_{2n}\}$.

We say that an algorithm P is *insensitive* after i rounds on V_1, V_2 in the family $\mathcal{G}$ if the content of messages transmitted by $v_1, \ldots, v_{2n}$ is the same for all graphs from $\mathcal{G}$ after i rounds. In other words, the whiteboard does not give any information about edges between nodes of $V_1 \cup V_2$ (except of the fact that there is a perfect bipartite matching on (V_1, V_2)), provided the input graph belongs to $\mathcal{G}$.

Lemma 2. *Assume that an adaptive whiteboard algorithm P with b-bit messages is insensitive on V_1', V_2' of size p after i rounds for constant $i \geq 0$. Then, there exist $V_1'' \subset V_1'$ and $V_2'' \subset V_2'$ of size $\frac{\log \log p}{2b}$ such that P is insensitive on V_1'', V_2'' after $i + 1$ rounds, provided n and p are large enough.*

Proof. Let $q \leq p$ be a natural number. Moreover, let $U \subset V_1'$ be an arbitrary subset of V_1' of size q, $U = \{u_1, \ldots, u_q\}$. We split V_2' into buckets labeled with bq-bit words in the following way. A node $v \in V_2'$ is assigned to the bucket with the label $(P_{u_1}(v), P_{u_2}(v) \ldots P_{u_q}(v))$, where $P_{u_j}(v)$ is the message written in the $(i + 1)$-st round on the whiteboard by the node u_j if it is connected with v in the matching on sets V_1', V_2'. Using the pigeonhole principle, we can deduce that there exists a bucket containing at least $\frac{p}{2^{bq}}$ nodes.

Now consider an r-element subset U' of the largest bucket, $U' = \{u_1', \ldots, u_r'\}$ and set $r = \frac{p}{2^{bq}}$. Then, let us assign labels to the nodes of U, defined by messages of nodes from U'. The node $u \in U$ is assigned the label $(P_{u'_1}(u), P_{u'_2}(u) \ldots P_{u'_p}(u))$, where $P_{u'_j}(u)$ is the message written on the whiteboard by the node u'_j in the $(i+1)$st round if it would be matched with u in the input graph. Next, we split the nodes of U into buckets with respect to the labels defined by the elements of U'. Thus there are at least $\frac{q}{2^{br}}$ nodes in a largest bucket, by the pigeonhole principle. Let U'' be a subset of a largest bucket of size $\frac{q}{2^{br}}$. Observe that the sets U' and U'' are chosen such that P is insensitive on V''_1, V''_2 after $i+1$ rounds for each V''_1, V''_2 such that $V''_1 \subset U''$, $V''_2 \subset U'$ and $|V''_1| = |V''_2|$. Indeed, according to the definitions of U' and U'', for each $u' \in U'$ ($u'' \in U''$, resp.) connected with a node from U'' (U', resp.), the messages sent by u' up to the round $i+1$ do not depend on the fact which element of U'' (U', resp.) is a neighbor of u' (u'', resp.). Thus, in order to prove the lemma, it is sufficient to show that a it is possible to choose choose U', U'' satisfying the above constraints such that

$$\min\{|U'|, |U''|\} = \min\{r, \frac{q}{2^{br}}\} \geq \frac{\log \log p}{2b},$$

where $r = \frac{p}{2^{bq}}$. Let us consider q as a variable (we chosen arbitrary $q \leq p$). Note that $|U'| = r$ is a decreasing function of q, thus $|U''| = \frac{q}{2^{br}}$ is a growing function of q. Thus, $\min\{|U'|, |U''|\}$ is maximized when $|U'| = |U''|$ which gives the relationship $r = \frac{p}{2^{bq}} = \frac{q}{2^{br}}$. The relationship $r = \frac{q}{2^{br}}$ implies that $q = r2^{rb}$ which combined with $r = \frac{p}{2^{bq}}$ gives $p = r2^{br2^{rb}}$. The right-hand side expression $r2^{br2^{rb}}$ of the above relationship is the growing function of r. Thus, in order to finish the proof of the lemma, it is sufficient to check whether r satisfying the above equality is larger than $\frac{\log \log m}{2b}$, i.e., whether $r2^{br2^{rb}} < p$ for $r = \frac{\log \log m}{2b}$:

$$r2^{br2^{rb}} = \frac{\log \log p}{2b} \cdot 2^{b\frac{\log \log p}{2b}2^{\frac{\log \log p}{2b}b}} = \frac{\log \log p}{2b}2^{\frac{\log \log p}{2}2^{\frac{\log \log p}{2}}}$$
$$= 2^{\log \log \log p - \log(2b)} \cdot 2^{\sqrt{\log p}\frac{\log \log p}{2}} = 2^{\Theta(\sqrt{\log p}\log \log p)} < p$$

where the last inequality holds if p is large enough. One may argue that r and q should be natural numbers, while our setting does not guarantee that property. However, the inequality in the above estimation can be guaranteed also e.g. when $|U'| = |U''|$ is equal to $\lceil r \rceil$ if p is large enough. $\square$

4.2   Grid Graphs with Gadgets and Shuffles

Now, we discuss a kind of embedding of some families of graphs on a grid and specific rearrangements of such embeddings called shuffles.

Let m, d, k, δ, g be positive natural numbers such that $g + k\delta < d$. Moreover, let $\tau = (\tau_1, \ldots, \tau_{md})$ be a sequence of pairwise different natural numbers. We define the family of *grid graphs with gadgets (ggg)* $\mathcal{G}^\tau_{m,d,g,k,\delta}$ as follows.

Consider m-column, d-row rectangular grid and the set of nodes $v_{\tau(1)}, \ldots, v_{\tau(md)}$. Then, we put consecutive nodes from the sequence $v_{\tau(1)}, \ldots, v_{\tau(md)}$ in consecutive rows of the grid, i.e., the node $v_{\tau(a)}$ is located in the row $i = \lfloor 1 + (a - 1)/m \rfloor$ and the column $j = a - (i - 1)m$. For brevity of notation, the node located in the row i and the column j will be denoted by $u_{i,j}$. Then, we distinguish k *gadgets* in the grid. The jth gadget for $j \in [1, k]$ consists of the $(g + (j - 1)\delta)$th row $W_j = \{u_{g+(j-1)\delta,1}, \ldots, u_{g+(j-1)\delta,m}\}$ (the top row of gadget j) and the $(g+1+(j-1)\delta)$th row $W'_j = \{u_{g+1+(j-1)\delta,1}, \ldots, u_{g+1+(j-1)\delta,m}\}$ (the bottom row of gadget j). We say that m is the *width* of a gadget and δ is the *distance* between gadgets.

Finally, an md-node graph G put on the grid in the above described way belongs to the family $\mathcal{G}^\tau_{(m,d,g,k\delta)}$, if the set of edges of G satisfies the following conditions:

1. For each rows j and $j+1$ which do not form a gadget and each $i \in [m]$, there is a "vertical" edge connecting the ith node in the jth row and the ith node in the $(j+1)$st row. Formally, for each $i < d$ such that $i \notin \{g+l\delta \,|\, 0 \le l < k\}$, and each $j \in [m]$, there is an edge $(u_{i,j}, u_{i+1,j})$.
2. For each gadget $j \in [k]$ consisting of rows W_j and W'_j (i.e., the rows $g+(j-1)\delta$ and $g + 1 + (j - 1)\delta$ – see above), G contains a perfect bipartite matching between W_j and W'_j.
3. There are no more edges connecting nodes of G, except of those described above.

Thus, all graphs from the family share the same set of "vertical" edges outside of gadgets, while they can have arbitrary bipartite matchings "inside" gadgets. We also distinguish the family of *cyclic grid graphs with gadgets (cggg)* $\mathcal{C}^\tau_{m,d,g,k,\delta}$ which satisfy requirements of $\mathcal{G}^\tau_{m,d,g,k,\delta}$ with additional set of "vertical" edges connecting nodes in the last row (row d) and the first row: for each $j \in [m]$, there is an edge $(u_{d,j}, u_{1,j})$.

We will use the *shuffle* operation on a graph G defined as follows. Let $j \in [k - 1]$ and let τ_j be a permutation of $\{1, \ldots, m\}$. In the (j, τ_j)-shuffle of the graph G, we use the permutation τ_j in all rows starting in the bottom row W'_j of the gadget j and finishing at the top row W_{j+1} of the gadget $j + 1$. Formally, for each row $i \in [g + 1 + (j - 1)\delta, \ldots, g + j\delta + 1]$, the nodes $u_{i,1}, \ldots, u_{i,m}$ (nodes from the row i) are permuted according to τ_j. That is, after the (j, τ_j)-shuffle, the position (i, l) of the grid for $i \in [g + 1 + (j - 1)\delta, \ldots, g + j\delta]$ is occupied by the node $u_{i,\tau_j(l)}$.

If G is a ggg (i.e., without cycles), the $(0, \tau_0)$-shuffle $((k, \tau_k)$-shuffle, resp.) is just the application of the permutation τ_0 (τ_k, resp.) to the top row (bottom row, resp.) of the first (last, resp.) gadget and to all rows preceding (following, resp.) it.

If G is a cggg (with cycles), the above definition of shuffles is extended for (k, τ_k)-shuffle by assuming that the gadget 1 is preceded by the last gadget k and gadget k is followed by the gadget 1. Below, we state a simple but useful observation that ggg and cggg are closed under shuffles.

Lemma 3. *Let $\mathcal{G}^\tau_{m,d,g,k,\delta}$ ($\mathcal{C}^\tau_{m,d,g,k,\delta}$) be a family of grid graphs with gadgets (cyclic grid graphs with gadgets, respectively). Moreover, let $0 \le j_1 < j_2 < \cdots < j_l$ and let $\tau_{j_1}, \ldots, \tau_{j_1}$ be permutations of $\{1, \ldots, m\}$. Then, after applying the set*

of shuffles $S = \{(j_1, \tau_{j_1}), \ldots, (j_l, \tau_{j_l})\}$, *we obtain the family of graphs* $\mathcal{G}^{\tau'}_{m,d,g,k,\delta}$
$(\mathcal{C}^{\tau}_{m,d,g,k,\delta},$ *resp.*), *where* τ' *is determined by* τ *and* $\tau_{j_1}, \ldots, \tau_{j_1}$. *Moreover, for
each* $j \in \{j_1, \ldots, j_l\}$ *the top row of gadget* j *and the bottom row of gadget* $j + 1$
are obtained by the application of τ_j *to these rows on graphs from* $\mathcal{G}^{\tau}_{m,d,g,k,\delta}$
$(\mathcal{C}^{\tau}_{m,d,g,k,\delta},$ *resp.*).

4.3   Adaptive Complexity of the HAM problem

In this section we show that any adaptive algorithm solving HAM in the n-party adaptive whiteboard model with message size $b = O(1)$ requires $\Omega(\log^* n)$ rounds. More generally, we prove the following theorem.

Theorem 3. *Any adaptive whiteboard protocol* P *with* b-*bit messages requires* $\Omega(\log^* n - \log^* b)$ *rounds to solve HAM problem.*

In the remaining part of this section we prove Theorem 3 and give the final conclusion in Corollary 1. In order to prove Theorem 3 we will show that there exist (P, i)-fooling family of graphs for $i \in [\Theta(\log^* n - \log^* b)]$ and a protocol P. Let $n, m, k, \delta \in \mathbb{N}$ such that n is divisible by m and $n \geq mk\delta$. Moreover, let σ be a permutation of $\{1, \ldots, n\}$ and let $d = n/m$. We define the family of graphs $\mathcal{H}^{\sigma}_{(m,k,\delta)}$ as the family of cyclic grid graphs with gadgets $\mathcal{C}^{\sigma}_{m,d,1,k,\delta}$. That is, $\mathcal{H}^{\sigma}_{(m,k,\delta)}$ is the specific instantiation $\mathcal{C}^{\sigma}_{m,d,1,k,\delta}$ of a family of cyclic grid graphs, where σ is just a permutation of $\{1, \ldots, n\}$ (thus, the grid contains the nodes with indices $\{v_1, \ldots, v_n\}$) and the top row of the first gadget is just the top row of the grid.

Observe that, as long as $k \geq 1 \wedge m \geq 2$, there are both disconnected and connected graphs in $\mathcal{H}^{\sigma}_{(m,k,\delta)}$. Our goal is to build families of graphs defined above for consecutive $i = 1, 2, 3, \ldots$ with various parameters such that the considered protocol P is insensitive on all gadgets W_j, W'_j after i rounds. Then, as long as there is at least one gadget, the family contains both connected and disconneccted graphs and therefore it is (P, i)-fooling.

Let $\phi(n) = \frac{\log \log n}{2b}$, and let $\phi^{(i)}$ denote the i-fold composition of ϕ, i.e., $\phi^{(0)}(n) = n$ and $\phi^{(i)}(n) = \phi(\phi^{(i-1)}(n))$ for $i > 0$. Moreover, let $\phi^*(n) = \min\{r \mid \phi^{(r)}(n) < 2\}$.

Lemma 4. *Let* P *be an adaptive whiteboard algorithm with* b-*bit messages. Then, for each* $i \in \mathbb{N}$, *there are infinitely many naturals* n *satisfying the following condition: there exists a permutation* σ *of* $\{1, \ldots, n\}$ *such that* $\mathcal{H}^{\sigma}_{(\phi^{(i)}(\log n), \frac{2^{\phi^*(\log n)}}{2^i}, 2^{i+1})}$ *is* (P, i)-*fooling.*

Proof. We will prove Lemma 4 by induction. Note that there is nothing on the whiteboard before the first round. Therefore, any set of graphs containing both connected and disconnected graphs is $(P, 0)$-fooling. In particular the set $\mathcal{H}^{\mathrm{Id}}_{(\log n), 2^{\phi^*(\log n)}, 2}$ is $(P, 0)$-fooling, where Id denotes the identity permutation. This observation (applied for all large enough n) gives the base step of the inductive proof.

Now, let $m = \phi^{(i)}(\log n)$, $2k = \frac{2^{\phi^*(\log n)}}{2^i}$, and $\delta = 2^{i+1}$. For the inductive step, assume that $\mathcal{H}^{\sigma}_{(\phi^{(i)}(\log n), \frac{2^{\phi^*(\log n)}}{2^i}, 3 \cdot 2^i - 1)} = \mathcal{H}^{\sigma}_{(m, 2k, \delta)}$ is a (P, i)-fooling set for $i > 0$ and some permutation σ. To prove the inductive step, it is sufficient to show that there exists a family $\mathcal{H}^{\sigma'}_{(m', k, \delta')}$ included in $\mathcal{H}^{\sigma}_{(m, 2k, \delta)}$ which is $(P, i+1)$-fooling, where

- σ' is a permutation of $\{1, \ldots, n\}$,
- $m' = \frac{\log \log m}{2b} = \phi^{(i+1)}(\log n)$, $\delta' = 2\delta = 2^{i+2}$ and $k = \frac{2^{\phi^*(\log n)}}{2^{i+1}}$.

In other words, we build a $(P, i+1)$-fooling subfamily of $\mathcal{H}^{\sigma'}_{(m', k, \delta')}$ of $\mathcal{H}^{\sigma}_{(m, 2k, \delta)}$ embedded in the $m' \times n/m'$ grid with k gadgets in distances δ'.

Observe that, according to the assumption that $\mathcal{H}^{\sigma}_{(m, 2k, \delta)}$ is a (P, i)-fooling set, the protocol P is insensitive on each gadget, i.e., on the sets (W_j, W'_j) for $j \in [2k]$, where W_j, W'_j are the rows $1 + \delta(j - 1)$ and $2 + \delta(j - 1)$, resp. Therefore, we can apply Lemma 2 on each gadget. In our construction, we apply Lemma 2 simultaneously to all gadgets with odd indices, i.e., (W_1, W'_1), $(W_3, W'_3) \ldots (W_{2k-1}, W'_{2k-1})$. Moreover, we fix edges in all gadgets with even indices and some edges in gadgets with odd indices such that we will eventually get a new family of graphs which is $(P, i+1)$-fooling.

Lemma 2 implies that the following property holds after round $i + 1$: for each $j \in \{1, 3, \ldots, 2k - 1\}$ there are sets $U_j \subset W_j$, $U'_j \subset W'_j$ of size $\frac{\log \log m}{2b}$ such that P is insensitive on U_j, U'_j after $i + 1$ rounds.

Given the sets U_j, U'_j for odd js, we shuffle the nodes in the grid and fix some edges in gadgets such that the elements of U_j form the leftmost $\frac{\log \log m}{2b}$ nodes of the row $(j - 1)\delta + 1$ (i.e., the top row of the jth gadget), and the elements of U'_j form the leftmost $\frac{\log \log m}{2b}$ nodes of the row $(j - 1)\delta + 2$ (i.e., the bottom row of the jth gadget). Simultaneously we maintain the property that the graph obtained after the shuffles is a cyclic grid graph with gadgets. Recall that $m' = \frac{\log \log m}{2b}$. In order to satisfy the above described property, we apply Lemma 3 as follows. For each odd gadget $j \in \{1, 3, \ldots, 2k - 1\}$, we choose permutations $\sigma_{j,\text{top}}$ and $\sigma_{j,\text{bottom}}$ of $\{1, \ldots, m\}$ such that all elements of U_j are on the first $|U_j| = m'$ positions of $\sigma_{j,\text{top}}$ and all elements of U'_j are on the first $|U'_j| = m'$ positions of $\sigma_{j,\text{bottom}}$. Then, we apply $(j - 1, \sigma_{j,\text{top}})$-shuffle and $(j, \sigma_{j,\text{bottom}})$-shuffle.

Finally, we set edges in all even gadgets and some edges in odd gadgets in order to build such a subfamily of $\mathcal{H}^{\sigma}_{(m, k, \delta)}$ that only edges connecting sets U_j and U'_j remain undetermined (though there must be a perfect matching between U_j and U'_j in order to guarantee that the graph is in $\mathcal{H}^{\sigma}_{(m, k, \delta)}$) and the nodes of graphs can be rearranged in the grid with $m' = \frac{\log \log m}{2b} = |U_j|$ (for each odd j) columns such that the pairs of the sets U_j, U'_j form new gadgets in distance $2\delta = 2^{i+2}$. Firstly, we describe how the edges in even gadgets of the graphs from $\mathcal{H}^{\sigma}_{(m, k, \delta)}$ are set. Let $j \in [k]$ be an even number of the gadget. Recall that we apply two shuffles affecting the nodes from gadget j: $(j - 1, \sigma_{j-1,\text{bottom}})$-shuffle and $(j + 1, \sigma_{j+1,\text{top}})$-shuffle. For each $i \in [m]$, we connect the node from the column $\sigma_{j-1,\text{bottom}}(i)$ in the top row of gadget j with the node from the column $\sigma_{j+1,\text{top}}(i)$ in the bottom row of gadget j. That is, we set the edge $(u_{(j-1)\delta+1,\sigma_{j-1,\text{bottom}}(i)}, u_{(j-1)\delta+2,\sigma_{j+1,\text{top}}(i)})$. (Note that this is a vertical edge in the family obtained through the above described shuffles.)

Secondly, consider an odd gadget j from $\mathcal{H}^\sigma_{(m,k,\delta)}$. As described above, the nodes in the top row W_j of gadget j are reordered according to the permutation $\sigma_{j,\text{top}}$ of $\{1,\ldots,m\}$ and the nodes in the bottom row W'_j of gadget j are reordered according to the permutation $\sigma_{j,\text{bottom}}$ of $\{1,\ldots,m\}$. Analogously to the above discussed case of odd gadgets, we would like to set vertical edges inside the gadget, i.e., connect the node from the column $\sigma_{j,\text{top}}(i)$ in the top row of gadget j with the node from the column $\sigma_{j,\text{bottom}}(i)$ in the bottom row of gadget j. That is, we set the edge $(u_{(j-1)\delta+1,\sigma_{j,\text{top}}(i)}, u_{(j-1)\delta+2,\sigma_{j,\text{bottom}}(i)})$. However, recall that the sets $U_j \subset W_j$ and $U'_j \subset W'_j$ form $m' = |U_j| = |U'_j|$ leftmost elements of the sequences $\sigma_{j,\text{top}}$ and $\sigma_{j,\text{bottom}}$. And, the goal is to build a family of graphs, where arbitrary perfect bipartite matchings between U_j and U'_j might appear. Therefore, we add the edges $(u_{(j-1)\delta+1,\sigma_{j,\text{top}}(i)}, u_{(j-1)\delta+2,\sigma_{j,\text{bottom}}(i)})$ only for

$$i > |U_j| = |U'_j| = m' = \frac{\log \log m}{2b}$$

and leave the possibility that the set of edges inside $U_j \cup U'_j$ forms an arbitrary bipartite matching between U_j and U'_j.

Given the above described setting of edges and rearrangement of nodes in the grid, the leftmost m' columns of the original $m \times d$ grid might either form a connected subgraph or not, depending on perfect matchings between U_{2j} and U'_{2j} for js in $[1,k]$. However, for each $i > m'$, the nodes from the column i form a separate connected component, i.e., a cycle consisting of vertical edges connecting nodes in consecutive rows. Thus, all graphs from the obtained family are disconnected. On the other hand, our goal is to build a subfamily of $\mathcal{H}^\sigma_{(m,k,\delta)}$ which is $(P, i+1)$-fooling, i.e., all nodes from the subfamily should give the same content of the whiteboard after $i+1$ rounds while the subfamily should contain both connected and disconnected graphs. Therefore we make a slight but *significant* change in the above described setting of edges. By choosing large enough appropriate n we can assure that m is divisible by m'. Then, we choose an even gadget, say W_{2k}, W'_{2k}. Applied shuffles in the original grid imply that the edges set in the gadget $2k$

$$(u_{(2k-1)\delta+1,\sigma_{2k-1,\text{bottom}}(i)}, u_{(2k-1)\delta+2,\sigma_{1,\text{top}}(i)}) \text{ for } i \in [m]$$

are just vertical edges connecting nodes in the same column. Instead of vertical edges, we make a cyclic shift by m', i.e., the $i$the node of the top row of gadget $2k$ will be connected by an edge with: the $(i+m')$th node of the bottom row of gadget $2k$ if $i+m' \le m$ and with the $(i+m'-m)$th node of the bottom row of gadget $2k$ if $i+m' > m$. Thus, we add the edges

$$(u_{(2k-1)\delta+1,\sigma_{2k-1,\text{bottom}}(i)}, u_{(2k-1)\delta+2,\sigma_{1,\text{top}}(i+m')}) \text{ for } i \in [m]$$

instead of $(u_{(2k-1)\delta+1,\sigma_{2k-1,\text{bottom}}(i)}, u_{(2k-1)\delta+2,\sigma_{1,\text{top}}(i)})$. In this way, for each $i \in [1, m']$, the nodes from the columns $i+m', i+2m', \ldots, i+m-2m', i+m-m'$

are connected by a path which is entered from the ith column of gadget $2k$ (after the shuffles in the original grid) and the last node of this path has an edge connecting it with the node of the ith column of the top row of the gadget 1.

Finally, the above construction gives us the way to put graphs from $\mathcal{H}^\sigma_{(m,k,\delta)}$ in the $m' \times n/m'$ grid and satisfy constraints of $\mathcal{H}^{\sigma'}_{(m',k,\delta')}$, where the permutation σ' is determined by σ, the above described shuffles (following from the choices of U_j, U'_j for odd $j < 2k$) and the next rearrangement of nodes from columns $i > m'$ into paths which can be put in the first m' columns. Therefore we obtained a family of graphs $\mathcal{H}^{\sigma'}_{(m',k,\delta')}$ with $k = \frac{2^{\phi^*(\log n)}}{2^{i+1}}$ gadgets of width $m' = \frac{\log\log m}{2b} = \phi(\phi^{(i)}(\log n)) = \phi^{(i+1)}(\log n)$ with distance between them $2\delta = 2^{i+2}$, which finishes the inductive step of the proof of Lemma 4. $\square$

Recall that non-empty family $\mathcal{H}^\sigma_{(m,k,\delta)}$ contains both connected and disconnected graphs, provided that $k \geq 1$ and $m \geq 2$. Therefore, by Lemma 4, any adaptive algorithm solving HAM using b-bit messages needs $\phi^*(\log n)$ rounds. As ϕ is actually the function of n and b, one can prove the following relationship.

Fact 2. $\phi^*(\log n) \in \Omega(\log^* n - \log^* b)$ *for the function* $\phi(n) = \frac{\log\log n}{2b}$.

Finally, Theorem 3 is a simple consequence of Lemma 4 and Fact 2. For any constant b, Fact 2 implies $\Omega(\log^* n)$ lower bound on the number of rounds necessary to solve HAM using b-bit messages. Moreover, for each $b(n)$ such that $(\log^* n - \log^* b(n)) \in \omega(1)$ we have a super constant lower bound on the number of rounds necessary to solve HAM using $b(n)$-bit messages. In particular, there are $b(n) \in \omega(1)$ such that $\log^* n - \log^* b \in \omega(1)$, which implies the following corollary.

Corollary 1. *Any adaptive whiteboard protocol for HAM with messages of size $b \in O(1)$ requires $\Omega(\log^* n)$ rounds. Moreover, there exists $b \in \omega(1)$ such that each whiteboard protocol for HAM with b-bit messages requires $\omega(1)$ rounds.*

4.4   Round Hierarchy Theorem by the Analysis of the PATH$_d$ Problem

In this section we prove a round hierarchy theorem for messages of size $b = O(1)$. The result is obtained by analysis of the PATH$_d$ problem. We say that a graph $G(V,E)$ belongs to PATH$_d$ iff V contains a connected component equal to a path of length d. First, we give a simple algorithm which solves PATH$_d$ using $\lfloor \frac{d}{2} + 1 \rfloor$ rounds. Then, a lower bound $\Omega(\log d)$ on the number of rounds necessary to solve the PATH$_d$ problem is given.

The algorithm for PATH$_d$ Our whiteboard adaptive algorithm for PATH$_d$ uses 1-bit messages and works in $\lfloor \frac{d}{2} + 1 \rfloor$ rounds. In the first round, the nodes of degree 1 are writing 1 on the whiteboard. In the round $i \in [2, \lfloor \frac{d}{2} + 1 \rfloor]$, a node writes 1 on the whiteboard if it did not write 1 before, its degree is equal to 2 and one of its neighbors wrote 1 in round $i - 1$. The last $\lfloor \frac{d}{2} + 1 \rfloor$th round looks slightly different for odd and even d. If d is odd, a node v writes 1 on the whiteboard if one of its neighbors and v wrote 1 in the previous round and the degree of v is equal to 2. For even d, a node v writes 1 if both its neighbors wrote 1 in the previous round and the degree of v is equal to 2. One can easily check that the input graph belongs to PATH_d iff at least one node writes 1 on the whiteboard in the last round.

Lemma 5. *For each $d \in \mathbb{N}$, the problem PATH_d can be solved in $\lfloor \frac{d}{2} + 1 \rfloor$ rounds in the adaptive whiteboard model using 1-bit messages, i.e., $\text{PATH}_d \in WB_{\lfloor \frac{d}{2}+1 \rfloor}(1)$*

The lower bound for PATH$_d$ The main results of this section establishes a lower bound on the number of rounds sufficient to solve PATH_d with constant-size messages.

Theorem 4. *Any oblivious whiteboard protocol P with b-bit messages requires $\Omega(\min(\log d, \log^* n - \log^* b))$ rounds to solve PATH_d problem.*

The proof of Theorem 4 is based on similar ideas to those from the proof of Theorem 3. The key difference with respect to the proof of Theorem 3 is in the inductive step. Before, we rearranged and restricted a family of grid graphs such that the number of rows was growing and the number of columns was decreasing. Now, the number of rows will correspond to the parameter d of the considered PATH_d problem. Therefore, we will keep the number of rows unchanged and conceptually "remove" the part of the graphs which is fixed and has no impact of the fact whether the graph belongs to PATH_d. Moreover, we assure that the new 'subfamily" contains both graphs from PATH_d and graphs which do not belong to PATH_d.

The following result establishes a more exact separation at the bottom level of the hierarchy.

Theorem 5. *Each one-round protocol solving PATH_1 requires messages of size at least $\frac{(\log n) - 3}{5}$.*

By combining Lemma 5 and Theorems 4, 5, we have the following corollary.

Corollary 2. *Let $b = O(1)$ be an arbitrary constant. Then, there exists $c > 0$ such that $WB_{r_i}(b) \subsetneq WB_{r_{i+1}}(b)$ and $WB_{r_{i+1}}(1) \not\subset WB_{r_i}(b)$ for each $i \in \mathbb{N}$, where $r_1 = 1$, $r_2 = 2$ and $r_{j+1} = \lceil c \cdot 2^{r_j} \rceil$ for $j > 1$.*

References

1. Ahn, K.J., Guha, S., McGregor, A.: Analyzing graph structure via linear measurements. In: Discrete Algorithms, SODA 2012, pp. 459–467 (2012)
2. Becker, F., et al.: Allowing each node to communicate only once in a distributed system: shared whiteboard models. Distrib. Comput. **28**(3), 189–200 (2015)
3. Becker, F., Montealegre, P., Rapaport, I., Todinca, I.: The simultaneous number-in-hand communication model for networks: private coins, public coins and determinism. In: Halldórsson, M.M. (ed.) SIROCCO 2014. LNCS, vol. 8576, pp. 83–95. Springer, Cham (2014). https://doi.org/10.1007/978-3-319-09620-9_8

4. Drucker, A., Kuhn, F., Oshman, R.: On the power of the congested clique model. In: PODC 2014, pp. 367–376. ACM (2014)
5. Ghaffari, M., Parter, M.: MST in log-star rounds of congested clique. In: PODC 2016, pp. 19–28. ACM (2016)
6. Hegeman, J.W., Pandurangan, G., Pemmaraju, S.V., Sardeshmukh, V.B., Scquizzato, M.: Toward optimal bounds in the congested clique: graph connectivity and MST. In: Distributed Computing, PODC 2015, pp. 91–100. ACM (2015)
7. Hegeman, J.W., Pemmaraju, S.V.: Lessons from the congested clique applied to mapreduce. Theor. Comput. Sci. **608**, 268–281 (2015)
8. Jurdzinski, T., Nowicki, T.: MST in O(1) rounds of congested clique. In: SODA 2018, SIAM, pp. 2620–2632 (2018)
9. Jurdzinski, T., Nowicki, K.: Brief announcement: on connectivity in the broadcast congested clique. In: DISC 2017, LIPIcs, pp. 54:1–54:4 (2017)
10. Kari, J., Matamala, M., Rapaport, I., Salo, V.: Solving the INDUCED SUBGRAPH problem in the randomized multiparty simultaneous messages model. In: Scheideler, C. (ed.) Structural Information and Communication Complexity. LNCS, vol. 9439, pp. 370–384. Springer, Cham (2015). https://doi.org/10.1007/978-3-319-25258-2_26
11. Karloff, H.J., Suri, S., Vassilvitskii, S.: A model of computation for mapreduce. In: SODA 2010, SIAM, pp. 938–948 (2010)
12. Klauck, H., Nanongkai, D., Pandurangan, G., Robinson, P.: Distributed computation of large-scale graph problems. In: SODA 2015, SIAM, pp. 391–410 (2015)
13. Korhonen, J.H., Suomela, J.: Brief announcement: towards a complexity theory for the congested clique. In: DISC 2017, pp. 55:1–55:3 (2017)
14. Montealegre, P., Todinca, I.: Brief announcement: deterministic graph connectivity in the broadcast congested clique. In: PODC 2016. ACM (2016)
15. Phillips, J.M., Verbin, E., Zhang, Q.: Lower bounds for number-in-hand multiparty communication complexity, made easy. In: SODA 2012, SIAM, pp. 486–501 (2012)
16. Yao, A.C.-C.: Some complexity questions related to distributive computing (preliminary report). In: STOC 1979, pp. 209–213. ACM (1979)

Time-Bounded Influence Diffusion
with Incentives

Gennaro Cordasco[1(✉)], Luisa Gargano[2], Joseph G. Peters[3],
Adele A. Rescigno[2], and Ugo Vaccaro[2]

[1] Department of Psychology, Università della Campania
"Luigi Vanvitelli", Caserta, Italy
gennaro.cordasco@unicampania.it
[2] Department of Computer Science, Università di Salerno, Fisciano, Italy
[3] School of Computing Science, Simon Fraser University, Burnaby, Canada

Abstract. A widely studied model of influence diffusion in social networks represents the network as a graph $G = (V, E)$ with an influence threshold $t(v)$ for each node. Initially the members of an initial set $S \subseteq V$ are influenced. During each subsequent round, the set of influenced nodes is augmented by including every node v that has at least $t(v)$ previously influenced neighbours. The general problem is to find a small initial set that influences the whole network. In this paper we extend this model by using *incentives* to reduce the thresholds of some nodes. The goal is to minimize the total of the incentives required to ensure that the process completes within a given number of rounds. The problem is hard to approximate in general networks. We present polynomial-time algorithms for paths, trees, and complete networks.

1 Introduction

The *spread of influence* in social networks is the process by which individuals adjust their opinions, revise their beliefs, or change their behaviours as a result of interactions with others (see [14] and references therein quoted). For example, *viral marketing* takes advantage of peer influence among members of social networks for marketing [13]. The essential idea is that companies wanting to promote products or behaviours might try to target and convince a few individuals initially who will then trigger a cascade of further adoptions. The intent of maximizing the spread of viral information across a network has suggested several interesting optimization problems with various adoption paradigms. We refer to [5] for a recent discussion of the area. In the rest of this section, we will explain and motivate our model of information diffusion, describe our results, and discuss how they relate to the existing literature.

1.1 The Model

A social network is a graph $G = (V, E)$, where the node set V represents the members of the network and E represents the relationships among members.

Z. Lotker and B. Patt-Shamir (Eds.): SIROCCO 2018, LNCS 11085, pp. 280–295, 2018.
https://doi.org/10.1007/978-3-030-01325-7_25

We denote by $n = |V|$ the number of nodes, by $N(v)$ the neighbourhood of v, and by $d(v) = |N(v)|$ the degree of v, for each node $v \in V$.

Let $t : V \to \mathbb{N} = \{1, 2, \dots\}$ be a function assigning integer thresholds to the nodes of G; we assume w.l.o.g. that $1 \leq t(v) \leq d(v)$ holds for all $v \in V$. For each node $v \in V$, the value $t(v)$ quantifies how hard it is to influence v, in the sense that easy-to-influence elements of the network have "low" $t(\cdot)$ values, and hard-to-influence elements have "high" $t(\cdot)$ values [16]. An *influence process in G* starting from a set $S \subseteq V$ of initially influenced nodes is a sequence of node subsets,[1]

$\mathsf{Influenced}_G[S, 0] = S$

$\mathsf{Influenced}_G[S, \ell] = \mathsf{Influenced}_G[S, \ell-1] \cup \Big\{ v : \big| N(v) \cap \mathsf{Influenced}_G[S, \ell-1] \big| \geq t(v) \Big\}$,

$\ell > 0$.

Thus, in each round ℓ, the set of influenced nodes is augmented by including every uninfluenced node v for which the number of *already* influenced neighbours is at least as big as v's threshold $t(v)$. We say that v is influenced *at* round $\ell > 0$ if $v \in \mathsf{Influenced}_G[S, \ell] \setminus \mathsf{Influenced}_G[S, \ell - 1]$. A target set for G is a set S such that it will influence the whole network, that is, $\mathsf{Influenced}_G[S, \ell] = V$, for some $\ell \geq 0$.

The classical *Target Set Selection* (TSS) problem having as input a network $G = (V, E)$ with thresholds $t : V \longrightarrow \mathbb{N}$, asks for a target set $S \subseteq V$ of *minimum* size for G [1,8]. The TSS problem has roots in the general study of the spread of influence in social networks (see [5,14]). For instance, in the area of viral marketing [13], companies wanting to promote products or behaviors might try to initially convince a small number of individuals (by offering free samples or monetary rewards) who will then trigger a cascade of influence in the social network leading to the adoption by a much larger number of individuals.

In this paper, we extend the classical model to make it more realistic. It was first observed in [12] that the classical model limits the optimizer to a binary choice between zero or complete influence on each individual whereas customized incentives could be more effective in realistic scenarios. For example, a company promoting a new product may find that offering one hundred free samples is far less effective than offering a ten percent discount to one thousand people.

Furthermore, the papers mentioned above do not consider the time (number of rounds) necessary to complete the influence diffusion process. This could be quite important in viral marketing; a company may want to influence its potential customers quickly before other companies can market a competing product.

With this motivation, we formulate our model as follows. An assignment of incentives to the nodes of a network $G = (V, E)$ is a function $p : V \to \mathbb{N}_0 = \{0, 1, 2, \dots\}$, where $p(v)$ is the amount of influence initially applied on $v \in V$. The effect of applying the incentive $p(v)$ on node v is to decrease its threshold, i.e., to make v more susceptible to future influence. It is clear that to start the process, there must be some nodes for which the initially applied influences are at least as large as their thresholds. We assume, w.l.o.g., that

---

[1] We will omit the subscript G whenever the graph G is clear from the context.

$0 \leq p(v) \leq t(v) \leq d(v)$. An influence process in G starting with incentives given by a function $p : V \to \mathbb{N}_0 = \{0, 1, 2, \ldots\}$ is a sequence of node subsets

$\text{Influenced}[p, 0] = \{v : p(v) = t(v)\}$

$\text{Influenced}[p, \ell] = \text{Influenced}[p, \ell-1] \cup \Big\{ v : \big| N(v) \cap \text{Influenced}[p, \ell-1] \big| \geq t(v) - p(v) \Big\}$,

$\ell > 0$.

The cost of the incentive function $p : V \longrightarrow \mathbb{N}_0$ is $\sum_{v \in V} p(v)$.

Let λ be a bound on the number of rounds available to complete the process of influencing all nodes of the network. The Time-Bounded Targeting with Incentives problem is to find incentives of minimum cost which result in all nodes being influenced in at most λ rounds:

TIME-BOUNDED TARGETING WITH INCENTIVES (TBI).

Instance: A network $G = (V, E)$ with thresholds $t : V \longrightarrow \mathbb{N}$ and time bound λ.

Problem: Find incentives $p : V \longrightarrow \mathbb{N}_0$ of minimum cost $\sum_{v \in V} p(v)$ s.t.

$$\text{Influenced}[p, \lambda] = V.$$

Example 1. Solutions to the TBI problem can be quite different from solutions to the TSS problem for a given network. Consider a complete graph K_8 on 8 nodes with thresholds shown in Fig. 1. The optimal target set is $S = \{v_8\}$ which results in all nodes being influenced in 4 rounds. The TBI problem admits different optimal solutions (with different incentive functions) depending on the value of λ, as shown in Fig. 1.

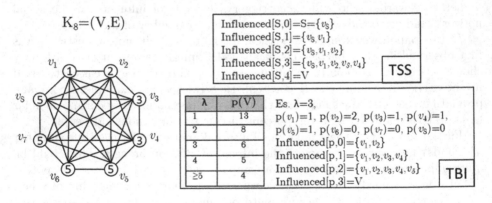

Fig. 1. A complete graph K_8. The number inside each circle is the node threshold. Optimal solutions for the TSS problem and the TBI problem, with various values of λ, are shown.

1.2   Related Work and Our results

The study of the spread of influence in complex networks has experienced a surge of interest in the last few years. Kempe *et al.* [17] introduced the Influence

Maximization (IM) problem, where the goal is to find a subset of nodes in a social network that has cardinality bounded by a certain budget β and that could maximize the spread of influence. However, they were mostly interested in networks with randomly chosen thresholds.

Chen [6] studied the TSS problem. He proved a strong inapproximability result that makes unlikely the existence of an algorithm with approximation factor better than $O(2^{\log^{1-\epsilon} |V|})$. Chen's result stimulated a series of papers including [1–4,7–9,15,20,22] that isolated many interesting scenarios in which the problem (and variants thereof) become tractable.

The problem of maximizing the number of nodes activated within a specified number of rounds has also been studied [8,9,23]. The problem of dynamos or dynamic monopolies in graphs is essentially the target set problem with every node threshold being half its degree [21].

The Influence Maximization problem with incentives was introduced in [12]. In this model the authors assume that the thresholds are randomly chosen values in $[0,1]$ and they aim to understand how a fractional version of the Influence Maximization problem differs from the original version. To that purpose, they introduced the concept of partial influence and showed that, in theory, the fractional version retains essentially the same computational hardness as the integral version, but in practice, better solutions can be computed using heuristics in the fractional setting.

The Targeting with Partial Incentives (TPI) problem, of finding incentives $p : V \longrightarrow \mathbb{N}_0$ of minimum cost $\sum_{v \in V} p(v)$ such that all nodes are eventually influenced, was studied in [11]. Exact solutions to the TPI problem for special classes of graphs were proposed in [10,11]. Variants of the problem, in which the incentives are modelled as additional links from an external entity, were studied in [18,19]. The authors of [23] study the case in which offering discounts to nodes causes them to be influenced with a probability proportional to the amount of the discount.

It was shown in [11] that the TPI problem cannot be approximated to within a ratio of $O(2^{\log^{1-\epsilon} n})$, for any fixed $\epsilon > 0$, unless $NP \subseteq DTIME(n^{polylog(n)})$, where n is the number of nodes in the graph. As a consequence, for general graphs, the same inapproximability result still holds for the time bounded version of the problem that we study in this paper.

Theorem 1. *The TBI problem cannot be approximated to within a ratio of $O(2^{\log^{1-\epsilon} n})$, for any fixed $\epsilon > 0$, unless $NP \subseteq DTIME(n^{polylog(n)})$, where n is the number of nodes in the graph.*

Our Results. Our main contributions are polynomial-time algorithms for path, complete, and tree networks. In Sect. 2, we present a linear-time greedy algorithm to allocate incentives to the nodes of a path network. In Sect. 3, we design a $O(\lambda n \log n)$ dynamic programming algorithm to allocate incentives to the nodes of a complete network. In Sect. 4, we give an $O(\lambda^2 \Delta n\})$algorithm to allocate incentives to a tree with n nodes and maximum degree Δ.

2   A Linear-Time Algorithm for Paths

In this section, we present a greedy algorithm to allocate incentives to nodes of a path network. We prove that our algorithm is linear-time.

We denote by $L(0, n-1)$ the path with n nodes $0, \ldots, n-1$ and edges $\{(i, i+1) : 0 \leq i \leq n-2\}$. Since the threshold of each node cannot exceed its degree, we have that $t(0) = t(n-1) = 1$ and $t(i) \in \{1, 2\}$, for $i = 1, \ldots, n-2$. For $0 \leq j \leq k \leq n-1$, we denote by $L(j, k)$ the subpath induced by the nodes $j, \ldots, k$.

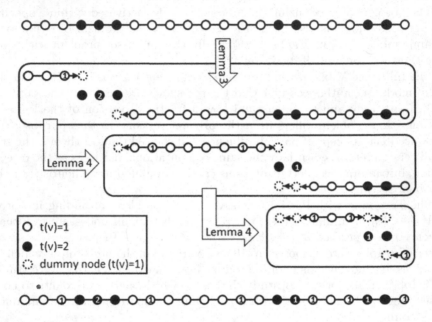

Fig. 2. An example of the execution of the Algorithm 1 on a path $L(0, 21)$ with a 2-path satisfying Lemma 3 and two 2-paths satisfying Lemma 4. Filled nodes represents nodes having threshold 2. Dashed nodes represents dummy nodes. The number inside the nodes represents the incentive assigned to the node.

Lemma 1. *Let $L(j, k)$ be a subpath of $L(0, n-1)$ with $t(j+1) = \cdots = t(k-1) = 2$ and $t(j) = t(k) = 1$. For any incentive function $p : V \to \{0, 1, 2\}$ that solves the TBI problem on $L(0, n-1)$ and for any λ,*

$$\sum_{i=j+1}^{k-1} p(i) \geq \begin{cases} k-j-2 & \text{if both } j+1 \text{ and } k-1 \text{ are influenced by } j \text{ and } k, \text{ resp.} \\ k-j-1 & \text{if either } j+1 \text{ or } k-1 \text{ is influenced by its neighbour } (j \text{ or } k) \\ k-j & \text{otherwise.} \end{cases}$$

Proof. Let p be an incentive function that solves the TBI problem on $L(0, n-1)$. For any node $i \in \{j+1, \ldots, k-1\}$, let $inf(i) \in \{0, 1, 2\}$ be the amount of influence that i receives from its neighbours in $L(0, n-1)$ during the influence process starting with p (that is, the number of i's neighbours that are influenced before round i).

For each $i = j + 1, \ldots, k - 1$, it must hold that $inf(i) + p(i) \geq t(i) = 2$. Hence,

$$\sum_{i=j+1}^{k-1} p(i) \geq \sum_{i=j+1}^{k-1} (2 - inf(i)) \geq 2(k - j - 1) - \sum_{i=j+1}^{k-1} inf(i). \qquad (1)$$

Noticing that each link in E is used to transmit influence in at most one direction, we have

$$\sum_{i=j+1}^{k-1} inf(i) \leq \begin{cases} k-j, & \text{if both } j+1 \text{ and } k-1 \text{ are influenced by } j \text{ and } k, \text{ resp.} \\ k-j-1, & \text{if either } j+1 \text{ or } k-1 \text{ is influenced by its neighbour } (j \text{ or } k) \\ k-j-2, & \text{otherwise.} \end{cases}$$

As a consequence, using Eq. (1) gives the desired result.

In the following we assume that $\lambda \geq 2$. The case $\lambda = 1$ will follow from the results in Sect. 4, since the algorithm for trees has linear time when both λ and the maximum degree are constant.

Definition 1. *We denote by $OPT(0, n - 1)$ the value of an optimal solution $p : V \rightarrow \{0, 1, 2\}$ to the TBI problem on $L(0, n-1)$ in λ rounds. For any subpath $L(j, k)$ of $L(0, n - 1)$, we denote by:*

(i) $OPT(j, k)$ *the value $\sum_{i=j}^{k} p(i)$ where p is an optimal solution to the TBI problem on $L(j, k)$;*

(ii) $OPT(j, k, \leftarrow)$ *the value $\sum_{i=j}^{k} p(i)$ where p is an optimal solution to the TBI problem on $L(j, k)$ with the additional condition that the node k gets one unit of influence from $k + 1$;*

(iii) $OPT(j, k, \overset{\ell}{\leftarrow})$ *the value $\sum_{i=j}^{k} p(i)$ where p is an optimal solution to the TBI problem on $L(j, k)$ with the additional condition that k is influenced by round $\lambda - \ell$ without getting influence from node $k + 1$;*

(iv) $OPT(\rightarrow, j, k)$ *the value $\sum_{i=j}^{k} p(i)$ where p is an optimal solution to the TBI problem on $L(j, k)$ with the additional condition that j gets one unit of influence from $j - 1$;*

(v) $OPT(\overset{\ell}{\rightarrow}, j, k)$ *the value $\sum_{i=j}^{k} p(i)$ where p is an optimal solution to the TBI problem on $L(j, k)$ with the additional condition that node j is influenced by round $\lambda - \ell$ without getting influence from $j - 1$.*

Lemma 2. *For any subpath $L(j, k)$ and for each $1 \leq \ell < \ell' \leq \lambda$:*

(1) If $t(k) = 1$ then $OPT(j, k, \leftarrow) \leq OPT(j, k) \leq OPT(j, k, \overset{\ell}{\leftarrow}) \leq OPT(j, k, \overset{\ell'}{\leftarrow}) \leq OPT(j, k, \leftarrow) + 1$.

(2) If $t(j) = 1$ then $OPT(\rightarrow, j, k) \leq OPT(j, k) \leq OPT(\overset{\ell}{\rightarrow}, j, k) \leq OPT(\overset{\ell'}{\rightarrow}, j, k) \leq OPT(\rightarrow, j, k) + 1$.

Proof. We first prove (1). We notice that each of the first three inequalities $OPT(j, k, \leftarrow) \leq OPT(j, k)$, $OPT(j, k) \leq OPT(j, k, \overset{\ell}{\leftarrow})$, $OPT(j, k, \overset{\ell}{\leftarrow}) \leq OPT(j, k, \overset{\ell'}{\leftarrow})$ is trivially true since each solution that satisfies the assumptions of the right term is also a solution that satisfies the assumptions of the left term. It remains to show that $OPT(j, k, \overset{\ell'}{\rightarrow}) \leq OPT(j, k, \leftarrow) + 1$. Let p be a solution that gives $OPT(j, k, \leftarrow)$. Consider p' such that $p'(i) = p(i)$, for each $i = j, \ldots, k-1$ and $p'(k) = 1$. Recalling that $t(k) = 1$, we get that the cost increases by at most 1 and p' is a solution in which node k is influenced at round $0 \leq \lambda - \ell'$. A similar proof holds for (2). □

Definition 2. $L(j, k)$, with $j + 1 \leq k - 1$, is called a 2-path if $t(j + 1) = \ldots = t(k - 1) = 2$ and $t(j) = t(k) = 1$.

Lemma 3. For any value of λ, if $L(j, k)$ is a 2-path with $m = k - j - 1 \neq 2$ then $OPT(0, n - 1) = OPT(0, j, \overset{1}{\rightarrow}) + k - j - 2 + OPT(\overset{1}{\leftarrow}, k, n - 1)$.

Outline of Proof. The proof shows that one can always find an optimal solution in which the $m = k - j - 1$ nodes in $L(j + 1, k - 1)$ receive the incentives using the sequence $0(20)^*$ when m is odd and $01(20)^*$ when $m > 2$ is even. See Fig. 2 for an example of the odd case. □

Lemma 4. For any time bound $\lambda > 1$, if $t(0) = t(1) = \ldots = t(j - 1) = 1$ and $L(j, j + 3)$ is a 2-path then $OPT(0, n-1)$ is equal to

$$1 + \min \left\{ OPT(0, j, \overset{1}{\rightarrow}) + OPT(\overset{2}{\leftarrow}, j+3, n-1), OPT(0, j, \overset{2}{\rightarrow}) + OPT(\overset{1}{\leftarrow}, j+3, n-1) \right\}.$$

Outline of Proof. The proof is a case analysis that shows that there is always an optimal solution in which either

1. $p(j + 1) = 0, p(j + 2) = 1$ or   2. $p(j + 1) = 1, p(j + 2) = 0$. □

Lemma 5. For any value of λ, the minimum cost for the TBI problem on a path of n nodes having threshold 1 is $\lceil n/(2\lambda + 1) \rceil$.

Outline of Proof. The basic idea is to break the path into subpaths of $2\lambda + 1$ nodes and assign an incentive of 1 to the middle node of each subpath and incentive 0 to the others. □

Remark 1. $OPT(j, k, \overset{\ell}{\rightarrow})$ can be obtained by solving the TBI problem on an augmented path $L(j, k + \ell)$ obtained from $L(j, k)$ by concatenating ℓ dummy nodes on the right of k with $t(k + 1) = t(k + 2) = \ldots = t(k + \ell) = 1$. Notice that, for $\ell \leq \lambda$, it is always possible to find an optimal assignment of incentives for the augmented path $L(j, k + \ell)$ in which all dummy nodes get incentive 0. Indeed it is possible to obtain such an assignment starting from any optimal assignment and moving the incentives from the dummy nodes to node k. An analogous observation holds for $OPT(\overset{\ell}{\leftarrow}, j, k)$.

Our algorithm iterates from left to right, identifying all of the 2-paths and, using Lemma 3 or 4 and Lemma 5, it optimally assigns incentives both to the nodes of threshold 2 and to the nodes (of threshold 1) on the left. It then removes them from the original path. Eventually, it will deal with a last subpath in which all of the nodes have threshold 1.

Algorithm 1.   TBI-Path($L(0, n-1)$)

Input: A Path $L(0, n-1)$, thresholds $t(i) \in \{1, 2\}$, $i = 0, \ldots, n-1$, and a time bound λ.

Output: A solution $p(i) : V \rightarrow \{0, 1, 2\}$ of the TBI problem.

1 $i = 0$
2 **while** *there exists a node j with $t(j) = 2$ for some $i < j < n-1$* **do**
3 $\quad$ Identify the leftmost 2-path in the current path $L(i, n-1)$; let it be $L(j, k)$.
4 $\quad$ **if** $L(j, k)$ *is a 2-path satisfying Lemma 3* **then**
5 $\quad\quad$ assign incentives to the nodes $j+1, \ldots, k-1$ as in Lemma 3;
6 $\quad\quad$ $t(j+1) = t(k-1) = 1$;
7 $\quad\quad$ obtain $p(i), \ldots, p(j)$ by using Lemma 5 on $L(i, j+1)$ with
   $\quad\quad\quad$ $t(i) = \cdots = t(j+1) = 1$;
8 $\quad\quad$ $i = k - 1$;

9 $\quad$ **else if** $L(j, k = j+3)$ *is a 2-path satisfying Lemma 4* **then**
10 $\quad\quad$ **if** $j - i + 2 = c(2\lambda + 1)$ *for some $c > 0$* **then**   // Case 1 of Lemma 4
11 $\quad\quad\quad$ $p(j+1) = 0;\quad p(j+2) = 1;\quad i' = j+1$;
12 $\quad\quad$ **else**   // Case 2 of Lemma 4
13 $\quad\quad\quad$ $p(j+1) = 1;\quad p(j+2) = 0;\quad i' = j+2$;
14 $\quad\quad$ $t(j+1) = t(j+2) = 1$;
15 $\quad\quad$ obtain $p(i), \ldots, p(j)$ by using Lemma 5 on $L(i, i')$ with
    $\quad\quad\quad$ $t(i) = \cdots = t(i') = 1$;
16 $\quad\quad$ $i = i'$;

17 Assign incentives to $L(i, n-1)$ (with $t(i) = \ldots = t(n-1) = 1$), using Lemma 5;
18 **return** p;

Theorem 2. *For any time bound $\lambda > 1$, Algorithm 1 provides an optimal solution for the TBI problem on any path $L(0, n-1)$ in time $O(n)$.*

Proof. We show that the algorithm selects an optimal strategy according to the length and the position of the leftmost 2-path $L(j, k)$ and then iteratively operates on the subpath $L(i, n-1)$ where $i = k - 1$ (one dummy node on the left) or $i = k - 2$ (two dummy nodes on the left). See Fig. 2.

Let $L(i, n-1)$ be the current path and $L(j, k)$ be the leftmost 2-path. If $L(j, k)$ satisfies the hypothesis of Lemma 3, then we have

$$OPT(i, n-1) = OPT(i, j, \tfrac{1}{\rightarrow}) + k - j - 2 + OPT(\tfrac{1}{\leftarrow}, k, n-1).$$

Hence, we can obtain optimal incentives for nodes $i, \ldots, j$ by using the result in Lemma 5 on $L(i, j+1)$ (where $j+1$ is a dummy node). Moreover, we assign $k - j - 2$ incentives to the nodes $j+1, \ldots, k-1$ as suggested in Lemma 3 (i.e., $0(20)^*$ when the length of the 2-path is odd and $01(20)^*$ otherwise) and the algorithm iterates on $L(k-1, n-1)$ (where $k-1$ is a dummy node).

Now suppose that $L(j, k = j+3)$ satisfies the hypothesis of Lemma 4. We have that $OPT(i, n-1)$ is equal to

$$1 + \min \left\{ OPT(i, j, \tfrac{1}{\rightarrow}) + OPT(\tfrac{2}{\leftarrow}, k, n-1), OPT(i, j, \tfrac{2}{\rightarrow}) + OPT(\tfrac{1}{\leftarrow}, k, n-1) \right\}. \quad (2)$$

We have two cases to consider, according to the distance between i and j.

First assume that $j-i+2 = c(2\lambda+1)$ for some $c > 0$. By Lemma 5 and Remark 1 we know that in this case $OPT(i,j,\xrightarrow{2}) = OPT(i,j,\xrightarrow{1}) + 1$ and since by (2) of Lemma 2 we know that $OPT(\xleftarrow{2},k,n-1) \leq OPT(\xleftarrow{1},k,n-1)+1$, we have that $OPT(i,j,\xrightarrow{1}) + OPT(\xleftarrow{2},k,n-1)$ corresponds to the minimum of Eq. (2) and hence the solution described by case 1 in Lemma 4 (i.e., $p(j+1) = 0, p(j+2) = 1$) is optimal. Incentives to $i, \ldots, j$ are assigned exploiting the result in Lemma 5 on $L(i, j + 1)$ (where $j + 1$ is a dummy node) and the algorithm iterates on $L(k - 2, n - 1)$ (where both $k - 1$ and $k - 2$ are dummy nodes).

Now assume that $j - i + 2 \neq c(2\lambda + 1)$ for some $c > 0$. In this case, we have $OPT(i,j,\xrightarrow{2}) = OPT(i,j,\xrightarrow{1})$.

By (1) of Lemma 2 we know that $OPT(\xleftarrow{1},k,n-1) \leq OPT(\xleftarrow{2},k,n-1)$. Hence, $OPT(i,j,\xrightarrow{2}) + OPT(\xleftarrow{1},k,n-1)$ corresponds to the minimum of Eq. (2) and the solution in case 2 in Lemma 4 (i.e., $p(j + 1) = 1, p(j + 2) = 0$) is optimal. Incentives to $i, \ldots, j$ are assigned using the result in Lemma 5 on $L(i, j + 2)$ (considering both $j + 1$ and $j + 2$ as dummy nodes) and the algorithm iterates on $L(k - 1, n - 1)$ (where $k - 1$ is a dummy node).

If there remains a last subpath of nodes of threshold one, this is solved optimally using Lemma 5.

Complexity. The identification of the 2-paths and their classification can be easily done in linear time. Then, the algorithm operates in a single pass from left to right and the time is $O(n)$.　　　　□

3　An $O(\lambda n \log n)$ Algorithm for Complete Graphs

In this section, we present an $O(\lambda n \log n)$ dynamic programming algorithm to allocate incentives to the nodes of a complete network $K_n = (V, E)$. We begin by proving that for any assignment of thresholds to the nodes of K_n, there is an optimal solution in which the thresholds of all nodes that are influenced *at* round ℓ are at least as large as the thresholds of all nodes that are influenced before round ℓ for every $1 \leq \ell \leq \lambda$.

Let K_m be the subgraph of K_n that is induced by $V_m = \{v_1, v_2, \ldots, v_m\}$. We will say that an incentive function $p : V_m \longrightarrow \mathbb{N}_0$ is ℓ-*optimal for* K_m, $1 \leq m \leq n, 0 \leq \ell \leq \lambda$, if $\sum_{v \in V_m} p(v)$ is the minimum cost to influence all nodes in V_m in ℓ rounds.

Lemma 6. *Given K_m, thresholds $t(v_1) \leq t(v_2) \leq \ldots \leq t(v_m)$, and $1 \leq \ell \leq \lambda$, if there exists an ℓ-optimal solution for K_m that influences $k < m$ nodes by the end of round $\ell-1$, then there is an ℓ-optimal solution that influences $\{v_1, v_2, \ldots, v_k\}$ by the end of round $\ell - 1$.*

Proof. Let p^* be an ℓ-optimal incentive function for K_m that influences a set $V_k^* = \{u_1, u_2, \ldots, u_k\}$ of k nodes of K_m by the end of round $\ell - 1$. We will show how to construct an ℓ-optimal incentive function for K_m that influences nodes $V_k = \{v_1, v_2, \ldots, v_k\}$ by the end of round $\ell - 1$ where $t(v_1) \leq t(v_2) \leq \ldots \leq t(v_k)$ and $t(v_j) \geq t(v_k)$ for $j = k + 1, k + 2, \ldots, m$.

Suppose that p is an incentive function for K_m that influences nodes $V_k = \{v_1, v_2, \ldots, v_k\}$ by the end of round $\ell - 1$. If V_k^* is different from V_k, then there is some $u_i \in V_k^* \backslash V_k$ and some $v_j \in V_k \backslash V_k^*$ such that $t(u_i) \geq t(v_j)$. Since v_j is influenced *at* round ℓ in the ℓ-optimal solution p^*, it must require the influence of $t(v_j) - p^*(v_j)$ neighbours. (If it required the influence of fewer neighbours, then p^* would not be ℓ-optimal.) Note that $t(v_j) - p^*(v_j) \geq 0$. Similarly, u_i requires the influence of $t(u_i) - p^*(u_i) \geq 0$ neighbours. Consider the set of nodes $V_k^* \cup \{v_j\} \backslash \{u_i\}$ and define p as follows. Choose $p(v_j)$ and $p(u_i)$ as

$$t(v_j) - p(v_j) = t(u_i) - p^*(u_i) \quad \text{and} \quad t(u_i) - p(u_i) = t(v_j) - p^*(v_j)$$

so that v_j is influenced at the same round as u_i was influenced in the ℓ-optimal solution and u_i is influenced at round ℓ. Set $p(v) = p^*(v)$ for all other nodes in K_m. The difference in value between p and p^* is

$$p(v_j) + p(u_i) - p^*(v_j) - p^*(u_i) = 0$$

We can iterate until we find an ℓ-optimal solution that influences $\{v_1, v_2, \ldots, v_k\}$ by the end of round $\ell - 1$. $\qquad\square$

By Lemma 6, our algorithm can first sort the nodes by non-decreasing threshold value w.l.o.g. The sorting can be done in $O(n)$ time using counting sort because $1 \leq t(v) \leq n - 1 = d(v)$ for all $v \in V$. In the remainder of this section, we assume that $t(v_1) \leq t(v_2) \leq \ldots \leq t(v_n)$.

Let $Opt_\ell(m)$ denote the value of an ℓ-optimal solution for K_m, $1 \leq m \leq n$, $0 \leq \ell \leq \lambda$. Any node v that is influenced at round 0 requires incentive $p(v) = t(v)$ and it follows easily that

$$Opt_0(m) = \sum_{i=1}^{m} t(v_i), \ 1 \leq m \leq n. \tag{3}$$

Now consider a value $Opt_\ell(m)$ for some $1 \leq m \leq n$ and $1 \leq \ell \leq \lambda$. If exactly j nodes, $1 \leq j \leq m$, are influenced by the end of round $\ell - 1$ in an ℓ-optimal solution for K_m, then each of the $m - j$ remaining nodes in V_m has j influenced neighbours at the beginning of round ℓ and these neighbours are $v_1, v_2, \ldots, v_j$ by Lemma 6. For such a remaining node v to be influenced at round ℓ, either $t(v) \leq j$ or v has an incentive $p(v)$ such that $t(v) - p(v) \leq j$. It follows that

$$Opt_\ell(m) = \min_{1 \leq j \leq m} \left\{ Opt_{\ell-1}(j) + \sum_{i=j+1}^{m} \max\{0, t(v_i) - j\} \right\}, \ 1 \leq m \leq n. \tag{4}$$

We will use $Ind_\ell(m)$ to denote the index that gives the optimal value $Opt_\ell(m)$, that is,

$$Ind_\ell(m) = \arg\min_{1 \leq j \leq m} \left\{ Opt_{\ell-1}(j) + \sum_{i=j+1}^{m} \max\{0, t(v_i) - j\} \right\}, \ 1 \leq m \leq n. \tag{5}$$

A dynamic programming algorithm that directly implements the recurrence in Eqs. (3) and (4) will produce the optimal solution value $Opt_\lambda(n)$ in time $O(\lambda n^3)$. We can reduce the complexity by taking advantage of some structural properties.

Lemma 7. *For any $1 \leq \ell \leq \lambda$, if $k < m$ then $Ind_\ell(k) \leq Ind_\ell(m)$, $1 \leq k \leq n-1$, $2 \leq m \leq n$.*

Outline of Proof. The lemma always holds when $k < Ind_\ell(m)$. Assuming that $Ind_\ell(k) > Ind_\ell(m)$ when $k \geq Ind_\ell(m)$ leads to a contradiction.

Theorem 3. *For any complete network $K_n = (V, E)$, threshold function $t : V \longrightarrow \mathbb{N}$, and $\lambda \geq 1$, the TBI problem can be solved in time $O(\lambda n \log n)$.*

Proof. Our dynamic programming algorithm computes two $n \times (\lambda + 1)$ arrays *VALUE* and *INDEX* and returns a solution p of n incentives. $VALUE[m, \ell] = Opt_\ell(m)$ is the value of an ℓ-optimal solution for K_m (for a given threshold function $t : V \longrightarrow \mathbb{N}$), and $INDEX[m, \ell] = Ind_\ell(m)$ is the index that gives the optimal value, $1 \leq m \leq n$, $0 \leq \ell \leq \lambda$.

The array entries are computed column-wise starting with column 0. The entries in column $VALUE[*, 0]$ are sums of thresholds according to (3) and the indices in $INDEX[*, 0]$ are all 0, so these columns can be computed in time $O(n)$. In particular, $VALUE[j, 0] = \sum_{i=1}^{j} t(v_i)$, $j = 1, \ldots, m$.

Suppose that columns $1, 2, \ldots, \ell - 1$ of *VALUE* and *INDEX* have been computed according to (4) and (5) and consider the computation of column ℓ of the two arrays. To compute $INDEX[m, \ell]$ for some fixed m, $1 \leq m \leq n$, we define a function

$$A(j) = Opt_{\ell-1}(j) + \sum_{i=j+1}^{m} \max\{0, t(v_i) - j\}, \quad 1 \leq j \leq m$$

and show how to compute each $A(j)$ in $O(1)$ time.
By (5), $Ind_\ell(m) = \arg\min\{A(j) \mid 1 \leq j \leq m\}$.

First we compute an auxiliary vector a where $a[j]$ contains the smallest integer $i \geq 1$ such that $t(v_i) \geq j$, $1 \leq j \leq n$. This vector can be precomputed once in $O(n)$ time because the nodes are sorted by non-decreasing threshold value. Furthermore, the vector a together with the entries in column $VALUE[*, 0]$ allow the computation of $\sum_{i=j+1}^{m} \max\{0, t(v_i) - j\}$ in $O(1)$ time for each $1 \leq j \leq n$. Since $Opt_{\ell-1}(j) = VALUE[j, \ell-1]$ has already been computed, we can compute $A(j)$ in $O(1)$ time. The values $Opt_\ell(m) = VALUE[m, \ell]$ can also be computed in $O(1)$ time for each $1 \leq m \leq n$ given $Ind_\ell(m) = INDEX[m, \ell]$, vector a, and column $VALUE[*, 0]$. The total cost so far is $O(\lambda n)$. It remains to show how to compute each column $INDEX[*, \ell]$ efficiently.

The following algorithm recursively computes the column $INDEX[m, \ell]$, $1 \leq m \leq n$ assuming that columns $0, 1, 2, \ldots, \ell - 1$ of *INDEX* and *VALUE* have already been computed. The algorithm also assumes that two dummy rows have been added to array *INDEX* with $INDEX[0, \ell] = 1$ and $INDEX[n + 1, \ell] = n$,

$0 \leq \ell \leq \lambda$, to simplify the pseudocode. The initial call of the algorithm is COMPUTE-INDEX$(1, n)$.

We claim that algorithm COMPUTE-INDEX$(1, n)$ correctly computes the values $INDEX[m, \ell]$ for $1 \leq m \leq n$. First, it can be proved by induction that when we call COMPUTE-INDEX(x, y), the indices $INDEX[x - 1, \ell]$ and $INDEX[y + 1, \ell]$ have already been correctly computed. By Lemma 7, $Ind_\ell(x - 1) \leq Ind_\ell(\lceil \frac{x+y}{2} \rceil) \leq Ind_\ell(y + 1)$, so the algorithm correctly searches for $INDEX[m, \ell]$ between $INDEX[x - 1, \ell]$ and $INDEX[y + 1, \ell]$.

Algorithm 2.   COMPUTE-INDEX(x, y)

Input: Indices x, y.
Output: The values $INDEX[i, \ell]$ for $i = x, \ldots y$.
1 **if** $x \leq y$ **then**   // Assume that $INDEX[0, \ell] = 1$ and $INDEX[n + 1, \ell] = n$
2 $\quad$ $m = \lceil \frac{x+y}{2} \rceil$;
3 $\quad$ $INDEX[m, \ell] =$
 $\quad\quad$ $\arg\min \{A(j) \mid INDEX[x - 1, \ell] \leq j \leq \min\{INDEX[y + 1, \ell], m\}\}$;
4 $\quad$ COMPUTE-INDEX$(x, m - 1)$;
5 $\quad$ COMPUTE-INDEX$(m + 1, y)$;

It is not hard to see that the height of the recursion tree obtained calling COMPUTE-INDEX$(1, n)$ is $\lceil \log(n + 1) \rceil$. Furthermore, the number of values $A(j)$ computed at each level of the recursion tree is $O(n)$ because the ranges of the searches in line 3 of the algorithm do not overlap (except possibly the endpoints of two consecutive ranges) by Lemma 7. Thus, the computation time at each level is $O(n)$, and the computation time for each column ℓ is $O(n \log n)$. After all columns of $VALUE$ and $INDEX$ have been computed, the value of the optimal solution will be in $VALUE[n, \lambda]$. The round during which each node is influenced and the optimal function p of incentives can then be computed by backtracking through the array $INDEX$ in time $O(\lambda + n)$. The total complexity is $O(\lambda n \log n)$. $\qquad\qquad\square$

4   A Polynomial-Time Algorithm for Trees

In this section, we give an algorithm for the TBI problem on trees. Let $T = (V, E)$ be a tree having n nodes and the maximum degree Δ. We will assume that T is rooted at some node r. Once such a rooting is fixed, for any node v, we denote by T_v the subtree rooted at v, and by $C(v)$ the set of children of v. We will develop a dynamic programming algorithm that will prove the following theorem.

Theorem 4. *For any $\lambda > 1$, the TBI problem can be solved in time $O(n\lambda^2 \Delta)$ on a tree having n nodes and maximum degree Δ.*

The rest of this section is devoted to the description and analysis of the algorithm that proves Theorem 4. The algorithm performs a post-order traversal of the tree T so that each node is considered after all of its children have been processed. For each node v, the algorithm solves some TBI problems on the subtree T_v, with some restrictions on the node v regarding its threshold and the round during which it is influenced. For instance, in order to compute some of these values we will consider not only the original threshold $t(v)$ of v, but also the reduced threshold $t'(v) = t(v) - 1$ which simulates the influence of the parent node.

Definition 3. *For each node $v \in V$, integers $\ell \in \{0, 1, \ldots, \lambda\}$, and $t \in \{t'(v), t(v)\}$, let us denote by $P[v, \ell, t]$ the minimum cost of influencing all of the nodes in T_v, in at most λ rounds, assuming that*

- *the threshold of v is t, and for every $u \in V(T_v) \setminus \{v\}$, the threshold of u is $t(u)$;*
- *v is influenced by round ℓ in T_v and is able to start influencing its neighbours by round $\ell + 1$.[2]*

Formally the value of $P[v, \ell, t]$ corresponds to $P[v, \ell, t] = \min\limits_{\substack{p: T_v \to \mathbb{N}_0,\ \text{Influenced}_{T_v}[p, \lambda] = T_v \\ |C(v) \cap \text{Influenced}_{F(v,d)}[p, \ell-1]| \geq t - p(v)}} \left\{ \sum\limits_{v \in T_v} p(v) \right\}$ We set $P[v, \ell, t] = \infty$ when the above problem is infeasible. Denoting by $p_{v,\ell,t} : V(T_v) \to \mathbb{N}_0$ the incentive function attaining the value $P[v, \ell, t]$, the parameter ℓ is such that:

1. if $\ell = 0$ then $p_{v,\ell,t}(v) = t$,
2. otherwise, v's children can influence v at round ℓ, i.e. $|\{C(v) \cap \text{Influenced}[p_{v,\ell,t}, \ell - 1]\}| \geq t - p_{v,\ell,t}(v)$.

Remark 2. It is worthwhile mentioning that $P[v, \ell, t]$ is monotonically non-decreasing in t. However, $P[v, \ell, t]$ is not necessarily monotonic in ℓ.

Indeed, partition the set $C(v)$ into two sets: $C'(v)$, which contains the c children that influence v, and $C''(v)$, which contains the remaining $|C(v)| - c$ children that may be influenced by v. A small value of c may require a higher cost on subtrees rooted at a node $u \in C'(v)$, and may save some budget on the remaining subtrees; the opposite happens for a large value of c.

The minimum cost to influence the nodes in T in λ rounds follows from decomposing the optimal solution according to the round on which the root is influenced and can then be obtained by computing

$$\min_{0 \leq \ell \leq \lambda} P[r, \ell, t(r)]. \qquad (6)$$

---

[2] Notice that this does not exclude the case that v becomes an influenced node at some round $\ell' < \ell$.

We proceed using a post-order traversal of the tree, so that the computations of the various values $P[v, \ell, t]$ for a node v are done after all of the values for v's children are known. For each leaf node v we have

$$P[v, \ell, t] = \begin{cases} 1 & \text{if } \ell = 0 \text{ and } t = t(v) = 1 \\ 0 & \text{if } 1 \leq \ell \leq \lambda \text{ and } t = t(v) - 1 = 0 \\ \infty & \text{otherwise.} \end{cases} \tag{7}$$

Indeed, a leaf v with threshold $t(v) = 1$ is influenced in the one-node subtree T_v only when either $p_{v,\ell,t}(v) = 1$ ($\ell = 0$), or for some $1 \leq \ell \leq \lambda$, it is influenced by its parent (i.e., the residual threshold $t = t(v) - 1 = 0$).

For any internal node v, we show how to compute each value $P[v, \ell, t]$ in time $O(d(v) \cdot t \cdot \lambda)$.

In the following we assume that an arbitrary order has been fixed on the $d = d(v) - 1$ children of any node v, that is, we denote them as $v_1, v_2, \ldots, v_d$, according to the fixed order. Also, we define $F(v, i)$ to be the forest consisting of the subtrees rooted at the first i children of v. We will also use $F(v, i)$ to denote the set of nodes it includes.

Definition 4. *Let v be a node with d children and let $\ell = 0, 1, \ldots, \lambda$. For $i = 0, \ldots, d$, $j = 0, 1, \ldots, t(v)$, we define $A_{v,\ell}[i, j]$ (resp. $A_{v,\ell}[\{i\}, j]$) to be the minimum cost for influencing all nodes in $F(v, i)$, (resp. T_{v_i}) within λ rounds, assuming that:*

(i) if $\ell \neq \lambda$, at time $\ell + 1$ the threshold of v_k is $t'(v_k)$, for each $k = 1, \ldots, i$;
(ii) if $\ell \neq 0$, at least j nodes in $\{v_1, v_2, \ldots, v_i\}$ (resp. $\{v_i\}$) are influenced by round $\ell - 1$, that is

$$|\{v_1, v_2, \ldots, v_i\} \cap \mathsf{Influenced}[\pi_{v,\ell,i,j}, \ell - 1]| \geq j,$$

where $\pi_{v,\ell,i,j} : F(v, i) \rightarrow \mathbb{N}_0$ denotes the incentive function attaining $A_{v,\ell}[i, j]$.

We also define $A_{v,\ell}[i, j] = \infty$ when the above constraints are not satisfiable.

By decomposing a solution according to how many nodes in $C(v)$ are influenced prior to the root v being influenced and denoting this number as j, the remaining cost to influence the root v is $t - j$ Hence, we can easily write $P[v, \ell, t]$ in terms of $A_{v,\ell}[d, j]$ as follows.

Lemma 8. *For each node v with d children, each $\ell = 0, \ldots, \lambda$ and each $t \in \{t(v), t'(v)\}$*

$$P[v, \ell, t] = \begin{cases} t + A_{v,0}[d, 0] & \text{if } \ell = 0 \\ \min_{0 \leq j \leq t} \{t - j + A_{v,\ell}[d, j]\} & \text{otherwise.} \end{cases} \tag{8}$$

Lemma 9. *For each node v, each $t \in \{t(v), t'(v)\}$, and each $\ell = 1, \ldots, \lambda$, it is possible to compute $A_{v,\ell}[d, t]$, as well as $A_{v,0}[d, 0]$, recursively in time $O(\lambda d t)$ where d is the number of children of v.*

Outline of Proof. The proof shows that the values $A_{v,0}[d,0]$ and $A_{v,\ell}[d,t]$ can be computed, in time $O(\lambda d)$ and $O(\lambda dt)$ respectively, using the following recursive equations. Let $M(\ell_1,\ell_2,t) = \min_{\ell_1 \leq \ell' \leq \ell_2}\{P[v_i,\ell',t]\}$ we have
$$A_{v,0}[d,0] = \sum_{v_i \in C(v)} \min\{P[v_i,0,t(v_i)], M(1,\lambda,t'(v_i))\},$$

$$A_{v,\ell}[i,j] = \begin{cases} 0, & \text{if } i = j = 0 \\ \infty, & \text{if } i < j \\ \min\Big\{A_{v,\ell}[i-1,j-1] + M(0,\ell-1,t(v_i)), \\ \quad A_{v,\ell}[i-1,j] + \min\Big\{M(0,\ell,t(v_i)), M(\ell+1,\lambda,t'(v_i))\Big\}\Big\}, & \text{otherwise.} \end{cases}$$

$\square$

Lemmas 8 and 9 imply that for each $v \in V$, for each $\ell = 0,\ldots,\lambda$, and $t \in \{t'(v), t(v)\}$, the value $P[v,\ell,t]$ can be computed recursively in time $O(\lambda d(v)t(v))$. Hence, the value in (6) can be computed in time $\sum_{v \in V} O(\lambda d(v)t(v)) \times O(\lambda) = O(\lambda^2 \Delta) \times \sum_{v \in V} O(d(v)) = O(\lambda^2 \Delta n)$, where Δ is the maximum node degree. Standard backtracking techniques can be used to compute the (optimal) influence function p^* that influences all of the nodes in the same $O(\lambda^2 \Delta n)$ time.

References

1. Ackerman, E., Ben-Zwi, O., Wolfovitz, G.: Combinatorial model and bounds for target set selection. Theor. Comput. Sci. **411**, 4017–4022 (2010)
2. Ben-Zwi, O., Hermelin, D., Lokshtanov, D., Newman, I.: Treewidth governs the complexity of target set selection. Discret. Optim. **8**, 87–96 (2011)
3. Chopin, M., Nichterlein, A., Niedermeier, R., Weller, M.: Constant thresholds can make target set selection tractable. In: Even, G., Rawitz, D. (eds.) MedAlg 2012. LNCS, vol. 7659, pp. 120–133. Springer, Heidelberg (2012). https://doi.org/10.1007/978-3-642-34862-4_9
4. Coja-Oghlan, A., Feige, U., Krivelevich, M., Reichman, D.: Contagious sets in expanders. In: Proceedings of SODA 2015, pp. 1953–1987 (2015)
5. Chen, W., Lakshmanan, L.V.S., Castillo, C.: Information and Influence Propagation in Social Networks. Morgan & Claypool, San Rafael (2013)
6. Chen, N.: On the approximability of influence in social networks. SIAM J. Discrete Math. **23**, 1400–1415 (2009)
7. Chiang, C.-Y., Huang, L.-H., Li, B.-J., Wu, J., Yeh, H.-G.: Some results on the target set selection problem. Journal of Comb. Opt. **25**(4), 702–715 (2013)
8. Cicalese, F., Cordasco, G., Gargano, L., Milanič, M., Peters, J., Vaccaro, U.: Spread of influence in weighted networks under time and budget constraints. Theor. Comput. Sci. **586**, 40–58 (2015)
9. Cicalese, F., Cordasco, G., Gargano, L., Milanič, M., Vaccaro, U.: Latency-Bounded target set selection in social networks. Theor. Comput. Sci. **535**, 1–15 (2014)
10. Cordasco, G., Gargano, L., Rescigno, A.A.: On finding small sets that influence large networks. Soc. Netw. Anal. Min. **6**(94) (2016)

11. Cordasco, G., Gargano, L., Rescigno, A.A., Vaccaro, U.: Optimizing spread of influence in social networks via partial incentives. In: Scheideler, C. (ed.) Structural Information and Communication Complexity. LNCS, vol. 9439, pp. 119–134. Springer, Cham (2015). https://doi.org/10.1007/978-3-319-25258-2_9
12. Demaine, E.D., et al.: How to influence people with partial incentives. In: Proceedings of WWW 2014, pp. 937–948 (2014)
13. Domingos, P., Richardson, M.: Mining the network value of customers. In: Proceedings of 7th ACM SIGKDD International Conference on Knowledge Discovery and Data Mining, pp. 57–66 (2001)
14. Easley, D., Kleinberg, J.: Networks, Crowds, and Markets: Reasoning About a Highly Connected World. Cambridge University Press, Cambridge (2010)
15. Gargano, L., Hell, P., Peters, J.G., Vaccaro, U.: Influence diffusion in social networks under time window constraints. Theor. Comput. Sci. **584**, 53–66 (2015)
16. Granovetter, M.: Thresholds models of collective behaviors. Am. J. Sociol. **83**(6), 1420–1443 (1978)
17. Kempe, D., Kleinberg, J.M., Tardos, E.: Maximizing the spread of influence through a social network. In: Proceedings of 9th ACM SIGKDD International Conference on Knowledge Discovery and Data Mining, pp. 137–146 (2003)
18. Lafond, M., Narayanan, L., Wu, K.: Whom to befriend to influence people. In: Suomela, J. (ed.) SIROCCO 2016. LNCS, vol. 9988, pp. 340–357. Springer, Cham (2016). https://doi.org/10.1007/978-3-319-48314-6_22
19. Narayanan, L., Wu, K.: How to choose friends strategically. In: Das, S., Tixeuil, S. (eds.) SIROCCO 2017. LNCS, vol. 10641, pp. 283–302. Springer, Cham (2017). https://doi.org/10.1007/978-3-319-72050-0_17
20. Nichterlein, A., Niedermeier, R., Uhlmann, J., Weller, M.: On tractable cases of target set selection. Soc. Netw. Anal. Min. **3**(2), 233–256 (2013)
21. Peleg, D.: Local majorities, coalitions and monopolies in graphs: a review. Theor. Comput. Sci. **282**, 231–257 (2002)
22. Reddy, T.V.T., Rangan, C.P.: Variants of spreading messages. J. Graph Algorithms Appl. **15**(5), 683–699 (2011)
23. Liu, X., Yang, Z., Wang, W.: Exact solutions for latency-bounded target set selection problem on some special families of graphs. Discret. Appl. Math. **203**(C), 111–116 (2016)

Balanced Allocations and Global Clock in Population Protocols: An Accurate Analysis

Yves Mocquard[1], Bruno Sericola[2], and Emmanuelle Anceaume[3(✉)]

[1] Université de Rennes 1 - IRISA, Rennes, France
yves.mocquard@irisa.fr
[2] INRIA Rennes - Bretagne Atlantique, Rennes, France
bruno.sericola@inria.fr
[3] CNRS - IRISA, Rennes, France
emmanuelle.anceaume@irisa.fr

Abstract. The context of this paper is the two-choice paradigm which is deeply used in balanced online resource allocation, priority scheduling, load balancing and more recently in population protocols. The model governing the evolution of these systems consists in throwing balls one by one and independently of each others into n bins, which represent the number of agents in the system. At each discrete instant, a ball is placed in the least filled bin among two bins randomly chosen among the n ones. A natural question is the evaluation of the difference between the number of balls in the most loaded bin and the one in the least loaded. At time t, this difference is denoted by $\mathrm{Gap}(t)$. A lot of work has been devoted to the derivation of asymptotic approximations of this gap for large values of n. In this paper we go a step further by showing that for all $t \geq 0$, $n \geq 2$ and $\sigma > 0$, the variable $\mathrm{Gap}(t)$ is less than $a(1 + \sigma)\ln(n) + b$ with probability greater than $1 - 1/n^\sigma$, where the constants a and b, which are independent of t, σ and n, are optimized and given explicitly, which to the best of our knowledge has never been done before.

1 Introduction

In this paper we address the important issue of the two-choice paradigm analysis [10]. To illustrate the multi-choice paradigm, suppose that we have a set of m balls which are sequentially throws into n bins, where each ball is placed in the least filled bin among $d \geq 1$ ones randomly chosen among the n bins. Azar et al. [5] have characterized this problem by those three values (m, n, d). A natural question is the analysis of the maximum load in any of the bins, or the maximal gap that may exist between the least loaded bin and the most loaded

This work was partially funded by the French ANR project SocioPlug (ANR-13-INFR-0003), and by the DeSceNt project granted by the Labex CominLabs excellence laboratory (ANR-10-LABX-07-01).

© Springer Nature Switzerland AG 2018
Z. Lotker and B. Patt-Shamir (Eds.): SIROCCO 2018, LNCS 11085, pp. 296–311, 2018.
https://doi.org/10.1007/978-3-030-01325-7_26

one. It has been proven that in the simplest case where $d = 1$ (see for example [14]), the maximum load is equal to $m/n + \Theta\left(\sqrt{(m/n)\ln n}\right)$, leading to a gap that increases with the square root of m. Now, instead of choosing a single bin at random, d bins, with $d \geq 2$, are independently and randomly chosen, and the least loaded bin one among those d ones receives a ball. Then Azar et al. [5] have shown that when $m = n$ the maximum load is $\ln(\ln(n))/\ln(2) + O(1)$, and the largest gap is also equal to $\ln(\ln(n))/\ln(2) + O(1)$. These results show that by simply introducing a small choice we get a drastically improved balanced load among all the bins. Citing Mitzenmacher et al. [10], "having just two random choices (i.e., $d = 2$) yields a large reduction in the maximum load over having one choice, while each additional choice beyond two decreases the maximum load by just a constant factor". Hence the name of the two-choice paradigm. Later Berenbrink et al. [7] have studied the case (m, n, d) for $d \geq 2$ and $m \gg n$, and proved that the maximum load is equal to $m/n + O(\ln(\ln(n)))$. Note that a simpler proof of this result has been recently found by Talwar and Wieder [15]. Very recently, Peres et al. [12,13], using a measurement based on the hyperbolic cosine, have generalized the problem in the $(1 + \beta)$-choice problem. The $(1 + \beta)$-choice consists, with probability $1 - \beta$, in choosing one bin uniformly at random and to throw a ball in it, and with probability β, in choosing two bins uniformly at random and to throw a ball in the least loaded one. The name comes from the fact that $\mathbb{E}\{d\} = 1 + \beta$. We can note that in their model, each ball is assigned with a random weight. They found a logarithmic bound for both the gap between the maximum loaded bin and the average one [12], and for the gap between the maximum loaded bin and the minimum one [13]. In both cases the gap is $O\left(\log(n)/\beta\right)$.

The two-choice paradigm can be used in a multitude of applications, including balanced online resource allocation (where jobs need to be dynamically allocated to the least loaded processor) [1,6,8], priority scheduling [4], load balancing [2,7,9], and very recently, population protocols [3]. In the later case, the model governing the evolution of these systems consists in throwing balls one by one and independently of each others into n bins, which represent the number of agents in the system. At each discrete instant, a ball is placed in the least filled bin among two bins randomly chosen among the n ones. A natural question is the evaluation of the difference between the number of balls in the most loaded and the one in the least loaded bin. At time t, this difference is denoted by $\mathrm{Gap}(t)$. A lot of work has been devoted to the derivation of asymptotic approximations of this gap for large values of n. In this paper we go a step further by showing that for all $t \geq 0$, $n \geq 2$ and $\sigma > 0$,

$$\mathbb{P}\left\{\mathrm{Gap}(t) \geq a(1 + \sigma)\ln(n) + b\right\} \leq \frac{1}{n^{\sigma}}, \tag{1}$$

where the constants a and b, which are independent of t, σ and n, are optimized and given explicitly, which to the best of our knowledge has never been done before.

The remaining of the paper is structured as follows. In Sect. 2 we present the addressed problem and a simple algorithm to solve it. Section 3 is the main contribution of our work which consists in providing an accurate bound of the distribution of the gap between any two nodes. Section 4 evaluates constants a and b obtained by our analysis and compares it to constants that we derived from the work of [4]. The gain in accuracy we obtained by our analysis is significant. Finally Sect. 5 provides a summary of simulations results.

2   Problem description

We consider a very large set of n nodes (also called agents), interconnected by a complete graph, that asynchronously start their execution in a given state. Agents do not maintain nor use identifiers (agents are anonymous and cannot determine whether any two interactions have occurred with the same agents or not). However, for ease of presentation the agents are numbered $1, 2, \ldots, n$. Each agent keeps a local counter, initialized at 0. Agents communicate through random pairwise interactions. On each interaction, the two interacting agents compare their counters, and the one with the lower counter value increments its local counter. The objective of this simple algorithm is the construction of a global clock by guaranteeing that the values of all agent counters are concentrated according to Relation (1). As interactions are uniformly random, this can be related to the classic two-choice load balancing process [13]. The goal of the paper is to evaluate the gap between any two agents, that is the maximal difference that may exist at any time t between any two local counters, by accurately evaluating constants a and b. By accurately estimating the maximal gap between any two local counters, other population protocols can use it as a *global clock* to perform actions in a probabilistic synchronized way.

We denote by $C_t^{(i)}$ the state of agent i at time t. The stochastic process $C = \{C_t, \ t \geq 0\}$, where $C_t = (C_t^{(1)}, \ldots, C_t^{(n)})$, represents the vector state of the system at time t.

The choice of the two agents which interact is made using a uniform distribution. Given the pair (i, j) of agents which interact at time t, we consider the following evolution of the agents states

$$
\left(C_{t+1}^{(i)}, C_{t+1}^{(j)}\right) = \begin{cases} \left(C_t^{(i)} + 1, C_t^{(j)}\right) & \text{if } C_t^{(i)} \leq C_t^{(j)} \\ \\ \left(C_t^{(i)}, C_t^{(j)} + 1\right) & \text{if } C_t^{(i)} \geq C_t^{(j)}. \end{cases}
$$

Note that in the case where agents i and j interact at time t with $C_t^{(i)} = C_t^{(j)}$ then either of two agents can be chosen to have its value increased by 1 at time $t + 1$. A particular choice is made below.

The state space of process C is thus $\mathbb{N}^n$ and a state of this process is also called a protocol configuration. At time 0, we set $C_t^{(i)} = 0$, for every $i = 1, \ldots, n$. At each instant the value of only one agent is increased by 1 which means that we have, for every $t \geq 0$,

$$\sum_{i=1}^{n} C_t^{(i)} = t.$$

For every $i = 1, \ldots, n$, we introduce the quantities $x_i(t) = C_t^{(i)} - t/n$, which leads, for every $t \geq 0$, to

$$\sum_{i=1}^{n} x_i(t) = 0.$$

The value $C_t^{(i)}$ maintained by agent i is its own view of the global clock t of the system divided by n. More precisely, the approximation of time t, provided by agent i, is $nC_t^{(i)}$.

At each discrete time $t \geq 0$, any two indices i and j are uniformly chosen to interact, independently of the vector state with probability $1/(n(n-1))$.

In order to simplify the presentation, we suppose without any loss of generality that at each instant t, the values of $x_i(t)$ are reordered in a decreasing way, assigning an arbitrary order to agents with the same value. More precisely, at time t the reordering gives

$$x_1(t) = \max_{i=1,\ldots,n} (C_t^{(i)} - t/n) \geq \cdots \geq x_n(t) = \min_{i=1,\ldots,n} (C_t^{(i)} - t/n).$$

We denote by X the rank of the agent whose value is incremented when interaction occurs between 2 agents. In the case where two agents interacting, say i and j, are such that $C_t^{(i)} = C_t^{(j)}$, we choose to increase by 1 the one with the highest rank. If X_1 and X_2 are the ranks of the successive agents which interact, then the probability p_ℓ that agent of rank ℓ is incremented is, for $\ell = 1, \ldots, n$,

$$p_\ell = \mathbb{P}\{X = \ell\} = \mathbb{P}\{X_1 = \ell, X_2 < \ell\} + \mathbb{P}\{X_1 < \ell, X_2 = \ell\} = \frac{2(\ell - 1)}{n(n-1)}. \quad (2)$$

As mentioned in the introduction, the goal of the paper is the evaluation of the distribution of difference between the maximum and the minimum of the entries of vector C_t. This difference is denoted by $\mathrm{Gap}(t)$ and is given, for $t \in \mathbb{N}$, by

$$\mathrm{Gap}(t) = \max_{1 \leq i \leq n} C_t^{(i)} - \min_{1 \leq i \leq n} C_t^{(i)} = x_1(t) - x_n(t).$$

In order to bound the complementary distribution of $\mathrm{Gap}(t)$, we introduce the following potential functions defined, for $\alpha \in \mathbb{R}$, by

$$\Phi(t) = \sum_{i=1}^{n} e^{\alpha x_i(t)}, \quad \Psi(t) = \sum_{i=1}^{n} e^{-\alpha x_i(t)} \quad \text{and} \quad \Gamma(t) = \Phi(t) + \Psi(t).$$

The use of these two functions has been proposed in a very clever way by Peres et al. in [13]. The potential function $\Gamma(t)$ is then related to function $\mathrm{Gap}(t)$ by the following lemma.

Lemma 1. *For every $t \geq 0$, we have*

$$\Gamma(t) \geq 2e^{\alpha Gap(t)/2}. \qquad (3)$$

Proof. The exponential function being convex, we have, for every $a, b \in \mathbb{R}$, $2e^{(a+b)/2} \leq e^a + e^b$. Recalling that $\text{Gap}(t) = x_1(t) - x_n(t)$, we obtain

$$\Gamma(t) = \sum_{i=1}^{n} e^{\alpha x_i(t)} + \sum_{i=1}^{n} e^{-\alpha x_i(t)} \geq e^{\alpha x_1(t)} + e^{-\alpha x_n(t)} \geq 2e^{\alpha \text{Gap}(t)/2},$$

which completes the proof.   ∎

This result will be used at the end of the paper for the evaluation of the distribution of $\text{Gap}(t)$ which is based on the evaluation of the one of $\Gamma(t)$, which forms the main part of the paper.

3   Analysis

We first need the two following technical lemmas which are proved in [11].

Lemma 2. *For all $x \in \mathbb{R}$, we have $1 + x \leq e^x$. For all $x \in (-\infty, c]$, we have $e^x \leq 1 + x + x^2$, where c is the unique positive solution to equation $e^c - 1 - c - c^2 = 0$. The value of c satisfies $1.79 < c < 1.8$.*

Lemma 3. *Let $u = (u_k)_{k \geq 1}$ and $v = (v_k)_{k \geq 1}$ be two monotonic sequences of real numbers and let m_n be the sequence of mean values of sequence v defined, for $n \geq 1$, by*

$$m_n = \frac{1}{n} \sum_{k=1}^{n} v_k.$$

If the sequences u and v are both non-decreasing or both non-increasing then we have

$$\sum_{k=1}^{n} u_k v_k \geq m_n \sum_{k=1}^{n} u_k.$$

If one of these two sequences is non-increasing and the other is non-decreasing then we have

$$\sum_{k=1}^{n} u_k v_k \leq m_n \sum_{k=1}^{n} u_k.$$

For every $t \geq 0$, we introduce the notation $x(t) = (x_1(t), \ldots, x_n(t))$.

Lemma 4. *For all $\alpha \in (-1, 1)$, we have*

$$\mathbb{E}\{\Phi(t+1) - \Phi(t) \mid x(t)\} \leq \left(\alpha + \alpha^2 \left(1 - \frac{2}{n}\right)\right) \sum_{i=1}^{n} p_i e^{\alpha x_i(t)} - \left(\frac{\alpha}{n} - \frac{\alpha^2}{n^2}\right) \Phi(t).$$

$$(4)$$

Proof. Since the $x_i(t)$ are ordered, they may change value at each time. We can thus define a permutation on $\{1, 2, \ldots, n\}$ named σ_t such that, for every $u = 1, \ldots, n$, if $x_i(t) = C_t^{(u)} - t/n$ then $x_{\sigma_t(i)}(t+1) = C_{t+1}^{(u)} - (t+1)/n$. Suppose that the rank of the agent (say agent u), whose value is incremented at time t, is equal to i. In this case, we have

$$x_{\sigma_t(i)}(t+1) = C_{t+1}^{(u)} - \frac{t+1}{n} = C_t^{(u)} + 1 - \frac{t+1}{n}$$

$$= C_t^{(u)} - \frac{t}{n} + 1 + \frac{t}{n} - \frac{t+1}{n} = x_i(t) + 1 - \frac{1}{n}.$$

This leads, for every $i = 1, \ldots, n$, to $x_{\sigma_t(i)}(t+1) = x_i(t) + 1_{\{X=i\}} - \frac{1}{n}$, where 1_A is the indicator function of event A. We then get

$$\Phi(t+1) - \Phi(t) = \sum_{i=1}^{n} \left(e^{\alpha x_i(t+1)} - e^{\alpha x_i(t)} \right) = \sum_{i=1}^{n} \left(e^{\alpha x_{\sigma_t(i)}(t+1)} - e^{\alpha x_i(t)} \right)$$

$$= \sum_{i=1}^{n} \left(e^{\alpha(1_{\{X=i\}} - 1/n)} - 1 \right) e^{\alpha x_i(t)}.$$

Using the fact that $e^x \leq 1 + x + x^2$ for $x \leq 1$, see Lemma 2, we obtain, since $\alpha(1_{\{X=i\}} - 1/n) \leq 1$,

$$e^{\alpha(1_{\{X=i\}} - 1/n)} - 1 \leq \alpha(1_{\{X=i\}} - 1/n) + \alpha^2(1_{\{X=i\}} - 1/n)^2$$

$$= \alpha(1_{\{X=i\}} - 1/n) + \alpha^2 \left(1_{\{X=i\}}(1 - \frac{2}{n}) + \frac{1}{n^2} \right)$$

$$= \left(\alpha + \alpha^2 \left(1 - \frac{2}{n} \right) \right) 1_{\{X=i\}} - \left(\frac{\alpha}{n} - \frac{\alpha^2}{n^2} \right).$$

Taking the expectation of $\Phi(t+1) - \Phi(t)$, given $x(t)$, we obtain since $\mathbb{E}\{1_{\{X=i\}}\} = p_i$,

$$\mathbb{E}\{\Phi(t+1) - \Phi(t) \mid x(t)\} \leq \sum_{i=1}^{n} \left[p_i \left(\alpha + \alpha^2 \left(1 - \frac{2}{n} \right) \right) - \left(\frac{\alpha}{n} - \frac{\alpha^2}{n^2} \right) \right] e^{\alpha x_i(t)}$$

$$= \left(\alpha + \alpha^2 \left(1 - \frac{2}{n} \right) \right) \sum_{i=1}^{n} p_i e^{\alpha x_i(t)} - \left(\frac{\alpha}{n} - \frac{\alpha^2}{n^2} \right) \Phi(t),$$

which completes the proof.   ∎

The following relations will be frequently used in the sequel. Since, for $i = 1, \ldots, n$, $p_i = 2(i-1)/(n(n-1))$, we have for all $\lambda \in (0, 1)$ with $\lambda n \in \mathbb{N}$,

$$\frac{1}{n}\sum_{i=1}^{n}p_i = \frac{1}{n} \tag{5}$$

$$\frac{1}{\lambda n}\sum_{i=1}^{\lambda n}p_i = \frac{\lambda n - 1}{n(n-1)} \leq \frac{\lambda}{n} \tag{6}$$

$$\frac{1}{(1-\lambda)n}\sum_{i=\lambda n+1}^{n}p_i = \frac{(1+\lambda)n - 1}{n(n-1)} \geq \frac{1+\lambda}{n} \tag{7}$$

Corollary 5. *For all $\alpha \in (0,1)$, we have*

$$\mathbb{E}\{\Phi(t+1) - \Phi(t) \mid x(t)\} \leq \frac{\alpha^2}{n}\left(1 - \frac{1}{n}\right)\Phi(t).$$

Proof. To prove this result, observe that sequence $\left(e^{\alpha x_i(t)}\right)_i$ is a non-increasing sequence and $(p_i)_i$ is an non-decreasing sequence, so using Relation (5) and applying Lemma 3 we obtain

$$\sum_{i=1}^{n}p_i e^{\alpha x_i(t)} \leq \frac{1}{n}\left(\sum_{i=1}^{n}p_i\right)\left(\sum_{i=1}^{n}e^{\alpha x_i(t)}\right) = \frac{\Phi(t)}{n}.$$

Putting this result in inequality (4), we get

$$\mathbb{E}\{\Phi(t+1) - \Phi(t) \mid x(t)\} \leq \left(\alpha + \alpha^2\left(1 - \frac{2}{n}\right)\right)\sum_{i=1}^{n}p_i e^{\alpha x_i(t)} - \left(\frac{\alpha}{n} - \frac{\alpha^2}{n^2}\right)\Phi(t)$$

$$\leq \left[\frac{\alpha}{n} + \frac{\alpha^2}{n}\left(1 - \frac{2}{n}\right) - \left(\frac{\alpha}{n} - \frac{\alpha^2}{n^2}\right)\right]\Phi(t)$$

$$= \frac{\alpha^2}{n}\left(1 - \frac{1}{n}\right)\Phi(t),$$

which completes the proof.   ∎

Lemma 6. *For all $\alpha \in (-1,1)$, we have*

$$\mathbb{E}\{\Psi(t+1) - \Psi(t) \mid x(t)\} \leq \left(-\alpha + \alpha^2\left(1 - \frac{2}{n}\right)\right)\sum_{i=1}^{n}p_i e^{-\alpha x_i(t)} + \left(\frac{\alpha}{n} + \frac{\alpha^2}{n^2}\right)\Psi(t). \tag{8}$$

Proof. It suffices to replace α by $-\alpha$ in the proof of Lemma 4.   ∎

Corollary 7. *For all $\alpha \in (0,1)$, we have*

$$\mathbb{E}\{\Psi(t+1) - \Psi(t) \mid x(t)\} \leq \frac{\alpha^2}{n}\left(1 - \frac{1}{n}\right)\Psi(t)$$

Proof. See [11].   ∎

The two previous lemmas, which give a bound of the increase of functions $\Phi(t)$ and $\Psi(t)$, will be used to prove Theorem 12. The proof of the results follow the clever ideas of the seminal paper [13] in which the authors prove that $\mathrm{Gap}(t)$ is less than $O(\ln(n))$ with high probability. In [4], Alistarh et al. provide a more rigorous proof from which we have extracted constants associated with this asymptotic behavior. Those constants are given at the end of Sect. 4. The main original idea of our paper is to parametrize as much as possible the proofs in order to obtain the smallest values of constants a and b used in Relation (1) which is proved in Theorem 14. The numerical evaluation of these constants, obtained in Sect. 4, shows that they are remarkably small with respect to the ones of [4].

In the following, we introduce two variable parameters $\mu, \rho \in (0, 1/2)$ (which are fixed to $1/4$ in [13] and [4]). Since $x_i(t)$ are non-increasing we have $x_{\rho n}(t) \geq x_{(1-\mu)n}(t)$. Lemmas 8 and 9 deal with the balanced conditions case that is $x_{\rho n}(t) \geq 0 \geq x_{(1-\mu)n}(t)$. The unbalanced conditions that are the complementary cases $x_{\rho n}(t) \geq x_{(1-\mu)n}(t) > 0$ and $0 > x_{\rho n}(t) \geq x_{(1-\mu)n}(t)$ are considered respectively in Lemmas 10 and 11. Theorem 12 examines systematically each case which lead to recurrence relation for $\mathbb{E}\{\Gamma(t)\}$. Theorem 13 uses this recurrence relation to bound $\mathbb{E}\{\Gamma(t)\}$. Finally, Theorem 14 gives a precise lower bound of $\Gamma(t)$ with high probability.

Lemma 8. *Let* $\alpha, \mu \in (0, 1)$ *with* $\mu n \in \mathbb{N}$ *and* $\mu > \alpha/(1+\alpha)$. *If* $x_{(1-\mu)n}(t) \leq 0$ *then we have*

$$\mathbb{E}\{\Phi(t+1) \mid x(t)\}$$
$$\leq \left(1 - \frac{\alpha}{n}\left[\mu - \alpha(1-\mu) + \frac{\alpha(1-2\mu)}{n}\right]\right)\Phi(t) + \alpha + \alpha^2\left(1 - \frac{2}{n}\right)$$
$$\leq \left(1 - \frac{\alpha}{n}[\mu - \alpha(1-\mu)]\right)\Phi(t) + \alpha + \alpha^2. \tag{9}$$

Proof. See [11]. ∎

An analogous result is obtained for $\Psi(t)$ in the following lemma.

Lemma 9. *Let* $\alpha, \rho \in (0, 1)$ *with* $\rho n \in \mathbb{N}$ *and* $\rho > \alpha/(1-\alpha)$. *If* $x_{\rho n}(t) \geq 0$ *then we have*

$$\mathbb{E}\{\Psi(t+1) \mid x(t)\} \leq \left(1 - \frac{\alpha}{n}\left[\rho - \alpha(1+\rho) + \frac{\alpha(1+2\rho)}{n}\right]\right)\Psi(t) + \alpha\rho(1+\rho)$$
$$\leq \left(1 - \frac{\alpha}{n}[\rho - \alpha(1+\rho)]\right)\Psi(t) + \alpha\rho(1+\rho). \tag{10}$$

Proof. See [11]. ∎

Lemma 10. *Let* $\alpha, \mu \in (0, 1/2)$ *with* $\mu n \in \mathbb{N}$ *and* $\mu \in (\alpha/(1+\alpha), (1-2\alpha)/(1-\alpha))$, *let* $\mu' \in (0, 1)$ *with* $\mu'n \in \mathbb{N}$ *and* $\mu' \in (\mu/(1-\mu), 1/(1+\alpha))$ *and let* $\gamma_1 \in (0, 1)$.

If $x_{(1-\mu)n}(t) > 0$ and $\mathbb{E}\{\Phi(t+1) - \Phi(t) \mid x(t)\} \geq -(1 - \mu'(\alpha+1))\dfrac{\alpha\gamma_1}{n}\Phi(t)$ and $\Phi(t) \geq \lambda_1\Psi(t)$ then we have $\Gamma(t) \leq c_1 n$, where

$$c_1 = \left(1 + \frac{1}{\lambda_1}\right)C_1\left(\frac{C_1}{\mu\lambda_1}\right)^{\mu/((1-\mu)\mu'-\mu)} \quad , \quad C_1 = \frac{(1-\mu')\,(2+\alpha)}{(1-\gamma_1)\,(1-\mu'(1+\alpha))}$$

and

$$\lambda_1 = \frac{1 - \mu - \alpha(2-\mu)}{2\alpha}.$$

The condition $\mu < (1-2\alpha)/(1-\alpha)$ is needed ta assure that constant $\lambda_1 > 0$. The value of λ_1 will be used in Theorem 12. The condition $\mu' > \mu/(1-\mu)$ is needed to assure that the power involved in constant c_1 is positive.

Proof. See [11].   ∎

Lemma 11. Let $\alpha, \rho \in (0, 1/2)$ with $\rho n \in \mathbb{N}$ and $\rho \in (\alpha/(1-\alpha), 1/(1+\alpha))$, let $\rho' \in (\rho/(1-\rho), (1-2\alpha)/(1-\alpha))$ with $\rho'n \in \mathbb{N}$ and let $\gamma_2 \in (0,1)$.

If $x_{\rho n}(t) < 0$ and $\mathbb{E}\{\Psi(t+1) - \Psi(t) \mid x(t)\} \geq -[1 - 2\alpha - \rho'(1-\alpha)]\dfrac{\alpha\gamma_2}{n}\Psi(t)$ and $\Psi(t) \geq \lambda_2\Phi(t)$ then we have $\Gamma(t) \leq c_2 n$, where

$$c_2 = \left(1 + \frac{1}{\lambda_2}\right)C_2\left(\frac{C_2}{\rho\lambda_2}\right)^{\rho/((1-\rho)\rho'-\rho)} \quad , \quad C_2 = \frac{(1-\rho')\,(2 - 2\alpha - \rho'(1-\alpha))}{(1-\gamma_2)\,(1 - 2\alpha - \rho'(1-\alpha))}$$

and

$$\lambda_2 = \frac{1 - \rho(1+\alpha)}{2\alpha}.$$

The condition $\rho < 1/(1+\alpha)$ is needed ta assure that constant $\lambda_2 > 0$. The value of λ_2 will be used in Theorem 12. The condition $\rho' > \rho/(1-\rho)$ is needed to assure that the power involved in constant c_2 is positive.

Proof. See [11].   ∎

Theorem 12. Let $\alpha, \mu, \rho \in (0, 1/2)$ with $\mu n, \rho n \in \mathbb{N}$, $\mu \in (\alpha/(1+\alpha), (1-2\alpha)/(1-\alpha))$ and $\rho \in (\alpha/(1-\alpha), 1/(1+\alpha))$. Let $\mu' \in (\mu/(1-\mu), 1/(1+\alpha))$ with $\mu'n \in \mathbb{N}$ and let $\rho' \in (\rho/(1-\rho), (1-2\alpha)/(1-\alpha))$ with $\rho'n \in \mathbb{N}$. Let $\gamma_1, \gamma_2 \in (0,1)$. We then have

$$\mathbb{E}\{\Gamma(t+1) \mid x(t)\} \leq \left(1 - c_4\frac{\alpha}{n}\right)\Gamma(t) + c_3,$$

where

$$c_4 = \min\left\{\mu - \alpha(1-\mu), \rho - \alpha(1+\rho), \gamma_1\,(1 - \mu'(\alpha+1)), \frac{\alpha\,(1 - \mu - \alpha(2-\mu))}{1 - \mu(1-\alpha)}, \right.$$
$$\left. \gamma_2\,(1 - 2\alpha - \rho'(1-\alpha)), \frac{\alpha\,(1 - \rho(1+\alpha))}{1 - \rho(1-\alpha) + 2\alpha}\right\}$$

and

$$c_3 = \max \left\{ \alpha \left(1 + \alpha + \rho(1 + \rho)\right), \alpha(1 - \mu)(2 - \mu), (\alpha + c_4)\alpha c_1, \alpha + \alpha^2, \right.$$
$$\left. (\alpha + c_4)\alpha c_2 \right\},$$

in which

$$c_1 = \left(1 + \frac{1}{\lambda_1}\right) C_1 \left(\frac{C_1}{\mu \lambda_1}\right)^{\mu/((1-\mu)\mu'-\mu)}, \quad C_1 = \frac{(1 - \mu')(2 + \alpha)}{(1 - \gamma_1)(1 - \mu'(1 + \alpha))},$$

$$c_2 = \left(1 + \frac{1}{\lambda_2}\right) C_2 \left(\frac{C_2}{\rho \lambda_2}\right)^{\rho/((1-\rho)\rho'-\rho)}, \quad C_2 = \frac{(1 - \rho')(2 - 2\alpha - \rho'(1 - \alpha))}{(1 - \gamma_2)(1 - 2\alpha - \rho'(1 - \alpha))},$$

$$\lambda_1 - \frac{1 - \mu - \alpha(2 - \mu)}{2\alpha}, \quad \lambda_2 = \frac{1 - \rho(1 + \alpha)}{2\alpha}.$$

Proof. See [11]. ∎

We are now able to give a upper bound of the expected value of $\Gamma(t)$.

Theorem 13. *For all $t \geq 0$, under the hypothesis of Theorem 12, we have $\mathbb{E}\{\Gamma(t)\} \leq c_3 n/(\alpha c_4)$.*

Proof. We prove this result by induction. For $t = 0$, we have $\Gamma(0) = 2n$. Moreover, we have

$$c_3 \geq \alpha \left(1 + \alpha + \rho(1 + \rho)\right) \geq \alpha \text{ and } c_4 \leq \mu - \alpha(1 - \mu) \leq \mu \leq 1/2,$$

which implies that $c_3/(\alpha c_4) \geq 2$. We thus have $\mathbb{E}\{\Gamma(0)\} = 2n \leq c_3 n/(\alpha c_4)$. Suppose that the result is true for a fixed $t \geq 0$. From Theorem 12, we have

$$\mathbb{E}\{\Gamma(t+1)\} = \mathbb{E}\left\{\mathbb{E}\{\Gamma(t+1) \mid x(t)\}\right\} \leq \mathbb{E}\left\{\left(1 - c_4 \frac{\alpha}{n}\right) \Gamma(t) + c_3\right\}$$
$$\leq \left(1 - c_4 \frac{\alpha}{n}\right) \frac{c_3}{\alpha c_4} n + c_3 = \frac{c_3}{\alpha c_4} n.$$

which completes the proof. ∎

Theorem 14. *For all $t \geq 0$ and $\sigma > 0$, under the hypothesis of Theorem 12, we have*

$$\mathbb{P}\left\{\text{Gap}(t) \geq \frac{2(1 + \sigma)}{\alpha} \ln(n) + \frac{2}{\alpha} \ln\left(\frac{c_3}{2\alpha c_4}\right)\right\} \leq \frac{1}{n^\sigma}$$

Proof. From Lemma 1 and Theorem 13, we have

$$\Gamma(t) \geq 2e^{\alpha \text{Gap}(t)/2} \text{ and } \frac{c_3 n}{\alpha c_4} \geq \mathbb{E}\{\Gamma(t)\}.$$

It follows that

$$2e^{\alpha \text{Gap}(t)/2} \geq n^\sigma \frac{c_3 n}{\alpha c_4} \implies \Gamma(t) \geq n^\sigma \frac{c_3 n}{\alpha c_4} \implies \Gamma(t) \geq n^\sigma \mathbb{E}\{\Gamma(t)\}.$$

Using Markov inequality, we obtain

$$\mathbb{P}\left\{\mathrm{Gap}(t) \geq \frac{2(\sigma+1)}{\alpha}\ln(n) + \frac{2}{\alpha}\ln\left(\frac{c_3}{2\alpha c_4}\right)\right\} = \mathbb{P}\left\{2e^{\alpha \mathrm{Gap}(t)/2} \geq n^\sigma \frac{c_3 n}{\alpha c_4}\right\}$$

$$\leq \mathbb{P}\left\{\Gamma(t) \geq n^\sigma \mathbb{E}\{\Gamma(t)\}\right\} \leq \frac{1}{n^\sigma},$$

which completes the proof. ∎

The following corollary shows that at any time, and for any agent, its local counter approximates the global clock with high probability.

Corollary 15. *For all $t \geq 0$ and $\sigma > 0$, under the hypothesis of Theorem 12, we have*

$$\mathbb{P}\left\{\left|C_t^{(i)} - \frac{t}{n}\right| < \frac{2(1+\sigma)}{\alpha}\ln(n) + \frac{2}{\alpha}\ln\left(\frac{c_3}{2\alpha c_4}\right), \; \forall i = 1,\ldots,n\right\} \geq 1 - \frac{1}{n^\sigma}$$

Proof. By definition, we have $x_i(t) = C_t^{(i)} - t/n$, and since $x_n(t) \leq 0 \leq x_1(t)$, we have $|x_i(t)| \leq x_1(t) - x_n(t) = \mathrm{Gap}(t)$. It follows, from Theorem 14, that

$$\mathbb{P}\left\{\left|C_t^{(i)} - \frac{t}{n}\right| \geq \frac{2(1+\sigma)}{\alpha}\ln(n) + \frac{2}{\alpha}\ln\left(\frac{c_3}{2\alpha c_4}\right), \; \forall i = 1,\ldots,n\right\}$$

$$\leq \mathbb{P}\left\{\mathrm{Gap}(t) \geq \frac{2(1+\sigma)}{\alpha}\ln(n) + \frac{2}{\alpha}\ln\left(\frac{c_3}{2\alpha c_4}\right)\right\} \leq \frac{1}{n^\sigma}$$

which completes the proof. ∎

4   Evaluation of the constants

This section is devoted to the evaluation of constants a and b of Relation (1) and, to compare them with the ones that we can derive from the analysis of Alistarh et al. [4].
From Theorem 14, we have

$$a = \frac{2}{\alpha} \text{ and } b = \frac{2}{\alpha}\ln\left(\frac{c_3}{2\alpha c_4}\right),$$

where c_3 and c_4 are given by Theorem 12. First of all, note that constraints given in Theorem 12 imply the following inequality: $\rho/(1-\rho) < (1-2\alpha)/(1-\alpha)$, that is, $\rho \leq (1-2\alpha)/(2-3\alpha)$, which combined with $\rho \geq \alpha/(1-\alpha)$, leads to $\alpha \leq (5-\sqrt{5})/10 \approx 0.276$.

For a fixed value of α, we have to determine the values of parameters μ, ρ, μ', $\rho', \gamma_1, \gamma_2$ that minimize constant b. This is achieved by applying a simple Monte-Carlo algorithm. Figure 1 shows several optimal values of the constants a and b, used in Theorem 14, and computed for several values of α.

Let us now evaluate constants a and b obtained in the paper of Alistarh et al. [4]. Note that the goal of their work was not necessarily focused on the

α	0.17	0.18	0.19	0.20	0.21	0.22	0.23	0.24	0.25	0.26	0.27
$a = 2/\alpha$	11.77	11.12	10.53	10	9.53	9.10	8.70	8.34	8	7.70	7.41
$b = (2/\alpha)\log(c3/(2\alpha c_4))$	59	63	68	74	82	93	109	134	179	281	739

Fig. 1. Optimal values of a and b in function of α

optimization of a and b constants. Nevertheless, as we will see, the evaluation of a and b constants is an important motivation of our work. From Relations (1) and (2) of [4] and as $\beta = 1$, we get $0 < \delta \leq \varepsilon = 1/16$ and thus we obtain, for $\gamma > 0$ and $c \geq 2$,

$$\frac{1 + \gamma + c\alpha(1+\gamma)^2}{1 - \gamma - c\alpha(1+\gamma)^2} \leq \frac{17}{16},$$

which gives,

$$\alpha \leq \frac{1}{33c(1+\gamma)^2} - \frac{1}{c(1+\gamma)^2} \leq \frac{1}{33c(1+\gamma)^2} \leq \frac{1}{66}.$$

Considering the difference between the lower bound and the upper bound of the inequality following (11), we obtain

$$\exp\left(\frac{\alpha B}{n}\left(3 - \frac{1}{1-\lambda}\right)\right) \leq \frac{16\lambda C(\varepsilon)}{\varepsilon},$$

which can also be written as

$$\exp\left(\frac{\alpha B}{(1-\lambda)n}\right) \leq \left(\frac{16\lambda C(\varepsilon)}{\varepsilon}\right)^{1/(2-3\lambda)}.$$

Using the last inequality obtained in the proof of Lemma 4.8, we get

$$\Gamma(t) \leq \frac{4+\varepsilon}{\varepsilon}\lambda n C(\varepsilon)\exp\left(\frac{\alpha B}{(1-\lambda)n}\right) \leq \frac{4+\varepsilon}{\varepsilon}\lambda n C(\varepsilon)\left(\frac{16\lambda C(\varepsilon)}{\varepsilon}\right)^{1/(2-3\lambda)}.$$

Using this result, we obtain from Lemma 4.11, $\mathbb{E}\{\Gamma(t)\} \leq 4Cn/(\hat{\alpha}\varepsilon)$, where

$$C = \frac{4+\varepsilon}{\varepsilon}\lambda C(\varepsilon)\left(\frac{16\lambda C(\varepsilon)}{\varepsilon}\right)^{1/(2-3\lambda)}, \quad C(\varepsilon) = \frac{(1+\delta)/\lambda - 1 + 3\varepsilon}{3\varepsilon - \varepsilon/3}$$

and $\hat{\alpha} = \alpha(1 - \gamma - c\alpha(1+\gamma)^2)$.

Following the same ideas we used to prove Theorem 14, we get

$$a = \frac{2}{\alpha} \text{ and } b = \frac{2}{\alpha}\ln\left(\frac{2C}{\hat{\alpha}\varepsilon}\right).$$

Since $\alpha \leq 1/66$, we have $a \geq 132$. Moreover, since $0 \leq \delta \leq \varepsilon = 1/16$, $\lambda = 2/3 - 1/54 = 35/54$, $\gamma > 0$ and $c \geq 2$, we obtain

$$C(\varepsilon) = \frac{(1+\delta)/\lambda - 1 + 3\varepsilon}{3\varepsilon - \varepsilon/3} \geq \frac{1/\lambda - 1 + 3\varepsilon}{3\varepsilon - \varepsilon/3} = \frac{1227}{280}$$

which leads to

$$C = \frac{4+\varepsilon}{\varepsilon}\lambda C(\varepsilon)\left(\frac{16\lambda C(\varepsilon)}{\varepsilon}\right)^{1/(2-3\lambda)} \geq \frac{26585}{144}\left(\frac{6544}{9}\right)^{18}.$$

Regarding $\widehat{\alpha}$, we have $\widehat{\alpha} = \alpha(1-\gamma-c\alpha(1+\gamma)^2) \leq \alpha \leq 1/66$. Therefore, we have

$$b = \frac{2}{\alpha}\ln\left(\frac{2C}{\widehat{\alpha}\varepsilon}\right) \geq 132\ln\left(\frac{1169740}{3}\left(\frac{6544}{9}\right)^{18}\right) \geq 17354.$$

It follows that constants a and b obtained from [4] satisfy $a \geq 132$ and $b \geq 17354$, which are at least two orders of magnitude larger than the ones we derived (see Fig. 1).

5   Simulations

We complete this paper by giving a summary of the experiments we have carried out to illustrate the performances of our protocol. Recall that n is the number of nodes in the system, and $T = t/n$ is the total number of interactions divided by n, which is often called the parallel time. We have conducted two types of experiments, the first one illustrates the expected proportion of nodes $Y_T(n,k)$ whose counter is equal to $T+k$ at time nT, for different values of n and k. More precisely, $Y_T(n,k)$ is defined by

$$Y_T(n,k) = \frac{1}{n}\sum_{i=1}^{n}1_{\{C_{nT}^{(i)}=T+k\}}.$$

We show in Fig. 2(a) the expected value of $Y_T(n,k)$, for $n = 1000$ and $k = -2, -1, 0, 1$, as a function of the parallel time T. These results have been

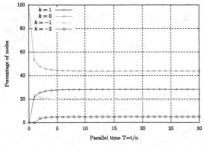

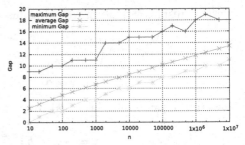

(a) Expected proportion $Y_T(n,k)$ of nodes as a function of parallel time T, for $n = 1000$, and $k = -2, -1, 0, 1$, from bottom to the top.

(b) Minimum, average and maximum gap as a function of n.

Fig. 2. Expected proportion and gap

Table 1. Expectation of $Y_{50}(n, k)$ from number of nodes n and shift k

k	n				
	10^3	10^4	10^5	10^6	10^7
-13	0.0	0.0	0.0	1.4E-9	1.42E-9
-12	0.0	2.0E-8	8.0E-9	9.0E-9	6.14E-9
-11	2.0E-7	4.0E-8	2.2E-8	2.8E-8	3.048E-8
-10	2.0E-7	8.0E-8	1.88E-7	1.436E-7	1.4814E-7
-9	4.0E-7	8.0E-7	7.7E-7	7.438E-7	7.2784E-7
-8	3.0E-6	3.6E-6	3.586E-6	3.48E-6	3.6029E-6
-7	1.42E-5	1.8E-5	1.8222E-5	1.7767E-5	1.7758E-5
-6	8.98E-5	8.602E-5	8.7176E-5	8.7372E-5	8.72753E-5
-5	4.372E-4	4.2706E-4	4.2957E-4	4.2901E-4	4.29349E-4
-4	0.0021144	0.0021023	0.0021071	0.0021092	0.0021086
-3	0.0102474	0.0102890	0.0102777	0.0102800	0.0102810
-2	0.0481626	0.0483366	0.0483382	0.0483465	0.0483437
-1	0.1930704	0.1932864	0.1933165	0.1933143	0.1933182
0	0.4389352	0.4380932	0.4380715	0.4380374	0.4380346
1	0.2824746	0.2827344	0.2826797	0.2827057	0.2827070
2	0.0243744	0.0245499	0.0245973	0.0245953	0.0245949
3	7.6E-5	7.224E-5	7.2248E-5	7.27752E-5	7.27974E-5
4	0.0	0.0	0.0	4.0E-10	3.6E-10

obtained after running 10, 000 independent experiments. Figure 2(a) shows that the expected value of $Y_T(n, k)$ seems to converge when T goes to infinity, and this convergence is reached very quickly. Note that for other values of k, proportions of nodes are too close to 0 to be depicted, as shown in Table 1. Table 1 shows the expected proportion of nodes $Y_T(n, k)$ whose counter is equal to $T + k$ at time $T = 50$, for different values of $n = 10^3, 10^4, 10^5, 10^6, 10^7$ and $k = -13, \ldots, 4$. These results have been obtained after running 5, 000 independent experiments, for each value of n. The expected value of $Y_{50}(n, k)$ seems to be almost independent of n for large values of n.

The second experiment illustrates the gaps (i.e., the maximal, average, and minimal) for different values of the size n of the system. Let $B = 2 \times 10^9$ be the total number of interactions considered. The maximal gap is computed as $\max_{100n \leq t \leq B} \mathrm{Gap}(t)$, the minimal one is given by $\min_{100n \leq t \leq B} \mathrm{Gap}(t)$, and the average gap is given by

$$\frac{1}{B - 100n} \sum_{t=100n}^{B-1} \mathrm{Gap}(t).$$

Figure 2(b) shows respectively the minimal, average and maximal gap in a system of size n over the interval $[100n, B]$ of interactions. As one may expect, the logarithmic progression of the Gap is clearly shown.

6 Conclusion

In this article we have gone a step further in the study of the two-choice paradigm by providing an accurate analysis of the gap problem. An important application of this study would be the improvement of leaderless population protocols. Indeed, we have shown in this paper that agents can construct a global clock by guaranteeing that the values of all agent counters are concentrated according to Relation (1), and thus can locally use this global clock to determine the instants at which some specific actions need to be triggered, or the instants from which all the agents of the system have converged to a given state. In the former case, this would allow agents to solve more complex problems by triggering a series of population protocols, whereas in the latter case this would allow agents to determine the instant from which all the agents have successfully computed a given feature of the population. The construction of efficient leaderless population protocols inspired from this orchestration is left for future work.

References

1. Adler, M., Berenbrink, P., Schröder, K.: Analyzing an infinite parallel job allocation process. In: Bilardi, G., Italiano, G.F., Pietracaprina, A., Pucci, G. (eds.) ESA 1998. LNCS, vol. 1461, pp. 417–428. Springer, Heidelberg (1998). https://doi.org/10.1007/3-540-68530-8_35
2. Adler, M., Chakrabarti, S., Mitzenmacher, M., Rasmussen, L.: Parallel randomized load balancing. Random Struct. Algorithms **13**(2), 159–188 (1998)
3. Alistarh, D., Aspnes, J., Gelashvili, R.: Space-optimal majority in population protocols. In: Czumaj, A. (ed.) Proceedings of the Annual ACM-SIAM Symposium on Discrete Algorithms (SODA), pp. 2221–2239 (2018)
4. Alistarh, D., Kopinsky, J., Li, J., Nadiradze, G.: The power of choice in priority scheduling. In: Proceedings of the ACM Symposium on Principles of Distributed Computing (PODC) (2017)
5. Azar, Y., Broder, A.Z., Karlin, A.R., Upfal, E.: Balanced allocations (extended abstract). In: Proceedings of the ACM Symposium on Theory of Computing (STOC) (1994)
6. Berenbrink, P., Czumaj, A., Friedetzky, T., Vvedenskaya, N.D.: Infinite parallel job allocation (extended abstract). In: Proceedings of the ACM Symposium on Parallel Algorithms and Architectures (SPAA), pp. 99–108 (2000)
7. Berenbrink, P., Czumaj, A., Steger, A., Vöcking, B.: Balanced allocations: the heavily loaded case. SIAM J. Comput. **35**(6), 1350–1385 (2006)
8. Berenbrink, P., Meyer auf der Heide, F., Schröder, K.: Allocating weighted jobs in parallel. Theory Comput. Syst. **32**(3), 281–300 (1999)
9. Mitzenmacher, M.: Load balancing and density dependent jump Markov processes. In: Proceedings of International Conference on Foundations of Computer Science (1996)

10. Mitzenmacher, M., Richa, A.W., Sitaraman, R.: The power of two random choices: a survey of techniques and results. In: Handbook of Randomized Computing, pp. 255–312. Kluwer (2000)
11. Mocquard, Y., Sericola, B., Anceaume, E.: Balanced allocations and global clock in population protocols: An accurate analysis (Full version), Technical report (2018). https://hal.archives-ouvertes.fr/hal-01790973
12. Peres, Y., Talwar, K., Wieder, U.: The $(1+\beta)$-choice process and weighted balls into bins. In: Proceedings of the ACM-SIAM Symposium on Discrete Algorithms (SODA) (2010)
13. Peres, Y., Talwar, K., Wieder, U.: Graphical balanced allocations and the $(1 + \beta)$-choice process. Random Struct. Algorithms **47**(4), 760–775 (2015)
14. Raab, M., Steger, A.: "Balls into Bins" — a simple and tight analysis. In: Luby, M., Rolim, J.D.P., Serna, M. (eds.) RANDOM 1998. LNCS, vol. 1518, pp. 159–170. Springer, Heidelberg (1998). https://doi.org/10.1007/3-540-49543-6_13
15. Talwar, K., Wieder, U.: Balanced allocations: a simple proof for the heavily loaded case. In: Esparza, J., Fraigniaud, P., Husfeldt, T., Koutsoupias, E. (eds.) ICALP 2014. LNCS, vol. 8572, pp. 979–990. Springer, Heidelberg (2014). https://doi.org/10.1007/978-3-662-43948-7_81

On Knowledge and Communication Complexity in Distributed Systems

Daniel Pfleger and Ulrich Schmid$^{(\boxtimes)}$

TU Wien, Treitlstrasse 3, 1040 Vienna, Austria
{dpfleger,s}@ecs.tuwien.ac.at

Abstract. This paper contributes to exploring the connection between epistemic knowledge and communication complexity in distributed systems. We focus on Action Models, a well-known variant of dynamic epistemic logic, which allows to cleanly separate the state of knowledge of the processes and its update due to communication actions: Exactly like the set of possible global states, the possible actions are described by means of a Kripke model that specifies which communication actions are indistinguishable for which process. We first show that the number of connected components in the action model results in a lower bound for communication complexity. We then apply this result, in the restricted setting of a two processor system, for determining communication complexity lower bounds for solving a distributed computing problem $\mathcal{P}$: We first determine some properties of the action model corresponding to any given protocol that solves $\mathcal{P}$, and then use our action model communication complexity lower bounds. Finally, we demonstrate our approach by applying it to synchronous distributed function computation and to a simple instance of consensus in directed dynamic networks.

Keywords: Distributed systems · Dynamic epistemic logic
Communication complexity

1 Introduction

Our paper is concerned with the idea to infer the communication complexity for solving a general distributed computing problem $\mathcal{P}$ from the epistemic knowledge that must be attained by the processes to solve $\mathcal{P}$. More specifically, we take a first step to bridge Hintikka's *epistemic logic* [12], variants of which have very successfully been applied in distributed computing already [3,4,11], and Yao's communication complexity [18]. In this seminal work, Yao introduced methods for deriving communication complexity lower bounds for distributed function computation in a system of 2 processes.

Epistemic logic [12] allows to formally reason about knowledge and belief in static multi-agent systems. It relies on a Kripke model M that describes the

This work has been supported by the Austrian Science Fund FWF under the projects ADynNet (P28182) and RiSE/SHiNE (S11405).

Z. Lotker and B. Patt-Shamir (Eds.): SIROCCO 2018, LNCS 11085, pp. 312–330, 2018.
https://doi.org/10.1007/978-3-030-01325-7_27

possible global states ("possible worlds") of the system, where certain atomic propositions (facts like "variable x_i of process p_i is zero") hold true or not, along with an indistinguishability relation $s \sim_a s'$ between global states s, s' for every agent (= process) a. Knowledge of some fact ϕ about the system in global state s is primarily captured by a modal *knowledge operator* K_a, used in formal expressions like $M, s \models K_a\phi$, which captures the intuition that, being in the global state s, process a knows ϕ if ϕ holds in every global state s' that is indistinguishable from s for a. We will use the term *epistemic model* to refer to this type of Kripke models, which focus entirely on a given "static" knowledge state of the processes.

Dynamic Epistemic Logic (DEL) [8,10] also allows to incorporate communication-induced knowledge gain into the formal reasoning. We will focus on a variant of DEL called *Action Models* [8], which are particularly suitable for our purpose. Action Models can be used to describe possible communication events that may occur at certain times in an execution of some algorithm. Formally, this is modeled by applying a sequence of (arbitrary) communication *action models* $\text{AM}_1, \text{AM}_2, \ldots$ to an initial epistemic model M_0, which results in a sequence of epistemic models $M_0, M_1, \ldots$ that describes the evolution of knowledge in the execution. Every AM_k is represented by an *independent* Kripke model here, which is orthogonal to the epistemic model M_{k-1} it is applied to. By means of a well-defined product $\otimes$ (see Sect. 2.1), this leads to the epistemic model $M_k = M_{k-1} \otimes \text{AM}_k$. This abstraction is particularly suitable for modeling synchronous systems, as AM_k can be used to express all the possible communication in round k here.

Our Contributions: (1) We first exploit the clean separation of epistemic models and action models to infer a natural lower bound for the number of bits that some process a must receive in any protocol that faithfully implements a given action model AM. It is closely related to the number of partitions in a's indistinguishability relation in AM. (2) Restricting our attention to systems of 2 processes, we then infer a communication complexity lower bound for solving a problem $\mathcal{P}$ by (i) determining the properties of the action model corresponding to any protocol that correctly solves $\mathcal{P}$, and (ii) inferring a communication complexity lower bound from this via the result of (1). We apply our approach to distributed function computation in synchronous systems [18] and sketch how it is applied to consensus under message adversaries [1,5].

Related Work: Van Ditmarsch et al. [8] provide a comprehensive introduction into Dynamic Epistemic Logic, including Action Models. Fagin et al. [9] introduces the powerful runs and systems framework, which allows to reason about knowledge in general distributed systems. Halpern and Moses [11] use this framework to reason the role of (various forms of) common knowledge in solving distributed consensus. They also elaborate on the *Muddy Children* problem, which is very closely related to the *Cheating Husbands* problem [15] that we use for illustrating Action Models in Sect. 4.

Communication complexity lower bounds deserve much to the seminal work [18] by Yao, which studies distributed function computation for two processes.

Indeed, [18] also sparked quite some interest in the distributed systems community, which led to very interesting lower bounds based on information-theoretic arguments. A few examples, among many possible others, are symmetry breaking in chains and rings [7] and lower bounds for all pair shortest paths [13]. However, unlike our results, these approaches are usually tied to the specific problem $\mathcal{P}$ at hand and do not use epistemic logic.

We are not aware of much work on the relation between communication complexity and epistemic models. Somewhat similar to our work is [6], which used dynamic epistemic logic and action models in a combinatorial way to find a lower bound on communication complexity for the Russian Cards problem. Alechina et al. [2] investigated bounds for a system of reasoning agents, where agents may have different knowledge and inferential capabilities and have to draw conclusions from received messages, which contain formulas. They established a framework to verify time, memory and communication bounds in such a system. Since the communication complexity in this work is defined as the number of formulas, rather than the number of bits, however, it cannot be compared to our approach.

Paper Organization: We start by defining our system model in Sect. 2, followed by an introduction to the relevant basics of Action Models (Sect. 2.1) and Communication Complexity (Sect. 3). In Sect. 4, we demonstrate how Action Models work by means of the well-known Cheating Husbands problem [15], and explain the connection between communication complexity and the number of partitions in action models. Sect. 5 elaborates further on the connection between Action Models and *Protocol Trees* introduced in [18]. Our main results can be found in Sect. 5.2, along with two applications in Sect. 5.3. Some conclusions and directions of future work in Sect. 6 round off our paper.

2   Model

We consider synchronous message passing systems only. Such systems are modeled as a set Π of n processes with unique identifiers, which are reliable and operate in lock-step rounds $r = 1, 2, \ldots$. The processes are modeled as state machines and connected by point-to-point communication links. We consider both reliable links and unreliable links controlled by a message adversary [1,5], which determines the links that successfully deliver the message sent over it in a round. More specifically, at the beginning of round r, all processes send out a message to every other process (and to themselves). Rounds are communication-closed, in the sense that each message sent in round r can only be delivered in r. The message adversary determines which message is indeed be delivered to the intended receiver. After this message exchange, all processes simultaneously perform an instantaneous local computation step that terminates round r.

Note that, in the case of reliable links, the guarantee that all the messages sent in round r will be delivered by the end of round r allows *communication by time*: if a process a did not receive the message from process b by the end of round r, then a *knows* that b did not send this message. Conversely, if a sends a message to b at the beginning of round r, a *knows* at the end of round r that b received this message. Note that Ben-Zvi and Moses [4] modeled communication by time via explicitly sending a virtual NULL-message instead of a real one.

Regarding the connection of action models and the synchronous model, we assume that a single action model AM_r is applied in each round r. For every possible communication pattern in round r, which is determined (i) by the protocol and (ii) by the message adversary, it contains a corresponding action. Clearly, two actions are indistinguishable for process a, if it receives the same messages. Note carefully that the synchrony assumption makes sure that every process knows that the action model AM_r is to be applied, even if it does not receive a single message in round r.

2.1   Knowledge and Action Models

Usually, distributed computing problems also involve global constraints, like agreement in distributed consensus (see Sect. 5.3). The actions of a single process in a distributed system depend solely on its local information, though, and the global behavior emerges from those local actions. Thus, defining and proving the correctness of distributed systems typically involves arguments about the behavior and interaction between individual processes. In such proofs, it is often argued that: "Once the synchronous round r begins, all processes *know* that all the sent messages have been delivered.", for example.

To formalize such arguments, frameworks like [8,9] allow to formally reason about knowledge in such systems. We utilize a variant of Dynamic Epistemic Logic [10,17], namely, Action Models, which are well-suited for the simple synchronous systems considered in our paper. The following Definition 1 to 4 and 7 will define epistemic models and action models, as well as the semantics of action model logic. Illustrating examples can be found in Sect. 4 (Figs. 2 and 3).

Definition 1 (Kripke model, see [8], Definition 2.6). *A Kripke model M is a tuple $\langle S, R, V \rangle$ on a set of processes A, where $S \neq \emptyset$ is a set of states, R is a set of accessibility relations: $R = \{R_a \mid a \in A\}$, with $R_a \subseteq S \times S$. A state $t \in S$ is accessible for process $a \in A$ from state $s \in S$, iff $sR_a t$. $V : P \to 2^S$, P a set of atomic propositions (also called atoms), is a valuation function for each $p \in P$. For any proposition $p \in P$, $V(p) \subseteq S$ is exactly the set of states in which p is true.*

In our context, the epistemic states $s \in S$, denoted (M, s), are the possible global states of the distributed system, and R_a is interpreted as an indistinguishability relation for process a, thus is denoted by $\sim_a$ in the sequel.

The following definition formally defines the Kripke model of the possible actions. It is independent of the epistemic model in Definition 1 it is applied to, except for the precondition function that governs which actions are applicable in which epistemic state.

Definition 2 (Action Model, see [8], Definition 6.2). *For given processes A and atomic propositions P and any logical language $\mathcal{L}$, the action model M is a* *structure* $\langle S, \sim, \textbf{pre} \rangle$ *such that* S *is a set of* actions, $\sim_a$ *is an equivalence relation on* S *for each* $a \in A$, *and* $\textbf{pre} : S \to \mathcal{L}$ *a preconditions function that assigns a* precondition $\textbf{pre}(s) \in \mathcal{L}$ *to each* $s \in S$. *A* pointed action model *is a structure* $(\textbf{M}, \textbf{s})$, *with* $\textbf{s} \in S$.

The following syntax is used to formally specify action model knowledge formulas. Basically, it consists of formulas ϕ related to the epistemic state (the "possible worlds") of the system, and formulas involving the application of some action α. The detailed semantics is given in Definition 7 below.

Definition 3 (Syntax of action model logic, see [8], Definition 6.3).
Given processes A *and atoms* P, *the language* $\mathcal{L}_{KC\otimes}(A, P)$ *is the union of formulas* $\phi \in \mathcal{L}_{KC\otimes}^{stat}(A, P)$ *and pointed action models* $\alpha \in \mathcal{L}_{KC\otimes}^{act}(A, P)$ *defined by:*

$$\phi ::= p \mid \neg\phi \mid (\phi \wedge \phi) \mid K_a\phi \mid E_B\phi \mid C_B\phi \mid [\alpha]\phi$$
$$\alpha ::= (\textbf{M}, \textbf{s}) \mid (\alpha \cup \alpha)$$

with $p \in P$, $a \in A$, $B \subseteq A$, *and* $(\textbf{M}, \textbf{s})$ *a pointed action model with finite domain* S *such that for all* $\textbf{t} \in S$ *the precondition* $\textbf{pre}(\textbf{t})$ *is a* $\mathcal{L}_{KC\otimes}^{stat}(A, P)$ *formula that has already been constructed in a previous stage of the inductively defined hierarchy.* $\alpha \cup \alpha'$ *denotes a* non-deterministic choice *between* α *and* α'.

Action models can be composed: To apply two different actions $(\textbf{M}, \textbf{s})$ and $(\textbf{M}', \textbf{s}')$ to some epistemic state (M, s) subsequently, one can either apply them one after the other or determine their composition $(\textbf{M}'', \textbf{s}'') = (\textbf{M}; \textbf{M}', (\textbf{s}, \textbf{s}'))$ and apply the resulting action $(\textbf{M}'', \textbf{s}'')$ in a single step.

Definition 4 (Composition of action models, see [8], Definition 6.7).
Let $\textbf{M} = \langle S, \sim, \textbf{pre} \rangle$ *and* $\textbf{M}' = \langle S', \sim', \textbf{pre}' \rangle$ *be two action models in* $\mathcal{L}_{KC\otimes}$. *Then their* composition $(\textbf{M}; \textbf{M}')$ *is the action model* $\textbf{M}'' = \langle S'', \sim'', \textbf{pre}'' \rangle$, *such that:*

$$S'' = S \times S'$$
$$(s, s') \sim_a'' (t, t') \text{ iff } s \sim_a t \text{ and } s' \sim_a' t'$$
$$\textbf{pre}''((s, s')) = \langle \textbf{M}, \textbf{s} \rangle \textbf{pre}'(s')$$

with $\langle \textbf{M}, \textbf{s} \rangle \textbf{pre}'(s')$ *denoting an abbreviation for* $\neg[\textbf{M}, \textbf{s}]\neg\textbf{pre}'(s')$.

To change the static epistemic status starting in an epistemic model M, one applies an action model $\textbf{M}$, resulting in a new epistemic modem M':

Definition 5 (Application of an action model). *We define the application of action model* $\textbf{M}$ *on epistemic model* M, *resulting in* M', $M' = (M \otimes \textbf{M})$, *as* $M' = \langle S', \sim', V' \rangle$ *with:*

$$S' = \{(s, \textbf{s}) \mid s \in S, \textbf{s} \in S, \text{ and } M, s \models \textbf{pre}(\textbf{s})\}$$
$$(s, \textbf{s}) \sim_a' (t, \textbf{t}) \text{ iff } s \sim_a t \text{ and } \textbf{s} \sim_a \textbf{t}$$
$$(s, \textbf{s}) \in V'(p) \text{ iff } s \in V(p)$$

Note that our complexity results will primarily rely on the axiom $(s, \textbf{s}) \sim_a' (t, \textbf{t})$ iff $s \sim_a t$ and $\textbf{s} \sim_a \textbf{t}$, which implies that the application of two *distinguishable* actions $s \not\sim_a t$ to indistinguishable epistemic states $s \sim_a t$ causes distinguishable epistemic states $(s, s) \not\sim'_a (t, t)$.

To define the semantics of common knowledge, the last ingredient of our Action Models, we need to introduce the *reflexive transitive closure* of a relation R. It allows to express facts and formulas ϕ that are commonly known to a subset of the processes, in the sense that "every process knows that every process knows that every process knows $\ldots \phi$.". We define "everybody in group B knows ϕ" $(E_B\phi)$, as a syntactic equivalence $E_B\phi = \bigwedge_{a \in B} K_a\phi$.

Definition 6. *The* reflexive transitive closure *of a relation R is the smallest relation R^* such that: (i) $R \subseteq R^*$, (ii) for all x, y, and z: $xR^*y \wedge yR^*z \Rightarrow xR^*z$ (transitivity), (iii) for all x, xR^*x (reflexivity).*

We can now give the formal semantics of Action Model logic. It specifies the meaning of both the operations for reasoning about knowledge in the epistemic model and the application of action models.[1]

Definition 7 (Semantics of action model logic, see [8], Definition 6.8).
Let $M = \langle S, \sim, V \rangle$ be an epistemic model with (M, s), $s \in S$, an epistemic state of this model, $\mathsf{M} = \langle \mathsf{S}, \sim, \mathbf{pre} \rangle$ an action model, and $\phi \in \mathcal{L}^{stat}_{KC\otimes}$ and $\alpha \in \mathcal{L}^{act}_{KC\otimes}$. Furthermore let A be a set of processes and P a set of atoms, while $a \in A$, $B \subseteq A$ and $p \in P$.

$$
\begin{aligned}
&M, s \models p &&\textit{iff } s \in V(p)\\
&M, s \models (\phi \wedge \psi) &&\textit{iff } M, s \models \phi \textit{ and } M, s \models \psi\\
&M, s \models \neg\phi &&\textit{iff } M, s \not\models \phi\\
&M, s \models K_a\phi &&\textit{iff for all } t \in S \textit{ such that } s \sim_a t: \; M, t \models \phi\\
&M, s \models E_B\phi &&\textit{iff for all } t \in S \textit{ such that } s \sim_{E_B} t: \; M, t \models \phi\\
&M, s \models C_B\phi &&\textit{iff for all } t \in S \textit{ such that } s \sim^*_{E_B} t: \; M, t \models \phi\\
&M, s \models [\alpha]\phi &&\textit{iff for all } M', s' \textit{ such that } (M, s)[\![\alpha]\!](M', s'): \; M', s' \models \phi\\
&(M, s)[\![\mathsf{M}, \mathsf{s}]\!](M', s') &&\textit{iff } M, s \models \mathbf{pre}(\mathsf{s}) \textit{ and } (M', s') = (M \otimes \mathsf{M}, (s, \mathsf{s}))\\
&[\![\alpha \cup \alpha']\!] &&= [\![\alpha]\!] \cup [\![\alpha']\!]
\end{aligned}
$$

with $\sim_{E_B} = \bigcup_{b \in B} \sim_b$.

In Sect. 4, we will use the Cheating Husbands problem [15] to exemplify how the Action Model semantics works in practice; particular instantiations can be found in Sect. 5.3.

3 Communication Complexity Basics

In [18], Yao considers two processes p_0 and p_1, which jointly solve the problem of evaluating the non-constant function $f : X \times Y \to Z$, where X, Y and Z are arbitrary finite sets. Herein, the input $x \in X$ is only known to p_0, whereas the

---

[1] Please observe the different fonts in our notation: in $s \sim_a t$, $\sim_a$ is taken from the epistemic model M, while in $\mathsf{s} \sim_a \mathsf{t}$, $\sim_a$ is from the action model M.

input $y \in Y$ is only known to p_1. Clearly, p_0 and p_1 have to communicate with each other in order to solve the problem. Yao's communication model assumes that all communication links are reliable and that the processes send information to each other alternatingly: one bit is sent by p_0 then one bit is sent by p_1 and so on, according to some protocol $\mathcal{P}$. The communication complexity of computing f is the least number of bits that need to be exchanged between p_0 and p_1 by any deterministic protocol $\mathcal{P}$ in order to determine $f(x, y)$ at p_0 or p_1. In fact, Yao assumes that the process that can compute $f(x, y)$ first sends a special NULL message to the other process and stops. It is also assumed that the processes both know the identity of themselves and the other process a priori. Note carefully that this allows the design of an asymmetric protocol, where some agreed-upon process, say, p_0 sends the first bit.

The following definition introduces the convenient notion of *protocol trees*, which uniquely describe the possible executions of a given protocol.

Definition 8 (Protocol trees, see [14], Definition 1.1). *A protocol $\mathcal{P}$ over $X \times Y$ with range Z is a binary tree, where each internal node v is labeled either by a function $g_v : X \to \{0, 1\}$ or by a function $h_v : Y \to \{0, 1\}$, and each leaf is labeled with an element z of Z. The root r is labeled by $g_1 : X \to \{0, 1\}$. Intuitively, g_k (resp. h_k) gives the bit sent by p_0 (resp. p_1) in round k.*

The cost of the protocol $\mathcal{P}$ on input (x, y) is the length of the path taken on input (x, y), denoted by $\mathcal{D}_\mathcal{P}(f)$. As the longest such path is the height of the protocol tree, the maximal cost over all inputs is the height of this protocol tree.

The cost of a problem f is the minimal cost of any protocol $\mathcal{P}$ that computes f, denoted by $\mathcal{D}(f)$.

Every root-leaf path in this tree corresponds to an execution of $\mathcal{P}$ on some input (x, y): At each internal node, the process that is the next to send a bit is computing g_v (resp. h_v), to determine the value of the next communicated bit. Figure 1 shows the definition of a function $f : \{x_0, x_1, x_2, x_3\} \times \{y_0, y_1, y_2, y_3\} \to \{0, 1\}$, which is computed by the protocol tree next to it.

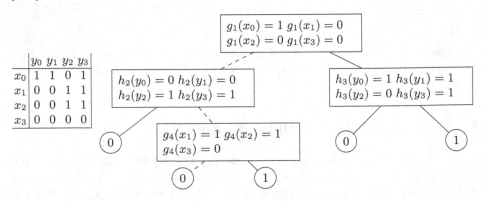

Fig. 1. An example of a function $f(x, y)$ and one possible protocol tree for computing it. The dashed path corresponds to some input in $\{x_3\} \times \{y_2, y_3\}$. The left (resp. right) branch from each node v corresponds to sending bit 0 (resp. 1).

As witnessed by the leaves in the protocol tree, a protocol $\mathcal{P}$ can be seen as a way to partition the set of possible inputs $X \times Y$ to multiple subsets leading to the same communication pattern. The left (resp. right) child of a node corresponds to the case where the bit sent by the node's corresponding process is 0 (resp. 1). For example, the sequence of communicated bits for any input pair in $\{x_3\} \times \{y_2, y_3\}$, corresponding to the dashed path in Fig. 1, is $(0, 1, 0)$. This leads to the crucial notion of *rectangles*:

Definition 9 (Rectangles, see [14], Proposition 1.13). *A partition* $R \subseteq X \times Y$ *is a rectangle iff:* $(x, y) \in R$ *and* $(x', y') \in R \Rightarrow (x, y') \in R$. *A rectangle is* f-*monochromatic iff for all* $(x, y) \in R$ *the result of* $f(x, y) = z$ *is the same.*

By exploiting the close relation between monochromatic rectangles and leaf, one can prove the following Corollary 1 [14, 18]:

Corollary 1 (Lower Bound, see [14], Corollary 1.17). *If any set of* f-*monochromatic rectangles induced by* f *has size at least* t, *then* $\log_2 t \leq D(f)$.

In the remaining paper, we will use a similar model as in [18], with the following two main differences: (i) Each process can send an arbitrary number of bits in every round. (ii) We consider symmetric function computation (sometimes without communication by time), i.e., once an algorithm for computing $f(x, y)$ terminates, the result must be commonly known by both p_0 and p_1.

4   Communication Complexity of Action Models

We use the *Cheating Husbands* problem [15] to illustrate the connection between knowledge and communication complexity. Herein, the women of a city ruled by a queen want to get rid of unfaithful husbands. It is common knowledge that each of the women knows the fidelity-status of the husbands of all other women, but does not know whether or not her own husband is unfaithful. The women are not allowed to discuss their husbands fidelity with each other. The left model in Fig. 2 depicts the initial epistemic model M_{CH} for three women (a, b, c). Each state is labeled with the atomic proposition (abc), $a, b, c \in \{0, 1\}$, $i = 1(0)$ interpreted as "husband of i is unfaithful (faithful)".

In the original problem considered in [15], it is common knowledge among the women that the queen will publicly announce whether there is at least one unfaithful husband or not. We can model this announcement using the actions $\neg t$ ("the queen does not announce anything") and ≥ 1 ("the queen publicly announces that there is at least one unfaithful husband"). It is well known [15] that, in this scenario, the women can find all the unfaithful husbands by a synchronous protocol, which requires every woman who gets to know her husband is unfaithful some day must shoot him at midnight.

There is a variant of this problem, which also allows a correct solution: Here it is common knowledge that, iff there is exactly one unfaithful husband, the queen tells his wife privately that her husband is unfaithful on some a priori

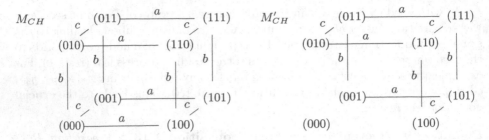

Fig. 2. Example Cheating Husbands: Left: The initial Kripke model M_{CH} for three wives a, b, c. Right: The epistemic model M'_{CH} reached after applying AM_{pub} or AM_{priv}.

known day. All the other women will never hear anything from the queen, and in no other case the queen announces anything. This can be modeled by the actions $\neg t$ ("the queen does not announce anything") and t_i for each woman i ("the queen tells woman i that her husband is unfaithful"). It can be shown that the women are also able to shoot all the unfaithful husbands, using the same protocol.

The two action models AM_{pub} for the public announcement and AM_{priv} for the private announcement are depicted in Fig. 3. Applying AM_{pub} and AM_{priv} on the initial epistemic model, the resulting epistemic model is the same, depicted in Fig. 2 (right). Still, in the public scenario AM_{pub}, the queen "sends out" a single bit ("There is no / at least one unfaithful husband.") to each woman, summing up to n bits in total for n women. In the private scenario AM_{priv}, though, the queen only needs to send a message ("You!") to a single woman in some special cases, and sends nothing in most other cases. Since we are a synchronous setting, however, every woman knows —via communication by time— that, if the queen sent her a message, she would have received it by midnight of the a priori known day. So, effectively, the queen sends a single bit ("Your husband is unfaithful") to at most one woman. Consequently, the communication complexity in the public scenario is higher than in the private one.

One may conjecture that this difference is related to the information complexity of the *a priori knowledge*: the communication complexity probably decreases with increasing a priori knowledge. Exploring this relation is a very interesting research question but still out of reach.[2] We therefore focus on the relation between communication complexity and the number of possible "knowledge-changing" events in an execution, which are neatly encapsulated in the action models: in essence, an action model just defines the possible observations of the global system state every single process can make.

Considering a single woman a, it is apparent from Fig. 3 that both of the action models are *partitioned* regarding the indistinguishability of a.

---

[2] We note, however, that our findings do support this claim, as the action model is common a priori knowledge and clearly more complex in the private than in the public scenario, cp. Fig. 3.

$$\neg t \qquad\qquad\qquad \geq 1 \qquad\qquad\Big| \qquad t_a \xrightarrow{\;\;b,c\;\;} \neg t \overset{a,b}{\underset{a,c}{\rightleftarrows}} \begin{matrix} t_c \\ t_b \end{matrix}$$

| Action model $\mathtt{AM}_{pub}$ modeling the public scenario of the cheating husbands problem. | Action model $\mathtt{AM}_{priv}$ modeling the private scenario of the cheating husbands problem. |

Fig. 3. Action models for the two scenarios of Cheating Husbands. The actions are denoted by: $\neg t$: the queen does not make a statement, ≥ 1: the queen publicly announces that there is at least one unfaithful husband, t_i: the queens tells i privately that her husband is unfaithful. The partitioning regarding a is depicted in red.

Definition 10 (Partitions of action models). *An Action Model $\mathtt{AM} = \langle S, \sim, pre\rangle$ is partitioned regarding process a if the underlying indistinguishability graph, consisting only of edges corresponding to $\sim_a$, is partitioned. I.e., there are sets of nodes V_i, such that for $i \neq j$ $V_i \cap V_j = \emptyset$, $\bigcup_i V_i = V$ and $v \in V_i$, $v' \in V_j \Rightarrow (v, v') \notin E$.*
The number of those partitions is denoted by $N^a_{\mathtt{AM}}$.

Our claim is that there is a strong connection between the communication complexity, more specifically, the number of bits received by process a, and the number of such partitions $N^a_{\mathtt{AM}}$ in the action model $\mathtt{AM}$. In Fig. 3, both action models $\mathtt{AM}_{pub}$ and $\mathtt{AM}_{priv}$ partition into two partitions, for each woman i. Indeed, if the queen does not announce anything in $\mathtt{AM}_{priv}$, woman a does NOT know that the action has been $\neg t$, she only knows that the action has been in the partition $\{\neg t, t_b, t_c\}$. In Fig. 3, if ≥ 1 (in $\mathtt{AM}_{pub}$) or t_a (in $\mathtt{AM}_{priv}$) occurs, woman a of course immediately knows the action itself, without receiving any additional information. In the case of $\neg t$ (in $\mathtt{AM}_{pub}$) or an action other than t_a (in $\mathtt{AM}_{priv}$), she learns the partition by not receiving anything, i.e., via communication by time.

Since every woman has to be able to identify the actual partition the current action is in, according to the semantics of $\otimes$ in Definition 5, the number of these partitions determines a lower bound on the number of bits received by a woman in some scenario, i.e., a worst-case lower bound: As both action models split into two partitions for each woman, the queen has to send one bit to each of them in BOTH scenarios.

Note that this does NOT contradict our above observation that, in $\mathtt{AM}_{priv}$, the queen sends a message to at most one woman. In more detail, in action t_a in $\mathtt{AM}_{priv}$, the queen *actively* sends a $bit = 1$ to woman a. In actions $\neg t$, t_b, t_c, the queen does not *actively* send anything to woman a, but it does so passively via communication by time: as in [4], we model this by virtually sending a NULL message.

Definition 11. *We define an active bit as a bit (i.e., 0 or 1) sent via explicit communication from some process a to some process b. A passive bit is defined as the bit "sent" in a NULL message from some process a to some process b (communication by time).*

Note that multiple active bits can be sent from a to b in a round, while a NULL message counts as a single passive bit only.

We are now ready to define the communication cost of the application of a single action model:

Definition 12. *The worst-case cost $D^a(AM)$ of the application of an action model AM for process a is the worst-case number of active bits received by a when the action model is applied, i.e., the maximum number of active bits received in some scenario.*

Note carefully that it is the particular protocol that actually determines the encoding used for communicating the occurrence of the actions to the processes. The number of active bits received by a may hence depend on which particular action occurs, which explains why we restrict our attention to the maximum number of active bits for defining $D^a(AM)$. Of course, this implies that we can only guarantee that $D^a(AM)$ bits are sent in *some* scenario, not in *any* scenario. Even worse, we cannot assume that the action causing the worst-case cost $D^a(AM)$ for process a is also causing the worst-case cost $D^b(AM)$ for process b. Therefore, defining the total worst-case cost $D(AM)$ of the application of an action model AM as the sum of $D^a(AM)$ over all processes a, would be overly conservative, and does hence not give a lower bound for the system-wide communication complexity. However, we can give a lower bound for $D^a(AM)$:

Lemma 1. *In a synchronous system with processes A, the worst-case cost $D^a(AM)$ of the application of an action model AM for process $a \in A$ satisfies $\log_2(N^a_{AM} - 1) \leq D^a(AM)$, where N^a_{AM} is the number of partitions regarding a in AM.*

Proof (Proof by contradiction). Suppose there exists an action model AM such that $D^a(AM) < \log_2(N^a_{AM} - 1)$ for some process $a \in A$. Then $2^{D^a(AM)} + 1 < N^a_{AM}$. Obviously, by receiving $D^a(AM)$ active bits with value 0 or 1, a can distinguish at most $2^{D^a(AM)} + 1$ (including the single passive bit) partitions of AM. Since $2^{D^a(AM)} + 1 < N^a_{AM}$, by a pigeonhole argument, there are at least two partitions P_0 and P_1 which cannot be distinguished by a.

Now assume that applying $(AM, \mathbf{s})$ at epistemic state (M, s) results in (M', s'), and consider the following scenario: (i) $s_0 \sim_a s_1$ in epistemic model M, (ii) $\mathbf{s}_0 \in P_0$, $\mathbf{s}_1 \in P_1$ in action model AM applicable to s_0 respectively s_1 ($\mathbf{s}_0 \not\sim_a \mathbf{s}_1$, but P_0 and P_1 indistinguishable by a). Such a scenario always exists, as one can choose $s_1 = s_0$ as well. By the semantics of Action Models (Definition 5 and 7), we must have $(s_0, \mathbf{s}_0) \not\sim_a (s_1, \mathbf{s}_1)$ in epistemic model M'. Since a cannot distinguish between (the actions in) P_0 and P_1, however, we inevitably have $(s_0, \mathbf{s}_0) \sim_a (s_1, \mathbf{s}_1)$, providing the required contradiction. Thus a has to receive at least $\log_2(N^a_{AM} - 1)$ active bits during the application of $(AM, \mathbf{s})$. ∎

Lemma 2 provides a lower bound for $D^a(AM)$ in the case in which communication by time cannot be used, e.g., when communication is unreliable. Its proof is almost identical to the proof of Lemma 1.

Lemma 2. *In a system with processes A, the worst-case cost of $D^a(AM)$ of the application of an action model AM for process $a \in A$ satisfies $\log_2(N^a_{AM}) \leq D^a(AM)$, where N^a_{AM} is the number of partitions regarding a in AM.*

So far, we only considered the application of a single action model. For the communication complexity of an algorithm $\mathcal{A}$ solving a specific problem $\mathcal{P}$ using multiple rounds of communication, the first thing that comes to mind is to sum up the communication complexity of single round action models. Unfortunately, this would not provide a tight lower bound on the overall communication complexity of $\mathcal{A}$: while the worst-case execution of $\mathcal{A}$ may include the worst-case scenario of some round r action model AM_r, it does not necessarily include the worst-case scenario of action model $AM_{r'}$ in round r'. Fortunately, however, Definition 4 provides a way to alleviate this problem: By computing the composition of the action models of rounds $1, 2, \ldots, k$, where k is the round in which $\mathcal{A}$ has terminated, we get a *single* action model for which we can compute the lower bound using the above method.

We conclude this section by stressing the fact that the worst-case cost $D^a(AM)$ given by Definition 12 is tied to the communication complexity for applying a given action model AM, i.e., of an algorithm $\mathcal{A}$ that *faithfully* implements a given AM. Obviously, this is not equivalent to the communication complexity for solving a specific problem, since the lower bounds for $D^a(AM)$ established in Lemma 1 and Lemma 2 are tied to a specific action model. In the following section, we will address the communication complexity of a given problem $\mathcal{P}$, by considering action models that are optimal for $\mathcal{P}$.

5 Action Models and Protocol Trees

In this section, we will restrict our attention to distributed function computation in synchronous systems of 2 processes. Clearly, all actions correspond to messages sent by one of the two processes here, and each action can be distinguished from any other action by both processes at the end of a round.

5.1 Action Models of Protocol Trees

As process a can only send some information x to process b if it knows that x is valid, an action corresponding to this sending process has to have a precondition containing $K_a x$. Consequently, even though process a can distinguish the action s "a sends x to b" in round r from the action t "a sends $\neg x$ to b" in round r, the application of one of those actions does not change a's view on the facts x and $\neg x$ in the resulting epistemic state compared to the original one, as a already knew x resp. $\neg x$.[3] On the other hand, since b can distinguish the actions s and

---

[3] To be precise, this is only true if x is a *preserved formula* (as introduced in [8]), which requires x to be propositional or positive knowledge (but not $x = \neg K_a \phi$, for example). Thus we will also restrict ourselves to algorithms in which preconditions of actions only involve preserved formulas, which is essentially a non-restriction for distributed algorithms.

t, b learns x resp. $\neg x$, which eliminates edges in $\sim_b$ in the resulting epistemic model, leading to a partitioning between the states where $K_b x$ and $K_b \neg x$.

Note, however, that the fact that all actions are distinguishable for every process is only valid because we have just two processes: In a system with e.g. three processes, it would be possible that p_0 doing actions s resp. t is sending 0 resp. 1 to process p_1 but nothing to process p_2, thus p_2 cannot distinguish actions s and t. Nevertheless, even in a system of n processes, the terminal epistemic model, in which the n processes all know the result of f, must be partitioned into several partitions that are separated for *all* processes: Each such partition consists of (potentially multiple) epistemic states in which the result of f must be the same. Otherwise, the result of f would not be common knowledge.

Since all the processes have the same initial knowledge (except for their own input value), the initial epistemic model $M = (S, \sim, V)$ is not partitioned, but rather a hypercube like in Fig. 2. The terminal epistemic model is the result of applying the composed action model (for all rounds) to the initial epistemic model. Thus, the required partitioning of the terminal epistemic model can only result from some partitioning of the composed action model.

Definition 13. *An algorithm $\mathcal{A}$ is defined by a set of action models $\{AM_1, \ldots, AM_k\}$, such that a single action model AM_i is applicable in round i of the synchronous execution. The action model AM_i partitions into $t_{AM_i} \geq 1$ disjoint partitions (identically for both processes p_0, p_1).*

Definition 14. *The composed action model of the first k rounds (CAM_k) is the composition of the action models $AM_1, \ldots, AM_k$, inductively defined as $CAM_1 = AM_1$ and $CAM_k = (CAM_{k-1}; AM_k)$. Every $CAM_k = (S_{CAM_k}, \sim_{CAM_k}, pre_{CAM_k})$ partitions into t_{CAM_k} disjoint partitions, where $P_{k,i}$ denotes the i-th partition of CAM_k, consisting of actions $S_{k,i} \subseteq S_{CAM_k}$.*

Clearly, if $\mathcal{A}$ computes f in m rounds, the relevant composed action model is CAM_m. Observe that applying the actions in $P_{k,i}$ to the initial epistemic model $M = (S, \sim, V)$ leads to a set of partitions of the epistemic model $M' = M \otimes CAM_k$, as required.

We can now define the protocol tree corresponding to the action model for algorithm $\mathcal{A}$:

Definition 15. *The protocol tree $\mathcal{T}_\mathcal{A} = (P, E)$ of an algorithm $\mathcal{A}$, starting at the root vertex v that represents the initial epistemic model $M = (S, \sim, V)$, is defined as:*

$P = \{v\} \cup \{P_{k,i} \mid P_{k,i} \text{ for some } i \text{ is a partition of } CAM_k, k \in \{1, m\}\}$
$E = \{(v, P_{1,j}) \mid P_{1,j} \text{ a partition of } CAM_1\} \cup \{(P_{k,i}, P_{k+1,j}) \mid \exists s \in S_{k,i}, t \in S_{AM_{k+1}} : (s, t) \in S_{k+1,j} \text{ for some } i \text{ and } j, \text{ and } k \in \{1, m-1\}\}.$

Informally, Definition 15 states that each partition of each composed action model CAM_k is a node in the protocol tree. All the nodes corresponding to the partitions of CAM_1 are connected to the root node v. There is a connection between two nodes $P_{k,i}$ and $P_{k+1,j}$ on levels k and $k+1$ if and only if there is an action

$s \in S_{k,i}$ which is a prefix of an action (s, t) of $S_{k+1,j}$, with $t \in S_{AM_{k+1}}$ an action of AM_{k+1}. The following Lemma 3 shows that $\mathcal{T}_{\mathcal{A}}$ is indeed a tree.

Lemma 3. $\mathcal{T}_{\mathcal{A}}$ *is a tree.*

Proof. For space reasons, we refer the reader to the full version [16]. $\square$

So far, we only considered the protocol tree $\mathcal{T}_{\mathcal{A}}$, which is solely defined in terms of the action models CAM_k. Now we turn our attention to the application of $\mathcal{T}_{\mathcal{A}}$ to the initial epistemic model $M = (S, \sim, V)$ that is a hypercube. As already said, this must induce a partitioning of the resulting epistemic model, i.e., the leaves in $\mathcal{T}_{\mathcal{A}}$, in order to correctly compute $f(x, y)$ at both processes. The following Lemma 4 shows that the CAM_m and the resulting $\mathcal{T}_{\mathcal{A}}$ of a correct solution must induce rectangles at the leafs of $\mathcal{T}_{\mathcal{A}}$.

Lemma 4. *Let $M = (S, \sim, V)$ be the hypercube describing the initial epistemic model of a solution algorithm for computing $f(x, y)$, defined by the action models $AM_1, \ldots, AM_m$ (resulting in the composed action model CAM_m) and the corresponding protocol tree $\mathcal{T}_{\mathcal{A}}$. Then, every rectangle corresponds to at least one partition in the final epistemic model $M' = M \otimes CAM_m$, i.e., at least one leaf, and every leaf corresponds to some (not necessarily maximal) rectangle.*

Proof. First, as $\mathcal{A}$ must compute $f(x, y)$ for every input (x, y), and $\mathcal{A}$ terminates only in leaves of $\mathcal{T}_{\mathcal{A}}$, every (x, y) leads to some leaf. Consequently, for every rectangle $\mathcal{R}$, which usually contains more than one input, say (x_1, y_1) and (x_2, y_2), we can assign the set of leafs $L_{\mathcal{R}}$ its constituent inputs lead to.

We now show that actually $|L_{\mathcal{R}}| = 1$, which implies that every leaf corresponds to some rectangle. Suppose that both inputs (x_1, y_1) and (x_2, y_2) allow the application of actions leading to the node ℓ of $\mathcal{T}_{\mathcal{A}}$, then also (x_1, y_2) and (x_2, y_1) lead to ℓ: The path through the tree has to be the same for all of the four input pairs. We start our inductive argument at level $k = 0$, the initial epistemic model. In the initial epistemic model, p_0 cannot distinguish the situation with input (x_1, y_1) from (x_1, y_2) resp. (x_2, y_1) from (x_2, y_2). A similar argument holds for p_1. Since (x_1, y_1) and (x_2, y_2) lead to the same node ℓ, the actions of AM_1 have to be in the same partition for both of them and since p_0 cannot distinguish (x_1, y_1) from (x_1, y_2), the action applied by p_0 has to be the same in both cases (similarly for (x_2, y_2) and (x_2, y_1)). Since p_1 cannot distinguish (x_1, y_1) from (x_2, y_1), the action applied by p_1 has to be the same in both cases (similarly for (x_2, y_2) and (x_1, y_2)). As p_0's action is the same for (x_1, y_1) and (x_1, y_2) and p_1's action is the same for (x_1, y_2) and (x_2, y_2), and the actions of AM_1 have to be in the same partition for (x_1, y_1) and (x_2, y_2), also the action for (x_1, y_2) has to be in the same partition in AM_1. By the analogous argument, it follows that also the action for (x_2, y_1) in AM_1 is in the very same partition of $AM_1 = CAM_1$.

For the induction step, assume that the execution of $\mathcal{A}$ for (x_1, y_1) resp. (x_2, y_2) reached some node $P_{k,i}$ on level k of $\mathcal{T}_{\mathcal{A}}$. By the induction hypothesis, also the executions for (x_1, y_2) and (x_2, y_1) have reached this node. Due to the initial premise of reaching the same leaf ℓ, the executions for (x_1, y_1) and (x_2, y_2) must reach some common node $P_{k+1,j}$ corresponding to a partition in $\mathtt{CAM}_{k+1} = (\mathtt{CAM}_k; \mathtt{AM}_{k+1})$. As already stated before, the epistemic model after round $k + 1$ can be derived in two ways: Applying action model by action model or once applying $\mathtt{CAM}_{k+1}$ on the initial epistemic model: the resulting epistemic models are equivalent. Thus, by the same argument as before (only using $\mathtt{AM}_{k+1}$ instead of $\mathtt{AM}_1$), it follows that all the actions on the inputs have to be in the same partition in $\mathtt{AM}_{k+1}$ and hence in $\mathtt{CAM}_{k+1}$. Consequently, all the inputs lead to the same node on level $k + 1$ as asserted.

5.2 Communication Complexity Lower Bounds

In this section, we will prove a lower bound on the number of bits received by the processes during the worst-case execution of a given algorithm $\mathcal{A}$ for computing a function $f(x, y)$, using the action model representation of Sect. 5.1. In the following, $\mathcal{D}^i = \mathcal{D}^i(\mathcal{A})$ denotes the maximum number of bits received by p_i in any execution of $\mathcal{A}$ for computing f, and $\mathcal{D} = \mathcal{D}(\mathcal{A})$ is the maximum total number of bits received system-wide.

The following Theorem 1 finally establishes a lower bound on $\mathcal{D}$ for a given algorithm $\mathcal{A}$ for computing f. It relies on a lower bound on the number of partitions $t_{\mathtt{CAM}_m}$ of the composed action model $\mathtt{CAM}_m$ for $\mathcal{A}$, and can hence be viewed as an action model analogon for Corollary 1.

Theorem 1. *The maximum total number of bits $\mathcal{D}$ received by the processes in any execution of any algorithm $\mathcal{A}$ that computes $f(x, y)$ in m rounds has the lower bound $\log_2 t_{\mathtt{CAM}_m} \leq \mathcal{D}$, where $t_{\mathtt{CAM}_m}$ is the number of partitions of the composed action model of $\mathcal{A}$ after m rounds. It satisfies $t_{\mathtt{CAM}_m} \geq t$, where t is the number of monochromatic rectangles of $f(x, y)$.*

Proof. Suppose $t_{\mathtt{CAM}_m} < t$, i.e., there are less leaves in $\mathcal{T}_\mathcal{A}$ than there are monochromatic rectangles of $f(x, y)$. Then, there are two rectangles $\mathcal{R}_1, \mathcal{R}_2$ that lead to the same leaf. However, this contradicts Lemma 4, as every leaf corresponds to a single rectangle.

By Lemma 2, a lower bound for the worst-case cost $\mathcal{D}^a = \mathcal{D}^a(\mathtt{CAM}_m)$ regarding process a is $\log_2 t_{\mathtt{CAM}_m}^a$, where $t_{\mathtt{CAM}_m}^a$ is the number of partitions of $\mathtt{CAM}_m$ regarding a. Additionally, every process has to be able to distinguish all the partitions of $\mathtt{CAM}_m$, else the result of $f(x, y)$ would not be common knowledge. Thus, $t_{\mathtt{CAM}_m}^a = t_{\mathtt{CAM}_m}$ and hence $\log_2 t_{\mathtt{CAM}_m} \leq \mathcal{D}^a$. Since trivially $\mathcal{D}^a \leq \mathcal{D}$, we can conclude that $\log_2 t \leq \log_2 t_{\mathtt{CAM}_m} \leq \mathcal{D}$.

5.3 Application Examples

We now demonstrate how to apply our approach by means of two simple examples. We first consider distributed function computation, using the function $f(x, y)$ and the protocol $\mathcal{A}$ given in Fig. 1, where the processes send a single bit in each round alternatingly. Recall that $\mathcal{A}$ is optimal in terms of communication complexity.

The corresponding action models, for the 3 rounds of algorithm $\mathcal{A}$, are given in Fig. 4. An action of the form e.g. (x_0, x_1) encodes that p_0 sends the information that its input value is either x_0 or x_1 to p_1. An expression like $K_i x_0$ in a precondition formula means that, in the appropriate epistemic state, p_i knows that x_0 is the input to p_0. Self-loops in the indistinguishability relation of the action models are denoted by *loops*, in the following.

AM_1: $\quad S_{AM_1} = \{(x_0), (x_1, x_2, x_3)\}$ $\sim_{p_i} = loops$ $\mathtt{pre}(x_0) = K_0 x_0$ $\mathtt{pre}((x_1, x_2, x_3)) = K_0(x_1 \vee x_2 \vee x_3)$	$(x_0) \qquad (x_1, x_2, x_3)$

AM_2: $\quad S_{AM_2} = \{(y_0, y_1, y_3), y_2, (y_0, y_1), (y_2, y_3)\}$ $\sim_{p_i} = loops$ $\mathtt{pre}((y_0, y_1, y_3)) = K_1(x_0 \wedge (y_0 \vee y_1 \vee y_3))$ $\mathtt{pre}(y_2) = K_1(x_0 \wedge y_2)$ $\mathtt{pre}((y_0, y_1)) = K_1((x_1 \vee x_2 \vee x_3) \wedge (y_0 \vee y_1))$ $\mathtt{pre}((y_2, y_3)) = K_1(x_1 \vee x_2 \vee x_3) \wedge (y_2 \vee y_3))$	$(y_0, y_1, y_3) \qquad (y_2)$ $(y_0, y_1) \qquad (y_2, y_3)$

AM_3: $\quad S_{AM_2} = \{x_3, (x_1, x_2)\}$ $\sim_{p_0} = loops$ $\mathtt{pre}(x_3) = K_0(x_3 \wedge (y_2 \vee y_3))$ $\mathtt{pre}((x_1, x_2)) = K_0((x_1 \vee x_2) \wedge (y_2 \vee y_3))$	$(x_3) \qquad (x_1, x_2)$

The resulting composed action model is CAM_3:

$S_{CAM_3} = \{s_0 = (x_0, (y_0, y_1, y_3)), s_1 = (x_0, y_2),$ $s_2 = ((x_1, x_2, x_3), (y_0, y_1)),$ $s_3 = (((x_1, x_2, x_3), (y_2, y_3)), x_3),$ $s_4 = (((x_1, x_2, x_3), (y_2, y_3)), (x_1, x_2))\}$ $\sim_{p_i} = loops$ $\mathtt{pre}((x_0, (y_0, y_1, y_2))) = K_0 x_0 \wedge K_1(y_0 \vee y_1 \vee y_3)$ $\mathtt{pre}((x_0, y_2)) = K_0 x_0 \wedge K_1 y_2$ $\mathtt{pre}((x_1, x_2, x_3), (y_0, y_1)) = K_0(x_1 \vee x_2 \vee x_3) \wedge K_1(y_0 \vee y_1)$ $\mathtt{pre}(((x_1, x_2, x_3), (y_2, y_3)), x_3) = K_0 x_3 \wedge K_1(y_2 \vee y_3)$ $\mathtt{pre}(((x_1, x_2, x_3), (y_2, y_3)), (x_1, x_2)) = K_0(x_1 \vee x_2) \wedge K_1(y_2 \vee y_3)$	$s_0 \qquad s_1$ $s_2 \qquad s_3$ s_4

Fig. 4. Precondition functions (left) and action models (right) for the optimal algorithm $\mathcal{A}$ for f, given in Fig. 1.

The corresponding protocol tree $\mathcal{T}_A$ and the rectangles corresponding to $\mathcal{T}_A$ are depicted in Fig. 5. It is apparent that there are 5 completely separated partitions in CAM_3 corresponding to 5 leaves in the protocol tree $\mathcal{T}_A$. Theorem 1 thus reveals that $\mathcal{D} \geq \log_2(5)$. Alternatively, since $\mathcal{A}$ follows the original Yao protocol, we can also directly apply Corollary 1. It confirms that $\mathcal{A}$ has to communicate at least $\log_2(5) \leq 3$ bits to compute f. And indeed, in the corresponding protocol tree (of height 3), there are paths where 1 bit is sent/received in each of the 3 rounds.

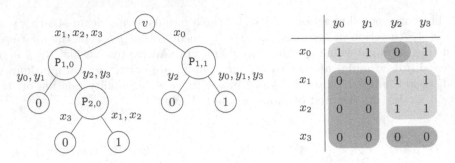

Fig. 5. The protocol tree $\mathcal{T}_A$ for function f defined in Fig. 1.

As our second application, we sketch how to use the approach developed in the previous sections to obtain lower bounds for the communication complexity $\mathcal{D}$ for distributed consensus in directed dynamic networks controlled by a message adversary.

In the *consensus* problem, each process p has an initial value x_p and a decision value y_p in its local state. The value y_p is written only once, and is undefined ($y_p = \perp$) initially. To solve consensus in our model, where processes cannot fail but communication is unreliable, an algorithm has to fulfill the following properties for each process $p, q \in \Pi$:

(Agreement) If p assigns value v_p to y_p and q assigns value v_q to y_q, then $v_p = v_q$.

(Termination) Eventually, every p assigns a value to y_p.

(Validity) If each process p has input $x_p = v$, then all processes q have to decide $y_q = v$.

We restrict our attention to the very simple directed dynamic network made up of two synchronous processes p_0, p_1. The communication graph $\mathcal{G}^r$ of each round r is controlled by an omniscient *message adversary* (MA) here [1,5], which determines which messages are delivered resp. lost in round r: it chooses a sequence $\sigma = \mathcal{G}^1, \mathcal{G}^2, \ldots$ of graphs $\mathcal{G}^r \in \{\leftarrow, \leftrightarrow, \rightarrow\}$ for any round r. We will focus on the message adversary $MA_{\leftrightarrow^2}$ here, which may generate all graph sequences not starting with $\mathcal{G}^1 = \mathcal{G}^2 = \leftrightarrow$. There is a simple algorithm $\mathcal{A}$ solving consensus under $MA_{\leftrightarrow^2}$.

Lacking space forced us to relegate the detailed modeling and analysis to [16]. In a nutshell, we pursued two different approaches there:

(1) One can consider consensus as the distributed computation of a function $f(x, y, \sigma)$, where the result depends on the inputs of p_0 (x) and p_1 (y) and on the particular graph sequence σ chosen by the MA.

(2) In order to directly apply our approach, we had to address the problem that consensus does not specify a unique function: Validity only fixes the outcome for all inputs being the same, but not in the remaining cases. Agreement, on the other hand, only requires the outputs at p_0 and p_1 to be the same. Consequently, the actual result of $f(x, y, \sigma)$ depends on x, y and σ, but also on the choices made by the algorithm $\mathcal{A}$. We solved this problem by partitioning the function $f(x, y, \sigma)$ into multiple functions $f_i(x, y)$, which can be treated independently.

Since every $f_i(x, y)$ led to the (trivial) lower bound $\mathcal{D} \geq 1$, we obtained the same (trivial) lower bound $\mathcal{D} \geq 1$ as in (1).

6   Conclusions

We established a relation between the number of partitions in the composed action model of a synchronous distributed algorithm $\mathcal{A}$ and the number of bits received by some process in a worst case execution. For the restricted case of deterministic distributed function computation among 2 processes, we also provided a lower bound for the total communication complexity of any correct solution algorithm. We provided two simple applications of our approach, which confirmed an already known communication complexity lower bound for distributed function computation and even reached out to consensus in directed dynamic networks under a message adversary. Part of our current work is devoted to the shortcomings of our current approach, most notably, the restriction to 2 processes.

References

1. Afek, Y., Gafni, E.: Asynchrony from synchrony. In: Frey, D., Raynal, M., Sarkar, S., Shyamasundar, R., Sinha, P. (eds.) Distributed Computing and Networking. Lecture Notes in Computer Science, vol. 7730, pp. 225–239. Springer, Heidelberg (2013). https://doi.org/10.1007/978-3-642-35668-1_16
2. Alechina, N., Logan, B., Nguyen, H.N., Rakib, A.: Verifying time, memory and communication bounds in systems of reasoning agents. Synthese **169**(2), 385–403 (2009)
3. Ben-Zvi, I., Moses, Y.: Beyond lamport's *Happened-Before*: on the role of time bounds in synchronous systems. In: Lynch, N.A., Shvartsman, A.A. (eds.) DISC 2010. LNCS, vol. 6343, pp. 421–436. Springer, Heidelberg (2010). https://doi.org/10.1007/978-3-642-15763-9_42
4. Ben-Zvi, I., Moses, Y.: Beyond Lamport's happened-before: On time bounds and the ordering of events in distributed systems. J. ACM **61**(2), 13:1–13:26 (2014). https://doi.org/10.1145/2542181
5. Biely, M., Robinson, P., Schmid, U., Schwarz, M., Winkler, K.: Gracefully degrading consensus and k-set agreement in directed dynamic networks. Theoretical Computer Science **726**, 41–77 (2018). https://doi.org/10.1016/j.tcs.2018.02.019, http://www.sciencedirect.com/science/article/pii/S0304397518301166
6. Cyriac, A., Krishnan, K.M.: Lower bound for the communication complexity of the russian cards problem. CoRR arXiv:abs/0805.1974 (2008)
7. Dinitz, Y., Moran, S., Rajsbaum, S.: Bit complexity of breaking and achieving symmetry in chains and rings. J. ACM **55**(1), 3:1–3:28 (2008). https://doi.org/10.1145/1326554.1326557
8. van Ditmarsch, H., van der Hoek, W., Kooi, B.: Dynamic Epistemic Logic. Springer, Netherlands (2008). https://doi.org/10.1007/978-1-4020-5839-4
9. Fagin, R., Moses, Y., Halpern, J., Vardi, M.: Reasoning About Knowledge. MIT Press, Cambridge (2003). https://books.google.at/books?id=xHmlRamoszMC

10. Gerbrandy, J.D.: Dynamic epistemic logic. Institute for Logic, Language and Computation (ILLC), University of Amsterdam (1997)
11. Halpern, J.Y., Moses, Y.: Knowledge and common knowledge in a distributed environment. J. ACM **37**(3), 549–587 (1990). https://doi.org/10.1145/79147.79161
12. Hintikka, J.: Knowledge and belief: an introduction to the logic of the two notions. Contemporary philosophy, Cornell University Press (1962). https://books.google.de/books?id=N28OAAAAIAAJ
13. Holzer, S., Wattenhofer, R.: Optimal distributed all pairs shortest paths and applications. In: Proceedings of the 2012 ACM Symposium on Principles of Distributed Computing, pp. 355–364. PODC 2012. ACM, New York (2012). https://doi.org/10.1145/2332432.2332504
14. Kushilevitz, E., Nisan, N.: Communication Complexity. Cambridge University Press (1997). https://books.google.at/books?id=yiV6pwAACAAJ
15. Moses, Y., Dolev, D., Halpern, J.Y.: Cheating husbands and other stories: a case study of knowledge, action, and communication. Distrib. Comput. **1**(3), 167–176 (1986). https://doi.org/10.1007/BF01661170
16. Pfleger, D., Schmid, U.: On knowledge and communication complexity in distributed systems. Technical report TUW-269752, Technische Universität Wien, Institute of Computer Engineering (2018). http://publik.tuwien.ac.at/files/publik_269752.pdf
17. Plaza, J.: Logics of public communications. Synthese **158**(2), 165–179 (2007). https://doi.org/10.1007/s11229-007-9168-7
18. Yao, A.C.: Some complexity questions related to distributive computing (preliminary report). In: Proceedings of the 11h Annual ACM Symposium on Theory of Computing, 30 April–2 May 1979, Atlanta, Georgia, USA, pp. 209–213 (1979). https://doi.org/10.1145/800135.804414

Connectivity and Minimum Cut Approximation in the Broadcast Congested Clique

Tomasz Jurdziński$^{(\boxtimes)}$ and Krzysztof Nowicki

Institute of Computer Science, University of Wrocław, Wrocław, Poland
{tju,knowicki}@cs.uni.wroc.pl

Abstract. In this paper we present two graph algorithms in the Broadcast Congested Clique model. In this model, there are n players, which communicate in synchronous rounds. Each player represents a single node of the input graph; initially each player knows the set of edges incident to his node. In each round of communication each node can broadcast a single b–bit message to all other nodes; usually $b \in \mathcal{O}(\log n)$. The goal is to compute some function of the input graph.

The first result we present is the first sub-logarithmic determinis-tic algorithm finding a maximal spanning forest of an n node graph in the Broadcast Congested Clique, which requires only $\mathcal{O}(\log n / \log \log n)$ rounds. The second result is a randomized $1 + \epsilon$ approximation algorithm finding the minimum cut of an n node graph, which requires only $\mathcal{O}(\log n)$ maximal spanning forest computations. In the Broadcast Congested Clique this approach, combined with the new maximal spanning forest algorithm, yields an $\mathcal{O}(\log^2 n / \log \log n)$ round algorithm. Additionally, it may be applied to different models, i.e. in the multi-pass semi-streaming model it allows to reduce required memory by $\Theta(\log n)$ factor, with only $\mathcal{O}(\log^* n)$ passes over the data stream.

1 Introduction

In this paper we study graph connectivity related problems in the Broadcast Congested Clique model. In this model there are n players, which communicate in synchronous rounds. Each player represents a single node of the input graph; initially each player knows the set of edges incident to his node. In each round of communication each node can broadcast a single b–bit message to all other nodes, usually $b \in \mathcal{O}(\log n)$. The main complexity measurement in this model is the number of rounds.

We propose the first algorithm in the Broadcast Congested Clique model, which requires only a sublogarithmic number of rounds, which is an improvement

This work was supported by the National Science Centre, Poland grant 2017/25/B/ST6/02010. Results from Sect. 3 were presented as Brief Announcement at DISC 2018 [8] and their initial version was obtained with support from the National Science Centre, Poland grant DEC-2012/07/B/ST6/01534.

Z. Lotker and B. Patt-Shamir (Eds.): SIROCCO 2018, LNCS 11085, pp. 331–344, 2018.
https://doi.org/10.1007/978-3-030-01325-7_28

over an $\mathcal{O}(\log n)$ round implementation of Borůvka's algorithm, which so far was the best known algorithm (our algorithm has been presented as a brief announcement at DISC 2017 [8]). Moreover, we study the minimum cut problem, which may be considered as a generalization of the graph connectivity problem, as finding the size of a minimum cut of G is equivalent to finding a maximal value c, such that G is c–connected. We propose an algorithm, which finds an $(1 + \epsilon)$ approximation of the minimum cut using only $\mathcal{O}(\log n)$ spanning forest computations.

Moreover, this approach can be applied to the Multi–Pass Semi–Streaming model. This model of computation is a version of the Semi–Streaming model [5], in which we allow multiple passes over the data stream in order to determine the output. In this model our approach yields an algorithm, which uses memory smaller by a $\Theta(\log n)$ factor, than a single pass semi streaming algorithm, but requires $\mathcal{O}(\log^* n)$ passes over the data stream. Moreover, if we allow the improvement factor to be $\Theta(\log n / \log^{(k)} n)$, then we require only $\mathcal{O}(k)$ passes over the data stream.

1.1   Congested Clique: Broadcast and Unicast

There are two version of the Congested Clique model which are usually considered: Broadcast and Unicast, however, for the latter we usually skip the Unicast part, and call it just Congested Clique. For both models, the main complexity measurement is the number of rounds. Both models differ only by the fact that in the Unicast Congested Clique each player is allowed to send different messages to different players. This difference has large impact on the complexity in both models – in particular there are no non–trivial lower bounds in the Congested Clique model, where in the Broadcast Congested Clique we can show some quite strong lower bounds [3, 4, 11].

The other relevant difference is in the complexity of the connected components problem in both models. The very first paper [12] regarding the Congested Clique model established that it is possible to solve the MST problem (which implies solving the connected components problem) deterministically in $\mathcal{O}(\log \log n)$ rounds. There are no known $o(\log \log n)$ deterministic algorithms solving this problem. On the other hand, quite recently, there was a sequence of randomized algorithms for the MST problem. The first proved that if we allow randomization, we can actually be faster and solve the problem in $\mathcal{O}(\log \log \log n)$ rounds [7]. It was improved to an $\mathcal{O}(\log^* n)$ round algorithm in [6], and in [9] the complexity of MST was established to be $\mathcal{O}(1)$ rounds.

In the Broadcast Congested Clique model, with a size of a message $\mathcal{O}(\log n)$ we had the opposite situation – not a single improvement over Borůvka's algorithm [15] was known. However, if we allow message size to be larger, then it is known that one can

- compute a maximal spanning forest deterministically in $O(\epsilon^{-1})$ rounds, for $b \in \mathcal{O}(n^\epsilon \log n)$

– compute a maximal spanning forest using public randomness in a single round, for $b \in \mathcal{O}(\log^3 n)$ [1][1]

For the min-cut problem in the Broadcast Congested Clique model we could also apply the result from the streaming model [2], but a straightforward implementation requires $\mathcal{O}(\log^2 n)$ spanning forest computations, which with Borůvka's algorithm as subroutine yields an $\mathcal{O}(\log^3 n)$ round algorithm. This implies that our results give an improvement by a $\Theta(\log n \log \log n)$ factor in the number of rounds.

1.2 Problems

In this paper we study graph connectivity related problems. More precisely, we provide an algorithm finding a maximal spanning forest for a given graph. It may be considered as a generalization of the graph connectivity problem, which is one of the fundamental graph problems. Moreover, procedure finding a maximal spanning forest is used as a subroutine in many more complex algorithms.

The second problem we study, is the minimum cut (min-cut) problem, in which we have to find a division of the set of nodes V into two parts V_1, V_2, such that the set of edges between V_1 and V_2 is as small as possible. It is equivalent to asking about the smallest set of edges such that removing them makes a graph disconnected. In this sense, it may be considered as a generalization of the connectivity problem, as we say that a graph is c–connected if its min-cut is not smaller than c.

2   Graph Terminology

Given a partition $\mathcal{C}$ of a graph $G(V, E)$ into connected components and $v \in V$, C^v denotes the component containing v. We define $\deg_{\mathcal{C}}(v)$ for a vertex v wrt a partition $\mathcal{C}$ as the number of components connected with v, i.e., $\deg_{\mathcal{C}}(v) = |N_{\mathcal{C}}(v)|$, where $N_{\mathcal{C}}(v) = \{C \in \mathcal{C} \mid \exists u \in C$ such that $(v, u) \in E$ and $C \neq C^v\}$. For a component $C \in \mathcal{C}$, we define $\deg_{\mathcal{C}}(C) = \max_{v \in C} \{\deg_{\mathcal{C}}(v)\}$. Note that, according to this definition, the degree of a component C might be smaller than the actual number of components containing nodes connected by an edge with nodes from C. Our definition of degree is adjusted to make it possible that degrees of components can be determined in $O(1)$ rounds. Given a partition $\mathcal{C}$ of the graph into components, we define the linear ordering $\succ$ of components, where $C \succ C'$ iff $\deg_{\mathcal{C}}(C) > \deg_{\mathcal{C}}(C')$ or $\deg_{\mathcal{C}}(C) = \deg_{\mathcal{C}}(C')$ and $\mathrm{ID}(C) > \mathrm{ID}(C')$. A component C is a *local maximum* if all its neighbors are smaller wrt the $\succ$ ordering.

By $G(p)$ we denote a random subgraph of G such that each edge is sampled to $G(p)$ with probability p. By *size of a cut* V_1, V_2 we understand the number of edges between nodes in the sets V_1 and V_2. Then, the *expected size of a cut* V_1, V_2 in $G(p)$ is the expected number of edges between V_1, V_2 in $G(p)$.

---

[1] In this paper authors present an algorithm for semi–streaming model, however the result applies to the Broadcast Congested Clique.

3   Spanning Forest in the Broadcast Congested Clique

First, we recall a distributed implementation of the well known Borůvka's algorithm for MST. Then, we design a new algorithm for connectivity which (unexpectedly?) shows that the $\log n$ bound on round complexity can be broken in the broadcast congested clique.

3.1   Minimum Spanning Forest in the Broadcast Congested Clique

The minimum spanning forest can be computed using a distributed version of the classical Borůvka's algorithm. The algorithm works in *phases*. At the beginning of phase i a partition $\mathcal{F}$ into fragments of size $\geq 2^i$ is given. During the phase i new fragments of size $\geq 2^{i+1}$ are determined, based on the lightest edges incident to all fragments.

In the distributed implementation of the Borůvka's algorithm each node knows the set of fragments at the beginning of a phase. During the phase each node v announces (broadcasts) the lightest edge connecting v with a node $u \notin F^v$. Using those edges, each node can individually (locally) perform the next phase of the Borůvka's algorithm and determine new (larger) fragments.

Theorem 1. *Borůvka's algorithm can be implemented in* Broadcast Congested Clique, *so that it requires* $O(\log n)$ *rounds.*

3.2   Connected Components Algorithm

To compute the connected components we could use the standard Borůvka's algorithm as well. However, we are not forced to select the lightest edge incident to each component. Our general idea is to prefer those edges which connect nodes to the components of large degree. And the intended result of a phase should be that each component either has a small degree or it is connected to some "host" of large degree (directly or by a path of length larger than one). As the number of such "hosts" will be relatively small, we obtain a significant reduction of the number of components of large degree in each phase. Moreover, we separately deal with components of small degree by allowing them to broadcast all their neighbours at the final stage of the algorithm. In particular, we show that using this approach, we can prove the following.

Theorem 2. *For a given graph, it is possible to find its maximal spanning forest in the* Broadcast Congested Clique *in* $O\left(\frac{\log n}{\log \log n}\right)$ *rounds.*

The remaining part of this section is dedicated to prove Theorem 2.

Algorithm. Our algorithm consists of the main part and the *playoff*. The main part is split into *phases*. At the beginning of phase 1 each node is *active* and it forms a separate component. The algorithm is parametrized by a natural number s which (intuitively) sets the threshold between components of small degree (smaller than s) and large degree (at least s).

During an execution of the algorithm, nodes from non growable components and components of small degree (smaller than s) are *deactivated*.

At the beginning of a phase, a partition of the graph of active nodes is known to the whole network.

First, each node v determines $N(v)$ and announces its degree $\deg(v)$ wrt the current partition of the set of active nodes into components (Round 1). With this information, each node v knows the ordering of components of the graph of active nodes according to $\succ$. Then, each *active* node v (except of members of local maxima) broadcasts its incident edge to the largest active component from $N(v)$ according to $\succ$ relation (Round 2).

Next, each node v of each local maximum C checks whether edges connecting C to all components containing neighbors of v (i.e., to components from $N(v)$) have been already broadcasted. If it is not the case, an edge connecting v to a new component C' (i.e., to such C' that no edge connecting C and C' was known before) is broadcasted by v (Round 3).

Based on broadcasted edges, new components are determined and their degrees are computed (Round 4). Each new component with degree smaller than s is *deactivated* at the end of a phase.

The playoff lasts s rounds in which each node v of each deactivated component broadcasts edges going to all components connected to v (there are at most s such components for each deactivated node).

More precise description of this strategy is presented as Algorithm 1. The key property for an analysis of complexity of our algorithm is that each active component C of large degree is either connected during a phase to all its neighbors or to a component which is larger than C according to $\succ$.

Theorem 3. *Algorithm 1 solves the spanning forest problem in $O(s + \log_s n)$ rounds for an n-node graph.*

Proof. First, consider round complexity of the algorithm. It is clear that Playoff has s rounds. To show the claimed complexity we show that the number of active components is decreased at least s times in each phase. An intuition is that all components join with (some) local maxima and thus each local maximum of large degree "combines" at least s components in a new, larger component. However, the situation is not that simple, as there might be many local maxima.

In order to formalize the intuition, consider a directed graph G_{phase} of components active at the beginning of a phase, where (C_1, C_2) is an edge in G_{phase} **iff** a node from C_1 broadcasts an edge connecting it with C_2 in step 6 of the phase (edge of type 1) **or** C_1 is a local maximum, a node from C_1 broadcasts an edge connecting it with some C' in step 12, while a node from C' broadcasts an edge connecting it with C_2 in step 6 (we call it edge of type 2).

The algorithm guarantees that: (a) G_{phase} is acyclic. Indeed, each edge (C_1, C_2) resulted from broadcasts in step 6 satisfies $C_1 \prec C_2$. Moreover, an edge is broadcasted from C_1 to C' in step 12 iff all nodes from C' broadcasted connections to components larger than C_1 wrt $\succ$ ordering.

(b) Each connected component C (i.e., each node of G_{phase}) is either a sink of G_{phase} connected with (at least) $\deg(C)$ nodes in G_{phase} or has out-degree at

---

Algorithm 1. BroadcastCC(v, s)   ▷ s is the threshold between small/large degree

---

1: **while** there are active components **do**   ▷ execution at a node v
2:     **Round 1:** v broadcasts $\deg(v)$
3:     **if** $\deg(v) > 0$ **then**
4:         $C_{max}(v) \leftarrow$ the largest element of $N(v)$ wrt the ordering $\succ$
5:         **Round 2:**
6:         **if** C^v is not a local maximum **then** v broadcast an edge (u, v) such that $u \in C_{max}(v)$
7:             **Round 3:**
8:             **if** C^v is a local maximum **then**
9:                 $N_{lost}(v) \leftarrow \{C \mid C \in N(v)$ and no edge connecting C and C^v was broadcasted$\}$
10:                **if** $N_{lost}(v) \neq \emptyset$ **then**
11:                    $u \leftarrow$ a neighbor of v such that $u \in C$ for some $C \in N_{lost}(v)$
12:                    v broadcasts an edge (u, v)
13:                **end if**
14:            **end if**
15:     **end if**
16:     v computes the new partition into components, using broadcasted edges
17:     **Round 4:** v broadcasts $\deg(v)$ ▷ degrees wrt the new components!
18:     **if** $\deg(C^v) < s$ **then** deactivate v
19: **end while**
20: **Playoff (s rounds):** deactivated nodes broadcast edges to neighboring components.

---

least one. This property follows from the fact that only nodes of local maxima are candidates for sinks, as only they do not broadcast in step 6. Moreover, assume that C is a local maximum and there is a neighbor C' of C whose nodes have not broadcasted connections with C in step 6. Then a node(s) from C broadcast in step 12 which implies that out-degree of C is at least one.

(c) Each connected component of a partition obtained at the end of a phase contains at least one sink of G_{phase}.

If one ignores that edges of G_{phase} are directed then certainly new components at the end of the phase correspond to connected components of G_{phase}. This follows from the fact that edges of G_{phase} correspond to connections between components (by an edge or a path of two edges in the original graph) broadcasted during the phase. As G_{phase} is acyclic, each connected component contains a sink.

Let $\mathcal{C}$ be a partition into components at the beginning of a phase and $\mathcal{C}'$ be the partition into components at the end of that phase, before deactivating components of small degree.[2] The above observations imply that each component of $\mathcal{C}'$ either contains only components of $\mathcal{C}$ of small degree (smaller than s) or it contains at least $s + 1$ components from $\mathcal{C}$. Contrary, assume that a component

---

[2] Note that deactivation of components of degree $< s$ at the end of a phase does not guarantee that degrees of all components are $\geq s$ at the beginning of the next phase. This is caused by the fact that deactivation of some components might decrease degrees of components which remain active (degrees are calculated only among active nodes).

C' of $\mathcal{C}'$ contains a component $C \in \mathcal{C}$ of degree $\geq s$, while C' contains altogether at most s components of $\mathcal{C}$. Then, there is a directed path from C to a sink C_{sink} of degree at least $\deg(C) \geq s$. Property (b) implies that at least s components of $\mathcal{C}$ have edges towards C_{sink} in G_{phase}. This contradicts the contrary assumption that C' contains altogether less than s components of $\mathcal{C}$.

Summarizing, assume that we have p active components at the beginning of a phase. Then, at the end of the phase, there are at most p/s new components which contain at least one component whose degree at the beginning of the phase was $\geq s$. It remains to consider the final components of the phase which are composed only from components whose degree was $< s$ at the beginning of the stage. However, as the degree of a node cannot increase during the algorithm, the degrees of these new components are $< s$ and they are deactivated at the end of the phase. Thus, each phase decreases the number of active components at least s times – there are at most $\log_s n$ phases.

Correctness of the algorithm follows from the fact that each node of each deactivated component can broadcast its connections with all other components during Playoff. Moreover, active components are connected subgraphs of G at each stage.

If node is deactivated, its degree must be lesser than s. Therefore, announcing edges of inactive components takes $O(s)$ rounds. In order to show that whole algorithm needs $O(s + \log_s n)$ rounds, we will show, that there are at most $O(\log_s n)$ iterations of while loop (called stages). In particular we will show following lemma

Lemma 1. *Number of components which are not disabled is decreasing at least s times in one stage.*

Proof. Each node announces edge to a component containing a node of the highest degree. For each component let us consider node v with the highest degree. C^v is a component of node v before stage. After the 3-rd line there are four cases:

1. all neighbours of v announced edges connecting them to C^v and $deg_{old}(v) \geq s$
2. not all neighbours of v announced edges connecting them to C^v and $deg_{old}(v) \geq s$
3. all neighbours of v announced edges connecting them to C^v and $deg_{old}(v) < s$
4. not all neighbours of v announced edges connecting them to C^v and $deg_{old}(v) < s$

In the first case v is merged with at least $s + 1$ other components.

In the second case, some neighbours did not announced edge connecting it with C^v. Thus, it must have a neighbour in component of higher degree. Edge announced in the line 7 will connect v with such neighbour, thus with component of higher degree. If node of higher degree in this component also was in case 2 we can repeat this reasoning. At the end we will end up in some node in case 1, therefore all old components are now part of larger new component composed from at least $s + 1$ old components.

In the third case all nodes in component in new component of v had non larger degree than v, thus whole component will be disabled in the 8th line.

In the fourth case we can repeat reasoning from case two, but at the end we will end up either in case 1 or case 3. Thus, C^v will be disabled or connected with at least $(s+1)$ other components.

Therefore all components were disabled or connected with $s+1$ other components, thus number of non deactivated components decreased at least $(s+1)$ times. Therefore Lemma 1 is correct.

After $O(\log_s n)$ stages number of active components will drop to 0, which implies, that E will be empty. Each phase required a constant round number, thus so far algorithm required $O(\log_s n)$ rounds. $E = \emptyset$ implies, that we will end while loop and move to announcing edges from disabled nodes. This part requires $O(s)$ rounds. Therefore Theorem 3 is correct.

The minimum of $s + \log_s n$ is obtained for $s = \frac{\log n}{\log \log n}$. For such s, Algorithm 1 works in $O(\log n/\log \log n)$ rounds, which ends the proof of Theorem 2. Moreover, we want to emphasize following remark.

Remark 1. It is possible to solve connectivity problem in the broadcast congested clique using $\mathcal{O}(\log_d n)$ rounds, if we allow in one round to send $d \log n$ bit messages. Moreover, each node transmits only $O(\log n(d + \frac{\log n}{\log d}))$ bits in total.

Proof. If we take $s = d$ in Algorithm 1, we get $\log_d n$ phases, each requiring $O(\log n)$ bits per node. Edges from deactivated nodes are broadcasted during Playoff in one round, using $O(d \log n)$ bits. This gives $O(\log_d n)$ round algorithm, with total number of bits per player in $O(\log n(d + \frac{\log n}{\log d}))$.

Remark 1 in a sense gives an improvement over a result from [13], where the total number of bits per node is $O(d\frac{\log^2 n}{\log d})$ in $O(\log_d n)$ rounds. Moreover, our algorithm is simpler than that in [13], since it does not require number theoretic techniques as d-pruning and deterministic sparse linear sketches. Also, we find Remark 1 interesting on its own as it concerns relation between the total number of bits communicated in the protocol, and the number of rounds in the protocol, which says something about communication complexity of this problem in Broadcast Congested Clique.

4   Minimum Cut Approximation

In this section we propose a min-cut approximation algorithm, which finds an $(1 + \epsilon)$ approximation of the min-cut using only $\mathcal{O}(\log n)$ maximal spanning forest computations. Together with our spanning forest algorithm, it yields the following.

Theorem 4. *For a given n node graph G, it is possible to find $(1+\epsilon)$ approximation of the min-cut in $\mathcal{O}(\epsilon^{-2} \log^2 n/\log \log n)$ rounds of the* Broadcast Congested Clique, *for any $\epsilon \in (0,1)$.*

The remaining part of this section is dedicated to prove Theorem 4. In the first place we recall usage of c–connectivity certificates, together with Karger's sampling approach. Then we show how to improve on this approach. In particular in Subsect. 4.2 we provide analysis of sampling with small probabilities, which in Subsect. 4.3 we apply, together with regular sampling, to get our improved version of the algorithm.

4.1   Connectivity Certificates and Karger's Sampling

The algorithm we propose may be considered as extension of the algorithm presented in [2]. This algorithm is based on c–connectivity certificates introduced in [14,16]. In those papers, authors provide a lemma which is useful for min-cut approximation in the Broadcast Congested Clique and Semi–Streaming models.

Lemma 2 [14,16]. *Let G be a graph and F_0 an empty set. For $i > 0$, let F_i be a maximal spanning forest of $G \setminus \bigcup_{j=1}^{i-1} F_j$. Let $G_c = \bigcup_{i=1}^{c} F_i$. Then G_c is c-connected iff G is c-connected. Moreover, if the min-cut of G is smaller than c, then G_c has exactly the same min-cut as G.*

Therefore, if we compute G_c, we can verify whether G is c connected, and if it's not, we can find the min-cut of G. In the remaining part of this section, we will address to this process of verifying c connectivity and finding the min-cut just by 'verifying c–connectivity'. If we combine Lemma 2 with Karger's sampling approach [10], it gives us a min-cut approximation algorithm. If by $G(p)$ we denote random subgraph of G such that each edge is sampled to $G(p)$ with probability p, then lemma proposed by Karger is following

Lemma 3 [10]. *Let G be any graph with min-cut λ and let $p = \frac{6 \ln n}{\epsilon^2 \lambda}$. Then the probability that the value of some cut in $G(p)$ is larger than $(1 + \epsilon)$ or smaller than $(1 - \epsilon)$ times its expected value is $\mathcal{O}(1/n)$.*

A straightforward implementation of the min-cut approximation algorithm using Lemma 3 and Lemma 2 is following. Let consider set $P = \{p_0, p_1 \ldots, p_{\log n}\}$, such that $p_i = 1/2^i$. Let p^* be the smallest member of P such that p^* is large enough to apply Lemma 3 and small enough, to have the expected value of the min-cut of $G(p^*)$ in $\mathcal{O}(\log n)$, with high probability. Let consider $G(p^*)_c$. For properly chosen $c \in \mathcal{O}(\log n)$, graph $G(p^*)_c$ is not c–connected. Thus, by Lemma 2 it has the same min-cut as $G(p^*)$. By Lemma 3 this cut corresponds to a cut of G which was at most $1 + \epsilon$ times larger than the min-cut of G. Since we do not know the value of p^* beforehand, straightforward approach to the problem is to compute $G(p)_c$ for all $p \in P$, which requires finding $\Theta(\log^2 n)$ spanning forests.

In Subsects. 4.2 and 4.3, we propose a way of finding p^*, which requires computing only $\mathcal{O}(\log n)$ spanning forests. Knowing p^* allows us to compute $G(p^*)_c$ using c additional spanning forest computations.

4.2 Sampling with Small Probabilities

In order to explain the algorithm, we have to understand what is the behaviour of Karger's sampling Lemma 3, if the expected value of the min-cut is sublogarithmic. Since for graphs which are not $\Theta(\log n)$–connected, we can find the min-cut using $\Theta(\log n)$ –connectivity verification, we consider here only graphs which are $\Theta(\log n)$–connected.

If we sample graph with probability p, significantly smaller than p^*, the expected size of the min-cut in $G(p)$ may be sublogarithmic. Which means that in order to have high probability of success, we could set ϵ to be large. Then, we do not have any reasonable bound on the probability of a cut appearing smaller than its expected value, but we still can guarantee that if the expected size of the min-cut in $G(p)$ was small, the actual size of the min-cut in $G(p)$ is not too large with high probability. We formalize this in Lemma 4.

Lemma 4. *Let G be any graph with the size of a min-cut λ and let $p \leq \frac{6\ln n}{\alpha \lambda}$. Then the size of the min-cut of $G(p)$ is $\mathcal{O}(\log n/\sqrt{\alpha})$, with high probability.*

Proof. For $p = \frac{6\ln n}{\alpha \lambda}$, the claimed relationship follows from Lemma 3, if we set $\epsilon = \sqrt{\alpha}$. Therefore, for each $p' < p$, the size of min-cut of $G(p')$ is $\mathcal{O}(\log n/\sqrt{\alpha})$, with high probability. □

We can use Lemma 4 in a following way: if we check $G(p)$ for c–connectivity, for some $c \in \mathcal{O}(\log n/\sqrt{\alpha})$, and it is not c connected, then the min-cut of G is smaller than $\Theta(p^{-1}\log n/\alpha)$, with high probability. Pushing this idea a little bit forward, if we find probability p', such that $G(p')$ is not c connected, and $G(2p')$ is, it gives us some information about p^*.

Lemma 5. *Let consider some some $c \in \mathcal{O}(\log n/\sqrt{\alpha})$, and set of probabilities $P = \{p_1, p_2, \ldots, p_k\}$, such that*

- *$p_1 = 1/2^x$, for some nonnegative integer x,*
- *for all $i > 1$, $p_i = p_{i-1}/2$,*
- *l is the index of p^* in P.*

For any r' such that graph $G(p_{r'})$ is not c–connected and $G(p_{r'-1})$ is c–connected, $l \in \{r', r'-1, \ldots, r' - \log \alpha - \mathcal{O}(1)\}$.

Proof. Let consider graphs $G(p_i)_c$, for some $c \in \mathcal{O}(\log n/\sqrt{\alpha})$. There exists r, such that $\forall i \geq r$ $G(p_i)$ are not c connected, and $G(p_{r-1})$ is c–connected. Let consider graphs $G(p_i)_{c'}$, for some $c' \in \Theta(\log n)$. Then, by Lemma 3, there exists the largest l, such that $\forall i \leq l$ $G(p_i)$ are c'–connected. Since p_r is the largest probability, for which $G(p_r)$ is not c connected, the expected value of the min-cut of $G(p_r)$ is $\Omega(\log n/\alpha)$. Since expected value of $G(p_l)$ is $\Theta(\log n)$ we have $p_l/p_r = \Theta(\log n)/\Omega(\log n/\alpha) = \mathcal{O}(\alpha)$. Since for each i $p_i = p_{i-1}/2$, we have $r - l \leq \log(\Theta(\alpha)) = \log \alpha + \mathcal{O}(1)$.

Thus, if we find value r such that $\forall i \geq r$ $G(p_i)$ are not c–connected, we know that l is in the set $\{r, r-1, \ldots, r - \log \alpha - \mathcal{O}(1)\}$. Which shows that if we would verify c connectivity for all $G(p_i)$, we narrow the number of possible values of

l to $\log \alpha + \mathcal{O}(1)$. Moreover, if we find r' such that $G(p_{r'})$ is not c–connected and $G(p_{r'-1})$ is c connected, then for sure $r' \in \{r, r-1, \dots, r - \log \alpha - \mathcal{O}(1)\}$, which implies that $l \in \{r', r'-1, \dots, r' - \log \alpha - \mathcal{O}(1)\}$.

4.3   Algorithm

In this subsection we present the algorithm, which finds a constant number of probabilities, such that p^* is among them with high probability. The algorithm uses only $\mathcal{O}(\log n)$ spanning forest computations.

Our algorithm is performed in phases. We maintain feasible set of probabilities – set of probabilities P, such that $p^* \in P$ with high probability, and in each phase we narrow this set exponentially. In particular, if we denote by P_i the set probabilities which are feasible (i.e. $p^* \in P_i$ with high probability) after ith phase, then we provide procedure which finds subset $P_{i+1} \subseteq P_i$, such that $p^* \in P_{i+1}$ and $|P_{i+1}| \in \Theta(\log |P_i|)$.

More precisely, we show the following

Lemma 6. *Let $\log^{(i)}$ be an i-fold composition of $\log$ function. After ith phase the size of the set of feasible probabilities is $\Theta(\log^{(i+1)} n)$. Moreover, ith phase require c–connectivity verification for $\mathcal{O}((\log^{(i)} n)^{1/3})$ graphs, for $c \in \Theta(\log n/(\log^{(i)} n)^{1/2})$.*

Proof. In a single phase of our algorithm, we use Lemma 5, together with regular sampling, to reduce the size of set of feasible probabilities. Let consider the ith phase of the algorithm. Let $c_i \in \Theta(\log n/(\log^{(i)} n)^{1/2})$. By applying c_i–connectivity verification to $G(p)$, for all $p \in P_{i-1}$, we can provide P_i of size $\log \log^{(i)} n + \mathcal{O}(1) = \mathcal{O}(\log^{(i+1)} n)$, which, by Lemma 5 would provide set of feasible probabilities of size $\Theta(\log(\log^{(i)} n))$. This approach is not efficient, but we can replace it with either binary search or regular sampling. Here, we describe version which employs the regular sampling approach, as it covers both Broadcast Congested Clique and Multi–Pass Semi–Streaming models. Also, very similar analysis applies to the version which uses binary search, thus including both versions seems to be counterproductive.

Regular Sampling. The problem we have to solve is following. We are given set of probabilities P of size k. Let consider elements of P as a descending sequence $(p_1, p_2, \dots, p_k)$. Let $f_c(p_i) = 1$ if $G(p_i)$ is c-connected, 0 otherwise. The sequence $(p_1, p_2, \dots, p_k)$ has following properties

1. $\exists x . \forall i \geq x . f_c(p_i) = 0$
2. $\exists y . \forall i \leq y . f_c(p_i) = 1$

Our goal is to find any z such that $f_c(p_z) = 1 \wedge f_c(p_{z+1}) = 0$.

Lemma 7. *It is possible to find z, such that $f_c(p_z) = 1 \wedge f_c(p_{z+1}) = 0$, using $\mathcal{O}(k^{1/3})$ c-connectivity verifications.*

Proof. To solve this problem, we use the approach known as the regular sampling. In the first step we compute $f_c(p_i)$ (verify c connectivity of a proper graph) for every $k^{2/3}$th element from the sequence $(p_1, p_2, \ldots, p_k)$. Among indices, for which we computed value of $f_c(p_i)$, we select two consecutive indices $i_1 < i_2$, such that $f_c(i_1) = 0 \wedge f_c(i_2) = 1$. Those exists by properties 1 and 2. Therefore, by verifying c–connectivity $k/k^{2/3} + \mathcal{O}(1) = k^{1/3} + \mathcal{O}(1)$ times we reduced our problem to some subset of indices, which also has properties 1 and 2, but its size is $k^{2/3} + \mathcal{O}(1)$. Then we can do exactly the same thing, but use every $k^{1/3}$th element. This step require verifying c–connectivity $k^{2/3}/k^{1/3} + \mathcal{O}(1) = k^{1/3} + \mathcal{O}(1)$ times, and leaves us with set of probabilities of the size $k^{1/3} + \mathcal{O}(1)$, in which we can verify all possible probabilities using $k^{1/3} + \mathcal{O}(1)$ c–connectivity verification. In total we executed $\mathcal{O}(k^{1/3})$ c–connectivity verifications.

In the ith phase, we have the size of feasible probabilities P_{i-1} of size $\log^{(i)}(n)$. Our goal is to find r' from Lemma 5, i.e. value r' such that graph $G(p_{r'})$ is not c_i–connected, and $G(p_{r'-1})$ is. By Lemma 7, to find r' it is enough to use $(\log^{(i)}(n))^{1/3}$ c_i-connectivity verifications.

4.4 Complexity in Broadcast Congested Clique

By Lemma 6, the ith phase of our algorithm requires c–connectivity verification for $\mathcal{O}((\log^{(i)} n)^{1/3})$ graphs, for $c \in \Theta(\log n/(\log^{(i)} n)^{1/2})$, which give $\mathcal{O}(\log n/(\log^{(i)} n)^{1/6})$ spanning forest computations. Moreover, the number of phases before we are left with set of feasible probabilities is $\Theta(\log^* n)$.

In the Broadcast Congested Clique this results with total number of spanning forest computations

$$\sum_{i=1}^{\Theta(\log^* n)} \mathcal{O}(\log n/(\log^{(i)} n)^{1/6}) = \mathcal{O}(\log n).$$

Since, by Theorem 2 each spanning forest requires $\mathcal{O}(\log n/\log\log n)$ rounds, we can find p^* in $\mathcal{O}(\log^2 n/\log\log n)$ rounds of Broadcast Congested Clique, which finishes the proof of Theorem 4.

5   Application to Multi–Pass Semi–Streaming Model

The presented algorithm can be also applied in the Multi–Pass Semi–Streaming model.

Theorem 5. *It is possible to find $1 + \epsilon$ of the min-cut in a* Multi–Pass Semi–Streaming *model, in $k + 1$ passes and $\mathcal{O}(n \log^4 n(\epsilon^{-2} + \log^{(k)} n))$ space[3], for $\epsilon \in \Theta(1)$.*

---

[3] Authors of papers [1,2] have inconsistent way of defining space complexity, i.e. c connectivity in [1] requires $\mathcal{O}(cn \log^3 n)$ 'space', when referenced in [2] it only requires $\mathcal{O}(cn \log^2 n)$ 'space'. Here we go with the approach from [1], which seems to count bits.

Proof. Algorithm on the top level is the same, it is executed in phases, in ith phase, we verify c–connectivity verification for $\mathcal{O}((\log^{(i)} n)^{1/3})$ graphs, for $c \in \Theta(\log n/(\log^{(i)} n)^{1/2})$. Each phase is executed in a single pass over data stream. To do so, we use c-connectivity certificates from [1], which require $\mathcal{O}(nc\log^3 n)$ bits. Since, we verify c connectivity, required space is $\mathcal{O}(\log n/(\log^{(i)} n)^{1/6})$.

After $k-1$ passes/phases, by Lemma 6, the set of feasible probabilities has size $\mathcal{O}(\log^{(k)} n)$. Since we allow $\mathcal{O}(n\log^4 n \log^{(k)} n)$ memory, we can verify $\Theta(\log n)$–connectivity for all remaining feasible probabilities, as it requires $\mathcal{O}(n\log^4 n)$ space per each of remaining $\mathcal{O}(\log^{(k)} n)$ verifications.

At this point we have constant number of potential values of p^*, in the next pass for all of them simultaneously we can use naive approach. Since there are only $\mathcal{O}(1)$ graphs to verify, we need $\mathcal{O}(\epsilon^{-2}n\log^4 n)$ space.

Corollary 1. *If in Theorem 5 we take $k = \log^* n$, we get $\mathcal{O}(\log^* n)$ passes, $\mathcal{O}(\epsilon^{-2}n\log^4 n)$ space* Multi–Pass Semi–Streaming *algorithm, which finds a $1 + \epsilon$ min-cut approximation.*

6   Conclusions

In this paper we presented $\mathcal{O}(\log n/\log\log n)$ round algorithm for the spanning forest (connectivity/connected components) problem in the Broadcast Congested Clique model. This is the first algorithm for this problem, with sublogarithmic number of rounds. The natural questions which arise are:

- can we use this approach to find *minimum* spanning forest in this model?
- is it possible to find a spanning forest in $o(\log n/\log\log n)$ (even if we allow randomized algorithms)?
- are there any lower bounds on the number of rounds, if a message size is $\mathcal{O}(\log n)$?

Moreover, we presented approximation algorithm for the min-cut problem, which finds $1 + \epsilon$ approximation for the min-cut.

- in the Broadcast Congested Clique it requires $\mathcal{O}(\epsilon^{-2}\log^2 n/\log\log n)$ rounds;
- in the Multi–Pass Semi–Streaming model it requires $k + 1$ passes, and uses $\mathcal{O}(n\log^4 n \log^{(k)} n)$ space, where $\log^{(k)}$ is an k-fold composition of log function.

This, rises the following questions:

- is it possible to c connectivity certificates in a way, which does not use c spanning forest computations?
- is it possible to find exact min-cut in Broadcast Congested Clique in polylogarithmic number of rounds?

References

1. Ahn, K.J., Guha, S., McGregor, A.: Analyzing graph structure via linear measurements. In: Proceedings of the Twenty-Third Annual ACM-SIAM Symposium on Discrete Algorithms, SODA 2012, Kyoto, Japan, 17–19 January 2012, pp. 459–467 (2012)
2. Ahn, K.J., Guha, S., McGregor, A.: Graph sketches: sparsification, spanners, and subgraphs. In: PODS 2012, pp. 5–14. ACM (2012)
3. Becker, F., Montealegre, P., Rapaport, I., Todinca, I.: The simultaneous number-in-hand communication model for networks: private coins, public coins and determinism. In: Halldórsson, M.M. (ed.) SIROCCO 2014. LNCS, vol. 8576, pp. 83–95. Springer, Cham (2014). https://doi.org/10.1007/978-3-319-09620-9_8
4. Drucker, A., Kuhn, F., Oshman, R.: On the power of the congested clique model. In: PODC 2014, pp. 367–376 (2014)
5. Feigenbaum, J., Kannan, S., McGregor, A., Suri, S., Zhang, J.: On graph problems in a semi-streaming model. Theor. Comput. Sci. **348**(2), 207–216 (2005)
6. Ghaffari, M., Parter, M.: MST in log-star rounds of congested clique. In: Proceedings of PODC 2016 (2016)
7. Hegeman, J.W., Pandurangan, G., Pemmaraju, S.V., Sardeshmukh, V.B., Scquizzato, M.: Toward optimal bounds in the congested clique: graph connectivity and MST. In: Proceedings of the 2015 ACM Symposium on Principles of Distributed Computing, PODC 2015, Donostia-San Sebastián, Spain, 21–23 July 2015, pp. 91–100 (2015)
8. Jurdzinski, T., Nowicki, K.: Brief announcement: on connectivity in the broadcast congested clique. In: 31st International Symposium on Distributed Computing, DISC 2017, Vienna, Austria, 16–20 October 2017, pp. 54:1–54:4 (2017)
9. Jurdziński, T., Nowicki, K.: MST in O(1) rounds of congested clique. In: SODA 2018, pp. 2620–2632 (2018)
10. Karger, D.R.: Random sampling in cut, flow, and network design problems. In: ACM Symposium on Theory of Computing (STOC), pp. 648–657 (1994)
11. Kari, J., Matamala, M., Rapaport, I., Salo, V.: Solving the INDUCED SUBGRAPH problem in the randomized multiparty simultaneous messages model. In: Scheideler, C. (ed.) Structural Information and Communication Complexity. LNCS, vol. 9439, pp. 370–384. Springer, Cham (2015). https://doi.org/10.1007/978-3-319-25258-2_26
12. Lotker, Z., Patt-Shamir, B., Pavlov, E., Peleg, D.: Minimum-weight spanning tree construction in o(log log n) communication rounds. SIAM J. Comput. **35**(1), 120–131 (2005)
13. Montealegre, P., Todinca, I.: Brief announcement: deterministic graph connectivity in the broadcast congested clique. In: Proceedings of PODC 2016 (2016)
14. Nagamochi, H., Ibaraki, T.: Computing edge-connectivity in multigraphs and capacitated graphs. SIAM J. Discret. Math. **5**(1), 54–66 (1992)
15. Nešetřil, J., Milková, E., Nešetřilová, H.: Otakar Boruvka on minimum spanning tree problem translation of both the 1926 papers, comments, history. Discret. Math. **233**(1), 3–36 (2001)
16. Nishizeki, T., Poljak, S.: Highly connected factors with a small number of edges. Preprint (1989)

Biased Clocks: A Novel Approach to Improve the Ability To Perform Predicate Detection with $O(1)$ Clocks

Vidhya Tekken Valapil$^{(\boxtimes)}$ and Sandeep Kulkarni

Michigan State University, East Lansing, MI 48823, USA
{tekkenva,sandeep}@cse.msu.edu

Abstract. In this paper, we present the notion of biased hybrid logical clocks ($BHLC$). These clocks are intended to improve the ability of a distributed system to perform predicate detection with just $O(1)$ sized clocks. In traditional logical clocks (or hybrid logical clocks, their extension), the only way to *guarantee* that two events are concurrent is by checking if their clock values are *equal*. By contrast, biased clocks provide a *window* where this guarantee is provided. We validate our intuition that these biased clocks substantially improve the ability to successfully detect a given predicate with just $O(1)$ sized clock. In particular, for many scenarios, we show that biased clocks improve the ability to detect predicates by 100–200 times when compared to *standard hybrid logical clocks*.

1 Introduction

Debugging distributed systems is essential to ensure their correctness and reliability. More specifically, analyzing and debugging distributed systems to detect violations or to ensure satisfaction of their system requirements is critical as well as challenging. One of the basic types of debugging is predicate detection, where the goal is to detect whether a given condition or predicate P (e.g., that captures the violation of an invariant condition) is true. The challenge in performing such predicate detection lies in the fact that in distributed systems there is no single total ordering of events. In other words, due to the underlying nondeterminism, an observer can order the system events (i.e. events that happened at the involved processes) in several different ways. Hence, to detect violations, one has to analyze and evaluate all possible event orderings, because any of these orderings may correspond to the actual order of events that happened in the underlying system.

To deal with the uncertainty in the ordering of events in a distributed system, the notion of happened-before [14] is introduced. In a distributed system, an event e_1 is said to have happened before another event say e_2, denoted as $e_1 \rightarrow e_2$ iff one of the following conditions hold: (i) if event e_1 occurred before event e_2 at

---

This work is supported by NSF XPS 1533802.

Z. Lotker and B. Patt-Shamir (Eds.): SIROCCO 2018, LNCS 11085, pp. 345–360, 2018.
https://doi.org/10.1007/978-3-030-01325-7_29

the same process, (ii) if e_1 is a message sending event and e_2 is the corresponding message receive event, or (iii) if event e_1 happened before event x and event x happened before event e_2. Furthermore, two events e_1, e_2 are concurrent (denoted as $e_1||e_2$) iff $((e_1 \nrightarrow e_2) \wedge (e_2 \nrightarrow e_1))$. Thus, the problem of predicate detection in distributed systems requires us to find a global snapshot (consistent cut) that contains a local state of each process such that (1) the states (i.e., local events created at those states) are concurrent with each other, and (2) the predicate of interest is true in the corresponding global state.

Another requirement for predicate detection arises from the fact that modern distributed systems provide partially synchronized clocks. If clocks are synchronized to be within ϵ then the time (clock values) of the local states in the snapshot are required to be within ϵ. Thus, the problem of predicate detection requires us to find a global snapshot consisting of one local state per process such that their corresponding local events are concurrent and within ϵ of each other, and then evaluate if the predicate of interest is true in that global snapshot.

One can utilize vector clocks [6,11] – consisting of vectors with n elements– to perform predicate detection. With Vector Clocks, an event a happened before event b, iff $vc.a < vc.b$. And, events a, b are concurrent iff $(vc.a \not< vc.b) \wedge (vc.b \not< vc.a)$. However, detection using Vector Clocks suffers from two main challenges: their size can be large and they do not take into account features (specifically clock synchronization) that are available in many distributed systems. The former can make the overhead of predicate detection unacceptable and the latter could result in false positives where the algorithm falsely concludes that the predicate of interest is true.

To deal with these limitations, we can utilize logical clocks (LC) [10] or hybrid logical clocks (HLC) [9] that combine logical clocks and physical clocks. Size of LC/HLC is $O(1)$ and, hence, it permits efficient implementation. Also, predicate detection with LC/HLC does not have any false positives. However, the cost of getting these features is that it can result in substantial false negatives where the predicate of interest is true but is not detected. In other words, HLC can miss snapshots where the predicate is true. However, without paying $O(n)$ cost, it is unavoidable.

With this motivation, in this paper, we consider an alternate implementation of clocks that does not have false positives, reduces the number of false negatives and maintains $O(1)$ size.

Contributions of the Paper.

- We introduce a new type of $O(1)$ sized clocks called Biased Hybrid Logical clocks ($BHLC$) and discuss timestamping and ordering of system events using $BHLC$. Compared to logical clocks (or hybrid logical clocks), they provide a larger window where we can conclude that the given events are concurrent.
- We compare HLC and $BHLC$ (and its variations) in their ability to detect whether a given predicate is true. We find that $BHLC$ is able to find 100–200 times as many instances as HLC where the given predicate is true. Furthermore, in many scenarios, $BHLC$ is able to find more than 50% of the total actual instances in the system where the given predicate is true. As a point

of comparison, $O(n)$ sized timestamps are needed to find all snapshots where the predicate is true and for many scenarios, HLC is able to find less than 1% of instances.

- We present extensions of $BHLC$ where we allow clocks to be reset periodically and to adapt to the underlying system behavior.
- We discuss the effectiveness of $BHLC$ and its extensions under different conditions.

2   System Model

We consider a distributed system of n processes. Each process has a local physical clock associated with it and the clocks in the system are guaranteed to be synchronized within ϵ of each other.[1] The physical clock value of a process j (where $0 \leq j < n$) is denoted as $pt.j$. The events on a process are categorized into local events, message send events and message receive events. As the name suggests, at a send event, a process sends one or more messages to other processes. And, at a receive event, the process receives one or more messages from other processes. Our algorithm does not depend on whether the underlying network is FIFO or lossy. Each event e is assigned a timestamp by the owner process, i.e. by the process where e occurs.

Since our goal is to provide an algorithm for timestamping, we first describe the naive HLC algorithm from [9]. This algorithm forms the basis of our biased clocks presented in this paper.

2.1   Naive HLC

In this algorithm, each process maintains a variable $l.j$ that captures the logical time associated with an event. Intuitively, $l.j$ maintains a logical clock subject to the constraint that $l.j$ is always at least as large as $pt.j$, the physical time at j. Hence, for a send event, rather than just increasing $l.j$ by 1, we set $l.j$ to be $max(l.j + 1, pt.j)$. And, for a receive event, instead of setting $l.j$ to be $max(l.j + 1, l.m + 1)$, we set it to $max(l.j + 1, l.m + 1, pt.j)$. Thus, the naive HLC algorithm is shown in Algorithm 1.

---

[1] Our implementation of $BHLC$ does not use the value of ϵ. It is used only during monitoring to rule out snapshots that are not feasible. Hence, analysis in this paper can be extended for the case where ϵ is determined at run time and provided to the monitor. However, this issue is outside the scope of the paper.

Algorithm 1 Naive HLC Algorithm from [9]	Algorithm 2 HLC Algorithm from [9]

At node j
1: Initially $l.j := 0$
Send/Local event
2: $l.j := max(l.j + 1, pt.j)$
3: Timestamp with $l.j$
Receive event of message m
4: $l.j := max(l.j + 1, l.m + 1, pt.j)$
5: Timestamp with $l.j$

Send/Local Event
1: $l'.a := l.a$
2: $l.a := max(l'.a, pt.a)$ //tracking maximum time event, $pt.a$ is physical time at a
3: If $(l.a = l'.a)$ then $c.a := c.a + 1$ //tracking causality
4: Else $c.a := 0$
5: Timestamp event with $l.a, c.a$
Receive Event of message m
6: $l'.a := l.a$
7: $l.a := max(l'.a, l.m, pt.a)$ //$l.m$ is l value in the timestamp of the message received
8: If $(l.a = l'.a = l.m)$ then $c.a := max(c.a, c.m) + 1$
9: Elseif $(l.a = l'.a)$ then $c.a := c.a + 1$
10: Elseif $(l.a = l.m)$ then $c.a := c.m + 1$
11: Else $c.a := 0$
12: Timestamp event with $l.a, c.a$

Key properties of naive HLC are:

$$e \longrightarrow f \Rightarrow l.e < l.f \quad \text{and} \quad l.e = l.f \Rightarrow e \| f$$

A key disadvantage of naive HLC is that the drift between $l.j$ and $pt.j$ could grow unbounded. In [9], an alternate algorithm is proposed to deal with this issue. Specifically, in this alternate algorithm, instead of adding 1 to $l.j$ or $l.m$ when an event is created, we allow l value to remain unchanged. This creates the possibility that two successive events on a process (or a send event and the corresponding receive event) can have the same timestamp. To deal with this situation, we maintain a variable $c.j$ that is a counter (shown to be provably bounded). Specifically, the algorithm of HLC is as shown in Algorithm 2. HLC timestamps are of the form $\langle l, c \rangle$ and they provide causality information by lexicographical comparison. Specifically, HLC satisfies the following property.

$$e \longrightarrow f \Rightarrow (l.e < l.f) \vee ((l.e = l.f) \wedge (c.e < c.f))$$

The algorithm proposed in this paper is based on naive version described in Algorithm 1. HLC and naive HLC have the same capability in terms of being able to detect a given predicate.

3   An Idea to Increase Effectiveness of HLC

Consider the system execution shown in Fig. 1a. Here, we have 3 processes whose clocks are perfectly synchronized ($\epsilon = 0$). We assume that the physical clock increases by 1 between every two events. Furthermore, assume that maximum clock drift ϵ is 10. (Note that this is the maximum clock drift. In this execution, clocks happen to increase in lock-step fashion. However, the processes do not know this.) There are no messages in this execution. This implies that $\langle a_i, b_j, c_k \rangle$ is a consistent snapshot for any $1 \leq i, j, k \leq 6$. In such a system, suppose that the predicate of interest is $\wedge pr_p$ where pr_p is a local predicate at process p.

In this execution, since there are no messages, the value of $l.e$ for any event e is same as the physical time at e. Thus, possible snapshots where HLC can detect

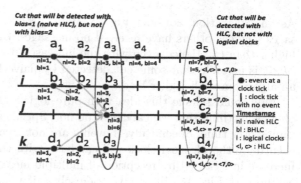

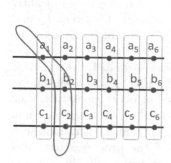

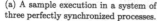

(a) A sample execution in a system of three perfectly synchronized processes.

(b) A scenario with non-uniform message distribution.

Fig. 1. Intuition behind $BHLC$

if $\wedge pr_p$ is true include $\{\langle a_1, b_1, c_1\rangle \; \langle a_2, b_2, c_2\rangle \; \langle a_3, b_3, c_3\rangle \; \langle a_4, b_4, c_4\rangle, \; \langle a_5, b_5, c_5\rangle, \; \langle a_6, b_6, c_6\rangle \}$. Even though there are several consistent snapshots, HLC does not have enough information to allow us to conclude that. Specifically, HLC does not allow us to conclude a_1 and b_2 are concurrent.

Now, suppose that we change the implementation of receive in Algorithm 1 as follows:

Upon receiving message m to create receive event b,
$$l.j = max(l.j + 1, l.m+ 2)$$
$$l.b = l.j$$

Observe that in this implementation, we have a bias for messages. Specifically, we add one to differentiate the new event from the previous event on the process. However, for the message, we added 2 (instead of 1 as done in Algorithm 1). This implementation, of HLC will guarantee that if e and f are events on two different processes then $e \longrightarrow f \Rightarrow l.e + 1 < l.f$. It follows that if e and f are events on two different processes and $|l.e - l.f| \leq 1$ then e and f are concurrent.

From this discussion, it follows that with this new implementation, even consistent cuts such as $\langle a_1, b_2, c_2\rangle$ (cf. Fig. 1a) are valid consistent cuts and we can use them to evaluate the predicate, i.e. to detect whether the given predicate $\wedge pr_p$ is true. We view the above implementation as a solution with bias of 2. The default implementation in Sect. 2.1 corresponds to a bias of 1. We also denote it as *unbiased* implementation.

With a bias of 2, in Fig. 1a, we can see that there is an increased potential to find a cut where the given predicate is true. Furthermore, the effectiveness of predicate detection in the scenario considered in Fig. 1a will increase as the value of bias increases. In particular, if we use a bias of 6 then all consistent cuts will be detected.

We note that the addition of bias is not free of cost. In particular, there is a potential that some cuts that were detected with bias of 1 (Algorithm 1 in Sect. 2.1) but are not detected by an algorithm with higher bias value.

As an illustration, consider Fig. 1b. Here, process j has received several messages whereas others have received no messages. In this case, $l.j$ is significantly higher than $l.k$. Hence, even with the increased drift permitted between $l.j$ and $l.k$, it is possible that some cut detected by naive HLC will not be detected by an algorithm with higher bias.

One could pursue this idea with pure logical clocks as well. However, HLC offers something that logical clocks do not. Specifically, HLC *smooths* the timestamps over time. Specifically, if events are not created at every clock tick, then eventually the physical clock will catch up with the l value (even after a rapid increase in l caused by reception of multiple messages). Hence, situations such as those in Fig. 1b will resolve themselves if none of the processes have any new events for a short duration. By contrast, (pure) logical clocks do not have this ability.

From the above discussion, we can see that biased hybrid logical clocks ($BHLC$) have the potential to increase the effectiveness of monitoring/debugging, without affecting the overall complexity of the detection algorithm. However, there is a potential of missing some cuts found by the unbiased algorithm.

4 Algorithm for Biased Clocks ($BHLC$)

Our first algorithm for biased hybrid logical clocks ($BHLC$) is based on the idea discussed in Sect. 3. Specifically in this work, each process j maintains a variable $l.j$ to keep track of its (biased) clock value. This value is initialized to 0. We also utilize physical clock $pt.j$ for process j. This value is updated automatically. As far as our algorithm is concerned, it is a read-only value. However, the underlying system will ensure that clocks of any two processes differ by at most ϵ. As an input, $BHLC$ takes a parameter B, that denotes the bias value.

When a new event is created on process j, $l.j$ is updated. And, the new value is assigned as a timestamp to the newly generated event. If the event generated is a send event or a local event then the algorithm works same as the naive HLC algorithm. It increases the value of $l.j$ by 1 and sets it to a value that is at least as large as $pt.j$. If this is a send event then the value of $l.j$ is sent with the message. If the event generated is a receive event, where message m is received, then $l.j$ is set to $max(l.j+1, l.m+\mathbf{B}, pt.j)$. In other words, the algorithm biases its clock to be at least $l.m + B$. Thus, the algorithm is as shown in Algorithm 3.

Algorithm 3. Algorithm $BHLC$ with Input Parameter B

At node j
  1: Initially $lc.j := 0$
  2: Set B to bias value
Send/Local event
  3: $l.j := max(l.j + 1, pt.j)$
  4: Timestamp with $l.j$
Receive event of message m
  5: $l.j := max(l.j + 1, l.m + B, pt.j)$
  6: Timestamp with $l.j$

From this algorithm, we can show that the following properties are satisfied.

Lemma 1. *Let e and f be events on two different processes and let l.e and l.f be the timestamps assigned to them by Algorithm 3. Then, we have*

$$e \longrightarrow f \ \Rightarrow \ l.e + B \le l.f$$

Proof. This proof follows from Line 5 of Algorithm 3.

Lemma 2. *Let e and f be events on two different processes and let l.e and l.f be the timestamps assigned to them by Algorithm 3. Then, we have*

$$|l.e - l.f| < B \ \Rightarrow e \| f$$

Proof. We consider two cases: $l.e \ge l.f$ and $l.f \ge l.e$. In the first case, clearly 'e happened before f' is false. Also, since $l.e - l.f < B$, i.e., $l.f + B > l.e$. From Lemma 1, 'f happened before e' is also false. Thus, $e\|f$. And, the analysis of the second case is identical.

4.1 Extension 1: Multiple Simultaneous Instances of $BHLC$

Algorithm $BHLC$ takes B as a parameter. We can run two versions of this algorithm, say with $B = B_1$ and $B = B_2$. Observe that if any one of them allows us to conclude that two events are concurrent then they are indeed concurrent. However, if we run two versions of the same algorithm, it would increase the storage cost and computational cost.

4.2 Extension 2: Algorithm $BHLC_r$: Resetting Clocks at Cut-Points

Based on the analysis of Fig. 1b, we can see that $BHLC$ will work effectively if the computation length is small so that the number of messages received by different processes are close. We consider two extensions to deal with these issues.

First, to deal with long computations, we introduce the notion of (periodic) cut-points with length C. Thus, the first interval is from $\langle 0..C - 1\rangle$. Next interval is from $\langle C..2C - 1\rangle$, and so on. Whenever, the clock of a process reaches its cut-point, we increase its l value to a large enough value that would not occur before the cut-point. This is straightforward to achieve since the computation length between cut-points has a fixed length. Note that this would create a problem in terms of comparing events when event on one process is just before the cut-point and one event is just after the cut-point. To deal with this issue, we can maintain another clock which resets at $\frac{1}{2}C$, $\frac{3}{2}C$, $\frac{5}{2}C$ and so on. As discussed in Sect. 4.1, if even one of these clocks would allow us to conclude concurrency of two events, it is sufficient. We use $BHLC_r$ to denote the resulting algorithm.

4.3   Extension 3: Algorithm $BHLC_a$: Adjusting Message Rate

Figure 1b also suggests that biased clocks would work most effectively if the number of messages received by each process is roughly the same. If this is not true then, we can achieve this by allowing a process to pretend to receive *fake messages*. This can be achieved as follows: Let $x(t)$ denote the number of messages that are expected to be received by a process by time t. If actual number of messages received by a process is smaller, we pretend to receive a message (and update the clock value). We denote the resulting algorithm as $BHLC_a$. We use $BHLC_{ra}$ to denote the algorithm that uses both extensions 2 and 3.

5   Comparison of HLC and $BHLC$ in Predicate Detection

5.1   Experimental Setup

To analyze the effectiveness of $BHLC$ (and its extensions), we use simulation of a system with 10 independent processes, where each process has a physical clock and a biased clock associated with it. The physical clocks of these processes are synchronized to be within ϵ of each other. This is achieved by executing the processes in a round-robin manner. When a process is given a chance to execute and increment its physical clock, it does so with a certain probability if incrementing it will not cause two clocks to differ by more than ϵ. If a process is able to increment its physical clock, it sends a message with probability α, one of the parameters in our simulation. The target is selected randomly with uniform distribution. These messages are received by the target processes after a specific message delay of δ. Every process i has a local boolean variable v_i which becomes true at a specific rate of β. When an event (message send or receive event/local event like change in v_i) occurs at a process, it updates its biased clock accordingly.

To understand the effectiveness of $BHLC$, we timestamp the events in this system using biased clock plus physical clock. When a snapshot is identified as a consistent snapshot using Lemma 2, the physical clock values are used to determine if the corresponding events are within ϵ window. The predicate being considered for our experimental analysis is conjunctive predicate, i.e. in the system under consideration, we use biased clocks to evaluate whether the predicate $\wedge v_i\ (1 \leq i \leq n)$ is satisfied.

We note that our experiments focus on identifying the number of global snapshots where the given predicate is true. Subsequently, we determine how many of these cuts are found by HLC/$BHLC$. In this sense, our analysis is independent of the predicate being considered. However, if we change the predicate under consideration, we will need to change the algorithm involved in detecting it. However, comparison of cuts identified by HLC and $BHLC$ will remain unaffected.

One issue in these experiments is caused by the fact that one consistent snapshot may have several other consistent snapshots that are *close* to it. For example, if we have two events per process where predicate v_i is true and there is no communication, there are 2^n consistent cuts. Counting all these cuts is not only expensive, but in fact they are not *independent* cuts. Hence, we require that two consistent cuts must occur at least ϵ apart. This ensures that these cuts are in fact disjoint.

Default Experimental Setup. In our simulations, we treat a clock tick to be 0.1 ms. Each simulation run is for a total of 100 (virtual) seconds, i.e., each process increments the clock from 0 to 1,000,000. The default set of parameters that we use are clock drift $\epsilon = 10$ ms (100 clock ticks), message delay $\delta = 1$ ms (10 clock ticks), $\beta = 10\%$ (the expected time before the variable becomes true is 1 ms), local predicate stays true for just 0.1 ms (1 clock tick) and an average communication frequency of 1000 messages per second (10% chance of sending a message every clock tick). To compare the effectiveness of biased clocks under different configurations, we vary one parameter at a time. The raw data from our experiments is available at http://cse.msu.edu/~tekkenva/biasedclocks/.

Organization of the Experimental Results. In Sect. 5.2, we present the algorithm that we use for finding whether the given conjunctive predicate is true. In Sect. 5.3 we analyze the effectiveness of $BHLC$ in detecting predicates (conjunctive predicate) as we vary the system parameters namely (a) clock drift ϵ (b) local predicate rate β (c) communication frequency α and (d) message delay δ. In Sects. 5.4 and 5.5, we analyze the effectiveness of $BHLC_r$, $BHLC_a$ and $BHLC_{ra}$ (discussed in Sects. 4.2 and 4.3).

5.2 Algorithm for Conjunctive Predicate Detection Using $BHLC$

As discussed above, the exact method for detecting a predicate is not relevant to our discussion since we are finding possible cuts where the predicate is true and comparing it with the ones that are detected by HLC or biased clocks with different bias value. However, for the sake of completeness, we present the algorithm we use for detecting conjunctive predicates used in our work.

When an event occurs at a process, the process updates its biased clock value and then timestamps the event with the updated biased clock value and current physical clock. Initially, v_i is false at every process i. Let e and f denote the (successive) events where v_i becomes true and false respectively. Let $\langle b.e, pt.e \rangle$ denote the value of $BHLC$ and physical clock timestamp of event e. Likewise, let the timestamp of f be $\langle b.f, pt.f \rangle$. Thus, v_i is true in the interval $[\langle b.e, pt.e \rangle, \langle b.f, pt.f \rangle)$. Hence, process i creates a candidate $[\langle b.e, pt.e \rangle, \langle b.f, pt.f \rangle)$ and adds it to its queue. The monitoring process uses these queues for the detection of conjunctive predicate i.e. in this instance to detect if the variable v was true at all processes at a same point in time ($\wedge v_i \ (1 \leq i \leq n)$).

The monitor forms a snapshot of the system by picking one candidate per process. It then evaluates if for any two candidates in the snapshot, say $[\langle b.e_i, pt.e_i \rangle, \langle b.f_i, pt.f_i \rangle)$ and $[\langle b.e_j, pt.e_j \rangle, \langle b.f_j, pt.f_j \rangle)$ if there is some point in time within a candidate interval that is possibly concurrent with some point in time within the other candidate interval. This is achieved by checking if $((|b.f_i - b.e_j| < B) \vee (|b.f_j - b.e_i| < B)) \wedge ((|pt.f_i - pt.e_j| \leq \epsilon) \vee (|pt.f_j - pt.e_i| \leq \epsilon))$.

If this evaluates to true for every pair of candidates in the snapshot then the snapshot forms a consistent snapshot. When the monitor detects a consistent snapshot, it reports a conjunctive predicate satisfaction.

5.3 Effectiveness of $BHLC$ Under Different System Parameters and Bias B

In this section, we analyze the effect of varying system parameters (clock drift, communication frequency, message delay, local predicate rate) and the amount of bias (B) on the ability of detecting predicates using $BHLC$ and HLC.

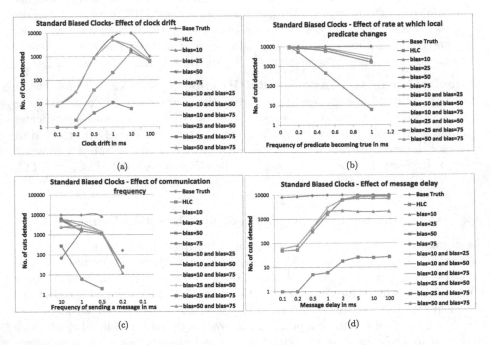

Fig. 2. (a) Effect of varying frequency of sending a message, (b) Effect of varying the rate at which local predicate becomes true, (c) Effect of varying frequency of sending a message, (d) Effect of varying Message delay on Standard biased clocks

Varying Clock Drift (ϵ). To analyze the effect of varying clock drift or clock synchronization of the processes in the system, we use the default set of system parameters and vary the clock drift ϵ. Specifically, we consider clock drifts of 0.1 ms, 0.2 ms, 0.5 ms, 1 ms, 10 ms and 100 ms. In Fig. 2a, we observe that $BHLC$ with any value of B performs (**orders of magnitude better**) than HLC. In terms of clock synchronization, $BHLC$ implementation performs better when the processes in the system are more tightly synchronized with each other. More specifically, when the processes in the system are tightly synchronized with each other, detection using $BHLC$ with $B = 10$ detects almost all cuts where the predicate is true.

Varying Local Predicate Rate (β). Starting from the default experimental setup, we consider cases where $\beta = 1$ (local predicate is always true) to $\beta = 0.1$ (probability that a local predicate is true is 0.1). The observed effect is as shown in Fig. 2b. As β decreases, the number of cuts satisfying the predicate in the system decrease. As expected, all methods of detection detect fewer cuts with smaller β. The earlier observation that $BHLC$ performs better than HLC continues to hold for Fig. 2b. Again, detection using $BHLC$ with $B = 10$ and a simultaneous implementation with $B = 10$ and $B = 75$ perform the best. More specifically, simultaneous implementation with $B = 10$ and $B = 75$ identifies approximately 70% of the actual cuts.[2]

Varying Communication Frequency (α). Starting from the default experimental setup, we vary α from 0.1 (roughly 100 messages per second) to 1 (roughly 10000 messages per second) As shown in Fig. 2c, all detection methods detect fewer cuts as the communication frequency increases; as the number of messages increase, there are fewer concurrent events.

We observe that $BHLC$ continues to perform better than the HLC. On an average, $BHLC$ detects about 50% of the total cuts. In general, predicate detection based on $BHLC$ performs better if communication frequency is low. This is expected, given that biased clocks were motivated by what happens when message communication frequency is very low. Hence, one may consider allowing for higher bias to improve performance. However, very high bias also means longer jumps in the $BHLC$. In turn, this may also result in rejection of more concurrent cuts. From Fig. 2c, we find that $BHLC$ with $B = 25$ and $B = 50$ works best.

Varying Message Delay (δ). For this case, we vary message delay to 0.1 ms, 0.2 ms, 0.5 ms, 1 ms, 2 ms, 5 ms, 10 ms and 100 ms. Remaining parameters are the same as in the case of default setup. From Fig. 2d, we can observe that with increase in message delay, predicate detection using $BHLC$ performs significantly better. When the message delay is small, $BHLC$ detects less that 10% of the cuts in the system. However, as the message delay increases the detection rate improves rapidly to 80%, specifically when message delay ≥ 2 ms.

5.4   Effectiveness of $BHLC_r$

In this section we analyze the effectiveness of $BHLC_r$, where clocks at the processes are reset periodically to overcome the issue of biased clocks growing far apart over time.

We perform the same set of experiments presented in Sect. 5.3 using $BHLC_r$ where clocks are reset every 1000 clock ticks, i.e. every 100 ms. We vary one system parameter at a time and present the results in Figs. 3a, b, c and d. We observe that the detection capability using $BHLC_r$ is similar to $BHLC$.

---

[2] Note that for the sake of comparison the analysis is done over the same set of execution traces and when we consider multiple bias amounts, common snapshots are counted only once.

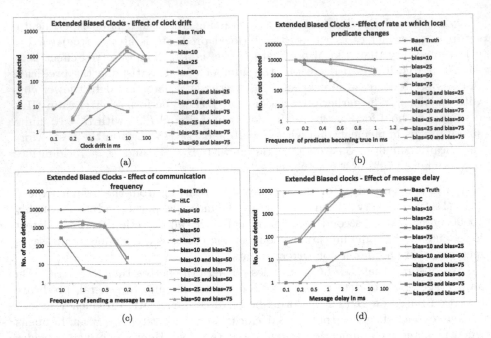

Fig. 3. (a) Effect of varying clock drift, (c) Effect of varying frequency of sending a message and (b) Effect of varying the rate at which local predicate becomes true and (d) Effect of varying Message delay on Extended biased clocks with reset every 100 ms.

5.5 Effectiveness of $BHLC$ Under Non-uniform Message Distribution

In this section we analyze the effectiveness of $BHLC$ and its extensions in a system with non-uniform message distribution. Specifically, we focus on the extension of $BHLC$ discussed in Sect. 4.3, where processes compensate for non-uniform message distribution.

We consider a scenario where we partition the set of 10 processes into two sets (5 in each). The first set (processes 1–5) receive messages at twice the rate of the second set (processes 6–10). Hence, processes 6–10 compensate by adding twice as much bias in the receive statement. In other words, if we instantiate $BHLC$ with $B = 10$, processes 1–5 add 10 on Line 5 and processes 6–10 add 20. In this scenario, we consider two subcases: (1) processes in one set only talk among themselves (cf. Fig. 4b) and (2) processes choose their destination randomly and, hence, they could send messages to processes in the other set (cf. Fig. 4a). For this version ($BHLC_a$), we utilize the following observation to decide if two events are related by happened-before relation.

$$e \longrightarrow f \quad \text{iff} \quad \begin{cases} l.e + 10 < l.f & f \text{ is on processes } 1..5 \\ l.e + 20 < l.f & f \text{ is on processes } 6..10 \end{cases}$$

From Figs. 4a and b, we find that $BHLC$, $BHLC_a$ and $BHLC_{ra}$ work better than HLC. However, $BHLC_a$ and $BHLC_r$ do not provide the desired improvement. Rather, (standard) $BHLC$ works better. In part, this happens because $BHLC_a$ does an abrupt jump of size $2B$ for processes 6..10. This abrupt jump makes it harder to find concurrent events at processes 1..5. That said, the addition of bias improves the predicate detection capability when compared with unbiased implementation (which corresponds to HLC).

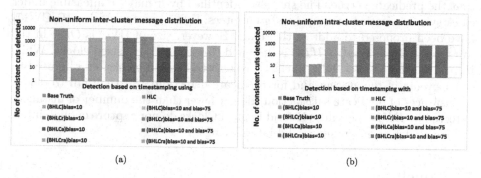

(a)   (b)

Fig. 4. No. of violations detected in (a) Non-uniform Message distribution 1 and (b) Non-uniform Message distribution 2

6 Related Work

The problem of predicate detection has been widely studied in the literature [4,7,13] for debugging distributed systems to identify errors that may happen due to race conditions. It is also generalized [1,12,15] to detect more complex properties. In general, the problem of predicate detection in distributed systems is NP-Complete [3]. This is mainly due to the underlying uncertainty in the order of events in the system, which in turn leads to the possibility of exponential number of valid interleavings that need to be examined for predicate satisfaction.

One approach is to order events in the system using physical time. However, the issue with using only physical time is that it fails to capture causality, especially when physical clocks at the processes are not perfectly synchronized with each other. For instance a send event may have a higher physical clock timestamp than the receive event, if the clock at the sending process is ahead of the receiver's clock due to clock drift. So using only physical time may result in consideration of invalid interleavings. In [16], Stoller showed that if approximately-synchronized real-time clocks are used for detecting global state predicates, and if the inter-event spacing, denoted as E, is more than the clock drift in the underlying system, then the number of possible states to be evaluated is only $O(En)$, where n is the number of processes in the system. However, in an asynchronous system the total number of possible states is $\Omega(E^n)$.

Another most prominently used approach to order events in distributed systems is the use of Vector Clocks for timestamping of events in the system. There are several existing works [2,3,5,8,19] that perform predicate detection using Vector Clocks. However the problem with using vector clocks is that vector clock timestamps are of size $O(n)$, where n is the number of processes in the system under consideration. Also, in modern systems where protocols can utilize the fact that the underlying system provides clock synchronization, detection using Vector Clocks can result in false positives, where the algorithm detects that the predicate is true. The snapshot identified by it may be infeasible under clock synchronization assumptions/guarantees. Hybrid vector clocks (HVC) [17] have been proposed to address this issue. However, size of HVC is $O(n)$ in the worst case. By contrast, $BHLC$ is designed to use only $O(1)$ sized clocks. While Hybrid Logical Clocks (HLC), which are also $O(1)$ sized clocks, can be used to order events in the system and for predicate detection [18], they fail to detect all instances of predicate satisfaction (detect fewer than the number of instances detected by $BHLC$) because Hybrid Logical Clocks do not capture enough information to detect all possible valid interleavings.

7   Conclusion

This paper presented a novel approach for predicate detection using biased clocks. It is well-known that if we want to detect all instances where predicate P is *possibly true* without finding any phantom instances then $O(n)$ size clocks are necessary. Logical clocks/Hybrid logical clocks can be used to perform such predicate detection. However, they cannot find all possible cuts. We considered the question: "Can we improve the effectiveness of predicate detection while still maintaining only $O(1)$ sized clocks".

Our analysis shows that with biased clocks presented in this paper, the chances of finding the predicate of interest being true increase substantially. On an average, in our experiments, biased clocks were able to find 100–200 times as many instances where the given predicate is true when compared to hybrid logical clocks. We find that this result is true for different communication frequencies, message delays, clock drifts and frequencies of the local predicate being true. Furthermore, for many scenarios, $BHLC$ was able to find more than half of the instances where the given predicate is true. Given that $BHLC$ performs predicate detection with $O(1)$ sized clocks and finds a substantial fraction of instances where the given predicate is true, we expect that $BHLC$ will provide an inexpensive and effective way to perform predicate detection.

One of the future work in this area is to develop a theory that will help identify the bias values that should be used to maximize the predicate detection capability of $BHLC$. Another future work is to learn from system behavior (e.g., number of messages received by a given process in a given time) and use it to automatically identify the bias values accordingly.

References

1. Bauer, A., Falcone, Y.: Decentralised LTL monitoring. Form. Methods Syst. Des. **48**(1–2), 46–93 (2016)
2. Charron-Bost, B.: Concerning the size of logical clocks in distributed systems. Inf. Process. Lett. **39**(1), 11–16 (1991)
3. Chase, C.M., Garg, V.K.: Detection of global predicates: techniques and their limitations. Distrib. Comput. **11**(4), 191–201 (1998)
4. Chauhan, H., Garg, V.K.: Fast detection of stable and count predicates in parallel computations. In: Aspnes, J., Bessani, A., Felber, P., Leitão, J. (eds.) 21st International Conference on Principles of Distributed Systems, OPODIS 2017, Lisbon, Portugal, 18–20 December 2017. LIPIcs, vol. 95, pp. 20:1–20:21. Schloss Dagstuhl - Leibniz-Zentrum fuer Informatik (2017)
5. Chauhan, H., Garg, V.K., Natarajan, A., Mittal, N.: A distributed abstraction algorithm for online predicate detection. In: Proceedings of the 2013 IEEE 32nd International Symposium on Reliable Distributed Systems, SRDS 2013, pp. 101–110. IEEE Computer Society, Washington, DC (2013)
6. Fidge, C.J.: Timestamps in message-passing systems that preserve the partial ordering. In: Proceedings of the 11th Australian Computer Science Conference, vol. 10(1), pp. 56–66 (1988)
7. Garg, V.K.: Brief announcement: applying predicate detection to the stable marriage problem. In: Richa, A.W. (ed.) 31st International Symposium on Distributed Computing, DISC 2017, Vienna, Austria, 16–20 October 2017. LIPIcs, vol. 91, pp. 52:1–52:3. Schloss Dagstuhl - Leibniz-Zentrum fuer Informatik (2017)
8. Garg, V.K., Waldecker, B.: Detection of weak unstable predicates in distributed programs. IEEE Trans. Parallel Distrib. Syst. **5**(3), 299–307 (1994)
9. Kulkarni, S.S., Demirbas, M., Madappa, D., Avva, B., Leone, M.: Logical physical clocks. In: Aguilera, M.K., Querzoni, L., Shapiro, M. (eds.) OPODIS 2014. LNCS, vol. 8878, pp. 17–32. Springer, Cham (2014). https://doi.org/10.1007/978-3-319-14472-6_2
10. Lamport, L.: Time, clocks, and the ordering of events in a distributed system. Commun. ACM **21**(7), 558–565 (1978)
11. Mattern, F.: Virtual time and global states of distributed systems. In: Parallel and Distributed Algorithms, pp. 215–226. North-Holland (1989)
12. Mostafa, M., Bonakdarpour, B.: Decentralized runtime verification of LTL specifications in distributed systems. In: 2015 IEEE International Parallel and Distributed Processing Symposium, IPDPS 2015, Hyderabad, India, 25–29 May 2015, pp. 494–503 (2015)
13. Natarajan, A., Chauhan, H., Mittal, N., Garg, V.K.: Efficient abstraction algorithms for predicate detection. Theor. Comput. Sci. **688**, 24–48 (2017)
14. Schwarz, R., Mattern, F.: Detecting causal relationships in distributed computations: in search of the holy grail. Distrib. Comput. **7**(3), 149–174 (1994)
15. Sen, K., Vardhan, A., Agha, G., Rosu, G.: Efficient decentralized monitoring of safety in distributed systems. In: 26th International Conference on Software Engineering (ICSE 2004), 23–28 May 2004, Edinburgh, United Kingdom, pp. 418–427 (2004)
16. Stoller, S.D.: Detecting global predicates in distributed systems with clocks. Distrib. Comput. **13**(2), 85–98 (2000)

17. Yingchareonthawornchai, S., Nguyen, D.N., Tekken Valapil, V., Kulkarni, S.S., Demirbas, M.: Precision, recall, and sensitivity of monitoring partially synchronous distributed systems. In: Falcone, Y., Sánchez, C. (eds.) RV 2016. LNCS, vol. 10012, pp. 420–435. Springer, Cham (2016). https://doi.org/10.1007/978-3-319-46982-9_26

18. Yingchareonthawornchai, S., Tekken Valapil, V., Kulkarni, S., Torng, E., Demirbas, M.: Efficient algorithms for predicate detection using hybrid logical clocks. In: Proceedings of the 18th International Conference on Distributed Computing and Networking, ICDCN 2017, pp. 10:1–10:10. ACM, New York (2017)

19. Zhu, W., Cao, J., Raynal, M.: Predicate detection in asynchronous distributed systems: a probabilistic approach. IEEE Trans. Comput. **65**(1), 173–186 (2016)

Gathering in the Plane of Location-Aware Robots in the Presence of Spies

Jurek Czyzowicz[1], Ryan Killick[2], Evangelos Kranakis[2(✉)], Danny Krizanc[3], and Oscar Morale-Ponce[4]

[1] Département d'informatique, Université du Québec en Outaouais, Gatineau, QC, Canada
[2] School of Computer Science, Carleton University, Ottawa, ON, Canada
kranakis@scs.carleton.ca
[3] Department of Mathematics and Computer Science, Wesleyan University, Middletown, CT, USA
[4] Department of Computer Science, California State University, Long Beach, CA, USA

Abstract. A set of mobile robots (represented as points) is distributed in the Cartesian plane. The collection contains an unknown subset of byzantine robots which are indistinguishable from the reliable ones. The reliable robots need to gather, i.e., arrive to a configuration in which at the same time, all of them occupy the same point on the plane. The robots are equipped with GPS devices and at the beginning of the gathering process they communicate the Cartesian coordinates of their respective positions to the central authority. On the basis of this information, without the knowledge of which robots are faulty, the central authority designs a trajectory for every robot. The central authority aims to provide the trajectories which result in the shortest possible gathering time of the healthy robots. The efficiency of a gathering strategy is measured by its competitive ratio, i.e., the maximal ratio between the time required for gathering achieved by the given trajectories and the optimal time required for gathering in the offline case, i.e., when the faulty robots are known to the central authority in advance. The role of the byzantine robots, controlled by the adversary, is to act so that the gathering is delayed and the resulting competitive ratio is maximized.

The objective of our paper is to propose efficient algorithms when the central authority is aware of an upper bound on the number of byzantine robots. We give optimal algorithms for collections of robots known to contain at most one faulty robot. When the proportion of byzantine robots is known to be less than one half or one third, we provide algorithms with small constant competitive ratios. We also propose algorithms with bounded competitive ratio in the case where the proportion of faulty robots is arbitrary.

J. Czyzowicz and E. Kranakis—Research supported in part by NSERC Discovery grant.
R. Killick—Research supported by OGS scholarship.
Due to space limitations all missing proofs can be found in the report [19].

Z. Lotker and B. Patt-Shamir (Eds.): SIROCCO 2018, LNCS 11085, pp. 361–376, 2018.
https://doi.org/10.1007/978-3-030-01325-7_30

Keywords: Byzantine · Competitive ratio · Gathering
Location aware · Reliable · Robots

1   Introduction

1.1   The Background

A collection of mobile robots need to meet at some point of the geometric environment. This task, known as *gathering* or *rendezvous*, has been extensively investigated in the past. The gathering may be necessary, e.g., to coordinate a future task or to exchange previously acquired information.

In most formerly studied cases, robots have limited knowledge about the environment and they do not know the positions of the other robots. In the present paper, the robots are distributed in the two-dimensional Cartesian plane. They are equipped with GPS devices and they can wirelessly communicate their positions to the central authority. The central authority then informs each individual robot of the trajectory it is to follow in order to meet. However, the team of reliable robots has been contaminated with "spies" - a subset of byzantine robots, indistinguishable from the original ones, controlled by an omnipotent adversary. The role of the faulty robots is simple – delay the gathering of the reliable ones for as long as possible. A byzantine robot may report a wrong position, fail to report any, or fail to follow its assigned route. As the central authority does not recognize which robots are byzantine, it sends the travel instructions to all of them.

Our goal is to design a strategy resulting in gathering of all reliable robots within the smallest possible time. We attempt to minimize the *competitive ratio* – the ratio of the time required to achieve gathering of the reliable robots, to the time required for such gathering to occur under the assumption that the reliable robots were known in advance.

1.2   The Model and the Problem

A collection S of n mobile robots move at maximum unit speed within the two-dimensional plane. It is assumed that each robot in S is equipped with a GPS device so it is aware of a pair of Cartesian coordinates representing its current location in the plane.

We consider the problem of gathering an unknown subset $\mathcal{N} \subseteq S$ of robots. The robots of $\mathcal{N}$ need to arrive at some time at a same point on the plane in order to complete some given task. We refer to this set $\mathcal{N}$ of at least $n - F$ robots as the set of reliable robots and define $\mathcal{F} = S \setminus \mathcal{N}$ of $f \leq F$ robots as the set of byzantine robots. We call this problem of gathering all reliable robots from a collection containing at most F byzantine robots the $Gather(n, F)$ problem.

At the beginning, all robots in S send a single message recording their starting positions to the central authority. In turn, the central authority computes a set of trajectories instructing each robot how to time their respective movements in order to achieve gathering. At this point the robots follow the trajectories provided.

The movement continues until all reliable robots meet for the first time. We imagine a successful gathering as a meeting of robots possessing pieces of information allowing them to solve some puzzle. As long as all pieces are disassembled, the puzzle remains unsolved, and the identification of useful or invalid information is not possible.

The byzantine robots may report incorrect initial locations, which can potentially adversely affect the robots' trajectories. Clearly, this results in byzantine robots not being able to follow the assigned trajectories. However, as long as all reliable robots complete their trajectories, the schedule must lead to their gathering.

The trajectories designed by the central authority are computed uniquely on the basis of the reported set of robot positions and possibly using the knowledge of the upper bound on the number of byzantine robots. Once the robots start their movements, no adaptation to our algorithm is ever possible as no extra information may be obtained. We assume that the adversary knows in advance our algorithm and it will put the byzantine robots in the positions which result in the worst possible competitive ratio.

We note that the requirement of a central authority may be removed by allowing the robots to instead broadcast their initial positions to all other robots. In this situation all robots compute the same set of trajectories using the same algorithms.

We are interested in developing algorithms solving the $Gather(n, F)$ problem which are optimal in terms of the competitive ratio for a given initial configuration S of n robots, at most F of which are byzantine. We define the competitive ratio $\mathrm{CR}_{n,F}(\mathcal{A}, \mathcal{N})$ of an algorithm $\mathcal{A}$ for the specific subset $\mathcal{N}$ of the input S as the ratio of the time $T_{\mathcal{A}}(\mathcal{N})$ – the time of the first gathering of all robots belonging to $\mathcal{N}$ – divided by $T_*(\mathcal{N})$ – the minimal time necessary to gather the robots in $\mathcal{N}$, i.e. $\mathrm{CR}_{n,F}(\mathcal{A}, \mathcal{N}) = \frac{T_{\mathcal{A}}(\mathcal{N})}{T_*(\mathcal{N})}$. We also define the *overall* competitive ratio $\widehat{\mathrm{CR}}_{n,F}(\mathcal{A}, S)$ of an algorithm $\mathcal{A}$ with input S as the maximal $\mathrm{CR}_{n,F}$ over any subset $\mathcal{N}$ of S, i.e. $\widehat{\mathrm{CR}}_{n,F}(\mathcal{A}, S) = \max_{\mathcal{N} \subset S} \mathrm{CR}_{n,F}(\mathcal{A}, \mathcal{N})$. We further define the *optimal* competitive ratio $\overline{\mathrm{CR}}_{n,F}(S)$ for an input S as the minimal $\widehat{\mathrm{CR}}_{n,F}(\mathcal{A}, S)$ for any algorithm $\mathcal{A}$, i.e. $\overline{\mathrm{CR}}_{n,F}(S) = \min_{\mathcal{A}} \widehat{\mathrm{CR}}_{n,F}(\mathcal{A}, S)$. For ease of presentation we will often drop the subscripts n and F when they are implied by context.

We define an optimal algorithm $\overline{\mathcal{A}}$ solving the $Gather(n, F)$ problem as any algorithm satisfying

$$\mathrm{CR}_{n,F}(\overline{\mathcal{A}}, S) = \overline{\mathrm{CR}}_{n,F}(S), \ \forall \ S. \tag{1}$$

1.3   Our Results

We provide algorithms with constant competitive ratio for all but a small bounded region in the space of possible n and F pairs. In doing so we demonstrate that having knowledge of the upper bound of the number of byzantine robots in the subset (represented by the parameter F) permits fine-tuning of the gathering algorithm, resulting in better competitive ratios. In Sect. 2 we consider the gathering problem for collections involving only a single byzantine robot. After developing insight into the problem we give a gathering algorithm that is optimal for any number of robots, at most one of which is byzantine. For the boundary case of three robots, one of which is byzantine, we give a closed form expression for the competitive ratio. Section 3 presents two algorithms with small constant competitive ratio when the number of byzantine robots is bounded by a small fraction of n. Specifically, we give algorithms with competitive ratios of 2 and $2\sqrt{2}$ when $F < \lceil n/3 \rceil$ and $F < \lceil n/2 \rceil$ respectively. Finally, in Sect. 4, we give two gathering algorithms solving the problem for any n and any F. The competitive ratio of one of these algorithms is constant, while the other is bounded by $F + 2$. We summarize the results of the paper in Table 1 and Fig. 1.

Table 1. Summary of competitive ratio bounds for various algorithms.

F	Upper-bound	Reference
1	Optimal	Algorithm 4
$\leq \lceil n/3 \rceil$	2	Algorithm 5
$\leq \lceil n/2 \rceil$	$2\sqrt{2}$	Algorithm 6
$> \lfloor 32\sqrt{2} \rfloor - 2$	$32\sqrt{2}$	Algorithm 7
$\leq \lfloor 32\sqrt{2} \rfloor - 2$	$F + 2$	Algorithm 8

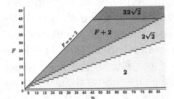

Fig. 1. Competitive ratio bounds for various regions of the space of possible n and F pairs.

1.4   Related Work

The gathering problem was originally introduced in [33] as a version of pattern formation (see also [20]). In operations research, Alpern [2,3] considers the gathering of two robots, referred to as the *rendezvous* problem (cf. [30]). Both problems are central in theoretical computer science. The rich related literature is due to the large variety of studied settings: deterministic and randomized, synchronous and asynchronous, for labeled and anonymous agents, in graphs and geometric environments, for same-speed or distinct maximal speed agents, etc. (cf. [4,9,13,21,24,28,31,35]). More recently, efficient solutions were proposed for the plane [15] and for grids [14].

In many papers on gathering the agents are a priori assumed to have limited knowledge of the environment. Moreover, most papers supposed that an agent is not aware of the positions in the environment of other agents. In the deterministic settings, one of the central studied questions was feasibility of gathering or rendezvous, cf. [20,21,31], which most often led to some form of the symmetry breaking problem, see [28,30]. Surprisingly, when agents were equipped with GPS devices, knowledge of the agent's own position in the environment permitted executing very efficient rendezvous algorithms (see [12,13]).

Fault tolerance in mobile agent algorithms has also been extensively studied in the past, but the failures were more often related to the static elements of the environment (network nodes or links), cf. [25,29]. The faults of the mobile agents were studied for the problems of convergence [10], flocking [34], searching [17,18] or patrolling [16]. Faults or imperfections arriving to mobile agents performing gathering were investigated in [1,11,22,26,32]. Research in [11], [26] and [32] considered the gathering problem in the presence of inaccurate or faulty robot perception components. In [1] the initial positions of the collection is known to all robots, which operate in so called *look-compute-move* cycle. The feasibility of the problem, as a function of faulty robots, is investigated in [1] for crash and byzantine faults. In [22], the gathering problem is studied in an unknown graph environment and the feasibility question for byzantine faults in the strong and weak sense are investigated. The results of [22] depend on the knowledge of the upper bound on the size of the graph environment (or the absence of such knowledge).

In [8] the authors studied, similar to ours, the online rendezvous problem using GPS-equipped robots on a line, where some robots may turn out to be byzantine. However the robot movements along the line are much easier to analyze than the setting studied in the present paper. Indeed, in the case of a line, the robots move inside a corridor forcing robots to meet.

1.5   Notation

We will use S to refer to a general collection of any robots (reliable and/or byzantine) and use $\mathcal{N}$ $(\mathcal{F})$ to represent a set of reliable (byzantine) robots only. We will represent the cardinality of a set S as $|S|$ and will always use $n = |S|$, and $f = |\mathcal{F}|$. We reserve the use of F for the upper bound on the number of byzantine robots in S (and, as such, it may be that $f \leq F$).

As we are dealing with robots in the plane we will use the term robot and point interchangeably. When it is required to refer to a particular robot/robots in a set we will use the capital letters A, B, and C. We use the capital letter D to refer to meeting points of robots.

We let the distance between any two points A and B be $|AB|$, and use $\overline{AB}$ to represent the directed line segment joining A and B. We will refer to the individual coordinates of a point using the subscripts x and y, e.g., $A = (A_x, A_y)$.

We define $\mathcal{MC}(S)$ as the minimum enclosing circle (MEC) of a set of points S, and let $\mathrm{Sup}[S]$ be the supporting set of $\mathcal{MC}(S)$. It is a well known property that $2 \leq |\mathrm{Sup}[S]| \leq 3$ [7]. We further define the radius $\mathrm{Radius}[S]$ and $\mathrm{Center}[S]$ of S to be the radius and center of the MEC of S respectively.

Finally, we let $\mathcal{FVD}(S)$ represent the furthest-point Voronoi diagram (FVD) of the point set S, and, for a point A in S, we let $\mathcal{FVR}(A)$ be the cell/region in $\mathcal{FVD}(S)$ belonging to the point A. See [5] for a description of the properties of the FVD.

2   One Byzantine Robot

In this section we develop optimal algorithms for the case that there is only a single byzantine robot within the collection $\mathcal{S}$. To do this we will need to consider subsets of $\mathcal{S}$ containing $n-1$ robots and we therefore introduce some convenient notation. We let $\mathcal{S}_i \subset \mathcal{S}$, $i \in [0, n-1]$ represent the n subsets of $n-1$ robots that can be formed from $\mathcal{S}$ and we define an ordering for the $\mathcal{S}_i$ in such a way that $\mathrm{Radius}[\mathcal{S}_i] \leq \mathrm{Radius}[\mathcal{S}_j]$ $\forall\, j \geq i$. For the sake of brevity, we use $r_{\mathcal{S}} = \mathrm{Radius}[\mathcal{S}]$ and $r_i = \mathrm{Radius}[\mathcal{S}_i]$ for the remainder of the section.

We start with the following (trivial) lemma concerning the optimal meeting time of any set of robots in the plane,

Lemma 1. *The minimal time needed to gather a set $\mathcal{S}$ of robots is $T_*(\mathcal{S}) = r_{\mathcal{S}}$.*

An immediate consequence of the above lemma is the following optimal algorithm for gathering a group of n reliable robots.

Algorithm 1. (Optimal *Gather*$(n, 0)$)

1: Set $D = \mathrm{Center}[\mathcal{S}]$;
2: All robots in $\mathcal{S}$ move at full speed towards D ;
3: The algorithm terminates when the last robot in $\mathcal{S}$ reaches D ;

To get an idea of how different the problem is when we consider the presence of even a single byzantine robot, let us run the above algorithm on the two inputs depicted in Fig. 2.

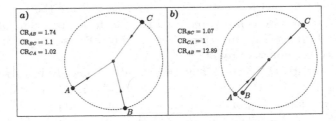

Fig. 2. Inputs for example analysis of competitive ratio. In both cases the robots A, B, and C move directly towards the center of the minimum enclosing circle of $\mathcal{S} = \{A, B, C\}$.

For a given input $\mathcal{S} = \{A, B, C\}$ the adversary can choose at most one of the robots A, B, and C to be byzantine. We assume that they will do so in such a way as to maximize the competitive ratio of our algorithm. Which robot would they choose? In the case (**a**) the choice is not so obvious, and, indeed, the competitive ratios for all three possibilities are not very different. In the case (**b**), however, there *is* an obvious choice: the adversary would make C byzantine since the robots A and B were initially very close but travelled far before meeting (Fig. 2).

This exercise, although simple, highlights an important observation – the "closest" robots should meet first. It turns out that, when $F = 1$, we can formalize this statement[1].

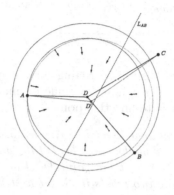

Fig. 3. Setup for the proof of Lemma 4.

Lemma 2. *Consider an optimal algorithm $\overline{\mathcal{A}}$ solving the Gather(n, 1) problem for the input S. Let S_i be the first group of $n - 1$ robots to meet. Then $S_i = S_0$, i.e. S_i is the group of $n - 1$ robots in S with the smallest enclosing circle.*

So, we now know that we have to make the smallest group of $n - 1$ robots meet first. What choice does this leave the adversary? Well, naturally, they would choose the byzantine robots in such a way that the second-smallest group of $n-1$ robots should have gathered. This observation leads us to the following:

Theorem 1. *The competitive ratio of any algorithm solving the Gather(n, 1) problem with input S is at least r_S/r_1.*

At this point we can make a useful observation: an optimal gathering algorithm ends either at the moment the first group of robots meet or the moment all robots meet. Furthermore, at the moment of the first meeting, all robots are located at either one of only two positions. Thus, in an optimal algorithm, we must send these remaining two groups of robots directly towards each other. We can claim the following:

Lemma 3. *An optimal algorithm $\overline{\mathcal{A}}$ solving the Gather(n, 1) problem can be completely described by the single point D at which the first $n - 1$ robots gather.*

Corollary 1 *(Lemma 3). There is an optimal algorithm solving the Gather(n, 1) problem following the strategy given in Algorithm 2.*

---

[1] When $F > 1$ there are cases when this is not true.

Algorithm 2. (General $Gather(n, 1)$)

1: All n robots start moving at full speed towards some point D ;
2: **if** The first $n - 1$ robots to arrive at D are all reliable; **then**
3:     The algorithm terminates ;
4: **else**
5:     Let D' be the midpoint of D and the position of the single robot that has not yet arrived at D (at the time the first group of robots gather at D) ;
6:     All robots move at full speed towards D'. The algorithm terminates once they meet ;

Corollary 1 reduces the task of searching for an optimal algorithm to the conceptually simpler task of searching for some optimal meeting point D. The following lemma tells us how to find this point:

Lemma 4. *Consider an optimal algorithm $\overline{A}$ solving the Gather(n, 1) problem for the input S parameterized by the point D. Let the group S_i represent the first group of $n - 1$ robots to gather at the point D. Then the point D lies on the perpendicular bisector of the two robots in S_i furthest from D.*

As a last step we derive an expression for the competitive ratio of an optimal $Gather(n, 1)$ algorithm.

Lemma 5. *An optimal algorithm following the strategy in Algorithm 2 solves the Gather(n, 1) problem for the input S with competitive ratio*

$$\widehat{CR} = \max\left\{ \frac{|AD|}{r_0}, \ \frac{|AD| + |CD|}{2r_1} \right\}$$

where A is one of the two points in S_0 furthest from D and C is the point in S that is not in S_0.

We are now ready to present our main result:

Algorithm 3. (Optimal $Gather(n, 1)$ point)

1: Set C as the single robot in S that is not in S_0;
2: Determine the Furthest-point Voronoi diagram $\mathcal{FVD}(S_0)$ of the point set S_0;
3: Set $CR_{min} = \infty$, and $D_{min} = NULL$;
4: **for** each edge E in $\mathcal{FVD}[S_0]$ **do**
5:     Set A and B as the two points such that the edge E separates $\mathcal{FVR}(A)$ and $\mathcal{FVR}(B)$;
6:     Determine the point D' on E that minimizes $CR(D') = \max\left\{ \frac{|AD'|}{r_0}, \frac{|AD'| + |CD'|}{2r_1} \right\}$.
7:     **if** $CR(D') < CR_{min}$ **then**
8:         Set $D_{min} = D'$ and $CR_{min} = CR(D')$
    return D_{min};

Algorithm 4. (Optimal *Gather*(n, 1))

1: The robots perform Algorithm 2 with the point D determined by Algorithm 3;

Theorem 2. *Algorithm 4 is an optimal algorithm solving the Gather(n, 1) problem with input S. The complexity of the algorithm is $\mathcal{O}(n \log n)$.*

It does not seem likely that a closed form expression can be derived for the competitive ratio of Algorithm 4 for arbitrary n. However, in the boundary case that $n = 3$ and $F = 1$ this is possible. The complete solution of the *Gather*(3, 1) is presented in [19] and the results are reproduced below:

Theorem 3. *Algorithm 2 optimally solves the Gather(3, 1) problem with input $\triangle ABC$ of side lengths $a \leq b \leq c$ and respective angles $\alpha \geq \beta \geq \gamma$ if the point D is chosen such that $D_x = \frac{1}{2}[(B_x + C_x) + a \tan \phi(B_y - C_y)]$, $D_y = \frac{1}{2}[(B_y + C_y) + a \tan \phi(C_x - B_x)]$, and $\tan \phi = \tan \beta$ if $\tan \beta \leq \sin \gamma$, otherwise $\tan \phi = \frac{2\sqrt{c^2 - (b-a)^2}}{\sqrt{(3b-a)^2 - c^2} + \sqrt{(b+a)^2 - c^2}}$. The competitive ratio of the algorithm equals c/b if $\tan \beta \leq \sin \gamma$, otherwise it is $1/\cos \phi$.*

3 Bounded Number of Byzantine Robots

We now consider instances of the *Gather*(n, F) problem when the value of F is a small constant fraction of n. We give two algorithms corresponding to the cases that $F < \lceil \frac{n}{3} \rceil$, and $F < \lceil \frac{n}{2} \rceil$. In both cases we show that a small constant competitive ratio is attainable. We start with the case that $F < \lceil \frac{n}{3} \rceil$.

Theorem 4. *Consider the Gather(n, F) problem with input S and for any $F < \lceil \frac{n}{3} \rceil$. Then, there is a gathering algorithm solving this problem with competitive ratio at most 2. The complexity of the algorithm is $\mathcal{O}(n)$.*

Proof (Theorem 4). We will make use of the centerpoint theorem (see [23] [Theorem 4.3]) which states that any finite set S of n points in $\mathbb{R}^d$ admits a point K (a centerpoint) such that any open half-space avoiding K contains at most $\lfloor \frac{dn}{d+1} \rfloor$ points of S. In particular, for $d = 2$, this implies that we can always determine a K such that any line L through K partitions S into two sets each with at least $F < \lceil \frac{n}{3} \rceil$ robots. This result inspires the following algorithm,

---
Algorithm 5. (Move to centerpoint)

1: The robots compute a centerpoint K of the set S of robots;
2: All robots move directly towards K;
3: The algorithm terminates once the final reliable robot reaches K;

Consider the reliable robot A that is initially furthest away from the point K determined in Algorithm 5. Draw a line L through K perpendicular to the line segment $\overline{AK}$ (as done in Fig. 4). Observe that, since K is a centerpoint, there are at least $\lceil \frac{n}{3} \rceil$ robots on either side of L. Furthermore, by assumption, F is strictly less than $\lceil \frac{n}{3} \rceil$ and we are thus guaranteed to have a reliable robot on either side of L. Consider any reliable robot B on the opposite side of L as A and note that the robot B is at least a distance $|AB| \geq |AK|$ away from the robot A. The competitive ratio of Algorithm 5 is therefore at most $\widehat{CR} \leq |AK|/(\frac{1}{2}|AB|) \leq 2$.

The complexity bound follows from the need to determine the centerpoint of the collection. The centerpoint of a set of n points can be determined in $\mathcal{O}(n)$ time using an algorithm by Jadhav [27].

The centerpoint theorem applies generally to any d-dimensional space and we thus have the following corollary,

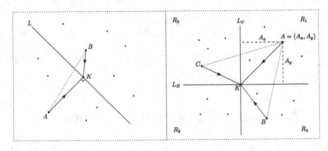

Fig. 4. Setup for the proofs of Theorem 4 (left) and Theorem 5 (right).

Corollary 2 *(Theorem 4). Consider the Gather(n, F) problem in $\mathbb{R}^d$ for any $F < \lceil \frac{n}{d+1} \rceil$. Then, there exists a gathering algorithm with competitive ratio at most 2.*

Now consider the case that $F < \lceil \frac{n}{2} \rceil$. We claim the following:

Theorem 5. *Consider the Gather(n, F) problem with input S and for any $F < \lceil \frac{n}{2} \rceil$. Then, there is a gathering algorithm solving this problem with competitive ratio at most $2\sqrt{2}$. The complexity of the algorithm is $\mathcal{O}(n)$.*

Proof (Theorem 5). The proof is based on the following algorithm,

---

Algorithm 6. (Move to intersection)

1: The robots compute a line L_H that partitions the robots into two disjoint sets each containing at least $\lceil \frac{n}{2} \rceil$ robots;
2: The robots compute a line L_V, perpendicular to L_H, that also partitions the robots into two disjoint sets each containing at least $\lceil \frac{n}{2} \rceil$ robots;
3: The robots move towards the point K that is the intersection of L_H and L_V.
4: The algorithm terminates once the final reliable robot reaches K;

First, we note that, in Algorithm 6, the existence of the lines L_H and L_V is ensured as a result of the ham-sandwich theorem (see [23] [Theorem 4.7]).

Now consider the four open regions R_1, R_2, R_3, and R_4 created by the intersection of L_H and L_V (as depicted in Fig. 4). Note that, by assumption, we have $F < \lceil \frac{n}{2} \rceil$ and we are therefore guaranteed to have at least one reliable robot in each of the regions R_1 and R_3, or in each of the regions R_2 and R_4.

Consider the reliable robot A that is furthest from K and assume without loss of generality that A is located in the region R_1. If there is a reliable robot B in R_3 then we have $|AB| \geq |AK|$ which implies that $\widehat{CR} \leq |AK|/(\frac{1}{2}|AB|) \leq 2$. If there is not a reliable robot in R_3 then there must be reliable robots B and C in R_2 and R_4 respectively. Let $d = \max\{|AB|, |AC|\}$ and let us adopt a coordinate system such that $K = (0,0)$ and $A = (A_x, A_y)$. Observe that $A_y \leq |AB| \leq d$ and $A_x \leq |AC| \leq d$. Thus, $|AK| = \sqrt{A_x^2 + A_y^2} \leq \sqrt{2}d$ and $\widehat{CR} \leq |AK|/(\frac{1}{2}d) \leq 2\sqrt{2}$.

The two lines L_H and L_V may be found in linear time by first choosing some line L' onto which we project the points in S. We then set L_H as the line perpendicular to L' dividing the points on L' in half (i.e. we need to find the median, $\mathcal{O}(n)$ time [6]). To find L_V we repeat with L' replaced with L_H.   □

4   Arbitrary Number of Byzantine Robots

In this section we consider algorithms that solve the $Gather(n, F)$ for any n and any F. We give two algorithms: the first, grid-rendezvous, is adapted from [12] and gives a constant competitive ratio independent of F. The second, shrinking-the-shortest-interval (SSI), gives a competitive ratio dependent on F.

4.1   Grid Rendezvous

We start with the grid-rendezvous algorithm which is a direct application of Algorithm 3 in [12]. The algorithm was originally designed to solve the rendezvous problem of two robots unaware of the other's position (but sharing a common coordinate system).

The idea of the algorithm is to calculate a hierarchy of grids $\Pi = \{\pi_0, \pi_1, ...\}$ which partition the plane into non-overlapping cells. The robots then travel through a series of potential meeting points located at the centers of ever larger cells from successive grids in Π.

In detail, each π_i exactly partitions the plane into square cells of side length 2^i such that one of the cells in π_i, the central cell, has its center at the origin. In order for the partition to be exact each cell is defined to include its top and right edges, as well as its top-right vertex (in addition to its interior).

We can nearly apply Algorithm 3 as given in [12]. We only need to specify the finest grid division that will be used by the robots. Let d_ϵ be the size of this finest grid cell. We present (the slightly modified) Algorithm 3 from [12] below.

Algorithm 7. (Grid-rendezvous [12])

1: The robots choose a d_ϵ much smaller than the closest pair of robots in the set;
2: The robots compute the hierarchy of grids Π;
3: **repeat** for $i = 1, 2, 3...$ and for each robot in S
4:     Set H equal to the cell of π_i containing your initial position p;
5:     Move to the center of H;
6:     Wait until $\sqrt{2} \cdot 2^{i-1}$ time has passed since the start of the current iteration;
7: **until** Gathering completed

The rendezvous time of the above algorithm is given by Corollary 9 in [12]. Using this time-bound we can state the following:

Theorem 6. *Consider the Gather(n, F) problem for the input S. Assume that the robots A and B are the closest pair of robots in S. Then the competitive ratio of Algorithm 7 is $\widehat{CR} \leq 2\sqrt{2}\left(16 + \frac{d_\epsilon}{|AB|}\right)$ where d_ϵ can be made as small as one chooses. The complexity of this algorithm is² $\mathcal{O}(n \log n)$.*

4.2   Shrink-Shortest-Interval

Consider the following algorithm, generalized from Algorithm 3 in [8]:

Algorithm 8. (Shrink-shortest-interval)

1: **repeat**
2:     Determine the two closest robots A and B in S that are not at the same position;
3:     Set D as the midpoint of A and B;
4:     Set $d = |AB|/2$.
5:     All robots move a distance d towards D;
6: **until** All robots in $\mathcal{N}$ gather.

Theorem 7. *Algorithm 8 solves the Gather(n, F) problem for the input S with competitive ratio at most $F + 2$. The complexity of the algorithm is $\mathcal{O}(n^2 \log n)$.*

To prove this we will need the following lemma:

Lemma 6. *Consider any point D and set of points S such that $A \in S$ is the closest point to D, and $C \in Sup[S]$ is the furthest point from D. Let S' be the positions of the points in S after moving them a distance $d \leq |AD|$ towards the point D. Then,*

$$Radius[S'] \leq \begin{cases} Radius[S] - d/2, & D \in \mathcal{MC}(S) \\ Radius[S], & otherwise \end{cases}.$$

---

² The complexity of the algorithm is entirely due to the determination of d_ϵ.

Proof (Theorem 7). Consider the *Gather*(n, F) problem for the input $\mathcal{S}$, let $\mathcal{N}$ be the subset of $\mathcal{S}$ that contains only reliable robots, and let f be the (actual) number of byzantine robots in $\mathcal{S}$.

Let $\mathcal{S}^{(i)}$ and $\mathcal{N}^{(i)}$ represent the unique positions of the robots in $\mathcal{S}$ and $\mathcal{N}$ after the ith iteration of the algorithm, and let $r_i = \text{Radius}[\mathcal{N}^{(i)}]$. We also let D_i be the midpoint and d_i be half the distance between the closest pair of points in $\mathcal{S}^{(i)}$. Finally, set $C_i \in \text{Sup}[\mathcal{N}^{(i)}]$ be the furthest point from D_i.

Now, if in the ith iteration the midpoint D_i lies within $\mathcal{MC}(\mathcal{N}^{(i)})$ then by Lemma 6 we have $r_{i+1} \leq r_i - d_i/2$. If we assume that their are m iterations of this kind then the time needed to complete these iterations is at most $T_m \leq \sum_{i=0}^{m} d_i \leq 2 \sum_{i=0}^{m} (r_i - r_{i+1})$. However, observe that $\sum_{i=0}^{m} r_i = r_0 + \sum_{i=1}^{m} r_i = r_0 + \sum_{i=0}^{m-1} r_{i+1}$ such that $T_m \leq 2r_0 - r_{m+1} \leq 2r_0$.

If D_i does not lie within $\mathcal{MC}(\mathcal{N}^{(i)})$, then we can only say that $r_{i+1} \leq r_i(1 - d_i/|C_i D_i|) \leq r_i$. However, observe that Algorithm 8 always gathers the two closest robots in $\mathcal{S}^{(i)}$ and we know that there is at least one pair of robots in $\mathcal{N}^{(i)}$ with separation no greater than $2r_i$. This tells us that $d_i \leq r_i$. Furthermore, since all reliable robots are, by definition, within $\mathcal{MC}(\mathcal{N})$, it is impossible for D_i to simultaneously be: (a) the midpoint of two reliable robots, and, (b) lie outside of $\mathcal{MC}(\mathcal{N})$. This implies that this type of iteration can occur at most f times (as it reduces the number of byzantine robots by one each time it occurs). Thus, the time needed to complete these iterations is at most $T_f = f \cdot r_0$.

Combining T_m and T_f gives us a bound on the total time necessary to complete the algorithm. We get $T \leq T_m + T_f = fr_0 + 2r_0 = (f + 2)r_0$. The bound on the competitive ratio follows from the fact that $f \leq F$, and $r_0 = \text{Radius}[\mathcal{N}]$ is the minimal time necessary to gather the robots in $\mathcal{N}$.

The complexity bound follows from the fact that we need to determine the closest pair of points $\mathcal{O}(n)$ times.   □

In the case that we have no knowledge of the number of byzantine robots in our collection (i.e. $F = n - 2$) the algorithm has a worst-case bound on the competitive ratio of n. This reflects the fact that an adversary, if allowed, would always choose $f = F$ robots in $\mathcal{S}$ to be byzantine. It is worth noting, however, that it was not necessary to know F in the proof of Theorem 7 and thus the algorithm has a competitive ratio that is bounded by the actual number of byzantine robots in $\mathcal{S}$. That is, for a particular instance $\mathcal{N} \subseteq \mathcal{S}$ such that $f = |\mathcal{S}| - |\mathcal{N}|$ we have $\text{CR}(\mathcal{N}) \leq f + 2 \leq F + 2$.

5   Conclusion

In this paper we analyzed the gathering problem for $n > 2$ robots in the plane at most F of which, $F \leq n - 2$, are byzantine. The robots were equipped with GPS and they could communicate their positions to a central authority. Several algorithms were designed with competitive ratio depending on the number of byzantine robots and the knowledge available to the robots.

In addition to improving the competitive ratio and/or complexity of our algorithms, several interesting open problems remain. In particular, one could consider models that allow the robots to communicate/exchange their positions at any time during the gathering process. Additionally, it would be interesting to consider robot gathering (in the presence of byzantine robots) under local (limited) communication range.

References

1. Agmon, N., Peleg, D.: Fault-tolerant gathering algorithms for autonomous mobile robots. SIAM J. Comput. **36**(1), 56–82 (2006)
2. Alpern, S.: The rendezvous search problem. SIAM J. Control. Optim. **33**(3), 673–683 (1995)
3. Alpern, S.: Rendezvous search: a personal perspective. Oper. Res. **50**(5), 772–795 (2002)
4. Alpern, S., Gal, S.: The Theory of Search Games and Rendezvous, vol. 55. Springer, New York (2003). https://doi.org/10.1007/b100809
5. de Berg, M., Cheong, O., van Kreveld, M., Overmars, M.: Computational Geometry: Algorithms and Applications, 3rd edn. Springer, Heidelberg (2008). https://doi.org/10.1007/978-3-540-77974-2
6. Blum, M., Floyd, R.W., Pratt, V., Rivest, R.L., Tarjan, R.E.: Time bounds for selection. J. Comput. Syst. Sci. **7**(4), 448–461 (1973)
7. Chrystal, G.: On the problem to construct the minimum circle enclosing n given points in the plane. Proc. Edinb. Math. Soc. **3**, 30–33 (1885)
8. Chuangpishit, H., Czyzowicz, J., Kranakis, E., Krizanc, D.: Rendezvous on a line of faulty, location-aware robots. In: Proceedings 13th International Symposium on Algorithms and Experiments for Wireless Networks, Vienna, Austria. LNCS. Springer (2017)
9. Cieliebak, M., Flocchini, P., Prencipe, G., Santoro, N.: Distributed computing by mobile robots: gathering. SIAM J. Comput. **41**(4), 829–879 (2012)
10. Cohen, R., Peleg, D.: Convergence properties of the gravitational algorithm in asynchronous robot systems. SIAM J. Comput. **34**(6), 1516–1528 (2005)
11. Cohen, R., Peleg, D.: Convergence of autonomous mobile robots with inaccurate sensors and movements. SIAM J. Comput. **38**(1), 276–302 (2008)
12. Collins, A., Czyzowicz, J., Gąsieniec, L., Kosowski, A., Martin, R.: Synchronous rendezvous for location-aware agents. In: Peleg, D. (ed.) DISC 2011. LNCS, vol. 6950, pp. 447–459. Springer, Heidelberg (2011). https://doi.org/10.1007/978-3-642-24100-0_42
13. Collins, A., Czyzowicz, J., Gąsieniec, L., Labourel, A.: Tell me where I am so I can meet you sooner. In: Abramsky, S., Gavoille, C., Kirchner, C., Meyer auf der Heide, F., Spirakis, P.G. (eds.) ICALP 2010. LNCS, vol. 6199, pp. 502–514. Springer, Heidelberg (2010). https://doi.org/10.1007/978-3-642-14162-1_42
14. Cord-Landwehr, A., Fischer, M., Jung, D., Meyer auf der Heide, F.: Asymptotically optimal gathering on a grid. In: Proceedings of the 28th ACM Symposium on Parallelism in Algorithms and Architectures, SPAA 2016, Asilomar State Beach/Pacific Grove, CA, USA, 11–13 July 2016, pp. 301–312. ACM (2016)
15. Courtieu, P., Rieg, L., Tixeuil, S., Urbain, X.: Certified universal gathering in $\mathbb{R}^2$ for oblivious mobile robots. In: Gavoille, C., Ilcinkas, D. (eds.) DISC 2016. LNCS, vol. 9888, pp. 187–200. Springer, Heidelberg (2016). https://doi.org/10.1007/978-3-662-53426-7_14

16. Czyzowicz, J., Gasieniec, L., Kosowski, A., Kranakis, E., Krizanc, D., Taleb, N.: When patrolmen become corrupted: monitoring a graph using faulty mobile robots. In: Elbassioni, K., Makino, K. (eds.) ISAAC 2015. LNCS, vol. 9472, pp. 343–354. Springer, Heidelberg (2015). https://doi.org/10.1007/978-3-662-48971-0_30

17. Czyzowicz, J., et al.: Search on a line by byzantine robots. In: 27th International Symposium on Algorithms and Computation, ISAAC 2016, Sydney, Australia, 12–14 December 2016. LNCS, pp. 27:1–27:12. Springer (2016)

18. Czyzowicz, J., Kranakis, E., Krizanc, D., Narayanan, L., Opatrny, J.: Search on a line with faulty robots. In: Proceedings of the 2016 ACM Symposium on Principles of Distributed Computing, PODC 2016, Chicago, IL, USA, 25–28 July 2016, pp. 405–414. ACM (2016)

19. Czyzowicz, J., Killick, R., Kranakis, E., Krizanc, D., Morale-Ponce, O.: Gathering in the plane of location-aware robots in the presence of spies. arXiv preprint arXiv:1712.02474 (2017)

20. Das, S., Flocchini, P., Santoro, N., Yamashita, M.: On the computational power of oblivious robots: forming a series of geometric patterns. In: Proceedings of the 29th PODC, Zurich, Switzerland, 25–28 July 2010, pp. 267–276. ACM (2010)

21. De Marco, G., Gargano, L., Kranakis, E., Krizanc, D., Pelc, A., Vaccaro, U.: Asynchronous deterministic rendezvous in graphs. Theor. Comput. Sci. 355(3), 315–326 (2006)

22. Dieudonné, Y., Pelc, A., Peleg, D.: Gathering despite mischief. ACM Trans. Algorithms (TALG) 11(1), 1 (2014)

23. Edelsbrunner, H.: Algorithms in Combinatorial Geometry, vol. 10. Springer, Heidelberg (2012)

24. Flocchini, P., Prencipe, G., Santoro, N., Widmayer, P.: Gathering of asynchronous robots with limited visibility. Theor. Comput. Sci. 337(1–3), 147–168 (2005)

25. Hromkovič, J., Klasing, R., Monien, B., Peine, R.: Dissemination of information in interconnection networks (broadcasting & gossiping). In: Du, D.Z., Hsu, D.F. (eds.) Combinatorial Network Theory. APOP, vol. 1, pp. 125–212. Springer, Boston (1996). https://doi.org/10.1007/978-1-4757-2491-2_5

26. Izumi, T., Souissi, S., Katayama, Y., Inuzuka, N., Défago, X., Wada, K., Yamashita, M.: The gathering problem for two oblivious robots with unreliable compasses. SIAM J. Comput. 41(1), 26–46 (2012)

27. Jadhav, S., Mukhopadhyay, A.: Computing a centerpoint of a finite planar set of points in linear time. Discret. Comput. Geom. 12(3), 291–312 (1994)

28. Kranakis, E., Krizanc, D., Rajsbaum, S.: Mobile agent rendezvous: a survey. In: Flocchini, P., Gąsieniec, L. (eds.) SIROCCO 2006. LNCS, vol. 4056, pp. 1–9. Springer, Heidelberg (2006). https://doi.org/10.1007/11780823_1

29. Lynch, N.A.: Distributed Algorithms. Morgan Kaufmann, Burlington (1996)

30. Pelc, A.: DISC 2011 invited lecture: deterministic rendezvous in networks: survey of models and results. In: Peleg, D. (ed.) DISC 2011. LNCS, vol. 6950, pp. 1–15. Springer, Heidelberg (2011). https://doi.org/10.1007/978-3-642-24100-0_1

31. Prencipe, G.: Impossibility of gathering by a set of autonomous mobile robots. Theor. Comput. Sci. 384(2–3), 222–231 (2007)

32. Souissi, S., Défago, X., Yamashita, M.: Gathering asynchronous mobile robots with inaccurate compasses. In: Shvartsman, M.M.A.A. (ed.) OPODIS 2006. LNCS, vol. 4305, pp. 333–349. Springer, Heidelberg (2006). https://doi.org/10.1007/11945529_24

33. Suzuki, I., Yamashita, M.: Distributed anonymous mobile robots: formation of geometric patterns. SIAM J. Comput. 28(4), 1347–1363 (1999)

34. Yang, Y., Souissi, S., Défago, X., Takizawa, M.: Fault-tolerant flocking for a group of autonomous mobile robots. J. Syst. Softw. **84**(1), 29–36 (2011)
35. Yu, X., Yung, M.: Agent rendezvous: a dynamic symmetry-breaking problem. In: Meyer, F., Monien, B. (eds.) ICALP 1996. LNCS, vol. 1099, pp. 610–621. Springer, Heidelberg (1996). https://doi.org/10.1007/3-540-61440-0_163

Formalizing Compute-Aggregate Problems in Cloud Computing

Pavel Chuprikov[1,2], Alex Davydow[1], Kirill Kogan[2], Sergey Nikolenko[1(✉)], and Alexander Sirotkin[1,3]

[1] Steklov Institute of Mathematics at St. Petersburg, St. Petersburg, Russia
pavel.chuprikov@imdea.org, adavydow@gmail.com, sergey@logic.pdmi.ras.ru,
avsirotkin@hse.ru
[2] IMDEA Networks Institute, Madrid, Spain
kirill.kogan@imdea.org
[3] National Research University Higher School of Economics, St. Petersburg, Russia

Abstract. Efficient representation of data aggregations is a fundamental problem in modern big data applications, where network topologies and deployed routing and transport mechanisms play a fundamental role to optimize desired objectives: cost, latency, and others. We study the design principles of routing and transport infrastructure and identify extra information that can be used to improve implementations of compute-aggregate tasks. We build a taxonomy of compute-aggregate services unifying aggregation design principles, propose algorithms for each class and analyze them.

Keywords: Compute-aggregate problems · Cloud computing

1 Introduction

Data centers store data at different interconnected locations. Modern big data applications are highly distributed, and requests need to satisfy various objectives: latency, cost efficiency, etc. [2,5,12]. *Compute-aggregate* problems, where several data chunks must be aggregated in a network sink, encompass an important class of big data applications implemented in modern data centers. Traditionally, applications have little control over how network transport handles the data. Latency optimization should account for properties of underlying transports in order to avoid, e.g., the *incast problem* [7,23], and optimizing latency for several compute-aggregate tasks can overload "fastest" (and more expensive) links. We believe that more fine-grained control is required to implement desired objectives transparently for applications.

In this work, we assume that each compute-aggregate task should conform to a budget constraint since different cloud tenants are able to invest different economic resources to compute their aggregations. To avoid oversubscription of

This work was supported by the Russian Science Foundation grant 17-11-01276.

Z. Lotker and B. Patt-Shamir (Eds.): SIROCCO 2018, LNCS 11085, pp. 377–391, 2018.
https://doi.org/10.1007/978-3-030-01325-7_31

"fastest" links, they can also have different costs of sending data over a link. The problem now divides into two completely decoupled phases: (1) find a "cheapest" plan given a distribution of data over the network, an aggregation function (that computes the size of aggregating two different pieces of data), and the cost of sending a unit of data over a link; and (2) actually redistribute aggregations computed in (1) while optimizing desired objectives. We can solve the first phase in a serial way, independently of the properties of underlying transport protocols, while the second phase can address such problems as incast. This is a natural generalization of traditional transports to implement efficient aggregations.

The first phase is of separate interest since it can represent various economic settings (e.g., energy efficiency) during aggregation; this phase can also lead to better utilization of network infrastructure since the cost to send a unit of data through the links can differ for different compute-aggregate instances. Hence, our primary goal is to identify universal properties of compute-aggregate tasks that allow for unified design principles of "perfect" aggregations on the first phase. Incorporating properties of aggregation functions into final decisions requires new insights on the model level and may lead to more efficient aggregation. There is definitely room for it: the average final output size jobs is 40.3% of the initial data sizes in Google [10], 8.2% in Yahoo, and 5.4% in Facebook [6].

In this work, we define a model for constructing an aggregation plan under budget constraints that requires applications to specify only one property: the (approximate) size of two data chunks after aggregation. Properties of aggregation functions can have a significant effect on the aggregation plan. We classify compute-aggregate tasks with respect to this property, propose algorithms for these optimization problems, and analyze their properties, proving a number of results on their performance and complexity, both positive (polynomial algorithms with good approximation ratios) and negative (inapproximability results).

The paper is organized as follows: Sect. 2 summarizes prior art, Sect. 3 introduces the model, Sect. 4 shows a classification of aggregate functions and their computational properties, i.e., hardness and approximability of optimization problems, and Sect. 5 concludes the paper.

2   Related Work

Various frameworks split computations into multiple phases: Camdoop [8] assumes that an aggregation's output size is a specific fraction of input sizes, Map-Reduce-Merge [19] extends MapReduce to implement aggregations, Astrolabe [15] collects large-scale system state and provides on-the-fly attribute aggregation, and so on. Like other data-flow systems [10,20,21], Naiad [14] offers the low latency of stream processors together with the ability to perform iterative and incremental computations. The work [21] introduces a distributed memory abstraction for fault-tolerant in-memory computation on large clusters, with orders of magnitude better latency than disk accesses. Other stream processing frameworks support low-latency dataflow computations over a static dataflow graph [1,13,16], while [9] explores optimal tree overlays to optimize latency of compute-aggregate tasks under specified budget constraints.

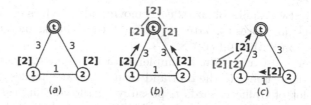

Fig. 1. Sample compute-aggregate task with 3 vertices, t (target), u, and v: (a) graph; (b) "move to root" plan with cost 12; (c) optimal aggregation plan with cost 11. Transmission cost of every edge is specified at the middle, e.g., $c(u, v) = 1$.

3   Motivation and Model Description

Our main objective is to use a network in the best possible way for a given compute-aggregate task. This is a problem with many variables. In this work, we leave most of them to the network transport layer (e.g., it chooses how to spread transmissions in time), concentrating on the *aggregation plan* that defines the order of aggregation and is fully decoupled from transport implementation.

3.1   Compute-Aggregate Tasks and "move to root" Plans

We model a network as an undirected connected graph $G = (V, E)$, where V is the set of computing nodes connected by links (edges) E. Since we operate on an application level, we can use any overlay topology in place of G that captures only information relevant to a specific compute-aggregate task. We model a task as a set of data chunks $C = \{ \boxed{x_0}, \boxed{x_1}, \ldots, \boxed{x_k} \}$, with each chunk $\boxed{x_i}$ characterized by its location $v(\boxed{x_i})$ and size $\mathrm{size}(\boxed{x_i})$. Many compute-aggregate tasks require to send the result to a specific node (e.g., to allow low-latency responses), so we introduce a root vertex $t \in V$ where all data chunks should be finally aggregated.

The hardest part to define is "the best possible way": application-specific objectives may include latency, throughput, or a more subtle objective such as congestion avoidance. We model them with a single per-link parameter, the *cost*, a flexible way to both freely combine objectives and keep the optimization problem clear. Formally, the cost function $c : E \to \mathbb{R}_+$ on the topology graph G maps each link e to its transmission cost per data unit $c(e)$; to transmit $\boxed{x}$ through e one must pay $c(e) \cdot \mathrm{size}(\boxed{x})$. A simple example of a compute-aggregate task is shown on Fig. 1a. Costs are shown on the edges, square brackets denote chunks, and the root vertex is marked by t.

We begin with the simplest form of an aggregation plan that we call "move to root": bring everything to the root node t (we say that an aggregation plan *moves* or *aggregates* for simplicity; in practice data transmission and aggregation are handled by the transport and application layers respectively). "Move to root" can be suboptimal with regard to transmission costs. Suppose that in the example on Fig. 1 the aggregation function chooses the best chunk, so the aggregated size does not exceed the maximal size of initial chunks. Now "move to root" has total cost 12 (Fig. 1b: two chunks of size 2 each moving along edges of cost 3), while on Fig. 1c one chunk moves to vertex 1 paying 2, then chunks merge, and chunk of size 2 moves to t with total cost 8.

Other concerns also arise. A naive implementation of the "move to root" plan that moves all data chunks to the root and then aggregates makes the transport layer direct a lot of traffic towards t, possibly overflowing ingress buffers and increasing latency due to the notorious TCP-incast problem. Moreover, in a low-latency application that aggregates in RAM [4,22] storage capacity can be exhausted when all data chunks are stored at t. This problem can be alleviated with intermediate aggregations: send data chunks to t sequentially in some order, aggregating some chunks immediately in the process. Recent studies [6,10] show that the result of a compute-aggregate task is often only a small fraction (usually less than half) of the total input size; e.g., in counting problems the aggregation result is just a few numbers. Thus, keeping in memory one intermediate chunk instead of initial data chunks can significantly reduce storage requirements.

In general, not every order can be used for intermediate aggregations because the final aggregation result might depend on this order (e.g., string field concatenation), and it is undesirable for an aggregation plan to affect the result [18]. Fortunately, most aggregation functions do not depend on the aggregation order, that is, they are *associative*: $\mathbf{aggr}(\boxed{x}, \mathbf{aggr}(\boxed{y}, \boxed{z})) = \mathbf{aggr}(\mathbf{aggr}(\boxed{x}, \boxed{y}), \boxed{z})$, and *commutative*: $\mathbf{aggr}(\boxed{x}, \boxed{y}) = \mathbf{aggr}(\boxed{y}, \boxed{x})$. Below we assume that aggregations are both associative and commutative; such systems as *MapReduce* already assume this for most *reduce* functions and allow aggregations of intermediate data chunks with combiner functions [10]. The TCP-incast problem, on the other hand, can be mitigated by spreading chunk transmissions in time (to reduce overlap), which requires complex synchronization on the part of the transport layer. Low-latency in-RAM applications also have to synchronize data transmissions to avoid too many data chunks "in the air" at the same time that cannot be aggregated; this is hard to implement in a distributed system, and if **aggr** is not commutative and associative this leads to more constraints since transmissions must occur in a specific order. All of the above suggests that it is hard for the "move to root" heuristic to reconcile network transport limitations with storage constraints and the distributed environment.

3.2   Moving Aggregation to Data and Aggregation Functions

The basic principle of *data locality optimization*, which lies at the heart of the *Hadoop* framework [17], is to *move computation to data* and as a result save on data transmission. We extend this strategy and try to *move aggregation to data* by allowing an aggregation plan to exploit intermediate nodes. Formally, an aggregation plan is a sequence P of operations $(o_0, o_1, \ldots, o_m)$, where each o_i is either $\mathbf{move}(\boxed{x}, v)$, which moves a chunk x to a vertex v, or $\mathbf{aggr}(\boxed{x}, \boxed{y})$, which merges chunks $\boxed{x}$ and $\boxed{y}$ located at the same vertex; the result is a new chunk $\boxed{xy}$ at that vertex. After all operations have been applied, the result must be a single data chunk $\boxed{z}$ at the root: $v(\boxed{z}) = t$. E.g., Figs. 1b and c show aggregation plans for the problem on Fig. 1a. Aggregation plans are fully decoupled from the transport layer, producing instructions and constraints that the transport layer must satisfy.

An aggregation plan has an associated transmission cost $\text{cost}(P)$, which is the sum of costs of all operations in P; here $\text{cost}(\textbf{aggr}(\boxed{x}, \boxed{y})) = 0$ (there is no data transmission), and $\text{cost}(\textbf{move}(\boxed{x}, v)) = \text{size}(\boxed{x}) \cdot d(\text{v}(\boxed{x}), v)$, where $d(u, v)$ is the total cost of the cheapest path from u to v.

This approach of "moving aggregation to data" has some important advantages over "move to root". First, the TCP-incast problem becomes less pronounced because inbound traffic is spread among different nodes, and fewer nodes need to be synchronized. Moreover, the total number of transmitted bits is reduced due to earlier aggregations (we usually expect an aggregation result to be smaller than the total input size). Second, storage capacity is now less of a constraint since less data has to be collected per node. Last but not least, data transmission cost is also reduced (cf. examples on Fig. 1). Note, however, that in practice not all nodes may be used for data aggregation. For example, we may be restricted to nodes where initial data chunks reside because it is expensive to allocate additional compute nodes; or it can be a security concern to perform computation on intermediate nodes (e.g., initial nodes belong to a private cloud, and the rest are transit nodes). This question must be carefully answered, and in what follows we assume that the overlay graph G reflects this answer and maps $\textbf{aggr}$ operations to appropriate nodes.

In order to formally define the optimization problem for aggregation, we have to know the following: given $\boxed{x}$ and $\boxed{y}$, what is the size of their aggregation result $\boxed{xy}$? This directly affects the cost of an aggregation plan, and different aggregation result sizes can lead to very different solutions. For example, if on Fig. 1 we assumed that the task is, e.g., sorting, where the size of an aggregated chunk is the sum of input sizes, the cost of the first plan would still equal 12, but the plan on Fig. 1c would now cost 14 and become suboptimal.

Unfortunately, the size of an aggregation result is application-specific, and in most cases the exact value depends on the actual content of $\boxed{x}$ and $\boxed{y}$; moreover, to determine this value we may need to actually perform aggregation (e.g., the number of key-value pairs in the counting problem cannot be predicted exactly unless we actually count). This is clearly infeasible since an aggregation plan must be constructed (and its cost evaluated) before the application performs any aggregations and the transport layer transmits any data. Therefore, we require each application to supply the *aggregation size function* $\mu : \mathbb{R}_+ \times \mathbb{R}_+ \to \mathbb{R}_+$ that would estimate this size using only sizes of the inputs, so that for the purposes of optimization $\text{size}(\boxed{xy}) = \mu(\text{size}(\boxed{x}), \text{size}(\boxed{y}))$. We do not expect these functions to be exactly correct, but they should provide the correct order of magnitude in order for the optimal solution to be actually good in practice. Since $\textbf{aggr}$ is assumed to be associative and commutative, μ should also have these properties. Some examples of μ for practical problems include: $\mu(a, b) = \text{const}$ for finding the top k elements in data with respect to some criterion; $\mu(a, b) = \min(a, b)$

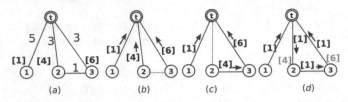

Fig. 2. Different μ lead to different plans: (a) sample task; (b) optimal plan for $\mu(a,b) = a + b$; (c) for $\mu(a,b) = \max(a,b)$; (d) for $\mu(a,b) = \min(a,b)$.

or $\mu(a,b) = \max(a,b)$ for choosing the best data chunk; $\mu(a,b) = a + b$ for concatenation or sorting; $\max(a,b) \leq \mu(a,b) \leq a+b$ for set union (word count).

Figure 2 shows how μ can affect the optimal aggregation plan. Figure 2a shows chunks of size 1 at vertex 1, of size 4 at vertex 2, and of size 6 at vertex 3, and the goal is to aggregate them at vertex 0. For $\mu(a,b) = a + b$, the optimal plan is to move each chunk to the root separately (Fig. 2b). For $\mu(a,b) = \max(a,b)$, it is cheaper to first move the chunk of size 4 along edge $2 \rightarrow 3$ and merge it, then move the resulting chunk of size 6 to the root (Fig. 2c). Finally, for $\mu(a,b) = \min(a,b)$ the optimal plan is to traverse the whole graph with the smallest chunk, merging larger ones along the way (Fig. 2d). Thus, even in a simple example the aggregation plan can change drastically depending on μ.

Problem 1. (CAM—*compute-aggregate minimization*). Given an undirected connected graph $G = (V, E)$, cost function c, a target vertex t, a set of initial data chunks C, and an aggregation size function μ, the CAM$[\mu]$ problem is to find an aggregation plan P such that cost(P) is minimized.

Interestingly, for general μ there is little we can do in the worst case.

Theorem 1. *Unless $P = NP$, there is no polynomial time constant approximation algorithm for* CAM *without associativity constraint on μ even if G is restricted to two vertices.*

Proof. We can encode an NP-hard problem in choosing the correct order of merging for a non-associative μ. For example, consider an instance of the knapsack problem with weights $w_1, \ldots, w_n$, unit values, and knapsack size W; then we have n chunks of size $w_1, \ldots, w_n$, and μ is defined as follows: if either $x = 0$, $y = 0$, or $x + y = W$ then $\mu(x,y) = 0$; else $\mu(x,y) = x + y$. This way, if we can fill the knapsack exactly the total resulting weight will be zero, and if not, it will be greater than zero, leading to unbounded approximation ratio unless we can solve the knapsack problem. □

In this work, we investigate two main degrees of freedom that the CAM problem has: network topology graph G and aggregation size function μ.

4   A Taxonomy of Aggregation Functions

There are different types of big data applications, a large variation in data-center network topologies, and countless data distributions, which collectively define

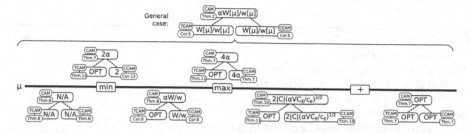

Fig. 3. The structure of our results in relation to aggregation size function μ and problem instance (CAM, TCAM, CCAM).

constraints for a compute-aggregate task. Handling each and every variation of these constraints separately does not scale, so a generalized decision procedure should be used to construct an aggregation plan. In this section, we present such a procedure and show worst-case guarantees for every choice.

Intuitively, stricter constraints may lead to better decisions, both in terms of the *cost of an aggregation plan*, which is our primary objective, and *performance* (running time). E.g., a better algorithm if the network graph is a tree or under certain constraints on the aggregation size function. Thus, a possible solution may have to account for *network topology, aggregation size function*, and *initial chunk distribution*. Chunk distribution varies from one instance to another and is unlikely to prove useful since it is expected to be roughly uniform (big data storage systems try to achieve even load distribution). Although there are common network topologies, such as *hypercube, fat-tree*, or *jellyfish*, there are plenty of variations and exceptions. So, while algorithms specifically tailored for, e.g., the hyper-cube topology [8] remain a valid topic for future study, in this work we mostly consider the aggregation size function, with only two special cases.

First, a tree is a topology that is both widespread and has a high potential for better algorithmic solutions; we call the CAM problem where G is a tree TCAM (tree CAM). Second, sometimes it is reasonable to limit aggregation to only those nodes that contain data chunks initially, either for security reasons or due to the need for additional resource provisioning on intermediate nodes that may significantly increase latency, while nodes with initial chunks usually already have computing resources for a preprocessing stage. If either security or provisioning impose the aforementioned restriction then a network graph G can be reduced to a complete graph over the nodes that contain chunks, and we call this special case CCAM (complete CAM). Our theoretical results are summarized in Fig. 3; the horizontal axis corresponds to how fast μ grows, and each small tree of results shows approximation ratios for CAM, TCAM, and CCAM, referring to specific theorems below.

4.1 General Case

In this subsection we assume no constraints on μ. As an example, consider a simpler setting where all chunks have size x, and $\mu(x, x) = x$. When paths of

---

Algorithm 1. steiner$(G, V' \subseteq V(G))$

1: **if** G is a tree **then**
2:     **return** unique subtree $T_{V'}$ covering V'
3: **else if** $V(G) = \{v(x) : x \in C\}$ **then**
4:     **return** min_spanning_tree(G)
5: **else**
6:     **return** steiner_tree_approx$(G, \{v(x) : x \in C\})$

Algorithm 2. steiner_rec(G, t, C)

1: $P \leftarrow ()$; $T \leftarrow$ steiner$(G, \{v(x) : x \in C\})$
2: **for** $v \in T$ in decreasing order of depth(v) **do**
3:   ▷ Denote $C(v) = \{x \in C : v(x) = v\}$
4:     **while** $|C(v)| > 1$ **do**
5:         P.append($\mathbf{aggr}(x, y)$), where $x, y \in C(v)$
6:         C.update($\mathbf{aggr}(x, y)$)
7:     **if** $\exists\, x \in C(v)$ **and** $\exists$ parent(v) **then**
8:         P.append($\mathbf{move}(x, \text{parent}(v))$)
9:         C.update($\mathbf{move}(x, \text{parent}(v))$)
10: **return** P

---

two chunks intersect, it is always better to merge at the intersection, so an optimal aggregation plan always proceeds along a tree subgraph of G, and it has cost equal to the weight of the tree multiplied by x. Thus, the problem reduces to finding a minimum weight tree that connects a given set of vertices, which is a well-studied minimum Steiner tree problem [11], MStT, that has many constant approximation algorithms. Using one of those we build our first aggregation plan construction algorithm steiner_rec (Algorithm 2). If there is a polynomial α-approximate algorithm for MStT, then steiner_rec provides an α-approximation for the special case when size$(x) = S$ for any $x \in C$, and $\mu(S, S) = S$. The steiner_rec algorithm has a number of interesting properties; e.g., it does not require any knowledge of μ or even chunk sizes. The infrastructure can run steiner_rec even before preprocessing (in map-reduce terminology, before a *map* phase). It turns out that in the general case, the price of using steiner_rec does not exceed the *ratio between the largest and smallest intermediate chunk*. We denote by $W_C[\mu]$ the maximal aggregate size of a subset of chunks from the set C, $W_C[\mu] = \max_{C' \subseteq C}\{\mu(C')\}$; it is well defined since μ is associative and commutative. We also denote by $w_C[\mu]$ the corresponding minimal aggregate size, $w_C[\mu] = \min_{C' \subseteq C}\{\mu(C')\}$.

Theorem 2. *If there exists a polynomial α-approximate algorithm for MStT, then there exists a polynomial algorithm that solves* CAM$[\mu]$ *with approximation factor* $\alpha \frac{W_C[\mu]}{w_C[\mu]}$.

Proof. First, note that any algorithm, even optimal, has to traverse at least the Steiner tree of G in total size, and has to carry at least weight w_c over each edge. The approximate algorithm begins by constructing the approximate Steiner tree with approximation ratio α, and then carries all chunks along this tree to the root, merging the chunks at first opportunity; in this process, the maximal possible chunk size is W_c, and it is carried over at most α times longer distance than in the actual Steiner tree, getting the approximation bound. $\square$

A well-known 2-approximation to MSTT is based on a minimum spanning tree (MST) of the distance closure G^* of G; the best known approximation ratio is $\ln 4 + \varepsilon \leq 1.39$ [3]. Although steiner_rec does not depend on either μ or chunk sizes, the approximation factor in Theorem 2 includes both. This result improves for special cases of TCAM and CCAM. E.g., if every vertex in G contains a data chunk, MSTT is equivalent to MST, so we get the following.

Theorem 3. *If every vertex in G contains a data chunk, then MSTT can be solved exactly in polynomial time.*

Theorem 4. *If G is a tree, MSTT can be solved exactly in polynomial time.*

Proof. There is only one subtree in G that connects a given set of vertices, and it can be found in polynomial time. $\square$

Theorems 3 and 4 essentially say that in these special cases we have 1-approximation algorithms for MSTT. Theorem 2 and this observation together imply the following.

Corollary 1. *There exist polynomial algorithms that solve CCAM$[\mu]$ and TCAM$[\mu]$ on a set of chunks C with approximation factor $\frac{W_C[\mu]}{w_C[\mu]}$.*

However, for many μ, including important ones (e.g., set union), Theorem 2 and Corollary 1 provide rather weak approximations; in particular, we would like to have approximation ratios independent of chunk sizes and specific values of μ since in practice $\frac{W_C}{w_C}$ may be very high. Unfortunately, it is impossible even for a restricted class of functions μ that reduce the weights.

Theorem 5. *There exists an aggregation size function μ such that $\forall a, b$ $\mu(a, b) \leq \min(a, b)$, and no polynomial time constant approximation algorithm for CCAM$[\mu]$ or TCAM$[\mu]$ exists unless $P = NP$.*

Proof. Consider a complete graph G where the root r contains an infinitely large chunk, all non-root vertices are terminals, edges between two terminal vertices cost 1, and edges between a terminal vertex and the root cost ∞. Given an instance of Set Cover, where a set S must be covered with a minimal number of m subsets $S_i \subseteq S$, we define $n(S_i)$ as the number with binary representation equivalent to $S \setminus S_i$ (for some fixed order of elements in the set). We encode S by a chunk of size 0 and any other subset $A \subset S$ by a chunk of size $n(A) + 4n \times 2^{|S|}$. The aggregation size function for two chunks corresponding to subsets A and B produces a chunk of size $n(A \cup B)$. Now, if there exists a set cover $S_{i_1}, S_{i_2}, \cdots, S_{i_k}$ then there is a solution to CAM$[\mu]$ of size $k \times 4n \times 2^{|S|} + c$, where $c \leq n \times 2^{|S|}$

(we can aggregate S_{i_j} in any order and then aggregate the rest with a zero chunk we obtained). On the other hand, if there exists a solution to CAM[μ] of size $g \times 4n \times 2^{|S|} + c$, where $c \leq n \times 2^{|S|}$, then there exists a solution to Set Cover of size g (to achieve this solution of CAM[μ] we have to obtain 0 in at most g aggregations). Thus, a constant approximation for CAM[μ] implies a constant approximation of Set Cover which is impossible unless $P = NP$. For TCAM[μ], consider the following transformation of G to a tree T_G: remove all the edges; introduce a new vertex c; connect c with r by an edge of weight ∞ and with the rest of G's vertices by edges of weight 1. Changing G to T_G does not increase cost more than twice (we traverse two edges now). Thus, the transformation preserves approximations, which again implies that TCAM[μ] does not have constant approximations unless $P = NP$.   □

Since CCAM is a strict subset of CAM, there is no constant-approximation solution for CAM either.

4.2   Range-Bounded Aggregation Size Functions

Depending on the application, the value of μ may be known to lie in a certain range. For example, if **aggr** represents *set union* then $\mu(x, y) \in [\max\{x, y\}, x+y]$, and if **aggr** represents *outer join* then $\mu(x, y)$ is likely to be always larger than $x + y$. We show a taxonomy of algorithms for different μ. Theorem 5 showed that aggregation size functions that reduce size too much are provably hard. On the other side of the spectrum, where $\mu(x, y) \geq x + y$, there is an optimal solution: bring all chunks to the sink.

Theorem 6. *If $\mu(a, b) \geq a+b$ for all a, b then there exists a polynomial optimal algorithm for CAM[μ], CCAM[μ], and TCAM[μ]; for TCAM[μ] the running time is $O(|C| + |G|)$.*

Proof. In this case, it does not make sense to merge chunks at all, it is optimal to bring all chunks separately to the sink. Formally, consider an optimal aggregation plan for CAM that merges two chunks not at the sink. Next, consider a transformed plan that carries both chunks separately and treats them separately until the final vertex. Since $\mu(a, b) \geq a+b$, the total cost does not increase in this transformation, so we can get an optimal plan without merging. The optimal strategy without merging is to move all chunks to the root along shortest paths, which can be computed in polynomial time. Because TCAM and CCAM are strict subsets of CAM, SPT is optimal for them too. For TCAM computing shortest paths is trivial, and the running time becomes linear.   □

We have found that for $\mu(x, y) \in (-\infty, \min\{x, y\}]$ the problem is inapproximable (Theorem 5), and for $\mu(x, y) \in [x + y, \infty)$ there is an optimal algorithm (Theorem 6). We split the remaining range $[\min\{x, y\}, x + y]$ at $\max\{x, y\}$ for two reasons. First, in practice max is a valid bound for many applications: set intersection, set union, outer join (symmetric or asymmetric); thus, the infrastructure often knows on which side of max μ lies. Second, theoretic results below show that max is an interesting demarcation line for worst-case guarantees: below max chunk sizes are a primary factor, and above max the graph structure starts to dominate. If $\mu(x, y) \in [\min(x, y), \max(x, y)]$, we can replace the ratio $W_C[\mu]/w_c[\mu]$ (Theorem 2), which depends on μ, with a simpler one that depends only on chunk sizes. In the next theorem $W_C = \max_{x \in C}\{\text{size}(x)\}$, $w_c = \min_{x \in C}\{\text{size}(x)\}$.

Theorem 7. *If* $\min\{a, b\} \leq \mu(a, b) \leq \max\{a, b\}$ *for all* a, b *and there exists a polynomial α-approximate algorithm for* MStT, *then there exists a polynomial algorithm that solves* CAM$[\mu]$ *with approximation factor* $\alpha\frac{W_C}{w_C}$.

Corollary 2. *If* $\min\{a, b\} \leq \mu(a, b) \leq \max\{a, b\}$ *then there exist polynomial algorithms that solves* CCAM$[\mu]$ *and* TCAM$[\mu]$ *with approximation factor* $\frac{W_C}{w_C}$.

For $\mu(x, y) \in [\max\{x, y\}, x + y]$, the last remaining range, we employ a mix of SPT and `steiner_rec`: merge chunks above a certain threshold with `steiner_rec`; below, with SPT.

Theorem 8. *If for all* a *and* b $\max(a, b) \leq \mu(a, b) \leq a + b$, *then there is an* $2NV^{1/2}\sqrt{\alpha\frac{c_{\max}}{c_{\min}}}$-*approximate polynomial algorithm for* CAM$[\mu]$, *which we call* RECH_MStTSplit, *where* V *is the number of vertices in* G, N *is the number of chunks,* $c_{\max}$ *is the cost of the most expensive edge in* G, $c_{\min}$, *of the cheapest edge, and* α *is an approximation factor for* MStT.

Proof. The idea of the algorithm is as follows. We split all chunks C into two sets: chunks with weight at least δM go into set C_1 and chunks with weight smaller than δM go into C_2, where M is the weight of the maximal chunk and δ is a constant to be defined later, so $C = C_1 \cup C_2$. Next we solve two separate CAM problems. For C_1 we run the general algorithm from Theorem 2, and for C_2 we run the algorithm from Theorem 6 that we used for μ such that $\mu(a, b) \geq a + b$. The first algorithm yields an $\frac{\alpha N}{\delta}$-approximate solution, and the total weight of the second solution does not exceed $\delta MVNc_{\max}$, where N is the number of chunks. Let W be the weight of the optimal solution. Now, since $\max(a, b) \leq \mu(a, b)$, and W is at least the weight of the optimal solution for C_1, we can conclude that the weight of the solution for C_1 is at most $\frac{\alpha V}{\delta}W$. On the other hand, since $W \geq Mc_{\min}$, the weight of the solution for C_2 is at most $\delta V^2\frac{c_{\max}}{c_{\min}}W$. Now if we choose $\delta = \frac{\sqrt{ac_{\min}}}{V^{1/2}\sqrt{c_{\max}}}$ to minimize the total result, the total weight of both solutions will be $2NV^{1/2}\sqrt{\alpha\frac{c_{\max}}{c_{\min}}}W$. $\square$

To improve the above theorem we cannot apply Theorem 3 to get rid of α for CCAM or TCAM since it uses Steiner tree only for a subset of chunks. But, remarkably, we can do better for TCAM: the following theorem proves that between *max* and "+" `steiner_rec` is optimal for TCAM. The algorithm is similar to Theorem 2.

Theorem 9. *There exists a polynomial optimal algorithm for the* TCAM$[\mu]$ *problem for any μ such that* $\forall a, b$ $\max(a, b) \leq \mu(a, b) \leq a + b$.

Proof. The algorithm is similar to Theorem 2: move chunks towards t, merging them in intermediate nodes. Consider an arbitrary subtree T of G. All data chunks from T have to be eventually moved upwards using parent edge e (if $T \neq G$). Minimal cost of this operation is $\leq s_T$, where s_T is the size of the aggregation result of all chunks from $C|_T$. Can it be less? Assume the opposite: consider a set of data chunks X that will be moved upwards through e s.t. $C_T \subseteq X^* = \bigcup_{x \in X} x^*$ and $\sum_{x \in X} \text{size}(x) < s_T$, where x^* is the set of initial chunks that contributed to x. If $C_T \subsetneq X^*$, we can throw away $X^* \setminus C_T$ without any increase in the cost because μ is at least max. Since μ does not exceed the sum, we can aggregate X with no cost increase. The resulting chunk has size s_T, but by construction each upward edge e from T will add exactly s_T. □

4.3   Specific Aggregation Size Functions

Sometimes we can improve performance further if we know μ exactly. This is especially interesting for the "junction points" between previous results.

Theorem 10. *If there exists a polynomial α-approximate algorithm for MSTT, then there exists a polynomial 2α-approximate algorithm for CAM[min].*

Proof. Given an instance (G, t, C) of the CAM[min] problem, first we find an α-approximation T to the MSTT instance $(G, V' = \{t\} \cup \{v(x) : x \in C\})$. Then, we construct an aggregation schedule by taking a data chunk with the smallest size and walking it through T. The resulting cost does not exceed $2m \cdot w(T)$, where m is the size of the smallest chunk. Similar to Theorem 13, an aggregation schedule defines a subgraph $H \supseteq V'$, and so incurs the cost of at least $m \cdot w(H)$. A sample solution for this algorithm is shown on Fig. 4. □

Corollary 3. *There exists a polynomial 2-approximate algorithm for CCAM[min].*

Theorem 11. *There is an optimal polynomial algorithm for TCAM[min].*

Proof. The optimal algorithm uses dynamic programming. Consider an instance $(G = (V, E), t, C)$ of the TCAM[min] problem, where G is a tree with a root t. For every vertex $v \in V$ we compute $\text{mc}(v)$, the size of the smallest chunk in a subtree T_v rooted at v, and for every $c \in C$ we compute $\text{dp}(v, \text{size}(c))$, an optimal solution for T_v with an additional chunk of size $\text{size}(c)$ at v. We can find $\text{mc}(v)$ for every vertex in linear time by running depth first search. If $\text{dp}(v, \text{size}(c))$ are known for every $u \in \texttt{children}(v)$ then $\text{dp}(v, \text{size}(c))$ can be computed as $\text{dp}(v, \text{size}(c)) = \sum_{u \in \texttt{ch}(v)} \min\{2 \cdot \text{size}(c) \cdot \text{d}(v, u) + \text{dp}(u, \text{size}(c)), \text{mc}(u) \cdot \text{d}(v, u) + \text{dp}(u, \text{mc}(u))\}$. Now $\text{dp}(t, \text{mc}(t))$ contains the cost of an optimal aggregation plan, which can be found with backtracking. □

Our approximate algorithm for CAM[max] is also based on MSTT, but with a different construction.

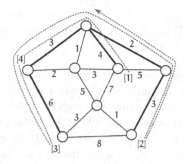

Fig. 4. Sample solution from Theorem 10. Steiner tree approximation is shown with bold lines, and the resulting path of the smallest chunk is shown by a dotted arrow.

Theorem 12. *If there exists a polynomial α-approximate algorithm for* MStT, *then there exists a polynomial 4α-approximate algorithm for* CAM[max], *which we call* RECH_MStTMax.

Proof. Let the maximal chunk size be equal to M. We break data chunks into several subsets: first with chunks with sizes in $(M/2, M]$, second in $(M/4, M/2]$, and so on. We build a solution for the first subset, extend it to a solution for the first two subsets, and so on. First, build a Steiner tree for the root and the first subset and solve the problem there. This solution is at least 2α-competitive, where α is the MStT approximation factor: edges used by a different solution must connect every chunk to the root, so their total cost is at least the cost of a minimum Steiner tree, and their sizes are at least $M/2$. Next, we merge the tree obtained on the first iteration into a single vertex, throw away the first subset, build a Steiner tree for the second subset, solve the problem for this tree, and so on. Suppose that there were k such subsets. Since we move chunks of size at most $M/2^{i-1}$, and merging vertices does not increase the weight of a Steiner tree, the cost of the ith subset does not exceed $M/2^{i-1}\alpha\mathrm{ST}(i, i-1)$, where $\mathrm{ST}(i,j)$ is the optimal Steiner tree weight for chunks with sizes in $(M/2^i, M/2^j]$. Thus, the total cost does not exceed $2\alpha M \sum_{i=1}^{k} \mathrm{ST}(i, i-1)/2^i$. For the lower bound, we count the cost of all data movements across every edge. The total cost of all edges with chunks of mass at least $M/2$ moved along them is bounded by $\mathrm{ST}(1,0)$, so the cost is bounded by $\frac{M}{2}\mathrm{ST}(1,0)$; repeating the process for $\frac{M}{4}$, $\frac{M}{8}$ and so on, we get in total $M \sum_{i=1}^{k} \mathrm{ST}(i,0)/2^i$. Some of the edges are counted more than once: an edge with the largest chunk of size $M/2^j$ moved along it has been counted once with a factor of $M/2^j$, once with $M/2^{j+1}$, and so on, but the real lower bound for this edge is $M/2^j$. Thus for every edge we have an extra factor of $1 + 1/2 + 1/4 + \ldots \leq 2$, and the optimal cost is at least $M/2 \sum_{i=1}^{k} \mathrm{ST}(i,0)/2^i$. Since $\mathrm{ST}(i,0) \geq \mathrm{ST}(i, i-1)$, the total approximation factor is $\frac{2\alpha M}{M/2} = 4\alpha$. □

Theorem 9 implies that TCAM[max] has an optimal solution since $\max\{x, y\}$ lies trivially in $[\max\{x, y\}, x + y]$. However, results for CAM[min] and CAM[max] cannot be significantly improved because both are NP-hard.

Theorem 13. *If there exists $a > 0$ s.t. $\mu(a,a) = a$ then* CAM$[\mu]$ *is NP-hard and does not have less than $\frac{19}{18}$-approximate polynomial algorithms even if all edge weights are equal, unless $P = NP$.*

Proof. The proof is by the reduction from the MStT problem. Given a MStT instance $(G, w, V' \subseteq V)$, we place data chunks of size a in each vertex of V' except one, which becomes the sink. Any aggregation schedule defines a connected subgraph H of G that contains all vertices from V'. The minimal cost is $a \cdot w(H)$, where $w(H) = \sum_{e \in E(H)} w(e)$, since all transmitted data chunks have size a, and we can always avoid transmitting more than one chunk across one link. Any spanning tree of H defines a Steiner tree for V', and vice versa, any Steiner tree T defines an aggregation schedule with cost $a \cdot w(T)$.   □

Theorem 14. *The* CAM$[\min]$ *problem is NP-hard even if all edge weights are equal, and each vertex is required to contain a data chunk.*

Proof. This time we reduce the Hamiltonian cycle problem. Given a Hamiltonian cycle problem instance G, we choose the sink arbitrarily, place a chunk of size 1 in the sink and chunks of size $|V|^2$ in every other vertex. The optimal solution travels with weight 1 along a Hamiltonian cycle.   □

5   Conclusion

In this work, we have introduced a model to find a schedule of aggregations that satisfies budget constraints rather than directly optimizing desired objectives such as latency or throughput. We believe that this approach will allow to decouple optimization problems from underlying transports and provide fine-grained control to exploit network infrastructure. Our primary contribution is a classification of aggregation functions together with extensive theoretical analysis that lead to unified design principles of "perfect" aggregations.

References

1. Akidau, T., et al.: MillWheel: fault-tolerant stream processing at internet scale. PVLDB **6**(11), 1033–1044 (2013)
2. Al-Fares, M., Radhakrishnan, S., Raghavan, B., Huang, N., Vahdat, A.: Hedera: dynamic flow scheduling for data center networks. In: USENIX, pp. 281–296 (2010)
3. Byrka, J., Grandoni, F., Rothvoß, T., Sanità, L.: An improved LP-based approximation for Steiner tree. In: Proceedings of the Forty-Second ACM Symposium on Theory of Computing, STOC 2010, pp. 583–592. ACM, New York (2010)
4. Carbone, P., Katsifodimos, A., Ewen, S., Markl, V., Haridi, S., Tzoumas, K.: Apache flink^{TM}: stream and batch processing in a single engine. IEEE Data Eng. Bull. **38**(4), 28–38 (2015)
5. Chang, F., et al.: Bigtable: a distributed storage system for structured data. In: OSDI, pp. 205–218 (2006)
6. Chen, Y., Ganapathi, A., Griffith, R., Katz, R.H.: The case for evaluating MapReduce performance using workload suites. In: MASCOTS, pp. 390–399 (2011)

7. Chen, Y., Griffith, R., Liu, J., Katz, R.H., Joseph, A.D.: Understanding TCP incast throughput collapse in datacenter networks. In: WREN, pp. 73–82 (2009)
8. Costa, P., Donnelly, A., Rowstron, A.I.T., O'Shea, G.: Camdoop: exploiting in-network aggregation for big data applications. In: NSDI, pp. 29–42 (2012)
9. Culhane, W., Kogan, K., Jayalath, C., Eugster, P.: Optimal communication structures for big data aggregation. In: INFOCOM, pp. 1643–1651 (2015)
10. Dean, J., Ghemawat, S.: MapReduce: simplified data processing on large clusters. Commun. ACM 51(1), 107–113 (2008)
11. Kaklamanis, C., Chlebk, M., Chlebkv, J.: Algorithmic aspects of global computing the steiner tree problem on graphs: inapproximability results. Theor. Comput. Sci. 406(3), 207–214 (2008)
12. Lakshman, A., Malik, P.: Cassandra: a decentralized structured storage system. Oper. Syst. Rev. 44(2), 35–40 (2010)
13. Malewicz, G., et al.: Pregel: a system for large-scale graph processing. In: SIGMOD, pp. 135–146 (2010)
14. Murray, D.G., McSherry, F., Isaacs, R., Isard, M., Barham, P., Abadi, M.: Naiad: a timely dataflow system. In: SIGOPS, pp. 439–455 (2013)
15. van Renesse, R., Birman, K.P., Vogels, W.: Astrolabe: a robust and scalable technology for distributed system monitoring, management, and data mining. ACM Trans. Comput. Syst. 21(2), 164–206 (2003)
16. Tucker, P.A., Maier, D., Sheard, T., Fegaras, L.: Exploiting punctuation semantics in continuous data streams. IEEE Trans. Knowl. Data Eng. 15(3), 555–568 (2003)
17. White, T.: Hadoop: The Definitive Guide, 1st edn. O'Reilly Media Inc., Sebastopol (2009)
18. Xiao, T., et al.: Nondeterminism in MapReduce considered harmful? An empirical study on non-commutative aggregators in MapReduce programs. In: Companion Proceedings of the 36th International Conference on Software Engineering, ICSE Companion 2014, pp. 44–53. ACM, New York (2014)
19. Yang, H., Dasdan, A., Hsiao, R., Parker, D.S.: Map-Reduce-Merge: simplified relational data processing on large clusters. In: SIGMOD, pp. 1029–1040 (2007)
20. Yu, Y., et al.: DryadLINQ: a system for general-purpose distributed data-parallel computing using a high-level language. In: OSDI, pp. 1–14 (2008)
21. Zaharia, M., et al.: Resilient distributed datasets: a fault-tolerant abstraction for in-memory cluster computing. In: NSDI, pp. 15–28 (2012)
22. Zaharia, M., et al.: Apache spark: a unified engine for big data processing. Commun. ACM 59(11), 56–65 (2016)
23. Zhang, Y., Ansari, N.: On architecture design, congestion notification, TCP incast and power consumption in data centers. IEEE Commun. Surv. Tutor. 15(1), 39–64 (2013)

Priority Evacuation from a Disk
Using Mobile Robots
(Extended Abstract)

Jurek Czyzowicz[1], Konstantinos Georgiou[2(⊠)], Ryan Killick[3],
Evangelos Kranakis[3], Danny Krizanc[4], Lata Narayanan[5], Jaroslav Opatrny[5],
and Sunil Shende[6]

[1] Départemant d'informatique, Université du Québec en Outaouais,
Gatineau, Canada
[2] Department of Mathematics, Ryerson University, Toronto, Canada
konstantinos@ryerson.ca
[3] School of Computer Science, Carleton University, Ottawa, ON, Canada
[4] Department of Mathematics and Computer Science,
Wesleyan University, Middletown, CT, USA
[5] Department of Computer Science and Software Engineering,
Concordia University, Montreal, QC, Canada
[6] Department of Computer Science, Rutgers University, Camden, USA

Abstract. We introduce and study a new search-type problem with
$(n+1)$-robots on a disk. The searchers (robots) all start from the center of
the disk, have unit speed, and can communicate wirelessly. The goal is for
a distinguished robot (the queen) to reach and evacuate from an exit that
is hidden on the perimeter of the disk in as little time as possible. The
remaining n robots (servants) are there to facilitate the queen's objective
and are not required to reach the hidden exit. We provide upper and lower
bounds for the time required to evacuate the queen. Namely, we propose
an algorithm specifying the trajectories of the robots which guarantees
evacuation of the queen in time always better than $2 + 4(\sqrt{2} - 1)\frac{\pi}{n}$ for
$n \geq 4$ servants. We also demonstrate that for $n \geq 4$ servants the queen
cannot be evacuated in time less than $2 + \frac{\pi}{n} + \frac{2}{n^2}$.

Keywords: Mobile robots · Priority · Evacuation · Exit
Group search · Disk · Wireless communication · Queen · Servants

1 Introduction

A fundamental research topic in mathematics and computer science concerns
search, whereby a group of mobile robots need to collectively explore an environment in order to find a hidden target. In the scenarios considered so far, the

J. Czyzowicz, K. Georgiou, E. Kranakis, L. Narayanan and J. Opatrny—Research
supported in part by NSERC Discovery grant.
R. Killick—Research supported by the Ontario Graduate Scholarship.
A full version of this work is available on the Computing Research Repository [14].

© Springer Nature Switzerland AG 2018
Z. Lotker and B. Patt-Shamir (Eds.): SIROCCO 2018, LNCS 11085, pp. 392–407, 2018.
https://doi.org/10.1007/978-3-030-01325-7_32

goal was to optimize the time when the first searcher reaches the target position. More recently, researchers studied the evacuation problem in which it is required to minimize the time of arrival to the target position of the last mobile robot in the group. In the work done on search so far, all robots are generally assumed to have exactly the same capabilities. However, it is quite natural to consider collaborative tasks in which the participant robots have different capabilities. For example, robots may have different maximum speeds, or have different communication capabilities. Robots with different speeds have been studied in the context of rendezvous [20] and evacuation [25]. In the context of search, a natural situation may be that only one of the robots has the capability to address an urgent need at the target, for example, performing an emergency procedure, or closing a breach in the perimeter. The remaining robots can help in searching for the target, but their arrival at the target does not accomplish the main purpose of finding the target. Therefore, the collective goal of the robots is to get the special robot to the target as soon as possible. In this paper, we are interested in such a type of search problem, which grants *priority* to a pre-selected participant. In other words, we assume that the collection of robots contains a leader, known in advance, and as long as the leader does not get to the target position, search is considered incomplete.

In this paper we propose and investigate the *priority evacuation* problem, a new form of group search in which a given selected searcher in the group is deemed more important than the rest. This distinguished robot is given priority over all other searchers during the evacuation process in that it should be evacuated as early as possible upon the exit being located by any searcher.

1.1 Model

In the priority evacuation, or PEvac_n problem, $n + 1$ robots (searchers) are placed at the center of a unit disk. There is a target (exit), placed at an unknown location on the boundary of the disk. The target can be discovered by any robot walking over it. A robot that finds the exit instantaneously broadcasts its current position. Among the robots there is a distinguished one called the *queen* and the remaining n robots are referred to as *servants*. The goal is to minimize the queen's *evacuation time*, i.e. the worst case total time until the queen reaches the target. We assume that all robots, including the queen, may walk using maximum unit speed. We note that the queen may or may not actively participate in the search of the exit.

1.2 Related Work

Search and exploration have been extensively studied in mathematics and various fields of computer science. If the environment is not known in advance, search implies exploration, and it usually involves mapping and localizing searchers within the environment [1,19,24,26]. However, even for the case of a known, simple domain like a line, there have been several interesting studies attempting to optimize the search time. These were initiated with the seminal works of

Bellman [6] and Beck [5], in which the authors attempted to minimize the competitive ratio in a stochastic setting. After the appearance of [3], where a search by a single robot was studied for infinite lines and planes, several other works on linear search followed (cf. [2]) and more recently the search by a single searcher was studied for different models, e.g., when the turn cost was considered [18], when a bound on the distance to the target is known in advance [8], and when the target is moving or for more general linear cost functions [7].

For the case of a collection of searchers, numerous scenarios have been studied, such as: graph or geometric terrains, known or unknown environments, stationary or mobile targets, etc. (cf. [21]). In many papers, the objective is to decide the feasibility of the search or to minimize its search time.

The evacuation problem from the disk was introduced in [12] where two types of robots' communication were studied – the wireless one and communication by contact (also called face-to-face). The bounds for evacuation of two robots communicating face-to-face were later improved in [16] and in [9]. The case of a disk environment with more than one exit was considered in [11,27]. Other variations included evacuation from environments such as regular triangles and squares [17], the case of two robots having different maximal speeds [25], and the evacuation problem when one of the robots is crash or byzantine faulty [13].

Group search and evacuation in the line environment were studied in [4,10]. The authors of [10] proved, somewhat surprisingly, that having many robots using maximal speed 1 does not reduce the optimal search time as compared to the search using only a single robot. However, interestingly, [10] shows that the same bound for group search (and evacuation) is achieved for two robots having speeds 1 and 1/3. For both types of robots' communication scenarios, [4] presents optimal evacuation algorithms for two robots having arbitrary, possibly distinct, maximal speeds in the line environment.

A priority evacuation-type problem has been previously considered in [22, 23] but with different terminology. Using the jargon of the current paper, an immobile queen is hidden somewhere on the unit disk, and a number of robots try to locate her, and fetch (evacuate) her to an exit which is also hidden. The performance of the evacuation algorithm is measured by the time the queen reaches the exit. Apart from these results, and to the best of our knowledge nothing is known about the priority evacuation problem. In this work we provide a general strategy for the case of $n \geq 4$ servants. When there are fewer than 4 servants more ad hoc strategies must be employed which do not fit with the general framework developed here and they are therefore treated elsewhere [15].

1.3   Results of the Paper

Section 2 introduces nomenclature and notation and discusses preliminaries. In Sect. 3 we provide an algorithm that evacuates the queen in time always smaller than $2 + 4(\sqrt{2} - 1)\frac{\pi}{n}$ for $n \geq 4$ servants (the exact evacuation times of our algorithm must be calculated numerically). In Sect. 4 we demonstrate that for $n \geq 4$ servants the queen cannot be evacuated in time less than

$1 + \frac{2}{n} \cdot \arccos(-\frac{2}{n}) + \sqrt{1 - \frac{4}{n^2}}$, or, asymptotically, $2 + \frac{\pi}{n} + \frac{2}{n^2}$. These results improve upon naive upper and lower bounds of $2 + \frac{2\pi}{n}$ and $2 + \frac{\pi}{n+1}$ respectively (see Sects. 2.2 and 4). A summary of the evacuation times for our algorithm (numerical results) as well as the upper and lower bounds (non-trivial and naive) is provided in Table 1 and in Fig. 1. We conclude the paper in Sect. 5 with a discussion of open problems. Many of our proofs are omitted from this extended abstract due to space limitations; see [14] for a full version of this paper.

Table 1. Evacuation times T of the queen using Algorithm 2 (numerical results). The upper bound of $2 + 4(\sqrt{2} - 1)\frac{\pi}{n}$ (Theorem 1), and the lower bound of $1 + \frac{2}{n}\cos^{-1}\left(\frac{-2}{n}\right) + \sqrt{1 - \frac{4}{n^2}}$ (Theorem 5) are also provided. For comparison, the naive upper bound and lower bound of $2 + \frac{2\pi}{n}$ (see Sect. 2.2) and $2 + \frac{\pi}{n+1}$ (see Sect. 4) are included.

n	T (Algorithm 2)	UB (Theorem 1)	LB (Theorem 5)	UB Naive	LB Naive
4	3.113	3.301	2.913	3.571	2.628
5	2.905	3.041	2.709	3.257	2.524
6	2.762	2.868	2.580	3.047	2.449
7	2.660	2.744	2.490	2.898	2.393
8	2.582	2.651	2.424	2.785	2.349

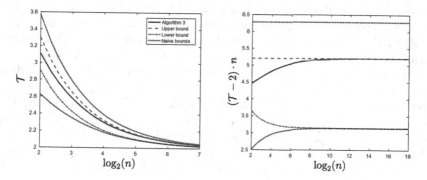

Fig. 1. Evacuation times T of Algorithm 2 for $n \in [4, 2^7]$ (left) and $n \in [4, 2^{18}]$ (right). The upper bound of $2 + 4(\sqrt{2} - 1)\frac{\pi}{n}$ (Theorem 1), the lower bound of $1 + \frac{2}{n}\cos^{-1}\left(\frac{-2}{n}\right) + \sqrt{1 - \frac{4}{n^2}}$ (Theorem 5) are also provided. For comparison, a naive upper bound and lower bound of $2 + \frac{2\pi}{n}$ (see Sect. 2.2) and $2 + \frac{\pi}{n+1}$ (see Sect. 4) are included.

2   Notation and Preliminaries

2.1   Notation

We denote by $\mathcal{U}$ the unit circle in $\mathbb{R}^2$ centered at the origin $O = (0, 0)$ which must be evacuated by the queen and we assume that all robots start from the origin.

We use n to denote the number of servants, and use $Q(t)$ and $S_k(t)$, $k = 1, \ldots, n$, to represent the trajectories of the queen and k^{th} servant respectively. The set of all servant trajectories is represented by $\mathcal{S} = \{S_k(t);\ k = 1, \ldots, n\}$. A trajectory will be given as a parametric function of time and, when referring to a robot's trajectory, it will be implied that we mean the path taken by the robot in the case that the exit has not been found.

2.2   Evacuation Algorithms

A priority evacuation algorithm $\mathcal{A}$ is specified by the trajectories of the queen and servants, $\mathcal{A} = \{Q(t)\} \cup \mathcal{S}$. We say that $\mathcal{A}$ solves the PEVAC_n problem if, in finite time, all points of $\mathcal{U}$ are visited/discovered by at least one robot. The evacuation time $\mathcal{T}$ of an algorithm solving the PEVAC_n problem is defined to be the worst-case time taken for the queen to reach the exit. As such, the evacuation time will be composed of two parts: the time taken until the exit is discovered plus the time needed for the queen to reach the exit once it has been found.

We will find it useful to define the restricted class of evacuation algorithms $\mathbb{S}$ containing all those algorithms in which: (a) the queen does not participate in searching for the exit, (b) the servants initially move as quickly as they can to the perimeter of $\mathcal{U}$, (c) each servant searches either counter-clockwise or clockwise along the perimeter of $\mathcal{U}$ at full speed, and, (d) each servant stops and is no longer used once it reaches an already discovered point of $\mathcal{U}$. Algorithms in this class can be defined by the trajectory of the queen $Q(t)$ together with the sets $\Phi = \{\phi_k \in [0, 2\pi];\ k = 1, \ldots, n\}$ and $\Sigma = \{\sigma_k = \pm 1;\ k = 1, \ldots, n\}$ which respectively specify the angular positions on $\mathcal{U}$ to which the servants initially move, and the directions in which each servant searches. We will enforce an ordering on the sets Φ and Σ such that for $\phi_k \in \Phi$, $1 \le k \le n-1$, we have $\phi_k \le \phi_{k+1}$. With this notation we can express the trajectory of the kth servant during the time it is searching as $S_k(t) = (\cos{(\phi_k + \sigma_k(t-1))},\ \sin{(\phi_k + \sigma_k(t-1))})$.

We additionally define the class of algorithms $\mathbb{S}_{sym} \subset \mathbb{S}$ containing those algorithms for which we can split the set of servants into two groups $\mathcal{S} = \mathcal{S}_+ \cup \mathcal{S}_-$ where: (a) servants in $\mathcal{S}_+$ follow trajectories which are reflections about the x-axis[1] of servants in $\mathcal{S}_-$, and, (b) all servants in $\mathcal{S}_+$ search counter-clockwise[2]. In the case that n is odd we permit one servant to follow a trajectory that is symmetric about the x-axis. For an algorithm in $\mathbb{S}_{sym}$ we may write $\Phi = \Phi_+ \cup \Phi_-$ where Φ_+ (resp. Φ_-) specifies the positions on $\mathcal{U}$ to which the servants above (resp. below) the x-axis initially move. Formally we may write $\Phi_+ = \{\phi_k \in [0, \pi]; k = 1, \ldots, \lceil \frac{n}{2} \rceil\}$ and $\Phi_- = -\Phi_+$ for even n and $\Phi_- = \{-\phi_k;\ k = 2, \ldots, \lceil \frac{n}{2} \rceil\}$ for odd n. In the class $\mathbb{S}_{sym}$ the directions in which the servants search are always counter-clockwise (resp. clockwise) for robots in Φ_+ (resp. Φ_-) and thus an algorithm $\mathcal{A} \in \mathbb{S}_{sym}$ is entirely specified by the set $\{Q(t)\} \cup \Phi_+$.

---

[1] The choice of the x-axis is arbitrary since we may always rotate $\mathcal{U}$. What is important is that a diameter of symmetry exists.

[2] Again, these choices of search directions are arbitrary since we can reflect $\mathcal{U}$ about the y-axis. What is important is that all servants within a group search in the same direction.

As a warm-up to the next section, and to demonstrate the intuitive nature of these definitions, consider the following trivial algorithm which achieves an evacuation time of $2 + \frac{2\pi}{n}$: the queen remains at the origin until the exit is found and the servants move directly to equally spaced locations on the perimeter of $\mathcal{U}$ each searching an arc of length $\frac{2\pi}{n}$ in the counter-clockwise direction. This algorithm can be seen to be in the class $\mathbb{S}$ and we can succinctly represent the algorithm as follows

---

Algorithm 1. Trivial Evacuation 1, $\mathcal{A} \in \mathbb{S}$

1: $Q(t) = (0, 0)$.
2: $\Phi = \{\frac{(k-1)}{n} 2\pi; \ k = 1, \dots, n\}$
3: $\Sigma = \{1; \ k = 1, \dots, n\}$

Observe that the above algorithm is not in $\mathbb{S}_{sym}$. We can, however, give an equivalent algorithm in $\mathbb{S}_{sym}$ which achieves the same evacuation time. This algorithm is depicted in Fig. 2 along with Algorithm 1 for the case that $n = 8$.

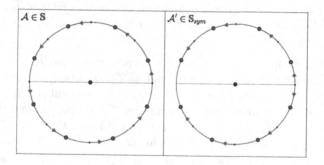

Fig. 2. Depiction of the two trivial algorithms each achieving an evacuation time of $2 + \frac{2\pi}{n}$. Both algorithms are in the class $\mathbb{S}$ and the algorithm on the right is also in the class $\mathbb{S}_{sym}$. The queen is indicated by the blue point and the servants by the red points. A red arc indicates points that have been discovered. (Color figure online)

3 Upper Bound

In the previous section we introduced two evacuation algorithms solving PEVAC_n with evacuation time $2 + \frac{2\pi}{n}$. We will show that this can be improved:

Theorem 1. *There exists an algorithm solving* PEVAC_n *for* $n \geq 4$ *with an evacuation time at most* $2 + 4(\sqrt{2} - 1)\frac{\pi}{n} \approx 2 + 1.657\frac{\pi}{n}$.

We will prove Theorem 1 constructively and present an evacuation algorithm in the class $\mathbb{S}_{sym}$ achieving the desired upper bound for $n \geq 4$ servants. For ease of presentation we will assume that n is even. Furthermore, as it will greatly simplify the algebra, we will redefine all times (including the evacuation time) to start from the moment the servants first reach the perimeter. To avoid confusion we will use $\mathcal{T}_p$ to represent the evacuation time of an algorithm as measured from the moment the servants reach the perimeter. The total evacuation time will thus be $\mathcal{T} = \mathcal{T}_p + 1$.

As we will describe an algorithm in the class $\mathbb{S}_{sym}$ we will only need to specify the queen's trajectory $Q(t)$ and the initial angular positions Φ_+ of the servants lying above the x-axis. We start by giving the trajectory for the queen which we parametrize using $\alpha > 0$:

$$Q(t) = \begin{cases} (0,\ 0), & 0 \le t < \alpha \\ (\alpha - t,\ 0), & \alpha \le t < \alpha + 1 \\ (-1,\ 0), & t \ge \alpha + 1, \end{cases} \tag{1}$$

In words, the queen waits at the origin until the time $t = \alpha$ at which moment she begins moving at full speed along the negative x-axis stopping when she arrives to the point $(-1,\ 0)$ at the time $t = \alpha + 1$. The crux of the algorithm will be in specifying the set Φ_+. In order to do this we consider the following simple observation:

Observation 2. *If the queen is to achieve an evacuation time of $\mathcal{T}_p$, then, for all $t < \mathcal{T}_p$, all of the undiscovered points of $\mathcal{U}$ must remain inside the disk centered on the queen with radius $\mathcal{T}_p - t$.*

Assume that we have an algorithm with evacuation time $\mathcal{T}_p$ and define $\mathcal{C}_Q(t)$ as the circle centered on the queen with radius $\mathcal{T}_p - t$. Then, in light of Observation 2, it is not so hard to imagine that the intersection points of the circles $\mathcal{C}_Q(t)$ and $\mathcal{U}$ will be of importance. Thus, assume that $\mathcal{T}_p$ is small enough that at some time $t \ge \alpha$ the circles $\mathcal{C}_Q(t)$ and $\mathcal{U}$ intersect. Considering the form of the queen's trajectory, we can conclude that the circles $\mathcal{U}$ and $\mathcal{C}_Q(t)$ will first intersect at the time $\gamma = \frac{\mathcal{T}_p + \alpha - 1}{2}$ at the point $(1,\ 0)$. For times $t > \gamma$ the circles will intersect at two points $A_\pm$ which are symmetric about the x-axis and which move from right to left along the perimeter of $\mathcal{U}$. The importance of the points $A_\pm$ is clear when one considers that $A_\pm$ mark the boundary between those points of $\mathcal{U}$ which must be discovered and those which may yet be undiscovered at the time t. Intuitively, we will want to position the servants such that they are searching only when they are to the left of A_+ and A_-. In particular, a servant will stop searching at precisely the moment the intercept A_+ or A_- catches up to it (with a small caveat to be described shortly). This condition will allow us to specify the set Φ_+.

At this time we will find it useful to re-express the evacuation time as $\mathcal{T}_p = 1 + \alpha + \rho$ where ρ is a parameter that will ultimately depend on α. Intuitively, ρ represents the radius of $\mathcal{C}_Q(t)$ at the moment the queen reaches the perimeter of $\mathcal{U}$ and its inclusion will greatly simplify algebra. Note that, with this definition, the circles $\mathcal{C}_Q(t)$ and $\mathcal{U}$ will first intersect at the time $\gamma = \alpha + \frac{\rho}{2}$.

As we only need to specify the set Φ_+ we will only consider the intercept A_+. The coordinates of A_+ for times $\gamma \le t \le \alpha + 1$ can be determined by simultaneously solving the implicit equations for $\mathcal{U}$ and $\mathcal{C}_Q(t)$, i.e. $\mathcal{U}: x^2 + y^2 = 1$

and $\mathcal{C}_Q(t): (x-\alpha+t)^2+y^2 = (1+\alpha+\rho-t)^2$. We find that $A_+(t) = (x_A(t),\ y_A(t))$ where

$$x_A(t) = \frac{\rho(2+\rho)}{2(t-\alpha)} - 1 - \rho \qquad (2)$$

and

$$y_A(t) = \frac{\sqrt{\rho(\rho+2)[2(t-\alpha)-\rho][\rho+2-2(t-\alpha)]}}{2(t-\alpha)} \qquad (3)$$

The angular position of A_+ will be represented as ϕ_A and is given by:

$$\phi_A(t) = \tan^{-1}\left(\frac{y_A(t)}{x_A(t)}\right). \qquad (4)$$

We define ν_A as the speed at which A_+ moves along the perimeter of $\mathcal{U}$. We can determine ν_A using $\nu_A(t) = \sqrt{\left(\frac{dx_A}{dt}\right)^2 + \left(\frac{dy_A}{dt}\right)^2}$ from which we find that:

$$\nu_A(t) = \frac{1}{t-\alpha}\sqrt{\frac{\rho(\rho+2)}{[\rho+2-2(t-\alpha)][2(t-\alpha)-\rho]}} \qquad (5)$$

Now consider the form of the function $\nu_A(t)$. For times just after $t = \alpha$ we can see that A_+ will move with a speed $\nu_A \gg 1$ and, as such, no single servant will be able to stay to the left of A_+ for long. What is not so obvious from (5) is that ν_A continuously decreases until some time τ at which $\nu_A = 1$.[3] Furthermore, starting at the time τ there will be an interval of time during which $\nu_A \leq 1$. Thus, if the intercept reaches a servant at exactly the time τ that servant does not have to stop searching. We will choose ρ to ensure that the servant $S_{n/2} \in \mathcal{S}_+$ satisfies exactly this property.

Therefore we can describe the following general overview of our algorithm: the servant S_1 begins at $\phi_1 = 0$ (for even n) and searches until the time t_1 at which $S_1(t_1) = A_+(t_1)$ or when $t_1 + \phi_1 = \phi_A(t_1)$. The servant S_2 will begin its search at the position $\phi_2 = \phi_1 + t_1$ and it will search for a time t_2 until $S_2(t_2) = A_+(t_2)$ or until $t_2 + \phi_2 = \phi_A(t_2)$. The servant S_3 will begin at the position $\phi_3 = \phi_2 + t_2 = \phi_1 + t_1 + t_2$, and so on. Continuing on like this we can see that the servant S_k will begin its search at the position $\phi_{k+1} = \phi_k + t_k = \phi_1 + \sum_{i=1}^{k} t_i$ with the t_k satisfying $t_k = \phi_A(t_k) - \phi_k$ or, equivalently, $\phi_1 + \sum_{i=1}^{k} t_i = \phi_A(t_k)$. We want the servant $S_{n/2}$ to be coincident with the intercept A_+ at exactly the time τ (recall that τ is the time at which the speed of A_+ is $\nu_A = 1$) and thus we will choose ρ to satisfy $\phi_{n/2} + \tau = \phi_A(\tau)$. In this case the servant $S_{n/2}$ will search for a total time $\pi - \phi_{n/2}$ after which all of $\mathcal{U}$ will have been discovered.

To extend this algorithm to the case that n is odd we will need to split the trajectory of the servant $S_1 \in \mathcal{S}_+$ between the upper and lower halves of $\mathcal{U}$.

---

[3] It is not guaranteed that for all $\rho > 0$ this intercept will reach a speed of one before the queen reaches the perimeter of $\mathcal{U}$. However, we will choose a ρ such that this does happen.

We will therefore start the servant S_1 at the position $\phi_1 = \frac{-t_1}{2}$. All of the other relevant equations remain unchanged.

We provide links [28, 29] to short animations of the algorithm for $n = 4$, 8. In these animations the queen is represented by the blue point, the servants by red points, and the intercepts $A_\pm$ by green points. A plot of the evacuation time as a function of the time at which the servants find the exit is also shown. Note that the servants stop searching at the exact moment the intercept reaches them (except for the two servants furthest to the left) and at these moments the evacuation time is maximized. The two servants that are last active will be coincident with the intercepts at the moment these intercepts reach a speed of one, and, again, at this moment the evacuation time is maximized. In total there will be n different locations for the exit (counting the top and bottom of $\mathcal{U}$) which will maximize the evacuation time. A keen eye will note that the queen reaches the perimeter of $\mathcal{U}$ before the servants have finished searching the perimeter and this would appear to hint that Algorithm 2 can be improved. We will argue in Sect. 5 that this is not the case.

Our algorithm is formally presented in Algorithm 2 where we have left α as a parameter. We claim that Algorithm 2 will always do better than the bound of Theorem 1 when the evacuation time is minimized over α. We will now prove this claim.

Algorithm 2. IntersectChase(α), $\mathcal{A}_\alpha \in \mathbb{S}_{sym}$

1:

$$Q(t) = \begin{cases} (0,\ 0), & 0 \le t < \alpha \\ (\alpha - t,\ 0), & \alpha \le t < \alpha + 1 \\ (-1,\ 0), & t \ge \alpha + 1, \end{cases}$$

2: $\Phi_+ = \{\phi_k;\ k = 1,\ \ldots,\ \lceil \frac{n}{2} \rceil\}$, where:

$$\phi_1 = \begin{cases} 0, & n \text{ even} \\ -\frac{t_1}{2}, & n \text{ odd} \end{cases}, \qquad \phi_k = \phi_1 + \sum_{i=1}^{k-1} t_i, \qquad \phi_{n/2} + \tau = \phi_A(\tau)$$

and,

$$\phi_1 + \sum_{i=1}^{k} t_i = \phi_A(t_k), \qquad \nu_A(\tau) = 1$$

Proof (Theorem 1). To simplify the algebra we will assume that n is even. Algorithm 2 specifies that we choose the t_k in order to satisfy $\sum_{i=1}^{k} t_i = \phi_A(t_k)$ where $\phi_A(t)$ is defined in (4). We note that each servant will be able to search for at least a time γ since this marks the first time at which $\mathcal{C}_Q(t)$ and $\mathcal{U}$ intersect. This motivates us to define the primed time coordinate $t' = t - \gamma$. In this primed coordinate the defining relation for the t'_k is $\sum_{i=1}^{k} t'_i = \phi_A(t'_k) - k\gamma$ (where we assume that ϕ_A is properly redefined for the primed time coordinate). We are interested in an asymptotic limit and thus we make the following claim:

Claim 3. *When we take the limit in large* n, *the sum* $\sum_{i=1}^{k} t_i'$ *becomes a definite integral* $\lim_{n\to\infty} \sum_{i=1}^{k} t_i' = \int_0^\kappa t'(u)du$ *where* $\frac{\kappa}{n}$ *is to be interpreted as the fractional servant number and* u *is a dummy integration variable.*

Due to the Claim 3, the asymptotic defining relation for $t'(\kappa)$ becomes an integral equation $\int_0^\kappa t'(u)du = \phi_A(t'(\kappa)) - \kappa\gamma$. Using the fundamental theorem of calculus we can rewrite this as a differential equation: $t'(\kappa) = \frac{d}{d\kappa}(\phi_A(t'(\kappa))) - \kappa\gamma) = \frac{d\phi_A(t'(\kappa))}{d\kappa} - \gamma$. Applying the chain rule we find that $\frac{d\phi_A(t'(\kappa))}{d\kappa} = \frac{d\phi_A(t'(\kappa))}{dt'} \cdot \frac{dt'(\kappa)}{d\kappa}$. Observe that $\frac{d\phi_A(t'(\kappa))}{dt'}$ is simply the speed of the intercept A_+ and we can therefore write the differential equation for $t'(\kappa)$ as $\frac{dt'}{d\kappa} = \frac{t'+\gamma}{v_A(t'(\kappa))}$. This ordinary differential equation can easily be solved for κ in terms of t' by separation of variables. We find that $\kappa(t') = \int_0^{t'} \frac{v_A(u)}{u+\gamma}du$. The equation for the speed v_A is given in (5), which, in the primed time coordinate takes the form $v_A(t') = \frac{1}{(2t'+\rho)}\sqrt{\frac{\rho(\rho+2)}{t'(1-1')}}$. Substituting this into the expression for $\kappa(t')$ yields $\kappa(t') = \int_0^{t'} \frac{1}{(u+\gamma)(2u+\rho)}\sqrt{\frac{\rho(\rho+2)}{u(1-u)}}du$. This integral has the closed form solution

$$\kappa(t') = \frac{1}{\alpha}\left[2\tan^{-1}\left(\frac{t'(2t'+\rho)}{\rho}v_A(t')\right)\right.$$
$$\left. - \sqrt{\frac{\rho(\rho+2)}{\gamma(\gamma+1)}}\tan^{-1}\left(t'(2t'+\rho)\sqrt{\frac{1+\gamma}{\gamma\rho(\rho+2)}}v_A(t')\right)\right].$$

We require that the servant $S_{n/2}$ be coincident with the intercept A_+ at the time $\tau' = \tau - \gamma$ and this implies that we need $\kappa(\tau') = \frac{n}{2}$ or

$$\frac{n}{2} = \frac{1}{\alpha}\left[2\tan^{-1}\left(\frac{\tau'(2\tau'+\rho)}{\rho}v_A(\tau')\right)\right.$$
$$\left. - \sqrt{\frac{\rho(\rho+2)}{\gamma(\gamma+1)}}\tan^{-1}\left(\frac{\tau'(2\tau'+\rho)}{\rho}\sqrt{\frac{\rho(\gamma+1)}{\gamma(\rho+2)}}v_A(\tau')\right)\right].$$

If we set $\alpha = a\frac{\pi}{n}$ and note that, by definition, $v_A(\tau') = 1$, we can simplify the above to obtain

$$\frac{\pi}{2} = \frac{1}{a}\left[2\tan^{-1}\left(\frac{\tau'(2\tau'+\rho)}{\rho}\right) - \sqrt{\frac{\rho(\rho+2)}{\gamma(\gamma+1)}}\tan^{-1}\left(\frac{\tau'(2\tau'+\rho)}{\rho}\sqrt{\frac{\rho(\gamma+1)}{\gamma(\rho+2)}}\right)\right].$$

Define $D(a, \rho)$ as the quantity

$$D(a, \rho) = \frac{\pi}{2} - \frac{1}{a}\left[2\tan^{-1}\left(\frac{\tau'(2\tau'+\rho)}{\rho}\right)\right.$$
$$\left. - \sqrt{\frac{\rho(\rho+2)}{\gamma(\gamma+1)}}\tan^{-1}\left(\frac{\tau'(2\tau'+\rho)}{\rho}\sqrt{\frac{\rho(\gamma+1)}{\gamma(\rho+2)}}\right)\right]$$

which we want to be zero. We now make the following claim:

Claim 4. *The asymptotic behaviour of τ is $\mathcal{O}\left(\rho^{1/3}\right)$.*

Using Claim 4, we have that $\lim_{n\to\infty} \frac{\tau'(2\tau'+\rho)}{\rho} = \lim_{n\to\infty} \mathcal{O}\left(\rho^{-1/3}\right) = \infty$ and thus

$$\frac{\pi}{2} = \lim_{n\to\infty} \tan^{-1}\left(\frac{\tau'(2\tau'+\rho)}{\rho}\right) = \lim_{n\to\infty} \tan^{-1}\left(\frac{\tau'(2\tau'+\rho)}{\rho}\sqrt{\frac{\rho(\gamma+1)}{\gamma(\rho+2)}}\right).$$

We can therefore write $\lim_{n\to\infty} D(a,\rho) = \frac{\pi}{a}\left(1 - \frac{a}{2} - \sqrt{\frac{\rho}{2\gamma}}\right)$. Now set $\rho = q\frac{\pi}{n}$ such that $\gamma = \alpha + \frac{\rho}{2} = \frac{\pi}{n}(a + \frac{q}{2})$. Using this notation we have $\lim_{n\to\infty} D(a,\rho) = \frac{\pi}{a}\left(1 - \frac{a}{2} - \sqrt{\frac{q}{2a+q}}\right)$. We want this limit to equal zero which implies that we need $1 - \frac{a}{2} - \sqrt{\frac{q}{2a+q}} = 0$ or $q = \frac{2(2-a)^2}{(4-a)}$.

Now, to optimize the algorithm we need to minimize the evacuation time $\mathcal{T}_p$. Since $\mathcal{T}_p$ increases with a we equivalently need to minimize $a + q = \frac{a^2 - 4a + 8}{4 - a}$. Taking the derivative of this with respect to a and setting the result equal to zero gives us the optimal value of a and q to be $a = 2(2-\sqrt{2})$ and $q = 2(3\sqrt{2}-4)$. The asymptotic cost of the algorithm is therefore $\mathcal{T}_p = 1 + \alpha + \rho = 1 + 4(\sqrt{2}-1)\frac{\pi}{n}$. The overall evacuation time is then $\mathcal{T} = 1 + \mathcal{T}_p$ which is the bound given in Theorem 1.

We note that, in the case that n is odd, the results of the proof will not change due to the fact that, as $n \to \infty$, we have $\phi_1 = -\frac{t_1}{2} \to 0$.   □

4   Lower Bound

In this section we develop a lower bound on the evacuation time of the queen. We first note that we can derive a naive lower bound of $2 + \frac{\pi}{n+1}$ since each robot can travel with a maximum speed of one and we have $n + 1$ robots in total. We will show that this can be improved:

Theorem 5. *In any algorithm with $n \geq 4$ the queen cannot be evacuated in time less than $1 + \frac{2}{n}\cos^{-1}\left(\frac{-2}{n}\right) + \sqrt{1 - \frac{4}{n^2}}$. In the limit of large n this bound approaches $2 + \frac{\pi}{n} + \frac{2}{n^2}$.*

The outline of the proof is as follows: we first demonstrate that the lower bound holds for any algorithm in which the queen does not participate in searching for the exit before some critical time. We will then show that the queen is not able to participate in the search for the exit before this critical time. We begin with a lemma first given in [12] which is reproduced here for convenience:

Lemma 1. *Consider a perimeter of a disk whose subset of total length $u + \epsilon > 0$ has not been explored for some $\epsilon > 0$ and $\pi \geq u > 0$. Then there exist two unexplored boundary points between which the distance along the perimeter is at least u.*

In the next two lemmas we demonstrate that the lower bound holds if the queen does not participate in the search.

Lemma 2. *For $n \geq 2$, any x satisfying $\frac{\pi}{n} \leq x < \frac{2\pi}{n}$, and any evacuation algorithm in which the queen does not participate in searching for the exit before the time $1 + x$, it takes time at least $1 + x + \sin\left(\frac{nx}{2}\right)$ to evacuate the queen.*

Lemma 3. *For any $n \geq 2$ and any evacuation algorithm in which the queen does not participate in searching for the exit before the time $t = 1 + \frac{2}{n}\cos^{-1}\left(\frac{-2}{n}\right)$ it takes time at least $1 + \frac{2}{n}\cos^{-1}\left(\frac{-2}{n}\right) + \sqrt{1 - \frac{4}{n^2}}$ to evacuate the queen.*

We will now demonstrate that the queen is not able to search before the time $1 + \frac{2}{n}\cos^{-1}\left(\frac{-2}{n}\right)$. This will be the goal of the next four lemmas and the following simple observation

Observation 6. *If the queen is to achieve an evacuation time of T, then, for any time $t \leq T$, she must remain in the region of intersection of all disks centered on the undiscovered points of $\mathcal{U}$ with radii $T - t$.*

Lemma 4. *Consider any two points A and B on the unit circle connected by a chord of length δ. Define the circles C_A and C_B as the circles centered on A and B with radii r. Then, if $r > \frac{\delta}{2}$, the circles intersect at two points C and D at distances $\sqrt{r^2 - \frac{1}{4}\delta^2} \pm \sqrt{1 - \frac{1}{4}\delta^2}$ from the origin.*

Lemma 5. *For a given $r > 0$ define the functions $f_{\pm}(x) = \frac{1}{2}\sqrt{4r^2 - x^2} \pm \frac{1}{2}\sqrt{4 - x^2}$. Then, for $0 \leq x \leq \min\{2, 2r\}$ f_+ is a decreasing function of x and f_- is an increasing function of x if $r > 1$ otherwise it is decreasing.*

Lemma 6. *Consider any $r > 0$ and assume that the unexplored subset of $\mathcal{U}$ has total length ϕ. Define $\mathcal{D}_P$ as the disk centered on an undiscovered point $P \in \mathcal{U}$ with radius r and define $\mathcal{G}$ as the region of intersection of all such disks. Then, if $r \geq \sin\left(\frac{\phi}{2}\right)$, $\mathcal{G}$ is completely contained inside of a disk centered on the origin with radius $R = \sqrt{r^2 - \sin^2\left(\frac{\phi}{2}\right)} + \cos\left(\frac{\phi}{2}\right)$. If $r < \sin\frac{\phi}{2}$ then $\mathcal{G} = \emptyset$.*

Lemma 7. *Consider an algorithm with evacuation time $T < 3$. Then if the queen is able to search the perimeter of $\mathcal{U}$ we must have*

$$R(t) = \sqrt{(T - t)^2 - \sin^2\left(\frac{n(t - 1)}{2}\right)} - \cos\left(\frac{n(t - 1)}{2}\right) > 1.$$

Armed with these lemmas we are now able to tackle our main result.

Proof (Theorem 5). Set $T_0 = 1 + \frac{2}{n}\cos^{-1}\left(\frac{-2}{n}\right) + \sqrt{1 - \frac{4}{n^2}}$ and assume we have an algorithm with an evacuation time $T < T_0$. By Lemma 3, this implies that

the queen must search the perimeter of $\mathcal{U}$ before the time $t_c = 1 + \frac{2}{n}\cos^{-1}\left(\frac{-2}{n}\right)$.[4]
Assume that at the time t_c the robots have collectively searched the perimeter
of $\mathcal{U}$ at a rate μ satisfying $n < \mu \le n+1$. Then at the time t_c the unexplored
subset of $\mathcal{U}$ has length $\phi(t) = 2\pi - \mu(t_c - 1) = 2\pi - 2\frac{\mu}{n}\cos^{-1}\left(\frac{-2}{n}\right) < \pi$. Since
$\phi(t_c) \le \pi$ we can use Lemma 6 to say that the queen must be located within a
distance of $R(t_c)$ of the origin at the time t_c. Furthermore, in order for the queen
to have searched the perimeter of $\mathcal{U}$ at the time t_c, we must have $R(t_c) \ge 1$.
However, observe that

$$
\begin{aligned}
R(t_c) &= \sqrt{(\mathcal{T} - t_c)^2 - \sin^2\left(\frac{n(t_c - 1)}{2}\right)} - \cos\left(\frac{n(t_c - 1)}{2}\right) \\
&\le \sqrt{(\mathcal{T}_0 - t_c)^2 - \sin^2\left(\frac{n(t_c - 1)}{2}\right)} - \cos\left(\frac{n(t_c - 1)}{2}\right) \\
&= \sqrt{1 - \frac{4}{n^2} - \sin^2\left(\cos^{-1}\left(\frac{-2}{n}\right)\right)} - \cos\left(\cos^{-1}\left(\frac{-2}{n}\right)\right) = \frac{2}{n}
\end{aligned}
$$

which is clearly less than one for $n \ge 4$. We have therefore arrived to a contradiction
and must conclude that the lower bound holds. To determine the asymptotic behaviour
of $\mathcal{T}_0$ we can compute a Taylor series of $\mathcal{T}_0$ about $n = \infty$. We find that the first few
terms in the series are $2 + \frac{\pi}{n} + \frac{2}{n^2}$. □

5   Conclusions

We studied an evacuation problem concerning priority search on the perimeter
of a unit disk where only one robot (the queen) needs to exit from an unknown
location. We focused on the case of $n \ge 4$ servants and showed in Sect. 3 that
for any $n \ge 4$ the queen can be evacuated in time at most $2 + 4(\sqrt{2} - 1)\frac{\pi}{n}$.
Furthermore, in Sect. 4, we demonstrated that the queen cannot be evacuated
in time less than $1 + \frac{2}{n}\cos^{-1}\left(\frac{-2}{n}\right) + \sqrt{1 - \frac{4}{n^2}} > 2 + \frac{\pi}{n} + \frac{2}{n^2}$. Thus, in the limit of
large n, we are left with a gap of $(4\sqrt{2} - 5)\frac{\pi}{n} \approx 0.657\frac{\pi}{n}$ between the best upper
and lower bounds. We conjecture that Algorithm 2 is in fact optimal. We will
now justify this conjecture.

As was previously mentioned, one might think from Algorithm 2 that, since
the queen is able to reach the perimeter of $\mathcal{U}$ before the servants have finished
their search, it would be possible to improve our algorithm. However, this is
not the case – similar to the proof of Theorem 5 there are critical times ($\frac{n}{2}$
of them) that occur before the queen reaches the perimeter and anything she
does after these critical times cannot improve the evacuation time. These critical
times result from a tradeoff between maximizing the rate at which the servants
search – for which the queen should remain near the origin – and minimizing
the distance of the queen from possible exits near the end of the algorithm – for
which the queen should be near the perimeter. Furthermore, in order to achieve

---

[4] Alternatively we can say that the robots must search at a collective rate $> n$ by the
time t_c. This is why we were able to ignore the "unreasonable case" in Lemma 7.

the best tradeoff, the queen should travel as fast as she can from the origin to the perimeter. In other words, between these critical times, the queen should maximize her radial velocity. If we could prove that the queen does not need to participate in searching then it would not be so difficult to conclude why Algorithm 2 would be optimal. Any other trajectory of the queen between the critical search times will result in the same or a reduced radial velocity of the queen. It therefore does not seem likely that, with a reduced radial velocity, we can reduce the evacuation time.

In addition to improving the bounds obtained in this paper there are several interesting open problems related to priority search and evacuation. In particular, we may define a *weighted evacuation* problem (for a given group of agents) as a generalization of the priority evacuation problem studied here. One can differentiate on agent preferences by assigning a weight w_i to each agent i and require to evacuate a subset of agents of total weight $\geq W$ in minimum time. With this formulation in mind, the regular evacuation problem (see [12]) is the case where $w_i = 1$ for all agents and $W = n$, while for the problem considered in this work $w_i = 0$ for all agents except the queen for which $w_{queen} = 1$ and $W = 1$.

References

1. Albers, S., Henzinger, M.R.: Exploring unknown environments. SIAM J. Comput. **29**(4), 1164–1188 (2000)
2. Alpern, S., Gal, S.: The Theory of Search Games and Rendezvous, vol. 55. Kluwer Academic Publishers, Dordrecht (2002)
3. Baeza Yates, R., Culberson, J., Rawlins, G., Rawlins, G.: Searching in the plane. Inf. Comput. **106**(2), 234–252 (1993)
4. Bampas, E., et al.: Linear search by a pair of distinct-speed robots. In: Suomela, J. (ed.) SIROCCO 2016. LNCS, vol. 9988, pp. 195–211. Springer, Cham (2016). https://doi.org/10.1007/978-3-319-48314-6_13
5. Beck, A.: On the linear search problem. Isr. J. Math. **2**(4), 221–228 (1964)
6. Bellman, R.: An optimal search. SIAM Rev. **5**(3), 274 (1963)
7. Bose, P., De Carufel, J.-L.: A general framework for searching on a line. In: Kaykobad, M., Petreschi, R. (eds.) WALCOM 2016. LNCS, vol. 9627, pp. 143–153. Springer, Cham (2016). https://doi.org/10.1007/978-3-319-30139-6_12
8. Bose, P., De Carufel, J.-L., Durocher, S.: Revisiting the problem of searching on a line. In: Bodlaender, H.L., Italiano, G.F. (eds.) ESA 2013. LNCS, vol. 8125, pp. 205–216. Springer, Heidelberg (2013). https://doi.org/10.1007/978-3-642-40450-4_18
9. Brandt, S., Laufenberg, F., Lv, Y., Stolz, D., Wattenhofer, R.: Collaboration without communication: evacuating two robots from a disk. In: Fotakis, D., Pagourtzis, A., Paschos, V.T. (eds.) CIAC 2017. LNCS, vol. 10236, pp. 104–115. Springer, Cham (2017). https://doi.org/10.1007/978-3-319-57586-5_10
10. Chrobak, M., Gasieniec, L., Gorry, T., Martin, R.: Group search on the line. In: Italiano, G.F., Margaria-Steffen, T., Pokorný, J., Quisquater, J.-J., Wattenhofer, R. (eds.) SOFSEM 2015. LNCS, vol. 8939, pp. 164–176. Springer, Heidelberg (2015). https://doi.org/10.1007/978-3-662-46078-8_14

11. Czyzowicz, J., Dobrev, S., Georgiou, K., Kranakis, E., MacQuarrie, F.: Evacuating two robots from multiple unknown exits in a circle. In: Proceedings of the 17th International Conference on Distributed Computing and Networking, Singapore, 4–7 January 2016, pp. 28:1–28:8 (2016)
12. Czyzowicz, J., Gąsieniec, L., Gorry, T., Kranakis, E., Martin, R., Pajak, D.: Evacuating robots via unknown exit in a disk. In: Kuhn, F. (ed.) DISC 2014. LNCS, vol. 8784, pp. 122–136. Springer, Heidelberg (2014). https://doi.org/10.1007/978-3-662-45174-8_9
13. Czyzowicz, J., et al.: Evacuation from a disc in the presence of a faulty robot. In: Das, S., Tixeuil, S. (eds.) SIROCCO 2017. LNCS, vol. 10641, pp. 158–173. Springer, Cham (2017). https://doi.org/10.1007/978-3-319-72050-0_10
14. Czyzowicz, J., et al.: Priority evacuation from a disk using mobile robots. CoRR, abs/1805.03568 (2018)
15. Czyzowicz, J., et al.: God save the queen. In: 9th International Conference on Fun With Algorithms (FUN18) (2018)
16. Czyzowicz, J., Georgiou, K., Kranakis, E., Narayanan, L., Opatrny, J., Vogtenhuber, B.: Evacuating robots from a disk using face-to-face communication (extended abstract). In: Paschos, V.T., Widmayer, P. (eds.) CIAC 2015. LNCS, vol. 9079, pp. 140–152. Springer, Cham (2015). https://doi.org/10.1007/978-3-319-18173-8_10
17. Czyzowicz, J., Kranakis, E., Krizanc, D., Narayanan, L., Opatrny, J., Shende, S.: Wireless autonomous robot evacuation from equilateral triangles and squares. In: Papavassiliou, S., Ruehrup, S. (eds.) ADHOC-NOW 2015. LNCS, vol. 9143, pp. 181–194. Springer, Cham (2015). https://doi.org/10.1007/978-3-319-19662-6_13
18. Demaine, E.D., Fekete, S.P., Gal, S.: Online searching with turn cost. Theor. Comput. Sci. **361**(2), 342–355 (2006)
19. Deng, X., Kameda, T., Papadimitriou, C.: How to learn an unknown environment. In: FOCS, pp. 298–303. IEEE (1991)
20. Feinerman, O., Korman, A., Kutten, S., Rodeh, Y.: Fast rendezvous on a cycle by agents with different speeds. Theor. Comput. Sci. **688**, 77–85 (2017)
21. Fomin, F.V., Thilikos, D.M.: An annotated bibliography on guaranteed graph searching. Theor. Comput. Sci. **399**(3), 236–245 (2008)
22. Georgiou, K., Karakostas, G., Kranakis, E.: Search-and-fetch with one robot on a disk (track: wireless and geometry). In: Chrobak, M., Fernández Anta, A., Gasieniec, L., Klasing, R. (eds.) ALGOSENSORS 2016. LNCS, vol. 10050, pp. 80–94. Springer, Cham (2017). https://doi.org/10.1007/978-3-319-53058-1_6
23. Georgiou, K., Karakostas, G., Kranakis, E.: Search-and-fetch with 2 robots on a disk - wireless and face-to-face communication models. In: Liberatore, F., Parlier, G.H., Demange, M. (eds.) ICORES 2017, Porto, Portugal, 23–25 February 2017, pp. 15–26. SciTePress (2017)
24. Hoffmann, F., Icking, C., Klein, R., Kriegel, K.: The polygon exploration problem. SIAM J. Comput. **31**(2), 577–600 (2001)
25. Lamprou, I., Martin, R., Schewe, S.: Fast two-robot disk evacuation with wireless communication. In: Gavoille, C., Ilcinkas, D. (eds.) DISC 2016. LNCS, vol. 9888, pp. 1–15. Springer, Heidelberg (2016). https://doi.org/10.1007/978-3-662-53426-7_1
26. Papadimitriou, C.H., Yannakakis, M.: Shortest paths without a map. In: Ausiello, G., Dezani-Ciancaglini, M., Della Rocca, S.R. (eds.) ICALP 1989. LNCS, vol. 372, pp. 610–620. Springer, Heidelberg (1989). https://doi.org/10.1007/BFb0035787

27. Pattanayak, D., Ramesh, H., Mandal, P.S., Schmid, S.: Evacuating two robots from two unknown exits on the perimeter of a disk with wireless communication. In: ICDCN 2018, Varanasi, India, 4–7 January 2018, pp. 20:1–20:4 (2018)
28. Animation of algorithm 2 for $n = 4$, 13 February 2018. https://drive.google.com/open?id=1OhmWeqFZLFLiwQalvPoZSTg9Ah860mMn
29. Animation of algorithm 2 for $n = 8$, 13 February 2018. https://drive.google.com/open?id=10ntWmekJr5pTywEfpTNAw6uyxxrfpHsA

Author Index

Printed in the United States
By Bookmasters